NANOTECHNOLOGY SCIENCE AND TECHNOLOGY

COMPREHENSIVE GUIDE FOR NANOCOATINGS TECHNOLOGY

VOLUME 1

DEPOSITION AND MECHANISM

NANOTECHNOLOGY SCIENCE AND TECHNOLOGY

Additional books in this series can be found on Nova's website under the Series tab.

Additional e-books in this series can be found on Nova's website under the e-book tab.

NANOTECHNOLOGY SCIENCE AND TECHNOLOGY

COMPREHENSIVE GUIDE FOR NANOCOATINGS TECHNOLOGY

VOLUME 1

DEPOSITION AND MECHANISM

MAHMOOD ALIOFKHAZRAEI
EDITOR

nova publishers
New York

Library of Congress Cataloging-in-Publication Data

ISBN: 978-1-63482-447-7

ISSN: 2377-7575

Published by Nova Science Publishers, Inc. † New York

CONTENTS

PREFACE

This is the first volume of a four volume set of the Comprehensive Guide for Nanocoatings Technology. Nanocoatings are a kind of thin film with at least one nanometric dimension. The main difference between nano and ordinary coatings is in their structure and properties. To recognize these differences, it is required to study the properties and structure of nanocoatings. The properties of every substance are controlled by their constituents. In this regard, properties of coatings depend on the constituent grains and particles. The chemical, physical, and mechanical properties of nanocoatings including their electrical resistance, permeability, corrosion and abrasion resistance, and magnetic properties, which are a function of structure and particle size of coatings. It is evident that use of these features creates many potential applications. This volume mainly discusses the deposition and growth mechanism of nanocoatings. There are 16 chapters in this volume; each one includes examples of these interesting materials, supported by appropriate figures for better clarification.

Chapter 1 - The Layer-by-Layer (LbL) method is a promising approach to fabricate functional nanostructured coatings with a large number of applications in different areas of Nanotechnology, from optical to biocompatible devices. This apparently simple and versatile methodology is based in complex physico-chemical principles and rules whose knowledge plays a main role in the design and fabrication of nanocoatings. Special attention must be paid in the control of the assembling conditions since these can modify dramatically the structure and properties of the materials, and therefore the coating functionality and their potential applications. This chapter presents a comprehensive review of the physico-chemical bases that govern the relationship between properties, structure and function of polymeric LbL films. Also, some general rules for the design and fabrication of film for specific applications. Finally some of the most recent applications of nanocoatings are discussed.

Chapter 2 - Performance properties of various (nano) particles are determined principally by their size and the chemical composition of its surface. Within this Chapter, the authorswill take an exhaustive look into the surface treatment of titanium dioxide – TiO_2 (nano) particles. Of the essential meaning is that particle surface is coated in a controlled manner to maximize its applicable value. Process conditions (e.g., temperature, pH value, the time used for certain process stage, the dosing time, the sequence of process stages) affect the coating morphology and finally applicable properties of TiO_2. Inorganic surface modifiers are most commonly precipitated coatings of alumina (Al_2O_3) and silica (SiO_2). These inorganic surface treatments provide improvements in one or more important performance properties such as:

dispersibility, hiding power efficiency, chalk resistance and resistance to discoloration by heat and/or photoreduction.

Chapter 3 - This scientific paper describes the structure, phase composition and mechanical properties of nanocrystalline coatings obtained using the ion beam-assisted deposition technology. The results of computer simulation of IBAD processes for different energy values of bombarding ions and for relationships between the densities of ion fluxes and metal deposition rates is given. The data of electron-microscope analysis of the initial stages of film structure formation (of 5nm thick and more) is presented. And finally, the authors demonstrate the influence produced by the ion energy and ion current density on the formation of nanocrystalline structure and its correlation with tribological properties of coatings.

Chapter 4 - The particle coating by rotating Plasma Chemical Vapor Deposition (PCVD) reactor will be discussed in this chapter. Rotating PCVD technique is used to synthesize nano-sized thin films with high uniformity. The rotating PCVD reactor allows to produce a single-layer, as well as double-layer thin films coated on small-size substrates of various shapes (spherical, cylindrical, oval, etc.) and materials (glass, polymer, zeolite, etc.) with using liquid precursors (TTIP (titanium tetra-isopropoxide), TEOS (tetraethyl orthosilicate), etc.) and/or gaseous precursors. This chapter will explain the principles of rotating PCVD process, discuss simulation and experimental results and show possible applications of this technique. Being flexible in changing of process parameters, rotating PCVD is the promising method for synthesis of various thin films with several desirable properties.

Chapter 5 - Nanocoatings are used in cell phones, solar power plants and electric cars. Nanocoatings are used in order to improve the light-to-electricity conversion efficiency in solar cells. Flake production, functionalization, particulation, sacrificial coating, specifications are critical in order to prepare lubricous, high performance, nanocoatings Nanocoatings are used in GMRs, giant magneto resistive thin film heads that go into the preparation of hard drives in desktop computers. They are also used in thermal barrier coating and corrosion-resistant ceramic coating applications. resistance. SSA, specific surface area can be used to characterize nanocoating such as the degree of nanostructuring. TBC can be used to prevent fire spread. For materials with relaxation times greater than ($R^2/15.33\alpha_{nr}$) the temperature in the naval warhead can be expected to undergo subcritical damped oscillations.

X-ray lithography and electroplating techniques were used in order to create a zone plate of thickness of 500 nm with zone dimensions less than 50 nm. PU coating can be grafted with nano-alumina in order to improve the performance of the coating by increasing the adhesion between nanoalumina and organic substrate. Coating machines and rollers can be used to manufacture solar cells at lower cost. A PV, photovoltaic device was created by casting a solution of conducting polymer and fullerenes into a thin film.

Nanostructured alumina/titania powders can be used to improve abrasive wear resistance of ceramic coatings by 400%. Process comprises of dispersion of nanopowders into a colloid. Binders and granules are spray dried using a Metro 9 MB plasma gun. Thermal spray robots are used to obtain nanocoating. Nanostructures can result from control of morphology of phase separated polymer blends. Nanostructuring can be controlled during preparation of metal chalcogenide nanostructured films by electrodeposition. These films can be formed within mesopores of silica.

The use of graphene layers as anti-reflection coatings in solar cells can result in the reduction in the cost of the solar cells. Light to electricity conversion was found to increase to

14.6%. Graphene sheets with less than 24,000 atoms/25 nm are metastable. Nanocoatings are used in the design and application of nanorobots for drug delivery. Principles of photodynamic therapy, fullerene chemistry, nanostructuring, x-rays, computers, pharmacokinetic and robotics are applied to design a nanorobot in order to treat Alzheimer's disease. Drug is inactive in its caged form. It is activated by irradiation at desired site in the human anatomy. The irradiation causes breakage of the fullerenes and the drug becomes available for action. Graphene is used in naval applications in order to demonstrate an infrared-transparent, electromagnetic shielding coating that can be applied to electro optic sensor windows and domes. Transmission greater than 90% in the 3-5 micron wavelength region can be achieved.

Protein folding, secondary structure of proteins, nucleic acid structure, ribosomes, chromosomes and phospholipid membranes are examples where self-assembly is critical.

Fluidized bed can be used to obtain the nanocoating. The precursor material is calcined in ammonia at 1000°C for 48 hours in a fluidized bed reactor. A red tri-tantallumpenta nitride, Ta_3N_5 coated mica pigment is produced. Bubbles and slugs and other instabilities formed during the operation of the fluidized bed are reduced by use of vibrators/agitators. The substrate particle size range is between 1 – 500 μm.

Chapter 6 - The inkjet printing of copper, silver and gold patterns using nanoparticle- and molecule-based inks will be discussed. Diverse applications including low-cost conductive structures for passive wiring, photovoltaic and near-field communication applications as well as microelectronic components are discussed.

Chapter 7 - Electroless nickel coatings are often used because they are resistant to scratch, abrasion and corrosion.

However, it is difficult to choose between electroless nickel-phosphorous and electroless nickel- boron coatings for any given application because their properties are not often compared.

Both types of coatings were completely compared: their structure, morphology, hardness, abrasion resistance, scratch test response, and corrosion resistance were investigated and the best candidate for each feature was determined.

Heat treated coatings were submitted to the same comparison.

Chapter 8 - Sol-gel is a wildly used method in many areas of science, engineering and technology, and has been gaining much attention worldwide. One of the most important applications of sol-gel technology is the thin film coating on the surfaces of materials against corrosion, degradation, other environmental effects, and decorations purposes. Sol-gel coating mainly contains four steps, including the gel formation, aging, drying and heat treatment, and can also be in various forms, sizes and shapes. Compared to the conventional coating methods, sol-gel coating process contains considerably less equipment and labor; thus, it is more economic and an easy process for numerous applications. Furthermore, the sol-gel technique is an environmentally friendly method which can be a potential replacement method for the toxic pretreatments of some other conventional methods (such as, hexavalent chromium conversion, surface acid/base treatment, etching, and sand blasting), and they can be in both organic (polymeric) and inorganic (metals, alloys, oxides and composites) forms. Among the sol-gel coating applications, the protective coatings at various thicknesses (from nanoscale to micro scale) have gained significant attention because of the excellent chemical stability at room and elevated temperatures, oxidation control, heat dissipation, and enhanced corrosion and UV resistance. In the present book chapter, the authors have analyzed all the latest developments in the sol-gel coatings and their applications on various materials (e.g., stainless steel, aluminum, magnesium, copper and their alloys), progressions and challenges, and future trends in the field. The authors believe that this study will greatly provide several benefits for the readers and companies in the coating and related businesses.

Chapter 9 - The use of antireflection coating and high reflection coating on laser diode facets provide the necessary passivation and change the reflectivity so that power emission is predominantly from one end. Design of antireflection coating is optimized for single layers of Al2O3 and high reflection coating is optimized for stacks of Al2O3andSi Bragg reflector. Electron beam evaporation has been used to deposit uniform Al2O3/Si film under high vacuum with pressure 10-6 mbar and temperature 150°C-1600°C.Experimental parameters like refractive index of the material and thickness of deposition are necessary to match with simulated AR/HR curves. LI measurements of a coated laser diode are compared with an uncoated one.

Chapter 10 - From the past few decades, the field of ultrathin films has grown remarkably. A material in the form of ultrathin film gains enormous surface-to-volume ratio and thereby enhances the properties like catalysis, reactivity and adhesion remarkably. The efficiency and sensitivity of the devices employing ultrathin films increases manifold. In this chapter, the authors will discuss three different types of ultrathin films viz. Langmuir monolayer, Langmuir-Blodgett films and self-assembled monolayer. Since the thickness of these films are mono-molecular, it can be considered as two-dimensional system. Here the authors will discuss the fabrication of the ultrathin films, its properties, effect of external parameters on its properties, its application, and challenges and opportunities in the field.

Chapter 11 - In this chapter, the authors attempt to summarize the available information about the electrodeposition of indium and indium alloys.Nowadays, indium is becoming an attractive alloying element enhancing the spectrum of properties of the basic metals. The authors focus our attention on the special features of electroplating of indium from aqueous solutions and its possible perspectives for practical application.

Indium can be electrodeposited from both acid and alkaline electrolytes. Precipitation of the electrolyte is possible by increasing of thepH value above 3.4. In order to prevent the creation of the hard-to-dissolveindium hydroxide under these conditions, ions of ammonia and sucrose (levulose and glucose) are used.Indium alloys have characteristics that depend on such specific to indium properties as low melting point, high plasticity, and corrosion resistance.

An improved corrosion resistance compared with nickel is observed in the nickel-indium alloy.The alloys of indium with bismuth, antimony, and tin are used for solders.

The co-deposited indium contributes to the significant decrease of the internal stress of the palladium and ruthenium coatings.

Decorative coatings of various colors are obtained by co-depositing of indium, palladium, andgold.

A number of separate intermetallic phases of the indium alloys with different metals demonstrate specific properties. For example,shape-memory effects have been registered in the indium-gold, cadmium-indium, and thallium-indium alloys. Intermetallic bismuth-indium compounds are used as zinc anodes in zinc-alkaline batteries; indium-antimonide is a photodetector for infraredandvisiblerays.

Self-organization phenomena are observed during electrodeposition of several indium alloys – with silver, cobalt and palladium. In this case, the increase in the content of the alloying element in electrodeposited alloys reflects the changes of their phase composition when the saturation limit of the lattice of the basic metal is reached.

Chapter 12 - Out of numerous synthesis methods of nano-coatings, electrochemical techniques have gained much interest, mainly because of their versatility. By means of advanced electrochemical strategies it is possible to obtain nano-coatings of diverse forms. Rapidity of the process, purity of products as well as production of coatings on widely varying surfaces are the most noticeable advantages of electrochemical synthetic methods. In this chapter the authors will present two principal electrochemical techniques: electrochemical deposition and electrophoretic deposition, together with less popular alternatives, such as plasma electrolytic deposition, laser-assisted electrodeposition and alternating current electrophoretic deposition. The fundamentals of each technique will be followed by specific examples illustrating details of the synthetic methods.

Chapter 13 - The discovery of graphene, a two-dimensional (2D) sp^2 hybridized carbon atoms arranged in a honeycomb structure brought tremendous attention as next generation material for various applications. Recent significant development on synthesis of large-area continuous graphene by the chemical vapor deposition (CVD) process opens up new possibilities of application. The CVD process can be a reliable technique for control synthesis of high quality and large-area graphene on metal substrates. Among CVD technique, plasma CVD has evolved as a novel technique for deposition of graphene film at a comparatively lower temperature, than that of thermal CVD process. A low-temperature deposition technique can be more feasible for industrial application and can pave the way for direct deposition on any arbitrary substrates. In a plasma CVD process, carbon radicals are formed with hydrogen termination from a hydrocarbon, which absorb continuously on the substrate surface to form sp^2 carbon and eventually graphene structure. Recent development of the plasma CVD process for synthesis of graphene based materials is elaborately discussed.

Chapter 14 - Friction stir processing (FSP) is a recent surface engineering method that allows obtaining in-situ ultrafine grains in surface layers. FSP is developed from friction stir welding, a thermo-mechanical welding process, now widely used due to technical, economic and environmental advantages. Friction stir basic principle is very simple and consists in moving a rotating tool along the workpiece surface while applying an axial load to increase generated heat.

Actually many studies are focused on friction stir processing as a surface modification method. The temperature and mechanical aspects must be balanced and optimal conditions need to be found in order to obtain the desired nanograin structure when processing different materials.

This chapter will present state-of-the art and future challenges in friction on stir as a method to obtain significantly refined nanograin structures at workpiece surface. The following topics are emphasized: friction stir processing - innovative surface engineering method, processing nanograin layers by friction stir processing of various materials and numerical control of friction stir processing.

Chapter 15 - The surface acts as an interface between the bulk of a material and its environment. Therefore, a modification in the surface properties plays an important role in performance characteristics of a component. Surface engineering involves bringing about desirable changes in the properties at or just below the surface of a substrate. It is generally employed for improving performance in various applications, such as tribological, electrical, optical, electronic, chemical or magnetic. Through surface engineering, it becomes possible to select the bulk of a component to meet general requirements, such as strength, stiffness, formability, cost, etc. and thereafter modify the surface up to a certain depth, viz. through heat treatment, or by adding a thin layer of another material on the substrate surface through some suitable deposition technique in order to induce the desired functional properties, viz. hardness, corrosion resistance, electrical conductivity etc. With rapid advancement in vacuum technology and control systems over the past few decades, surface engineering processes involving deposition of thin films have gained widespread importance in various fields – especially metal cutting and semiconductor chip fabrication industries.

Chapter 16 - Electrochemistry plays a significant role in nanotechnology. The electrochemical techniques help not only in synthesizing a variety of nanomaterials and coatings starting from metals to tubes, rods or wires of different dimensions but also in characterizing them for various applications including preventing global warming, which is a major threat to the entire world. The chapter reviews the importance of electrochemical techniques in synthesizing nanostructured metals, alloys, coatings, composites, Titania nanotubes; rods, wires and carbon nanotubes that exhibit remarkably enhanced characteristics over other materials/coatings apart from their classification, synthesis, properties and applications. Subsequently, the advantages of electrochemically prepared nano metals, alloys, coatings, tubes, wires, rods and the necessity of their use in effectively preventing global warming and in modern industries that allow the systems to be used safely at varied environmental conditions, which not only would enhance the efficiency of industrial systems but also their life span and consequently the profitability, has been stressed.

In: Comprehensive Guide for Nanocoatings Technology. Vol. 1 ISBN: 978-1-63482-447-7
Editor: Mahmood Aliofkhazraei

Chapter 1

FABRICATION OF POLYMERIC NANOCOATINGS BY THE LAYER-BY-LAYER METHOD: A PHYSICO-CHEMICAL APPROACH

***Eduardo Guzmán*[1,2,*], *Francisco Ortega*[2] *and Ramón G. Rubio*[2]**

[1]Istituto per l'Energetica e le Interfasi - U. O. S. Genova,
Consiglio Nazionale delle Ricerche, Genova, Italy
[2]Departamento de Química Física I-Facultad de Químicas,
Universidad Complutense de Madridm, Madrid, Spain

ABSTRACT

The Layer-by-Layer (LbL) method is a promising approach to fabricate functional nanostructured coatings with a large number of applications in different areas of Nanotechnology, from optical to biocompatible devices. This apparently simple and versatile methodology is based in complex physico-chemical principles and rules whose knowledge plays a main role in the design and fabrication of nanocoatings. Special attention must be paid in the control of the assembling conditions since these can modify dramatically the structure and properties of the materials, and therefore the coating functionality and their potential applications. This chapter presents a comprehensive review of the physico-chemical bases that govern the relationship between properties, structure and function of polymeric LbL films. Also, some general rules for the design and fabrication of film for specific applications. Finally some of the most recent applications of nanocoatings are discussed.

1. INTRODUCTION: ON THE WAY TO FUNCTIONAL COATINGS

The interest in the modification of surfaces at molecular level by polymeric layers has grown in recent years due to the multiple possibilities offered by polymers for tuning the physico-chemical properties of surfaces. [1] Polymer nanocoatings modify the properties and

* To whom correspondence should be sent: eduardogs@quim.ucm.es

interactions of the surfaces which is important in several technological fields: fabrication of biocompatible films, stabilization of colloidal solutions and particle dispersions, flocculation processes, surface treatments, cosmetics and many others. [2] Among the different types of polymer coatings built using interfacial engineering approaches, the electrostatic self-assembling or Layer-by-Layer (LbL) method is especially important due to their low cost, modularity and versatility. [3-5] Moreover, this method is suitable for scaling-up for industrial production.

The LbL approach was firstly introduced by Iler [6] in the 60´s of the last century, though did not attract the interest of researchers until the works of Decher et al. thirty years later. [7] The paradigm of this type of coatings are those fabricated from the alternate deposition of polyelectrolytes with opposite charge onto a solid substrate. [3] However, the list of potential molecules that can be involved in the fabrication of LbL coatings is not limited to polyelectrolytes, being possible to use a broad range of substances; such as colloidal particles, biopolymers, carbon nano-objects, clays, small molecules, living cells and many others. [5, 8, 9] Furthermore, the shape and size of the surfaces that can be coated by LbL multilayers are not limited to flat substrates, but also nano- or microparticles, vesicles, bubbles, drop emulsions or flat fluid surfaces have been used. [10-13] This is an advantage over other traditional methods of the Soft Nanotechnology for the fabrication of nanocoatings: Langmuir-Blodgett, [14] self-assembled monolayers. [15] Furthermore, the LbL approach can be used for assembling both strongly and weakly interacting blocks, which allows one to use intermolecular interactions other than the electrostatic ones: hydrogen bonding, [16] charge transfer, [17] molecular recognition [18] or coordination. [19] Nevertheless, it is necessary that the interactions between the components used to build the film are strong enough to ensure its integrity when exposed to chemical or mechanical stresses. This type of multilayers is thermodynamically unstable versus the formation of interpolyelectrolyte complexes. [20] As a consequence, the protocol followed for building them up is very important. Moreover, a rigorous control of different physico-chemical variables (solvent, temperature, pH, ionic strength, adsorption time) is necessary to obtain the desired structure and properties. [20]

Among the applications of the LbL coatings are: antibacterial and antimicrobial coatings, antireflection coatings, self-healing and superhydrophobic surfaces, etc. [5, 21-23] This has opened important challenges in the fabrication of nanocoatings with tailored properties which can be divided in two groups: passive and "smart" coatings. [5] In the following, a description of the physico-chemical bases behind the fabrication of LbL coatings, together with the possible methodologies used will be presented. Furthermore, the most promising applications developed for LbL coatings will be summarized at the end of the chapter.

2. Physico-Chemical Bases of the Fabrication of LBL Coatings

The understanding of the physico-chemical bases that govern the building of LbL nanocoatings requires a detailed description of the interactions involved in the multilayer formation (polymer–polymer, polymer–solvent and polymer-surface). [20] The entropic contributions in the formation of the multilayers play a key role, being the driving force in many cases.

Initially, the LbL method was developed using molecules in which the electrostatic interactions played a major role (hence electrostatic self-assembling). [24] In these systems the adsorption of a layer provoked a charge inversion, which for some time was considered the driving force of the assembling. Charge inversion is a counterintuitive phenomenon that implies the adsorption of material on the formed layers is not stopped when the neutralization of the surface charge is reached, but continues till a certain level of overcompensation of the charge is reached. Charge inversion plays a main role on the fabrication of polyelectrolyte multilayers, PEMs. [25] Different results based on the evaluation of the surface potential changes of surfaces coated by PEMs have pointed out the existence of this inversion mechanism during the assembly process, independently of the nature of the substrate or of the polyelectrolytes. [11, 26] This means that the surface potential shows positive or negative values depending of the charge of the last adsorbed layer. The most frequently used methods for evaluating the charge inversion is by measuring the ζ-potential of the coated colloidal templates, [11] or the streaming potential [27] of flat substrates coated with PEMs. A typical example of charge inversion is shown for $(PDADMAC + PSS)_n$ multilayers, being PDADMAC poly(diallyldimethylammonium chloride), and PSS poly(4-styrene sulfonate of sodium). In this system, ζ changes between values around (30 ± 10) mV for the adsorption of the polycation layers and values of (-30 ± 10) mV after exposition to PSS solutions. Other systems behave similarly in relation to the charge inversion (see Figure 1). [28]

Some authors have pointed out that LbL multilayers can be built in systems in which the electrostatic interactions are not involved and a charge inversion is not expected. [29] Furthermore, other systems such as $(PAA + CHI)_n$ multilayers (being PAA poly(acrylic acid)) do not show any charge inversion after the deposition of a certain number of layers as consequence of the interpolyelectrolyte complexation. [30] This scenario requires to take into consideration the existence of additional contributions in the multilayers assembling. [24, 26]

The charge inversion principle presents the PEMs as charged supramolecular architectures. However, the electroneutrality of the multilayers at the macroscopic level (beyond the Debye length) is needed to guarantee the stability on PEMs. [25, 26] This condition of zero net charge requires the establishment of some mechanisms to achieve the charge neutralization or compensation on the supramolecular system. In particular, two different mechanisms have been described in the literature (see Figure 2), which are correlated to the relative importance of entropic contributions in the formation of the PEMs. [24, 26]

The first mechanism (intrinsic compensation) considers a perfect matching between the charges in adjacent layers that leads to the formation of supramolecular architectures with charge stoichiometry 1:1. This mechanism implies the formation of interpolyelectrolyte structures with a high ionic cross-linking (see Figure 2). The driving force of the intrinsic compensation is mainly entropic, as a consequence of the release of a large amount of counterions during the PEMs formation that leads to a strong decrease on the system free energy, due to the entropic increase. In the second mechanism, called extrinsic, the charge neutralization is partly due to the co-adsorption of counterions together with the polymer chains (see Figure 2). The counterions incorporation into the PEMs leads to the formation of complexes with less defined stoichiometries, and the entropic contribution are less important than in the intrinsic mechanism. In general, extrinsic compensation is more frequent in PEMs than the intrinsic one, being the latter mainly limited to highly charged strong polyelectrolytes. [26, 31] It is worth mentioning that the prediction of the type of compensation involved in the PEMs is not easy because the charge compensation is strongly dependent on many physico-chemical parameters (pH, ionic strength, etc.). [25] A careful look on the type of compensation is essential to control the growth and properties of the PEMs.

Table 1. Main works in the field of the fabrication of nanocoating based in LbL films

Topic	Reference
First work dealing with the assembly of multilayers films using the LbL approach. Assembly of multilayers of particles with opposite charge.	Iler, 1966 [6]
Work revisiting the bases of the LbL assembly of multilayers. First step for the development of the fabrication of nanocoatings by the LbL assembly of polyelectrolytes.	Decher et al., 1991 [7]
First evidence of the LbL assembly using interactions different to the electrostatic ones.	Stockton et al., 1997 [16]
Discussion of the advancements in the field of the LbL films and the future routes of development.	Decher, 1997 [3]
First LbL assembly of polyelectrolytes onto non-flat substrates.	Caruso et al., 1998 [11]
Establishment of the physico-chemical bases that govern the LbL electrostatic assembly.	Dubas and Schlenoff, 1999 [24] Schlenoff and Dubas, 2001 [25]
First application of the spraying methodology for deposition of LbL multilayers.	Schlenoff et al., 2000 [46]
Revision of the main parameters that govern the LbL assembly.	von Klitzing, 2006 [20]
Analysis of the development of the LbL approach during the first twenty years of application.	Schlenoff, 2009 [5]
First comparison of the behavior of solid-supported multilayers and free-standing one. Importance of the substrate in the fabrication of multilayers.	Guzmán et al., 2009 [13]
Application of deposition methods with potential industrial application for the fabrication of LbL nanocoating.	Kolansinska et al., 2009 [44]
First evidence of the absence of stratification in LbL multilayers.	Guzmán et al., 2009 [26]
First systematic study related to the coating of vesicles/liposomes by LbL films. Important development for applications of LbL films in the drug delivery field.	Ruano et al, 2010 [12]
Revision of the applications in the biomedical field of LbL nanocoatings.	Boudou et al., 2010 [21]
Work dealing with the growth of LbL multilayers that do not follow any of the two traditional growth trends.	Cini et al., 2010 [38]
Development of the LbL assembly for the coating of emulsion drops.	Adamczak et al., 2013 [10]

Among the different parameters that affect the charge compensation in PEMs, the ionic strength, I, of the solutions is one of the most important due to its influence on the ionic equilibrium of the system. This takes place through the modification of the counterions condensation around the polyelectrolytes and consequently their charge densities. Schlenoff and Dubas [25] pointed out that changes on I allows one to modify the type of the charge compensation in (PDADMAC + PSS)$_n$ multilayers, which passes from a mainly intrinsic for low I conditions to strongly extrinsic for high I (Figure 3). For low I values, the release of small amounts of counterions increases significantly the entropy of the system. [26] This favours the PEMs with strong ionic pairing between the polyelectrolytes and low counterions content. Conversely, the release of counterions does not alter significantly the system entropy for high I solutions. This favours supramolecular architectures containing a high amount of adsorbed counterions, thus reducing significantly the ionic pairing between the adjacent layers in the PEMs.

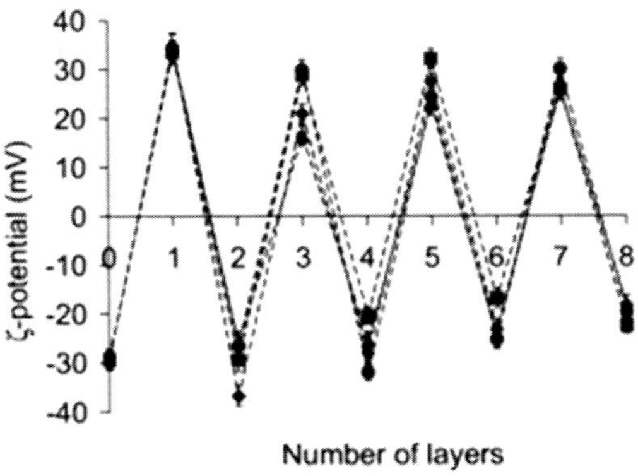

Figure 1. Charge inversion obtained as the ζ-potential dependence on the Number of Layers for different PEMs deposited onto colloidal particles of cellulose. The different set of symbols are data corresponding to different multilayers: ● (PDADMAC + PSS)$_n$, ■ (PDADMAC + Bovine gelatin)$_n$, ▲ (CHI + PSS)$_n$ and ♦ (CHI + Bovine gelatin)$_n$ (being CHI Chitosan). Note that in all the PEMs, the first polyelectrolyte is the polycation and the second one the polyanion due to the negative charge of the cellulose particles. Adapted from Reference 28. Copyright (2011), with permission from Elsevier.

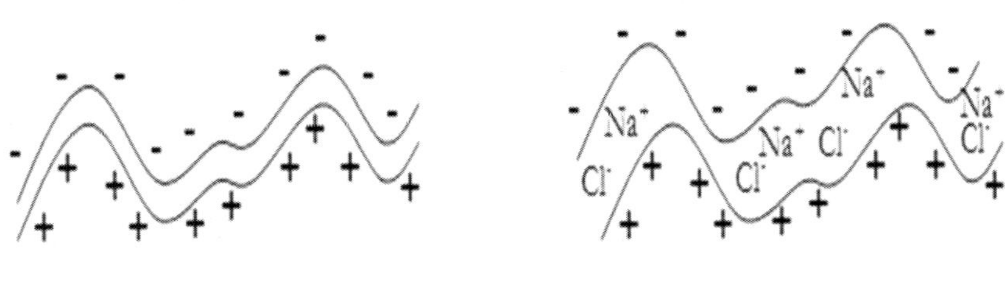

Figure 2. Types of charge compensation: Intrinsic vs. Extrinsic. Reproduced from Reference 26 with permission from The Royal Society of Chemistry.

Moreover, the type of compensation may be different for the two polyelectrolytes forming a multilayer, as reported by Lehaf et al. [32] for (PDADMAC + PSS)$_n$. They pointed out that the compensation of PDADMAC-capped films is strongly extrinsic, whereas for the case of PSS-terminated PEMs the compensation is mainly intrinsic. This is in good agreement with the existence of compensation ratio (the ratio between the charged units in PDADMAC layers to that in the PSS layers) higher than one. A different compensation mechanism of the capping layer of PEMs, modifies the structure and roughness of the film, which is very important for the design of nanocoatings. [32] In the particular case analyzed by Lehaf et al. [32], $(PDADMAC + PSS)_n$, the high ionic content of PDADMAC layers leads to the formation of fuzzier layers than those PSS-capped. This can be explained by the existence of a strong osmotic effect due to the presence of counterions. Analogous effects to those induced by changes in I are expected for other parameter that modify the ionic equilibrium of the systems, such as the type of polyelectrolyte, the charge density of the chains, [31] the solvent quality [24] or the pH. [30]

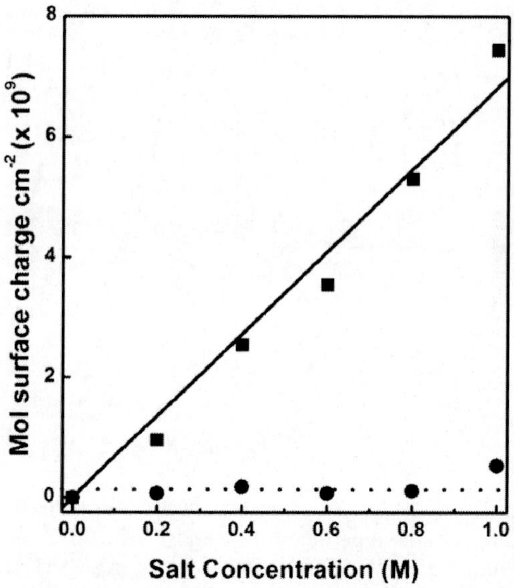

Figure 3. Surface charge per monomer monomers for polyanion (●) and polycation (■) layers in $(PDADMAC + PSS)_n$ multilayers adsorbed from solution of different I. Notice that the different amount of adsorbed PDADMAC and PSS evidences the existence of extrinsic compensation. Adapted with permission from Reference 25. Copyright (2001) American Chemical Society.

The above discussion has focused on the entropic contributions due mainly to the counterions release. However, other types of entropic contributions play a certain role in the multilayer assembling: entropy due to the release and reorientation of water molecules associated to polyelectrolyte chains and entropy penalty due to the attachment of the molecules to the surface. [33]

In summary, the assembling of PEMs must be considered a sum of several factors, mainly governed by electrostatic and entropic contributions. This latter is critical to build PEMs without limitation on the number of adsorbed layers, N. [4] In fact if only the electrostatic interactions are important, the adsorption may stop after the deposition of a reduced number of layers, which is especially important for components with low charge density or if the charges are strongly screened. The entropic contributions are especially relevant for multilayers of compounds with very low charge.

3. Methods for Building of LBL Coatings

Different techniques allow one to monitor the building of PEMs either during the assembling process or in a post-fabrication stage. The choice of the most adequate technique to characterize the multilayers depends on the specific characteristics of the substances and the substrate used. Flat multilayers can be studied by spectroscopic techniques (UV-Visible, Infrarred ...), [34] dissipative quartz crystal microbalance (D-QCM), [26] ellipsometry [30] and reflectivity techniques (neutrons or X-Rays). [26, 35] Other techniques such as dynamic light scattering, [12] Nuclear Magnetic Resonance (NMR) [36] are suitable for studying PEMs adsorbed onto colloidal templates. An example of a growth process followed by UV-visible spectroscopy is shown in Figure 4, where the absorbance of multilayers formed by poly(vinylpirrolidone) and PAA with different number of layers is shown. [37] The results point out clearly that the quantity of adsorbed material grows linearly with the number of adsorbed layers. For other systems or under different assembling conditions, it is possible to find a stronger dependence of the adsorbed amount in the number of layers.

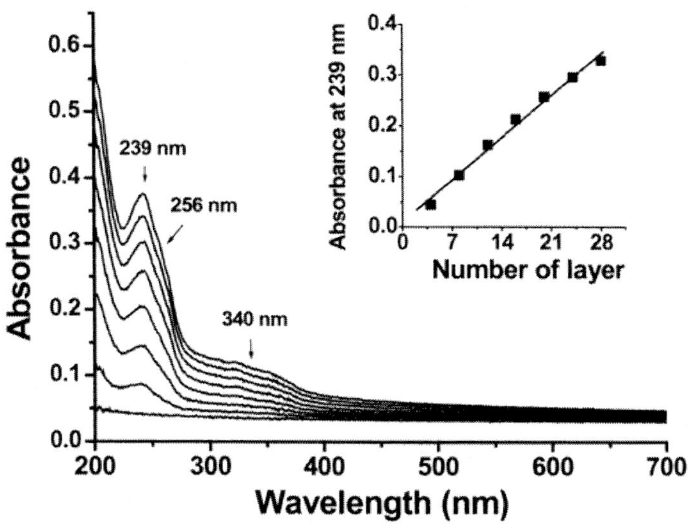

Figure 4. Changes on the UV-Visible spectrum of multilayers formed by poly(vinylpirrolidone) and PAA for films with 0, 2, 4, 6, 8, 10 and 12 bilayers (from bottom to top). The inserted figure represents the dependence of the absorbance for the band at 239 nm on the number of layers. Reprinted with permission from Reference 37. Copyright (2002) American Chemical Society.

One of the first issues to clarify when a PEM is being formed is the way its thickness increases with the number of layers deposited, N. Despite two growing mechanisms have been clearly identified, currently it is not possible to accurately predict which one will be found for two given compounds under different conditions of pH, ionic strength, or temperature. [20] In the first growing mechanism a linear dependence of the adsorbed amount on N is found, and it is usually known as linear-growth (see Figure 5). In the second type of multilayers, a supra-linear dependence of the adsorbed amount on N is found. This is the so called exponential growth mechanism despite that the dependence on N is not strictly in many cases. Some other mechanisms, presenting more complex dependences of the adsorbed amount on N have been recently described. [38] It is worth mentioning that the control of the type of growth of the PEMs is critical in the design of functional materials and coatings, since it affects strongly the film properties and roughness.

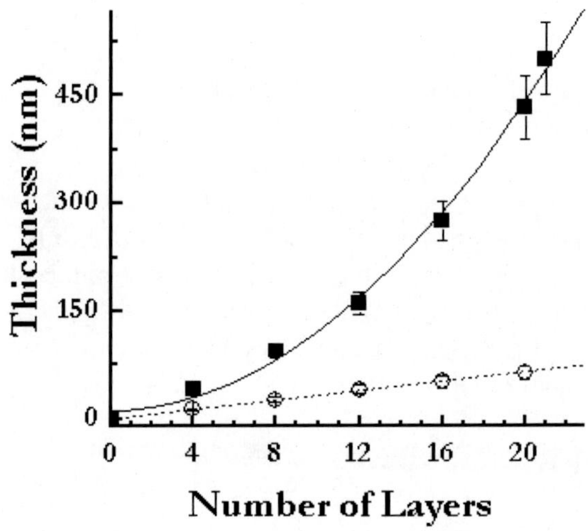

Figure 5. Linear vs. Non-Linear growth for PEMs deposited from NaCl solutions with concentration 500 mM. Linear growth correspond to $(PAH + PSS)_n$ multilayers (■, —) whereas non-linear growth to $(PDADMAC + PSS)_n$ multilayers (○, ···). Adapted with permission from Reference 39. Copyright (2012) American Chemical Society.

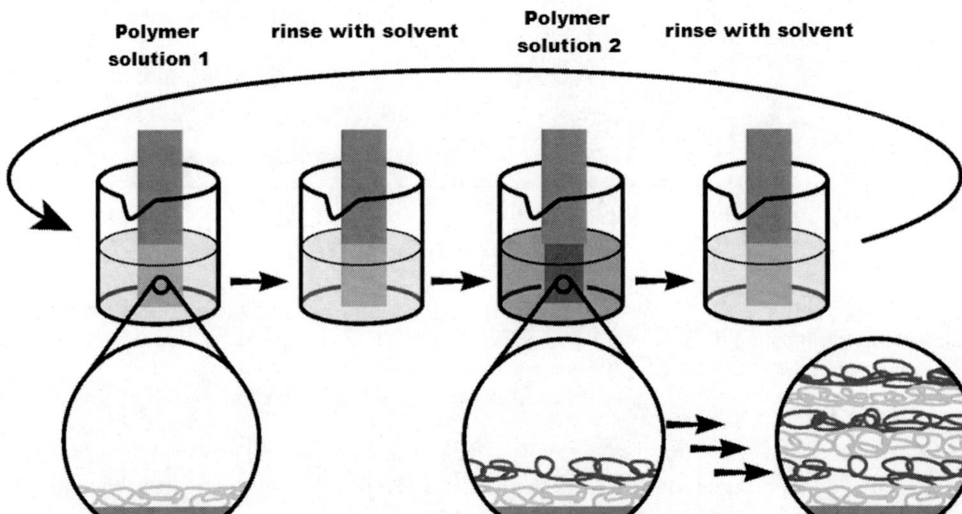

Figure 6. Building of LbL PEMs by dipping. Adapted from Reference 42. Copyright (2012), with permission from Elsevier.

The chemical and structural versatility of the LbL over other methods have stimulated the research to find the optimal conditions for the fabrication of LbL films with the desired properties. This has lead to the development of deposition methods that are suitable for automatization and scaling-up. [40] The most common method used is the dip-coating scheme illustrated in Figure 6, which is based on a direct material transference from the solution to the substrate by the immersion of this latter in the solution. The alternate immersion of the substrate in solutions of materials with different nature leads to the LbL nanocoatings. Intermediate steps of rinsing and drying of the supramolecular film can be included between the different adsorption steps in order to remove the molecules that are weakly adsorbed on the films. This allows avoiding the precipitation of complex on the film, which could alter the fabrication process and properties of the materials. [40] The fabrication of LbL multilayers by dipping is widely standardized and some automatic dippers have been recently developed for a well controlled and reproducible fabrication of coatings. [41, 42]

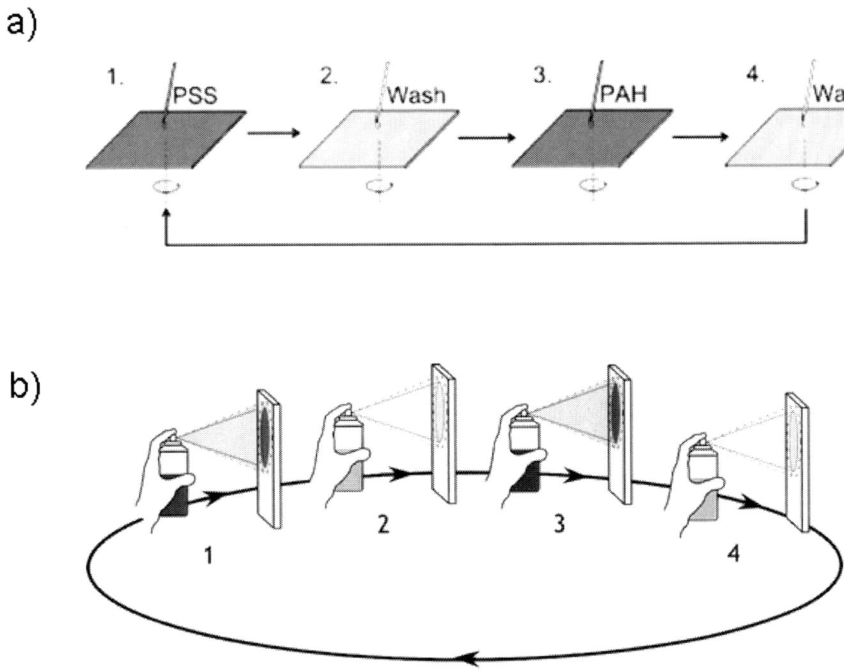

Figure 7. Alternative methods to replace the dip-coating assembling. (a) Spin-Assisted Assembling of LbL multilayers. Reprinted with permission from Reference 43. Copyright (2010) American Chemical Society. (b) Spray-Assisted Assembling. Reprinted with permission from Reference 44. Copyright (2005) American Chemical Society. In both Figures, steps 2 and 4 correspond to rinsing whereas steps 1 and 3 are deposition steps.

Despite the simplicity of the dipping method, it presents several problems, such as the time required for the film fabrication, and the difficulties in scaling-up that limits its application at industrial level. In order to overcome these difficulties, methods such as the spin-coating (Figure 7a) [43] and the spraying (Figure 7b) [44] have been developed. They allow one to decrease the deposition times and the amount of material needed, which is very important for scaling-up to the industrial production. [4]

In the spin-coating method a drop of solution is deposited onto a substrate that spins at controlled velocity till the total drying/evaporation of the solvent. [43] The steps involved in the spin coating fabrication of multilayers are the same than those involved in the fabrication by dipping. An important advantage of the fabrication of multilayers by spinning is that it leads to smaller roughness of the films than those obtained by dipping. [45] The main requirements for the correct fabrication of multilayers by spin-coating are the precise control of the spinning velocity and of the solution concentration. However, it is noteworthy that the LbL deposition following this methodology is strongly limited by the volatility of the solvent, which very frequently is water.

The second alternative to the application of the dipping in the fabrication of LbL films is the spray-assisted deposition, introduced by Schlenoff et al. [46] This method presents the same main rules than dipping and spin-assisted depositions, following a similar deposition-rinsing scheme. In this type of deposition, the solutions are directly sprayed onto the substrate, separating the successive step by a time gap long enough to allow for the liquid drainage. Therefore, two simultaneous processes have taken into account during the deposition/rinsing steps: the material adsorption and the solution drainage. The spray-assisted coating leads to materials with similar morphology, uniformity and chemical composition than those obtained by dipping, being the main advantages the significant reduction in the fabrication times [46] and the possibility to fabricate more stratified films than those obtained by dipping. [47] The most extended configuration for spray-assisted deposition implies to do the spraying perpendicular to the surface. This experimental conformation allows the adequate draining of the solution by gravity and the subsequent quick removal of the excess of liquid. Several parameters must be optimized to obtain good films, among them the distance between the spray and the surface, the times of spraying and draining, the concentration of the solutions and the volume and flow of the sprayed solution.

4. Control of Physico-Chemical Parameters in the Fabrication of Tailored Coatings

The fabrication of coatings for a certain application requires a good control over the structure and properties of the films which can be reached by the modification of different physico-chemical parameters during the LbL assembly. The understanding of the quantitative effect of these parameters is essential to control the fabrication processes. [20]

4.1. Structure of the Assembled Materials

The materials used for the fabrication of nanocoatings determine the final structure and properties of the films. [20] It was pointed out by von Klitzing et al., [48] that the hydrophilic/hydrophobic character of the PEMs components is one of the most important characteristics. These authors showed that the thickness increased with the hydrophobicity of the polyelectrolyte according to the following series: $(\text{PDADMAC} + \text{PAMS})_n < (\text{PAH} + \text{PSS})_n < (\text{PDADMAC} + \text{PSS})_n$, being PAMS poly(acrylamidesulfonic acid)). This dependence on the hydrohobicity is associated to the hydration and swelling of the adsorbed polymers. [24]

The flexibility/rigidity of the polymer chains plays also an important role due to its effect on the interactions along the multilayer. [39] The polymer nature can also modify other aspects such as the growth mechanism and the adsorption dynamics. This was observed in $(PAH + PSS)_n$ and $(PDADMAC + PSS)_n$ multilayers assembled from NaCl solutions of concentration 500 mM. The use of PDADMAC as polycation to replace the PAH changes both the growth trend of the multilayer and the dependence of the characteristic time for the polymer adsorption on N. [35, 39]

4.2. Effect of the Solution Concentration

The most adequate concentration for the fabrication of LbL films strongly depends on the nature of the materials to deposit. In general, the adsorbed amount in the multilayer increases with the solution concentration as was pointed out by Guzmán et al. [31] for $(PDMAEMA + PSS)_n$ multilayers (being PDMAEMA a copolymer of 2-(N,N-dimethylamino)ethyl-methacrylate and ε-caprolactone) and by Garg et al. [49] for multilayers of PAH and Poly{1-[*p*-(3′-carboxy-4′-hydroxyphenylazo)benzenesulfonamido]-1,2-ethandiyl} (see Figure 8). The minimum concentration needed to form a layer is closely correlated to the material's solubility and its charge density. Above a concentration threshold value further concentration increases do not modify the multilayer formation. [50]

4.3. Properties of the Solutions

The properties of the films are correlated to the characteristic of the solutions used for the assembling. The solution properties tune the charge density, conformation and solubility of the molecules used, and therefore the properties of the films. [20]

A most important characteristic is the solvent quality for each component. This can be controlled in different ways, e.g., changing pH, adding inert salts or co-solvents, changing the temperature, etc. In general, most materials used in the fabrication of LbL films, such as polyelectrolytes, are highly soluble in water and poorly soluble in organic solvents. [51] Therefore, adding small amounts of organic solvents (e.g., ethanol) changes the ionization degree and the solubility and, in most cases reduces the importance of the electrostatic stabilization of the multilayers.

Dubas and Schlenoff [24] showed that the addition of increasingly amounts of ethanol to the water solutions used for assembling $(PDADMAC + PSS)_n$ multilayers increases the average thickness of the film. This is due to the lower dielectric constant of the medium that decreased the solvent quality, thus favouring the coil conformation of the chains, which enhances the adsorption. The limited solubility of the polyelectrolytes in ethanol does not allow one to use solutions with alcoholic content higher than the 40 v/v%. Similar behaviours have been reported for $(PAH + PSS)_n$ in mixtures of water with chloroform and formamide. [52]

The solvent quality can also be modified by changes of I or by changes in the pH of the solutions. [24, 30, 53] These parameters modify the charge density of the polyelectrolytes, and consequently the strength and type of the interactions. The decrease of the charge density of the chains lowers the solvent quality, with the same consequence described above. The reduction of the charge density changes can be induced by increasing I (charge screening), or by changing the pH of the solutions (reduction of the ionization degree). [24, 26] The increase of the multilayer thickness as the charge density is reduced, due to modifications of I is illustrated in Figure 9 for $(PDADMAC + PSS)_n$ multilayers. [39] Reducing the charge density of the polymer chains below a threshold value can hinder completely the multilayer growth due to a high level of charge screening. [30] The influence of the reduction of the charge density is especially important in the case of weak polyelectrolytes, such as PAH and (PAA) or a vast number of biopolymers. [30, 53]

A last parameter that allows one to modify the solvent quality is the temperature. The effect of this variable in the multilayer assembling has been less explored than those previously described even though its influence can be stronger. This was pointed out by Salomäki et al. [54] who showed that changes in the temperature of the solutions alter the growth mechanism (from linear to non-linear) of both $(PDADMAC + PSS)_n$ and $(PAH + PSS)_n$ multilayers. The latter case is especially relevant because no changes in the growth mechanism have been found by modifying any of the variables discussed above. [39, 55]

4.4. Nature of the Supporting Electrolyte

The effect of the supporting electrolyte in the assembling of PEMs can be rationalized in terms of the well-known Hofmeister series. [56] In general, the anions affect strongly the building-up of the PEMs, whereas the cations have a smaller effect. [56] The increase of the hydrophobicity of the anions usually increases the thickness and roughness of the PEMs. Salomäki et al. [57] pointed out that for $(PDADMAC + PSS)_n$ the effect of sodium salts on the thickness decreases according to: $Br^- > NO_3^- > ClO_3^- > Cl^- > BrO_3^- > HCOO^- > F^-$, *as shown in Figure 10. This is correlated to the ability of the co-anions to decrease the hydration of the polyelectrolytes. Highly hydrophobic counterions bind more strongly to chains; thus reducing their hydration degree, and consequently the polyelectrolyte adopts a more coiled conformation. In the case of the cations, their influence can be ascribed to their different polarizability. Long* et al. *[58] pointed out that the thickness of* $(PDADMAC + PAMS)_n$ *multilayers increase with the ion polarizability following the series:* $Li^+ < Na^+ < K^+$.

4.5. Drying, Rinsing and Adsorption Time

Since PEMs are non-equilibrium structures, the protocol used in the fabrication of nanocoatings has a strong influence in their structure and properties. [35] A first aspect to be controlled is the time used to adsorb the layers. The internal structure and roughness of the coatings are correlated to the reorganization of the polymers chains along the whole multilayer structure. The use of long adsorption times can lead to multilayers in which the layers are not stratified because of the interdiffusion of the material through the multilayer structure. [26, 35] Furthermore, the interdiffussion can induce a significant increase in the multilayers roughness. [35] This allows one to draw a picture where the role of the contact time of the multilayers with the solution is critical to control the structure and properties of the films. As a consequence, it is convenient to use assembling techniques that require low contact time between the substrate and the solutions in case that the goal is to obtain stratified films with low roughness. [26, 59]

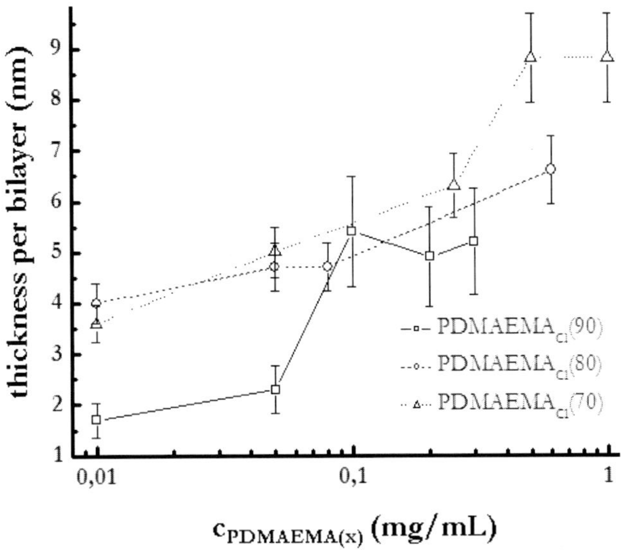

Figure 8. Concentration dependence of the average thickness per bilayer of (PDMAEMA + PSS)$_n$, where x indicates the weight content of 2-(N,N-dimethylamino)ethyl-methacrylate in the copolymer) in the copolymer. Adapted with permission from Reference 31. Copyright (2010) American Chemical Society.

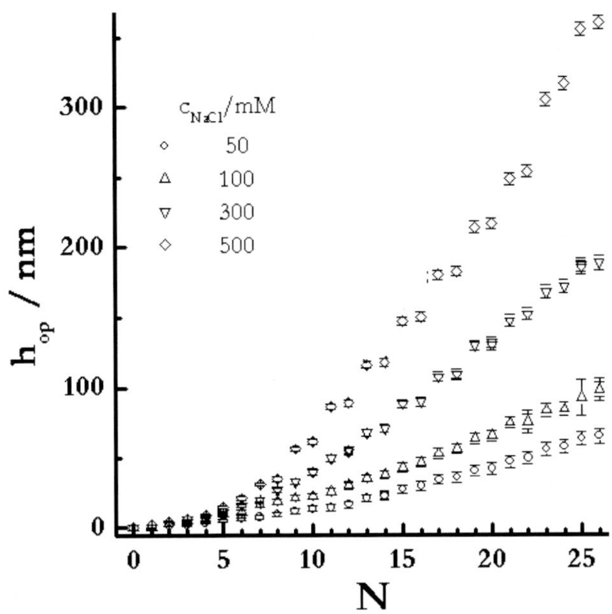

Figure 9. Effect of I on the growth of (PDADMAC + PSS)$_n$ multilayers. The optical thickness, h_{op}, was measured by ellipsometry. Adapted from Reference 26 with permission from The Royal Society of Chemistry.

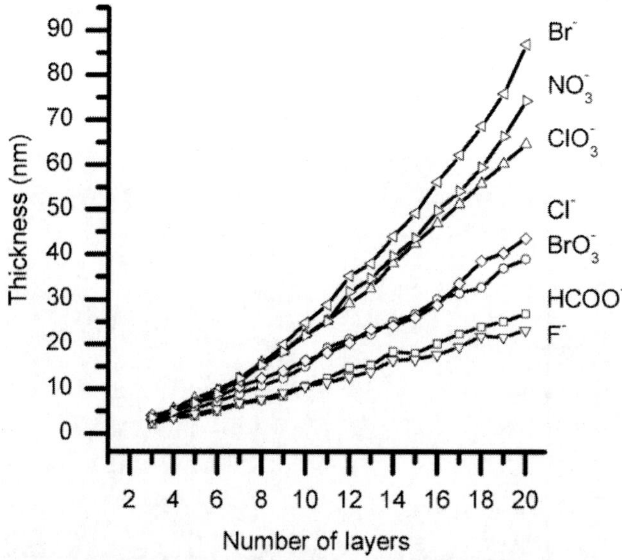

Figure 10. Counter-anion effect on the thickness of $(PDADMAC + PSS)_n$ multilayers. Reprinted with permission from Reference 57. Copyright (2004) American Chemical Society.

The influence of the rinsing step between the assembling of two consecutive layers depends strongly on the molecules used. In general, molecules strongly attached to the multilayers, such as strong polyelectrolytes, remain almost unaffected by rinsing, whereas weakly attached materials are generally removed by rinsing. [4]

Since LbL is a wet method, [4] it is expected that processes changing the layer hydration can modify the structure of the films. The drying step between the adsorption of adjacent layers can change multilayer growth trend and even it can block completely the growth. [60] The significant role of the drying is observed in multilayers of poly(o-methoxyaniline) and poly(vinylsulfonic acid) where the drying steps are needed to obtain an optimal multilayer growth. [60] It is also important to note that the way used to perform the drying plays also a noticeable role in the multilayer structure. The roughness of $(PAH + PSS)_n$ multilayers increase when the ambient air is replace by a nitrogen stream for drying. [61]

5. Applications

The number of applications of LbL coatings has undergone a spectacular grown in recent years due to the good properties associated to this type of films. This has allowed their use in the design and fabrication of both passive and active ("smart coatings") coatings for multiple applications. [5] In the following, some of these applications will be described.

5.1. Superhydrophobic and Superhydrophilic Coatings

The fabrication of superhydrophobic and superhydrophilic surfaces by the LbL approach is one of the fields with a more intense development. These coatings are generally composed by mixtures of polyelectrolytes and nanoparticles in order to control the roughness of the films to minimize the contact angle hysteresis. [62] Zhai et al. [23] showed that the combination of polyelectrolyte layers of PAA and PAH with silica nanoparticles and a fluorinated copolymer lead to superhydrophobic coatings with a contact angle around of 175°. Another example was given by Qi et al. [63] with multilayers formed by cellulose and chitin fibrils. Han et al. [64] built superhydrophobic surfaces based in (PAH + PAA-coated ZrO_2)$_n$. They pointed out the necessity to built PEMs with a minimum of twenty bilayers to obtain surfaces with good water-repellent properties. Furthermore, they showed that the water repellency of the PEMs was associated to the nature of the last layer (PAH-capped multilayers evidenced better repellency) due to the orientation of the polyelectrolytes in the formed films. Han et al. [65] showed that the exposure of multilayers formed by PAH and silica nanoparticles to UV-ozone plasma allows tuning reversibly the wettability properties of the coating from superhydrophobic (contact angle > 165°) to superhydrophilic (contact angle < 10°).

5.2. Anticorrosion and Self-Healing Coatings

LbL multilayers can be also used in the fabrication of anticorrosion protective coatings. Farhat and Schlenoff pointed out that the coating of stainless steel with ten bilayers of (PDADMAC + PSS)$_n$ reduced significantly the corrosion of the metal. [66] This anticorrosion performance was improved for multilayers with intrinsic compensation where the permeability to small ions and the water content of the PEMs were reduced due to the ionic pairing between polyelectrolytes in adjacent layers. An alternative way to enhance the anticorrosive properties is by the building of PEMs with hydrophobic polyelectrolytes: (Poly(N-octadecyl-2-ethynyl-pyridinum bromide) + sulfonate poly(ether-ether-ketone sulfonate))$_n$ or (poly(vinyl-pyridine) + PSS)$_n$.[66] Khaled et al. [67] pointed out that the increase of the polyanions acidity alters the anticorrosion properties of the PEMs onto steel. Faure et al. [68] demonstrated the possibility to protect galvanized steel by multilayers formed by a polyelectrolyte with catechol groups, operating as corrosion inhibitor, and clays, acting as protective barrier. These coatings did not change the surface aspect of the metal. Andreeva et al. [69] pointed out the good performance of (PEI + PSS)$_n$ (being PEI poly(ethylenimine)) in the corrosion protection of aluminium treated previously by ultrasonication.

Another active research field for the application of the LbL films is the fabrication of self-healing surfaces. [22, 70, 71] The possibility to introduce in the interior of the multilayers nano-reservoirs or nano-reactors allows one to design coatings with self-healing properties. Shchukin et al. [72] showed the good performance of LbL film in the fabrication of anticorrosion coatings onto Aluminium. Wang et al. [73] pointed out that multilayer formed only by polyelectrolyte can also present self-healing properties. These authors showed that (PEI + PAA)$_n$ multilayers can self-heal in contact with water due to the interdiffussion of the polyelectrolytes in the internal layers to the exterior region of the multilayers. Moreover, the self-healing properties of these multilayers depend on their structure and thickness. The existence of strong cross-linking in the multilayer structure reduces the ability to self-heal due to the blocked interdiffussion.

5.3. Biocompatible and Antifouling Coatings

The possibility of using biocompatible materials in the fabrication of multilayers has stimulated the research in the field of the biomaterials. Implantable biomaterials, such as stent coated by LbL layers have been described. [74] This type of LbL coatings can be used for purposes such as anticoagulation or as antibacterial materials. [21] Etienne et al. [75, 76] pointed out that PEMs coating functionalized with defensin and chromofungin inhibit the growth and proliferation of bacteria that attach to the coated surface. Some of these coatings have been applied for bone repairing, vascular engineering, tracheal prostheses or dental applications. [21, 77, 78] Kim et al. [77] show that the coating of stents with multilayers formed by polylysine and hyaluronic acid-grafted-poly(lactic-co-glycolic acid) loaded with heparin and paclitaxel avoid the proliferation of smooth muscle cell onto stent surfaces. Thierry et al. [79] pointed out that the coating of prosthesis with polysaccharides based PEMs enhance the thromboresistance.

A very promising application is the encapsulation of drugs for controlled delivery. [80, 81] Some of the drugs encapsulated are daunorubicin, doxorubicin, polyphenols, dextran, heparin, α-lactalbumin, lysozyme, ovalbumin, α-chymotrypsin, insulin and pronase.

The possibility to modulation the properties of LbL multilayers have allowed their application as coating for cellular culture media. Among the different properties that make special interesting this type of systems for this application is the possibility to control the mechanical properties of the films. This is particularly interesting in the control of the cell adhesion and attachment onto the substrate. For this purpose stiff films are requires, for which multilayers formed by synthetic polymers are important, whereas the lower stiffness of multilayers formed by biopolymers hinder the cellular attachment. [74]

Associated to the above cell adhesion, the antifouling properties of the surfaces play a key role. Antifouling coatings prevent the adsorption of protein and other organic materials. The modification of $(PAH + PSS)_n$ multilayers by grafting poly(ethylene-oxide) and phosphorylcholine reduces the adsorption of proteins onto the coated surfaces. [82] It is also possible to design surfaces with good anti-fouling properties by the coating with polysaccharide multilayers. In this case, the anti-fouling properties are conferred by the high water content of the layers. [78] Wong et al. [83] have demonstrated that the antifouling properties of $(PEI + PAA)_n$ were enhanced with the increase of N due to the formation of a coating with high coverage density.

5.4. Other Applications

Qi et al. [84] have showed that a classical multilayer such as $(PAH + PSS)_n$ shows a good performance in electroosmotic separation. Its behavior depends on the number of layers and the nature of the last adsorbed layer. The LbL coatings have also showed a good performance in the design of microfluidic devices. [85]

The possibility to introduce into the multilayers materials such as carbon nanotubes or clays allows one to design highly performing composite coatings. These coatings confer special mechanical, electrical and optical properties to the coated materials. Bertoni et al. [86] showed that the combination of PDADMAC with single-walled carbon nanotubes lead to highly conductive coatings.

Conclusion

The recent developments in the application of the LbL approach in the nanocoatings fabrication have opened new perspective in the design of new soft materials with enhanced properties. The multiple possibilities of control of the structure and properties of this type of systems offered by this methodology have allowed the design of functional coating with application in several nano-technological fields, e.g., optical and electrical materials, tissue engineering, drug delivery, etc. Furthermore, the versatility of the LbL leads to the fabrication of responsiveness materials with activity against different physico-chemical stimuli that have allowed the fabrication of both passive and active coatings. In addition, the progressive development of new approaches to perform the LbL deposition has lead to the scaling up of this approach from the laboratory scale to the industrial level open the application of LbL nanocoatings in real life applications. In summary, the LbL approach has become a reference methodology to design and fabricate nanocoatings.

Acknowledgments

This work was supported by MINECO (FIS2012-14008-C02-01), by ESA (FASES and PASTA), and by EU (Marie Curie ITN-COWET), and carried out in the framework of the COST actions CM1101. We are grateful to the UIRC of the CAI of Spectroscopy (Universidad Complutense de Madrid) for allow us the use of their facilities.

References

[1] Yoshida, M., Lendlein, R. L. A., Lahann, J., From Advanced Biomedical Coatings to Multi-Functionalized Biomaterials. *J. Macromol. Sci., Part C* 2006, 46, 347-375.

[2] Elbert, D. L., Herbert, C. B., Hubbel, J. A., Thin Polymer Layers Formed by Polyelectrolyte Multilayer Techniques on Biological Surfaces. *Langmuir* 1999, 15, 5355-5362.

[3] Decher, G., Fuzzy Nanoassemblies: Toward Layered Polymeric Multicomposites. *Science* 1997, 277, 1232-1237.

[4] Decher, G., Schlenoff, J. B., *Multilayer Thin Films-Sequential Assembly of Nanocomposite Materials*. Wiley-VCH Verlag: Berlin, 2003.

[5] Schlenoff, J. B., Retrospective on the Future of Polyelectrolyte Multilayers. *Langmuir* 2009, 25, 14007-14010.

[6] Iler, R. K., Multilayers of colloidal Particles. *J. Colloid Interface Sci.* 1966, 21, 569-594.

[7] Decher, G., Hong, J. D., Buildup of ultrathin multilayer films by a self-assembly process: II. Consecutive adsorption of anionic and cationic bipolar amphiphiles and polyelectrolytes on charged surfaces. *Ber. Bunsenges. Phys. Chem* 1991, 95, 1430-1434.

[8] Priolo, M. A., Gamboa, D., Holder, K. M., Grunlan, J. C., Super Gas Barrier of Transparent Polymer-Clay Multilayer Ultrathin Films. *NanoLett.* 2010, 10 4970-4974.

[9] Zhang, T., Yan, H., Peng, M., Wang, L., Ding, H., Fang, Z., Construction of flame retardant nanocoating on ramie fabric via layer-by-layer assembly of carbon nanotube and ammonium polyphosphate. *Nanoscale* 2013, 5, 3013-3021.

[10] Adamczak, M., Para, G., Simon, C., Warszynski, P., Natural oil nanoemulsions as cores for layer-by-layer encapsulation. *J. Microencapsul.* 2013, 30, 479-489.

[11] Caruso, F., Donath, E., Möhwald, H., Influence of Polyelectrolyte Multilayer Coatings on Förster Resonance Energy Transfer between 6-Carboxyfluorescein and Rhodamine B-Labeled Particles in Aqueous Solution. *J. Phys. Chem. B* 1998, 102, 2011-2016.

[12] Ruano, M., Ritacco, H., Rubio, R. G., Ortega, F., Layer-by-layer assembly of polyelectrolytes onto fluid micro-interfaces: Polyelectrolyte-coated liposomes. *Abs. Pap. Am. Chem. Soc.* 2011, 239, 28-COLL.

[13] Guzmán, E., Ritacco, H., Ortega, F., Svitova, T., Radke, C. J., Rubio, R. G., Adsorption Kinetics and Mechanical Properties of Ultrathin Polyelectrolyte Multilayers: Liquid-Supported versus Solid-Supported Films. *J. Phys. Chem. B* 2009, 113, 7128-7137.

[14] Blodgett, K. B., Films built by depositing successive monomolecular layers on a solid surface. *J. Am. Chem. Soc.* 1935, 57, 1007-1022.

[15] Love, J. C., Estroff, L. A., Kriebel, J. K., Nuzzo, R. G., Whitesides, G. M., Self-Assembled Monolayers of Thiolates on Metals as a Form of Nanotechnology. *Chem. Rev.* 2005, 105, 1103-1170.

[16] Stockton, W. B., Rubner, M. F., Molecular-Level Processing of Conjugated Polymers. 4. Layer-by-Layer Manipulation of Polyaniline via Hydrogen-Bonding Interactions. *Macromolecules* 1997, 30, 2717-2725.

[17] Shimazaki, Y., Mitsuishi, M., Ito, S., Yamamoto, M., Preparation of the Layer-by-Layer Deposited Ultrathin Film Based on the Charge-Transfer Interaction. *Langmuir* 1997, 13, 1385-1387.

[18] Bourdillon, C., Demaille, C., Moiroux, J., Savéant, J. M., Step-by-Step immunological construction of a fully active multilayer enzyme electrode. *J. Am. Chem. Soc.* 1994, 116, 10328-10329.

[19] Xiong, H. M., Cheng, M. H., Zhou, Z., Zhang, X., Shen, J. C., A new approach to the fabrication of a self-organizing film of heterostructured polymer/Cu_2S nanoparticles. *Adv. Mater.* 1998, 10, 529-532.

[20] von Klitzing, R., Internal Structure of polyelectrolyte multilayer assemblies. *Phys. Chem. Chem. Phys.* 2006, 8, 5012-5033.

[21] Boudou, T., Crouzier, T., Ren, K., Blin, G., Picart, C., Multiple Functionalities of Polyelectrolyte Multilayer Films: New Biomedical Applications. *Adv. Mater.* 2010, 22, 441-467.

[22] Shchukin, D. G., Möhwald, H., Self-Repairing Coatings Containing Active Nanoreservoirs. *Small* 2007, 3, 926-943.

[23] Zhai, L., Cebeci, F. C., Cohen, R. E., Rubner, M. F., Stable Superhydrophobic Coatings from Polyelectrolyte Multilayers. *NanoLett.* 2004, 4, 1349-1353.

[24] Dubas, S. T., Schlenoff, J. B., Factors Controlling the Growth of Polyelectrolyte Multilayers. *Macromolecules* 1999, 32, 8153-8160

[25] Schlenoff, J. B., Dubas, S. T., Mechanism of Polyelectrolyte Multilayer Growth: Charge Overcompensation and Distribution. *Macromolecules* 2001, 34, 592-598.

[26] Guzmán, E., Ritacco, H., Rubio, J. E. F., Rubio, R. G., Ortega, F., Salt-induced changes in the growth of polyelectrolyte layers of poly(diallyldimethylammoniumchloride) and poly(4-styrene sulfonate of sodium). *Soft Matter* 2009, 5, 2130-2142.

[27] Adamczyk, Z., Zembala, M., Warszynski, P., Jachimska, B., Characterization of Polyelectrolyte Multilayers by the Streaming Potential Method. *Langmuir* 2004, 20, 10517-10525.

[28] Strydom, S. J., Otto, D. P., Liebenberg, W., Lvov, Y. M., de Villiers, M. M., Preparation and characterization of directly compactible layer-by-layer nanocoated cellulose. *Int. J. Pharm.* 2011, 404, 57-65.

[29] Kharlampieva, E., Kozlovskaya, V., Ankner, J. F., Sukhishvili, S. A., Hydrogen-Bonded Polymer Multilayers Probed by Neutron Reflectivity. *Langmuir* 2008, 24, 11346-11349.

[30] Guzmán, E., Cavallo, J. A., Chuliá-Jordán, R., Gómez, C., Strumia, M. C., Ortega, F., Rubio, R. G., pH-Induced Changes in the Fabrication of Multilayers of Poly(acrylic acid) and Chitosan: Fabrication, Properties, and Tests as a Drug Storage and Delivery System. *Langmuir* 2011, 27, 6836-6845.

[31] Guzmán, E., Miguel, V. S., Peinado, C., Ortega, F., Rubio, R. G., Polyelectrolyte Multilayers Containing Triblock Copolymers of Different Charge Ratio. *Langmuir* 2010, 26, 11494-11502.

[32] Lehaf, A. M., Hariri, H. H., Schlenoff, J. B., Homogeneity, Modulus, and Viscoelasticity of Polyelectrolyte Multilayers by Nanoindentation: Refining the Buildup Mechanism. *Langmuir* 2012, 28, 6348-6355.

[33] Kotov, N. A., Layer-by-layer self-assembly: The contribution of hydrophobic interactions. *Nanostruct. Mat.* 1999, 12, 789-796.

[34] Lvov, Y., Yamada, S., Kunitake, T., Non-linear optical effects in layer-by-layer alternate films of polycations and an azobenzene-containing polyanion. *Thin Solid Films* 1997, 300, 107-112.

[35] Guzmán, E., Ritacco, H., Ortega, F., Rubio, R. G., Evidence of the influence of adsorption kinetics on the internal reorganization of polyelectrolyte multilayers. *Colloids Surf. A* 2011, 384, 274- 281.

[36] Schönhoff, M., Schwarz, B., Larsson, A., Kuckling, D., Dynamics in polymer layers investigated by NMR techniques. *Progr. Colloid Polym. Sci.* 2002, 121, 80-87.

[37] Fu, Y., Chen, H., Qiu, D., Wang, Z., Zhang, X., Multilayer Assemblies of Poly(4-vinylpyridine) and Poly(acrylic acid) Bearing Photoisomeric Spironaphthoxazine via Hydrogen Bonding. *Langmuir* 2002, 18, 4989-4995.

[38] Cini, N., Tulun, T., Decher, G., Ball, V., Step-by-Step Assembly of Self-Patterning Polyelectrolyte Films Violating (Almost) All Rules of Layer-by-Layer Deposition. *J. Am. Chem. Soc.* 2010, 132, 8264-8265.

[39] Guzmán, E., Ritacco, H. A., Ortega, F., Rubio, R. G., Growth of Polyelectrolyte Layers Formed by Poly(4-styrenesulfonate sodium salt) and Two Different Polycations: New Insights from Study of Adsorption Kinetics. *J. Phys. Chem. C* 2012, 116, 15474-15483.

[40] Li, Y., Wang, X., Sun, J., Layer-by-layer assembly for rapid fabrication of thick polymeric films. *Chem. Soc. Rev.* 2012, 41, 5998-6009.

[41] Gamboa, D., Priolo, M. A., Ham, A., Grunlan, J., Note: Influence of rinsing and drying routines on growth of multilayer thin films using automated deposition system. *Rev. Sci. Instr.* 2010, 81, 036103.

[42] Wohl, B. M., Engbersen, J. F. J., Responsive layer-by-layer materials for drug delivery. *J. Controlled Release* 2012, 158, 2-14.

[43] Kiel, M., Mitzscherling, S., Leitenberger, W., Santer, S., Tiersch, B., Sievers, T. K., Möhwald, H., Bargheer, M., Structural Characterization of a Spin-Assisted Colloid-polyelectrolyte Assembly: Stratified Multilayer Thin Films. *Langmuir* 2010, 26, 18499-18502.

[44] Kolansinska, M., Krastev, R., Gutberlet, T., Warszynski, P., Layer-by-Layer Deposition of Polyelectrolytes. Dipping versus Spraying. *Langmuir* 2009, 25, 1224-1232.

[45] Cho, J., Char, K., Hong, J.-D., Lee, K.-B., Fabrication of Highly Ordered Multilayer Films Using a Spin Self-Assembly Method. *Adv. Mat.* 2001, 13, 1076-1078.

[46] Schlenoff, J. B., Dubas, S. T., Farhat, T., Sprayed Polyelectrolyte Multilayers. *Langmuir* 2000, 16, 9968-9969.

[47] Félix, O., Zheng, Z., Cousin, F., Decher, G., Are sprayed LbL-films stratified? A first assessment of the nanostructure of spray-assembled multilayers by neutron reflectometry. *Comptes Rendus Chimie* 2009, 12, 225-234.

[48] von Klitzing, R., Wong, J. E., Jaeger, W., Steitz, R., Short range interactions in polyelectrolyte multilayers. *Curr. Opin. Colloid Interface Sci.* 2004, 9, 158-162.

[49] Garg, A., Heflin, J. R., Gibson, H. W., Davis, R. M., Study of Film Structure and Adsorption Kinetics of Polyelectrolyte Multilayer Films: Effect of pH and Polymer Concentration. *Langmuir* 2008, 24, 10887-10894.

[50] Okubo, T., Suda, M., Absorption of polyelectrolytes on colloidal surfaces as studied by electrophoretic and dynamic light-scattering techniques. *J. Colloid Interface Sci.* 1999, 213, 565-571.

[51] Dobrynin, A. V., Rubinstein, M., Theory of polyelectrolytes in solutions and at interfaces. *Progr. Polym. Sci.* 2005, 30, 1049-1118.

[52] Kamineni, V. K., Lvov, Y. M., Dobbins, T. A., Layer-by-Layer Nanoassembly of Polyelectrolytes Using Formamide as the Working Medium. *Langmuir* 2007, 23, 7423-7427.

[53] Bieker, P., Schönhoff, M., Linear and Exponential Growth Regimes of Multilayers of Weak Polyelectrolytes in Dependence on pH. *Macromolecules* 2010, 43 5052-5059.

[54] Salomaki, M., Vinokurov, I. A., Kankare, J., Effect of Temperature on the Buildup of Polyelectrolyte Multilayers. *Langmuir* 2005, 21, 11232-11240.

[55] Lavalle, P., Gergely, C., Cuisinier, F. J. G., Decher, G., Schaaf, P., Voegel, J. C., Picart, C., Comparison of the Structure of Polyelectrolyte Multilayer Films Exhibiting a Linear and an Exponential Growth Regime: An in Situ Atomic Force Microscopy Study. *Macromolecules* 2002, 35, 4458-4465.

[56] Record, M. T., Guinn, E., Pegram, L., Capp, M., Introductory Lecture: Interpreting and predicting Hofmeister salt ion and solute effects on biopolymer and model processes using the solute partitioning model. *Far. Disc.* 2013, 160, 9-44.

[57] Salomäki, M., Tervasmäki, P., Areva, S., Kankare, J., The Hofmeister Anion Effect and the Growth of Polyelectrolyte Multilayers. *Langmuir* 2004, 20, 3679-3683.

[58] Long, Y., Wang, T., Liu, L., Liu, G., Zhang, G., Ion Specificity at a Low Salt Concentration in Water-methanol Mixtures Exemplified by a Growth of Polyelectrolyte Multilayer. *Langmuir* 2013, 29, 3645-3653.

[59] Kharlampieva, E., Kozlovskaya, V., Chan, J., Ankner, J. F., Tsukruk, V. V., Spin-Assisted Layer-by-Layer Assembly: Variation of Stratification as Studied with Neutron Reflectivity. *Langmuir* 2009, 25 14017-14024.

[60] Raposo, M., Pontes, R. S., Mattoso, L. H. C., Oliveira, O. N., Kinetics of adsorption of poly(o-methoxyaniline) self-assembled films *Macromolecules* 1997, 30, 6095-6101.

[61] Silva, H. S., Miranda, P. B., Molecular Ordering of Layer-by-Layer Polyelectrolyte Films Studied by Sum-Frequency Vibrational Spectroscopy. *J. Phys. Chem B* 2009, 113, 10068-10071.

[62] Ghostine, R. A., Jisr, R. M., Lehaf, A., Schlenoff, J. B., Roughness and Salt Annealing in a Polyelectrolyte Multilayer. *Langmuir* 2013, 29, 11742-11750.

[63] Qi, Z.-D., Saito, T., Fan, Y., Isogai, A., Multifunctional Coating Films by Layer-by-Layer Deposition of Cellulose and Chitin Nanofibrils. *Biomacromolecules* 2012, 13, 553-558.

[64] Han, J. T., Zheng, Y., Cho, J. H., Xu, X., Cho, K., Stable Superhydrophobic Organic-Inorganic Hybrid Films by Electrostatic Self-Assembly. *J. Phys. Chem. B* 2005, 109, 20773-20778.

[65] Han, J. T., Kim, S., Karim, A., UVO-Tunable Superhydrophobic to Superhydrophilic Wetting Transition on Biomimetic Nanostructured Surfaces. *Langmuir* 2007, 23, 2608-2614.

[66] Farhat, T. R., Schlenoff, J. B., Corrosion Control Using Polyelectrolyte Multilayers. *Electrochem. Solid-State Lett.* 2002, 5, B13-B15.

[67] Khaled, M., Abu-Sharkh, B., Amr, E., Yilbas, B. S., Manda, A., Abulkibash, A., Corrosion properties of 316L stainless steel coated with polyelectrolyte multilayers of varying anionic acidity. *Corrosion Eng. Sci. Tech.* 2007, 42, 356-362.

[68] Faure, E., Halusiak, E., Farina, F., Giamblanco, N., Motte, C., Poelman, M., Archambeau, C., Weerdt, C. V. D., Martial, J., Jérôme, C., Duwez, A.-S., Detrembleur, C., Clay and DOPA Containing Polyelectrolyte Multilayer Film for Imparting Anticorrosion Properties to Galvanized Steel. *Langmuir* 2012, 28, 2971-2978.

[69] Andreeva, D. V., Fix, D., Möhwald, H., Shchukin, D. G., Buffering polyelectrolyte multilayers for active corrosion protectionn. *J. Mater. Chem.* 2008, 18, 1738-1740.

[70] Skorbab, E. V., Andreeva, D. V., Layer-by-Layer approaches for formation of smart self-healing materials. *Polym. Chem.* 2013, 4, 4834-4845.

[71] Shchukin, D. G., Container-based multifunctional self-healing polymer coatings. *Polym. Chem.* 2013, 4, 4871-4877.

[72] Shchukin, D. G., Zheludkevich, M., Yasakau, K., Lamaka, S., Ferreira, M. G. S., Möhwald, H., Layer-by-Layer Assembled Nanocontainers for Self-Healing Corrosion Protection. *Adv. Mater.* 2006, 18, 1672-1678.

[73] Wang, X., Liu, F., Zheng, X., Sun, J., Water-Enabled Self-Healing of Polyelectrolyte Multilayer Coatings. *Angew. Chem. Int. Ed.* 2011, 50, 11378 -11381.

[74] Picart, C., Polyelectrolyte Multilayer Films: From Physico-Chemical Properties to the Control of Cellular Processes. *Curr. Medicinal Chem.* 2008, 15, 685-697.

[75] Etienne, O., Gasnier, C., Taddei, C., Voegel, J. C., Aunis, D., Schaaf, P., Metz-Boutigue, M. H., Bolcato-Bellemin, A. L., Egles, C., Antifungal coating by biofunctionalized polyelectrolyte multilayered films. *Biomaterials* 2005, 26, 6704-6712.

[76] Etienne, O., Picart, C., Taddei, C., Haikel, Y., Dimarcq, J. L., Schaaf, P., Voegel, J. C., Ogier, J. A., Egles, C., Multilayer Polyelectrolyte Films Functionalized by Insertion of Defensin: a New Approach to Protection of Implants from Bacterial Colonization. *Antimicrob. Agents Chemother.* 2004, 48, 3662-3669.

[77] Kim, T. G., Lee, H., Jang, Y., Park, T. G., Controlled Release of Paclitaxel from Heparinized Metal Stent Fabricated by Layer-by-Layer Assembly of Polylysine and Hyaluronic Acid-g-Poly(lactic-co-glycolic acid) Micelles Encapsulating Paclitaxel. *Biomacromolecules* 2009, 10, 1532-1539.

[78] Crouzier, T., Boudou, T., Picart, C., Polysaccharide-based polyelectrolyte multilayers. *Curr. Opin. Colloid Interface Sci.* 2010, 15, 417-426.

[79] Thierry, B., Winnik, F. M., Merhi, Y., Silver, J., Tabrizian, M., Bioactive Coatings of Endovascular Stents Based on Polyelectrolyte Multilayers. *Biomacromolecules* 2003, 4, 1564-1571.

[80] Guzman, E., Ruano, M., Rubio, R. G., F.Ortega, Polyelectrolyte assemblies for drug storage and delivery: Multilayers, nanocapsules, and multicapsules. In *Inhaler devices: Fundamentals, design and drug delivery*, Prokopovich, P., Ed. Woodhead Publishing Limited: Sawston, Cambridge, UK 2013, pp 94-145.

[81] Shchukina, E. M., Shchukin, D. G., Layer-by-layer coated emulsion microparticles as storage and delivery tool. *Curr. Opin. Colloid Interface Sci.* 2012, 17, 281-289.

[82] Reisch, A., Hemmerlé, J., Chassepot, A., Lefort, M., Benkirane-Jessel, N., Candolfi, E., Mésini, P., Letscher-Bru, V., Voegel, J.-C., Schaaf, P., Anti-fouling phosphorylcholine bearing polyelectrolyte multilayers: Cell adhesion resistance at rest and under stretching. *Soft Matter* 2010, 6, 1503-1512.

[83] Wong, S. Y., Han, L., Timachova, K., Veselinovic, J., Hyder, M. N., Ortiz, C., Klibanov, A. M., Hammond, P. T., Drastically lowered protein adsorption on microbicidal hydrophobic/hydrophilic polyelectrolyte multilayers. *Biomacromolecules* 2012, 13, 719-726.

[84] Qi, S., Li, W., Zhao, Y., Ma, N., Wei, J., Chin, T. W., Tang, C. Y., Influence of the properties of layer-by-layer active layers on forward osmosis performance. *J. Membrane Sci.* 2012, 423-424, 536-542.

[85] Kumlangdudsana, P., Tuantranont, A., Dubas, S. T., Dubas, L., Polyelectrolyte multilayers coating for organic solvent resistant microfluidic chips. *Mat. Lett.* 2011, 65, 3629-3632.

[86] Bertoni, C., Skákalová, V., Roth, S., Layer-by-layer deposition of ultra-thin films of carbon nanotubes. *Physica E* 2008, 40, 2257-2262.

In: Comprehensive Guide for Nanocoatings Technology. Vol. 1 ISBN: 978-1-63482-447-7
Editor: Mahmood Aliofkhazraei

Chapter 2

SURFACE TREATMENT OF TiO_2 (NANO) PARTICLES

*Nika Veronovski**
Cinkarna Metallurgical and Chemical Industry Celje, Inc., Slovenia

ABSTRACT

Performance properties of various (nano) particles are determined principally by their size and the chemical composition of its surface. Within this Chapter, we will take an exhaustive look into the surface treatment of titanium dioxide – TiO_2 (nano) particles. Of the essential meaning is that particle surface is coated in a controlled manner to maximize its applicable value. Process conditions (e.g., temperature, pH value, the time used for certain process stage, the dosing time, the sequence of process stages) affect the coating morphology and finally applicable properties of TiO_2. Inorganic surface modifiers are most commonly precipitated coatings of alumina (Al_2O_3) and silica (SiO_2). These inorganic surface treatments provide improvements in one or more important performance properties such as: dispersibility, hiding power efficiency, chalk resistance and resistance to discoloration by heat and/or photoreduction.

1. INTRODUCTION

About Titanium Dioxide

Titanium dioxide (TiO_2) is produced in two major commercial forms worldwide: the traditional and quantity-wise essential production of pigment grade, and the innovative nano grade, which is increasingly being utilized in modern technologies. TiO_2 has been used for many years in a vast range of industrial and consumer goods including paints, coatings, adhesives, paper and paperboard, plastics and rubber, printing inks, coated fabrics and textiles, catalyst systems, ceramics, floor coverings, roofing materials, cosmetics and

* nika.veronovski@cinkarna.si.

pharmaceuticals, water treatment agents, food colorants and in automotive products, etc. As a pigment, TiO_2 has excellent light-scattering properties and is used in a variety of applications that require white opacity and brightness. New developments make use of the particle size dependency of light scattering of TiO_2 to produce »transparent« TiO_2 products. TiO_2 as a nanomaterial is engineered to have primary particles less than 100 nm in order to optimize such properties. Nano-scaled TiO_2 particles scatter visible light far less than TiO_2 pigments, yet still have the intrinsic property of UV-light absorption [1].

As viewed from crystalline structure, TiO_2 is a multi-crystal form substance. It includes brookite, rutile and anatase, although only rutile and anatase are widely used. Both rutile and anatase forms have a square crystal system structure. Anatase was found to be a photosensitiser that accelerates photooxidation, while rutile acted as an effective stabilizer retarding photooxidation [2].

Surface Treatment

Although high photocatalytic activity is required for TiO_2 nanoparticles used as photocatalysts, low photocatalytic activity is sometimes preferred for titania additives to avoid the degradation of a matrix. TiO_2 nanoparticles sometime cause photocatalytic degradation of organic molecules, especially when they are exposed to ultraviolet rays or sunlight [3]. TiO_2 with high photocatalytic activity facilitates the generation of reactive oxygen species, which can oxidize and degrade other ingredients.

The catalytic activity of TiO_2 nanoparticles can be enhanced or suppressed by coating them with appropriate materials. The insulating layers like silica (SiO_2) [4], alumina (Al_2O_3), or a polymer are often expected to isolate the nanoparticles from the catalyzed components to suppress the catalytic activity of TiO_2 nanoparticles [3, 5]. These treatments are most commonly precipitated in layers. However, some of the components can be co-precipitated to alter the pigment characteristics. Within this Chapter, silica and alumina surface treatments will be presented in detail.

The core TiO_2 particles are generally subjected to inorganic surface treatments. In this production step, the crude TiO_2 particles are subsequently covered with one or more layers of inorganic substances. Inorganic surface treatments play a role in improving weather resistance and photostability, as well as colloid chemical and optical properties of TiO_2 pigments. Surface treatments may be carried out either discontinuously in stirred, heated vessels, or otherwise, continuously in a number of consecutive tanks of a cascade. Both procedures lead to similar results, although the discontinuous procedure is easier to handle when the surface treatment is to be applied rather slowly for the sake of achieving a low specific surface area in the final product. The principle of this method of pigment modification is to first add dissolved inorganic precursors to a milled suspension of the crude pigment. For this, it is necessary to adjust the pH value of the pigment slurry to a value where the precursors remain in solution. From this mixture, the inorganic substances are finely precipitated by gradual adjustment of the pH value towards neutral. Due to high ionic strength in solution, these precipitates are not stable against flocculation, but agglomerate and attach to the pigment's surface. The art involved in these surface treatments lies in the ability to achieve as complete a coating as possible around the TiO_2 core particles [1].

Silica Coatings

Silica (SiO_2) is a functional material used to improve surfaces and mechanical properties of many materials. For many coating applications silica particles need to be smaller than 40 nm. Silica nanoparticles can be produced on the industrial scale in a “wet process”, which is based on chemical reaction between sodium silicate (Na_2SiO_3) and sulphuric acid (H_2SO_4) in aqueous solutions. Created in this reaction, monomer of silicic acid (SiH_4O_4), condenses to form larger oligomers, which link together to produce primary particles (nucleation). Depending on the process conditions, these particles can either take more monomers and grow or aggregate. Aggregation can finally lead to gelation of the colloidal suspension [6].

Silica treatment can produce very disparate pigment characteristics, depending on the silica precipitation conditions. Silica treatment can produce a durable pigment or a high dry-hiding pigment.

If silica is deposited at acidic or neutral pH, it is deposited in a form of submicroscopic particles joined together in a gel-like structure. This "fluffy" coating provides better spacing and optical efficiency, increases oil absorption and decreases gloss. These pigments are used in matte, dry-hiding formulations. Silica treatments can also be deposited as a dense, glassy shell that encapsulates the particle and provides the highest durability available in titanium dioxide pigments. Dense silica is deposited slowly at basic pH. The dense silica shell does not increase the oil absorption of the TiO_2. Durable silica treatment tends to reduce the gloss of pigments, probably by making it more difficult to develop gloss through the final milling step. The durability of the silica layer and the amount of energy needed to develop gloss are proportional to the amount of silica in the layer [7].

Alumina Coatings

Treatment with hydrous aluminum oxide - $Al(OH)_3$ - is probably the most common treatment on TiO_2 pigments. Hydrous aluminum oxide can be deposited from sodium aluminate ($NaAlO_2$), which is reacted with acid; or aluminum sulfate ($Al_2(SO_4)_3$), which is reacted with a base. Alumina can also be deposited from a combination of sodium aluminate and aluminum sulfate, which neutralize one another. The coating properties greatly depend on the structure and morphology of the coating layer which determine the electrical, magnetic, mechanical, and diffusive properties. Thus, controlling the morphology of coated particles is very important in industrial application. Aluminum oxide and hydrous alumina have many different structures, which can give many different properties to coated particles. It was reported that the structure of the alumina coating on a TiO_2 pigment varied with the pH, i.e., alumina precipitated under an acidic condition was amorphous, while alumina precipitated under a basic condition was crystalline, or boehmite under special conditions [8]. The transition pH is temperature-sensitive and tends to shift to lower pH at higher temperature. A layer of alumina oxide coating can increase the -OH groups on the particle surface, which will improve the dispersibility of the particles in aqueous solution and provide more active sites for further organic modification. The hydrous alumina particles on the pigment surface reduce the particle-particle attractive forces and improve dispersibility. The alumina also acts as a spacer for the particles and improves gloss and opacity of the pigment. The amorphous alumina can provide a moderate amount of durability and does not seem to provide as effective spacing as the pseudoboehmite alumina. When multiple treatment layers are deposited on a pigment, alumina is most often the final layer [7, 9]. The particles coated with alumina behave in some ways that are characteristic of alumina.

Binary Silica-Alumina Coatings

Binary silica/alumina-coated rutile TiO_2 composites were prepared by a liquid-phase deposition method starting from Na-silicate and Na-aluminate.

The coating extent and the morphology of the alumina coating layers are affected by the pH value of reaction solution. The IEPs of the naked TiO_2, SiO_2, and hydrated alumina reported are about 4.1 [10], 2.5 [11], and 6.8 [10], respectively. Under the experimental conditions, where the pH medium of reaction solutions are in a range from acidic to neutral, the silica-coated TiO_2 powders should be negatively charged and the hydrated alumina formed by hydrolysis of Na-aluminate should be positively charged. The positively charged alumina probably combines at the negatively charged silica-coated TiO_2 surfaces via electrostatic force during the coating process. In acidic medium, the formation rate of hydrated alumina aggregates rate via hydrolysis of Na-aluminate is quicker than the deposition rate of hydrated alumina on the silica-coated TiO_2 surfaces via electrostatic force, resulting in the formation of hydrated alumina aggregates between the binary silica/alumina-coated TiO_2 particles. When the pH values are raised to higher values, the amount of positive charges on the hydrated alumina formed by hydrolysis of Na-aluminate is decreased, resulting in a weak combination on the negatively charged silica-coated TiO_2 surfaces and giving the formation of floccular alumina coating layers. Furthermore, at high pH values, the hydrolysis rate of Na-aluminate becomes rapid because Na-aluminate has amphoteric property. The rapid hydrolysis of Na-aluminate favors the homogeneous nucleation of alumina rather than the heterogeneous nucleation on the silica-coated TiO_2 surfaces. When the coating process is carried out at a medium pH, continuous and dense alumina coating layers form on the silica-coated TiO_2 surfaces. Therefore, the formation of continuous and dense alumina coating layers is dependent on both electrostatic force between alumina and silica coated TiO_2 and hydrolysis rate of Na-aluminate [12].

According to the literature [12] the silica coating layers anchor at TiO_2 surfaces via Si–O–Ti bonds and the alumina coating layers probably anchor at the silica-coated TiO_2 surfaces via Al–O–Si bonds [12-14]. The formation of continuous and dense binary silica/alumina coating layers depended on the pH value of reaction solution and the alumina loading. However, alumina coating layers have a tendency to loosely anchor at TiO_2 surfaces [15, 16].

Organic Coatings

Many pigments are treated with an organic coating, but some pigments contain no organic treatment. This treatment can be added into the wet washed pigment before drying, onto the dry pigment, or onto the pigment during the final milling step. The organic materials used are most often polyols, amines or amine salts. Silicones and siloxanes are commonly used to produce hydrophobic pigments for plastics applications. Organic treatments are used to improve compatibility with the coating or plastic and to improve dispersion. Organic coatings also reduce agglomeration during storage of the TiO_2. They reduce water absorption on the TiO_2 surface, which leads to a lower volatile content for plastics processing.

TiO_2 pigment properties are determined by the base pigment and the finishing process. The base pigment is affected by the manufacturing process, the crystal structure, the particle size distribution and the additives during manufacture. During inorganic treatment, pigment properties are affected not only by the amounts of various chemicals but also by the conditions under which the treatment is deposited. Temperature, pH, deposition rate, source chemicals and other factors can influence the morphology or crystal structure of the treatment layers and finally applicable properties of TiO_2. The morphology of the treatment layers can, in turn, have an effect on the final properties of the pigment.

Different precursors as a source of silica and/or alumina used by the various authors are listed in Table I.

Table I. Different precursors as a source of silica and/or alumina used by the various authors

Formation of silica Lin et al. [9], Liu et al. [16], Veronovski et al. [17-19], Viravathana and Marr [20] used Na_2SiO_3 as a source of silica Mori et al. [4], Kalele et al. [21], Bechger et al. [22], El-Toni [23], Park and Kang [24] used TEOS as a source of silica Nur et al. [25] used OTS as a source of silica
Formation of alumina Pacewska et al. [26], Wu et al. [10, 27], Liu et al. [34] used $Al_2(SO_4)_3$ as a source of alumina Hayrapetyan et al. [28], Ozimek et al. [29] used $NaAlO_2$ as a source of alumina *Formation of silica-alumina*
Formation of silica-alumina Zhang et al. [15], Gao et al. [30] used Na_2SiO_3 as a source of silica and $NaAlO_2$ as a source of alumina van Dyk and Heyns [5] used Na_2SiO_3 as a source of silica and $Al_2(SO_4)_3$ or $NaAlO_2$ as a source of alumina

* Abbrevations for precursors used:
Sodium silicate (Na_2SiO_3), Tetraethyl orthosilicate (TEOS), Octadecyltrichlosilane (OTS), Aluminium sulfate ($Al_2(SO_4)_3$), Sodium silicate ($NaAlO_2$)

2. Coatings Formation and Characterization

2.1. Coatings Formation

Nano and pigmentary TiO_2 particles in rutile crystalline form, produced by the sulphate process in Cinkarna Celje were used in this research.

Suspensions of pigmentary TiO_2 were subsequently surface treaded with hydrated amorphous silica, alumina and binary silica/alumina coating using precursors sodium silicate solution (Na_2SiO_3), sodium aluminate solution ($NaAlO_2$) and aluminium sulphate solution ($Al_2(SO_4)_3$).

The effects of the pH value of the reaction solution and the precursor content on the morphology of the oxide coating layers on the TiO_2 surface were investigated. The pH of the reaction solutions varied from acidic to alkaline.

2.2. Coatings Characterization

2.2.1. Potentiometric Titration

The potentiometric titration is a basic method for exploring the conversion of Si and Al species in the hydrolysis-polymerization. By analysing the simplest titration curve (pH vs V) the study reported here focuses on the conversion trends of the species in the hydrolysis-polymerization of Si^{4+} and Al^{3+}.

Diluted Na-silicate, Na-aluminate and Al-sulphate solutions and 1.0 M acid or base solution were prepared with corresponding reagents dissolved in distilled water. The titrator 905 Titrando (Metrohm Swiss made, Switzerland) was used to inject titrant (acid/base) and record pH value automatically.

2.2.2. Surface Morphology Determination by SEM and TEM

The morphology of surface treated pigmentary TiO_2 particles can be identified directly only with utilization of scanning (SEM) and transmission microscope (TEM).

A drop of TiO_2 suspension was dropped on the adhesive carbon band on brass holders for the observation on a Zeiss SIGMA VP Scanning Electron Microscope (Carl Zeiss NTS GmbH, Germany), with a maximum resolution up to 1.5 nm at 20 kV. SEM images were collected and elaborated according to standard procedures.

In addition, morphology analysis of surface treated pigmentary TiO_2 was performed using transmission electron microscope (TEM, JEOL 2100F, Japan). For the analysis, a drop of suspension was dropped on Cu-grid.

The morphology of surface treated pigmentary TiO_2 was observed at different magnifications.

2.2.3. Electrochemical Properties Determination

Pure oxides have well defined iso-electric points (IEP) which can be used to verify the cleanliness of their surfaces. Although the IEP value for TiO_2 is usually quoted as pH 5.8, that value is rarely observed because commercial TiO_2 is normally covered with one or more layers of either silica (SiO_2), alumina (Al_2O_3) or zirconia (ZrO_2) or a combination of those oxides, in order to improve its performance in various situations. The IEP remains an important characteristic of the surface in this case because it helps identifying the coating layer and allows one to easily determine whether a supplier is properly controlling a particular product [31].

Mütek™ PCD-04 Particle Charge Detector (BTG Instruments GmbH, Germany) was used for charge identification of aqueous samples and for determination of their isoelectric point (IEP). If a liquid is forced along a particle surface under pressure, it will shear off the counter-ions of the diffuse layer and an electric field is built up. The resulting electric current is picked up by two electrodes and is either indicated by a streaming current signal or converted to a potential.

2.2.4. Microstructure Determination Using Fourier Transform Infrared (FTIR) Spectroscopy

The FTIR spectra of untreated and surface treated pigmentary TiO_2 were obtained on a Bruker spectrometer IFS 66/S with integration sphere Optosol. The spectra of studied samples were recorded over the range 1300-400 cm-1 in absorbance mode.

2.2.5. Microstructure Determination Using XRD

X-ray diffraction spectra of untreated and surface treated pigmentary TiO_2 particles were obtained on Cubix PRO PW 3800 (PANalyitcal).

3. ANALYSIS

3.1. Potentiometric Titration

Referring to the potentiometric titration experimental method, there are 3 characteristic points on the titration curve when titrating Si solution with acid under the medium rate of acid injection (Figure 1). These three critical points express different meanings, respectively: at the first critical point, a lot of mononuclear Si begins to convert into small/middle polynuclear Si (oligomer). At the second critical point, it is very likely that the small/middle polynuclear Si begin to form large polynuclear Si. At the third critical point, the amorphous gel or sediment appears. Generally, the critical point is the characteristic point of the titration curve related to Si salt solution in the hydrolysis-polymerization course, which is related to experiment conditions [32].

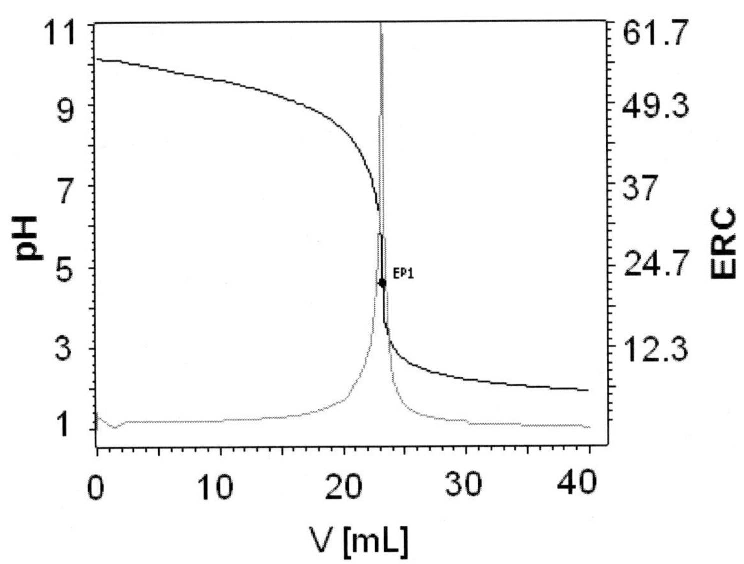

Figure 1. Na-silicate solution titration curve and ERC curve.

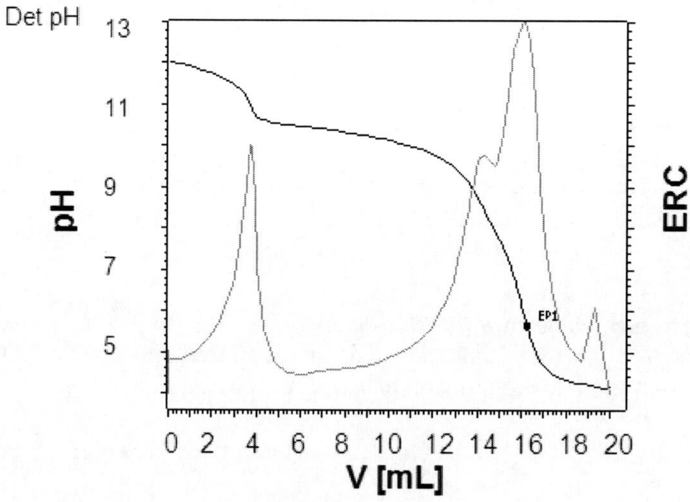

Figure 2. Na-aluminate solution titration curve and ERC curve.

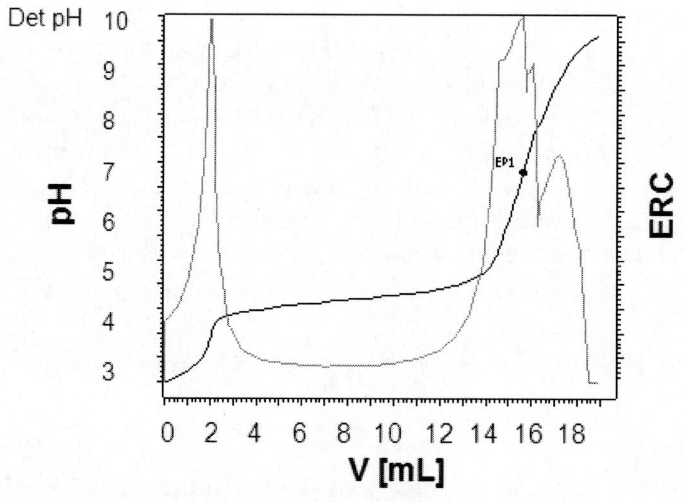

Figure 3. Al-sulphate solution titration curve and ERC curve.

According to the OH/Al mole ratio, the Al species can be divided into four regions quantitatively by three characteristic points on the titration curves. Referring to potentiometric titration experimental method, there are 3 characteristic points on the base titration curve when titrating Al solutions with base under the medium rate of base injection (Figure 2 and 3) [32]. At the first critical point a lot of mononuclear Al begins to convert into small/middle polynuclear Al (oligomer). At the second critical point it is very likely that the small/middle polynuclear Al begin to form large polynuclear Al. At the third critical point the amorphous gel or sediment appears [33].

In the standard S-titration curve (black curve) titrator determines the inflection point on the curve. It is displayed as a dot on the endpoint recognition criterion curve (ERC) (gray curve).

3.2. SEM and TEM Analysis

Formation of the Silica Coatings on the Surface of TiO_2 Nanoparticles

SEM and TEM images of TiO_2 nanoparticles in rutile crystalline structure are presented in Figure 4. Nanoparticles are non-spherical in shape, with approximately 80 nm in length and 25 nm in width. Sharp edges and crystal lattice fringes for rutile TiO_2 can be seen in TEM image (right).

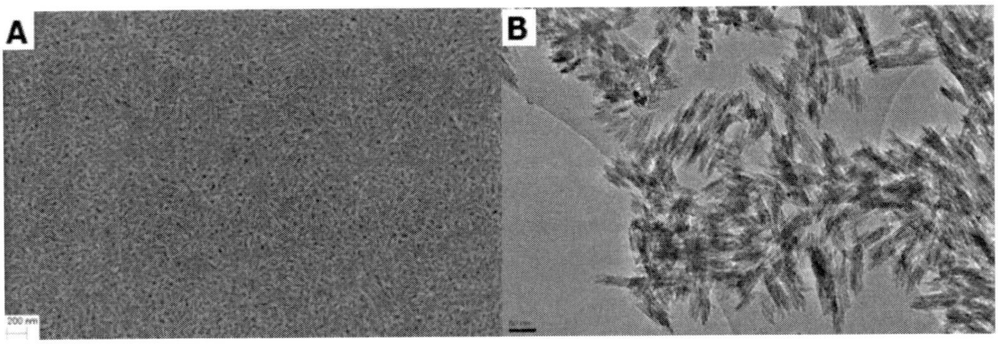

Figure 4. SEM and TEM images of TiO_2 nanoparticles in rutile crystalline structure.

Homogeneous amorphous coatings form in heterogeneous nucleation of silica molecules on the surface of TiO_2 nanoparticles (Figure 5B). Therefore, the surface of rutile TiO_2 nanoparticles became rougher due to amorphous silica precipitation. Another important observation is a uniform particles distribution. Figure 5A shows SEM image of evenly distributed untreated and surface-treated TiO_2 nanoparticles. No agglomerates are seen, which proves that during the surface treatment, individual particles were effectively kept separated. Prevention of agglomeration represented the greatest challlenge due to the fact that TiO_2 particles tend to agglomerate. However, the use of the proper synthesis conditions yielded efficiently surface-treated TiO_2 nanoparticles. The second aim to reach was homogeneous precipitation of amorphous silica on the surface of individual TiO_2 nanoparticles only. This goal was also achieved by heterogeneous silica nucleation on the surface of TiO_2 nanoparticles. No individual silica particles can be observed within the sample.

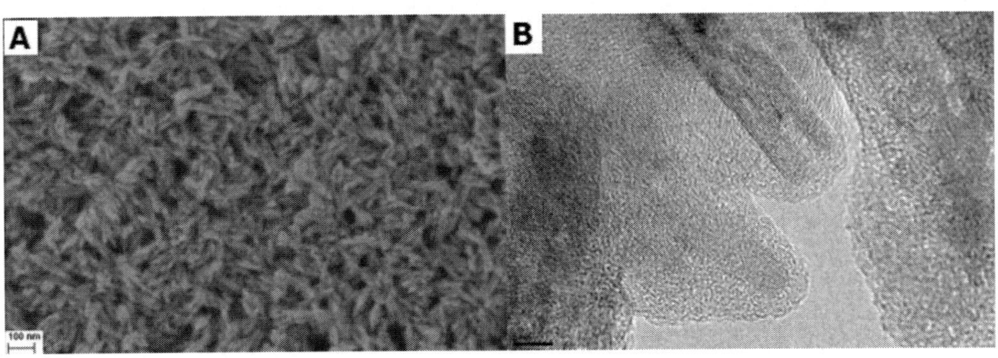

Figure 5. SEM and TEM images of TiO_2 nanoparticles in rutile crystalline structure, surface treated with 5 wt% silica.

Formation of the Silica Coatings on the Surface of Pigmentary TiO_2 Particles

Figure 6 shows the TEM image of the pigmentary TiO_2 particles. From the TEM image, it can be found that the naked TiO_2 powders have smooth surfaces and the average primary particle size is ~ 250 nm.

If silica is deposited at acidic or neutral pH, it is deposited as submicroscopic particles joined together in a gel-like structure. This "fluffy" coating provides better spacing and optical efficiency. Silica treatments can also be deposited as a dense, glassy shell that encapsulates the particle and provides the highest durability available in titanium dioxide pigments. Dense silica is deposited slowly at basic pH. [7].

Under the neutral pH conditions, the TiO_2 surface is negatively charged. Under neutral experimental conditions, Na-silicate rapidly hydrolyzes to form a large number of siliceous micelles. The resultant siliceous micelles anchor on the surface of TiO_2 powders via Ti–O–Si bonding to obtain the island-like coating layers. Increasing the pH value to alkaline pH region should lower the hydrolysis rate of Na-silicate resulting in the formation of small-sized or less aggregated siliceous micelles. First, the small-sized micelles anchor on the surface of TiO_2 powders via Ti–O–Si bonding. Then both the micelles present in the solution and the anchored micelles condense via Si–O–Si bonding to obtain continuous and uniform coating layers [34].

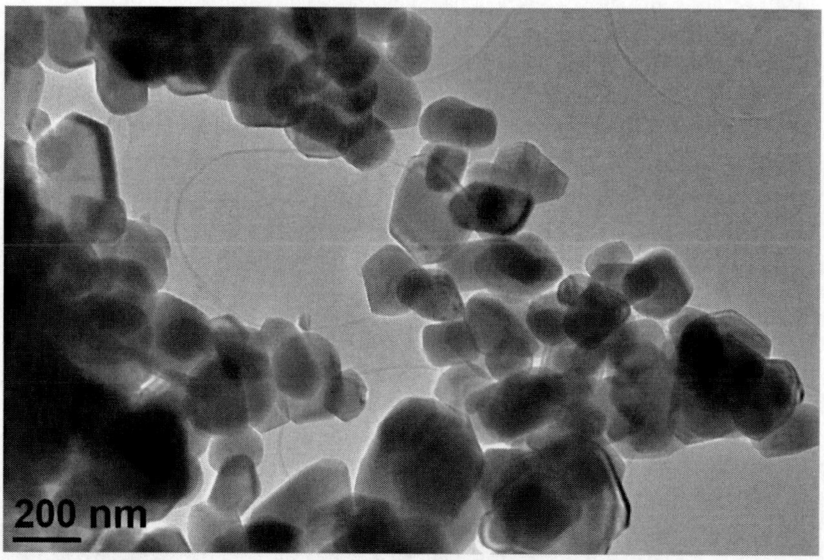

Figure 6. TEM image of pigmentary TiO_2 particles in rutile crystalline structure.

The TEM images of the silica coated TiO_2 samples (3 wt% silica) prepared at the different pH values are shown in Figure 7A. When the pH value of the reaction solution lies within neutral pH region, the densely island-like silica coating layers dominantly form. The thickness of the densely island-like coating layers is ~ 10 nm. While increasing the pH values from neutral to higher pH values, the morphologies of the silica coating layers changes from densely island-like coating to densely continuous and uniform coating (Figure 7B). The average thickness of the silica coating layer is ~ 3 nm.

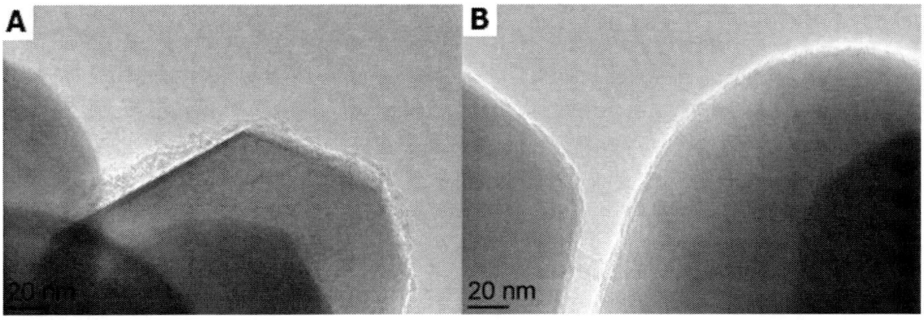

Figure 7. TEM images of pigmentary TiO_2 particles, surface treated with 5 wt% silica in neutral (A) and alkaline media (B).

The results show that the thickness of the silica coating layer is increased by increasing the mole ratio of Na-silicate to TiO_2 (10 wt.% silica) and that the densely continuous and uniform silica coating layers were formed in a large range of pH values when the mole ratio of Na-silicate to TiO_2 was higher (Figure 8). The average thickness of the silica coating layers formed via silica precipitation in neutral and alkaline media is ~ 25 and 10 nm, respectively.

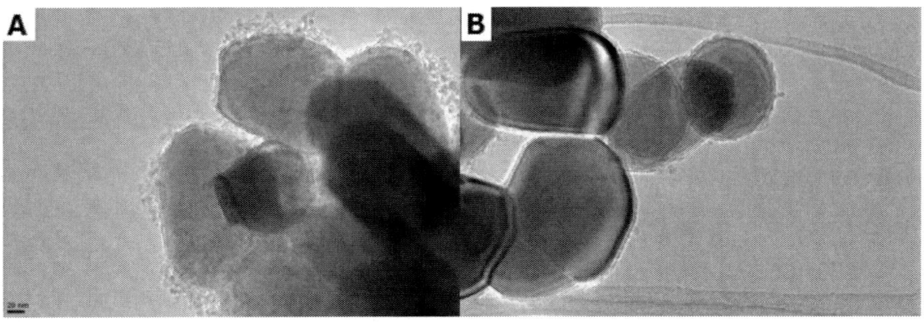

Figure 8. TEM images of pigmentary TiO_2 particles, surface treated with 10 wt% silica in neutral (A) and alkaline media (B).

Silica precipitation in acidic to neutral media resulted in fluffy coatings (Figure 9).

Formation of the Alumina Coatings on the Surface of Pigmentary TiO_2 Particles

The –OH groups on a particle surface affect the precipitation and condensation of the OH-Al species and determine whether homogeneous or heterogeneous nucleation dominates in the precipitation coating process.

The hydrous alumina coating on TiO_2 particles is produced through the collision and condensation of -OH groups on the core particle surface and OH-Al species [10, 33] and the -OH groups on the particle surface provide protons to trigger the condensation of OH-Al species. The ζ-potential of the TiO_2 particles varies with the pH, which indirectly reflects the amount of -OH groups on the particle surface that could provide protons. When the -OH groups can easily provide protons, the particles carry a negative charge; when it is difficult for the -OH groups to provide protons and easily accept protons, the particles carry a positive charge. Thus, at different pH, the ζ-potential reflects the different capacities of the TiO_2 particle surfaces to provide or accept protons. When the pH is lower than the isoeletric point

Figure 9. SEM image of pigmentary TiO_2 particles, surface treated in neutral media with 10 wt% silica.

(pH 4-5), the TiO_2 particles carry a positive charge [10], which reflects that the capacity of -OH groups providing protons is very weak, and the condensation between the OH-Al species and the -OH groups on the particle surface seldomly occurres. Since only a few -OH groups could provide protons, the condensed OH-Al species are usually isolated and the neighboring condensed OH-Al species hardly condensed each other, forming a dotted coating of nanoparticles on the surface, as shown in Figure 11B. When the pH is higher than the isoeletric point, the TiO_2 particles carry a negative charge [10, 33], their surface easily provides protons and facilitates the condensation between the OH-Al species and the –OH groups on the particle surface. Since a large amount of -OH groups on the particle surface could provide protons at the higher pH, there are many OH-Al species that condense with the -OH groups on the particle surface, and the neighboring condensed OH-Al species easily condense each other, forming a continuous film coating, as shown in Figure 10. With less -OH groups on the core particle surface, it is easier to cause homogeneous nucleation, i.e., a dotted coating of nanoparticles. However, with more -OH groups on the core particle surface, it is easier to cause heterogeneous nucleation, e.g., film coating.

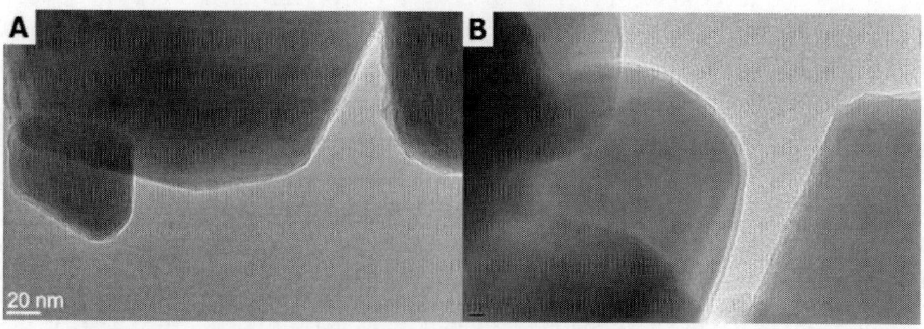

Figure 10. TEM images of pigmentary TiO_2 particles, surface treated with 4 wt% alumina continuous film coating under neutral pH conditions using Al-sulphate (A) and Na-aluminate precursor (B).

Flake coating with flakes almost vertically coated on the core particle surface is produced when the pH is higher than the isoeletric point ~ 5 (Figure 11A). Dotted coating with nanoparticles adhering on the core particle surface was produced when the pH was lower than the isoeletric point ~ 5. The alumina nanoparticles are usually isolated, which is shown in Figure 11B. Floccule coating with a loose layer, which covers the core particle surface nonuniformly, is shown in Figure 11C.

In coating and aging, the coating morphology with a continuous and uniform film forms only at low pH and temperature. In gel precipitation or coating, the temperature affects the direction and growth speed of the OH-Al species, while in the aging process, the temperature affects the self-assembly of the OH-Al species. [33]. The results are in agreement with the literature [33,35].

Formation of the Binary Silica-Alumina Coatings on the Surface of Pigmentary TiO_2 Particles

TEM image (Figure 12) shows that uniform continuous coating layers are formed on the TiO_2 surfaces, and the average thickness of the binary silica/alumina coating layers is ~ 8.5 nm.

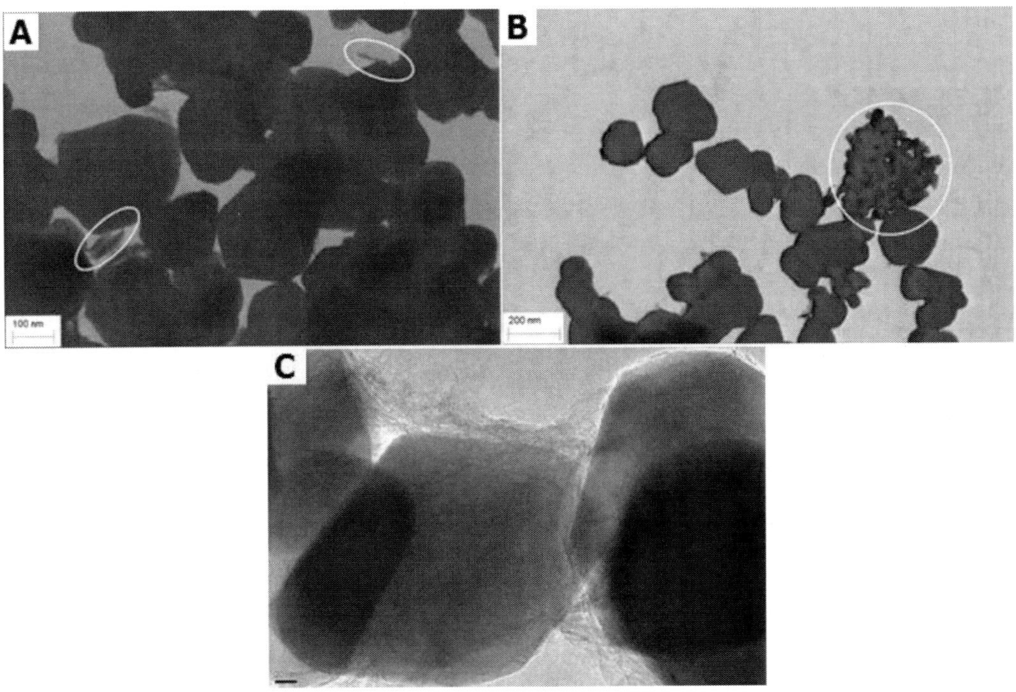

Figure 11. TEM images of pigmentary TiO_2 particles, surface treated with alumina flake (A), dotted (B) and floccule coating (C).

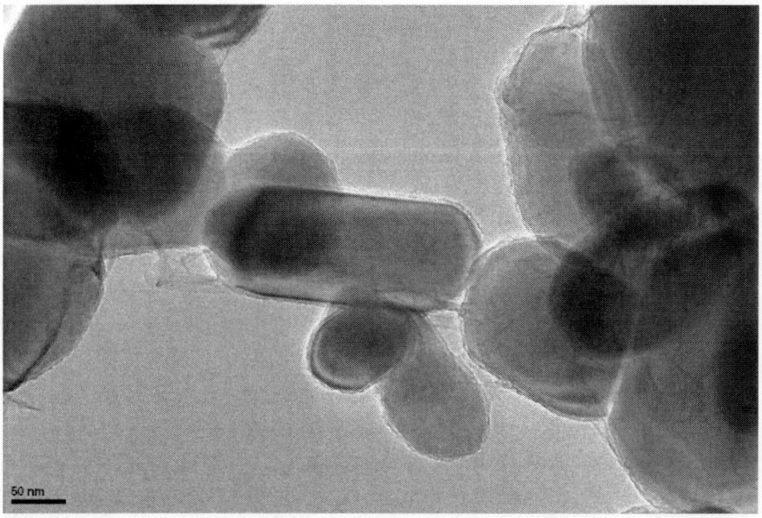

Figure 12. TEM images of pigmentary TiO2 particles, surface treated with binary silica (4 wt%)-alumina (4 wt%) coatings using Na-silicate and Na-aluminate precursor.

3.3. Electrochemical Properties of Untreated and Surface Treated TiO_2

Inorganic oxides carry on their surfaces OH-groups that differ in number and acidity [1]. Pure TiO_2 has its IEP at pH values between 4.5 and 6.5. The IEP of TiO_2 from the chloride process lies at pH values between 6.5 and 7, due to doping with rather appreciable amounts of Al_2O_3 that are partly on the surface of the particles. Sulfate grade TiO_2 core particles have their IEP at lower pH values. This is caused by the phosphates that are used as annealing salts for calcination. These form acidic titanium phosphates on the surfaces of the particles. The =PO-OH group, that is very acidic, then becomes potential determining [1].

Modifying the pigment surface with inorganic layers (for example, Al_2O_3, SiO_2, ZrO_2 and further TiO_2, phosphates of low solubility or mixed oxides), again changes the IEPs. Whereas, for example, Al_2O_3 (IEP of Al_2O_3 = 9) shifts the IEP to higher pH-values, other substances normally tend to reduce the IEP to lower pH-values (IEP of SiO_2 = 2, IEP of ZrO_2 = 4) [1].

The isoelectric point (IEP) of rutile nano and pigmentary TiO_2 lies around pH 4-5 [1, 11, 15, 36].

The results of streaming potential (U) as a function of pH value are graphically presented in Figure 13. This clearly indicates the presence of adsorbed silica on the surface of the TiO_2 particles, since the presence of silica causes the IEP of surface treated material to shift towards the acid region. Surface treated particles alter their electrokinetic behaviour in a way similar to silica, which suggests a complete silica coating layer is obtained on the TiO_2 surface. IEP of untreated TiO_2 lies around pH 4, while the presence of silica on the surface of TiO_2 shifts IEP of surface treated TiO2 to lower pH values (Figure 13 and 14). Complete coatings contribute to major shift of IEP (to pH value < 2), which proves a successful surface treatment, since IEP values of surface treated samples lie close to the IEP values, characteristic for silica. These results are in accordance with the literature [1, 3, 37, 38].

Alumina surface treated particles alter their electrokinetic behaviour in a way similar to Al-hydroxide, which suggests that a complete alumina coating layer is obtained on the TiO_2

surface. IEP of untreated TiO_2 lies around pH 4-5, while the presence of alumina on the surface of TiO_2 shifts IEP of surface treated TiO_2 to higher pH values (Figure 13 aand 14). The isoelectric point (IEP) of alumina in its hydroxide form ($Al(OH)_3$) is accepted as pH 7.4. This corresponds to the point of minimum solubility of aluminium hydroxide [39]. The U values, obtained for surface treated TiO_2, differ dependent on the surface coating effectiveness. Complete coatings contribute to shift of IEP to pH value 6.0 and 6.4, respectively, which proves a successful surface treatment, since IEP values of surface treated samples lie close to the IEP values, characteristic for Al-hydroxide. The results again are in agreement with the literature [10].

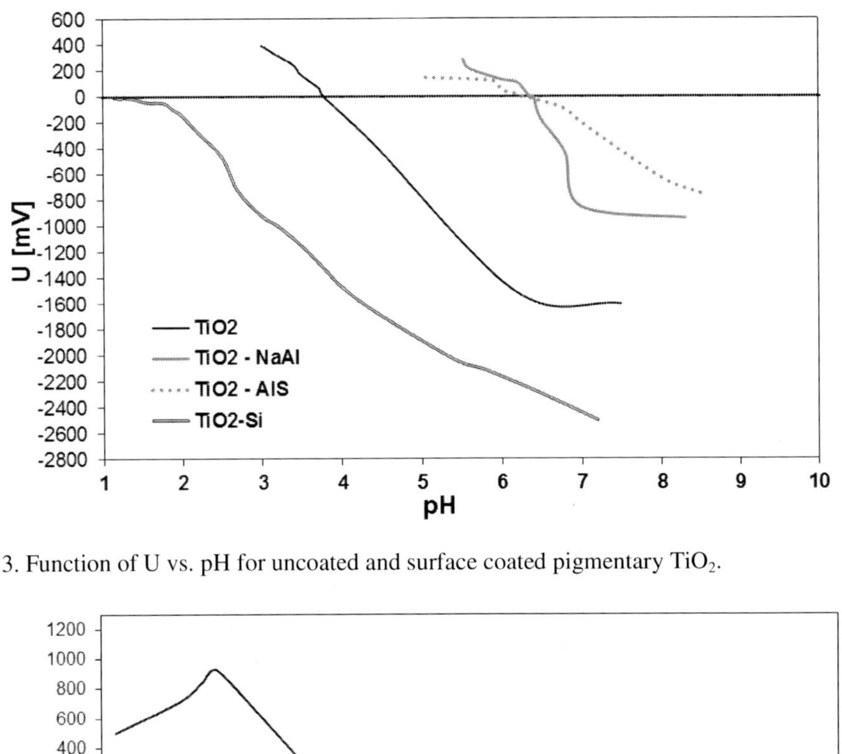

Figure 13. Function of U vs. pH for uncoated and surface coated pigmentary TiO_2.

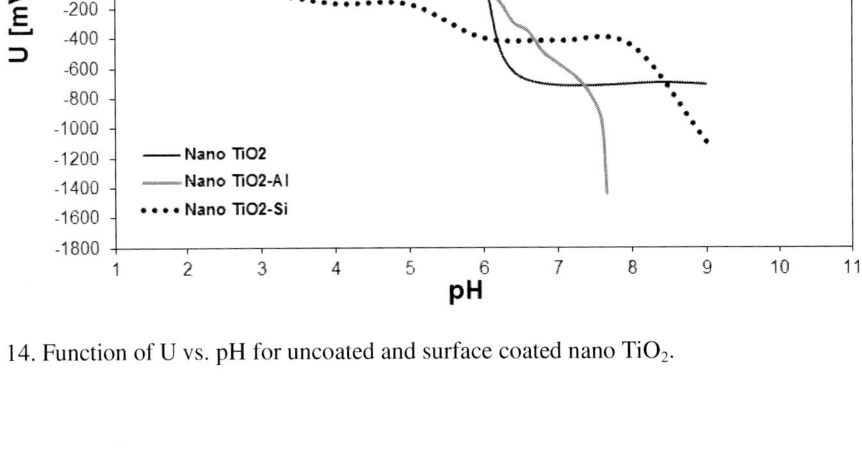

Figure 14. Function of U vs. pH for uncoated and surface coated nano TiO_2.

3.4. FTIR Analysis

In order to determine the surface microstructure, the FTIR analysis of uncoated and coated pigmentary TiO_2 was performed. The spectra of studied samples were recorded over the range of 1300 - 400 cm^{-1} in absorbance mode.

The FTIR spectra of uncoated and coated pigmentary TiO_2 are shown in Figure 15-17. The bands at 800–400 cm^{-1} assigned to Ti–O–Ti vibrations were observed over all of the samples.

In the spectra of the silica-coated TiO_2 sample, the bands appearing at 1077 cm^{-1} and 1237 cm^{-1} are assigned to the asymmetric stretching vibration modes of Si–O–Si bond [34]. The bands intensities at 1077 cm^{-1} and 1237 cm^{-1} are assigned to the asymmetric stretching vibration modes of the Si–O–Si bridge [13,14] increased when the mole ratio of Na-silicate to TiO_2 increased, revealing that the thickness of the silica coating layer increased along with the silica loading.

Figure 15 shows the FTIR spectra of the naked rutile TiO_2, (b) the hydrated silica coated TiO_2 sample (3 wt.%) and the hydrated silica coated TiO_2 sample (10 wt.%).

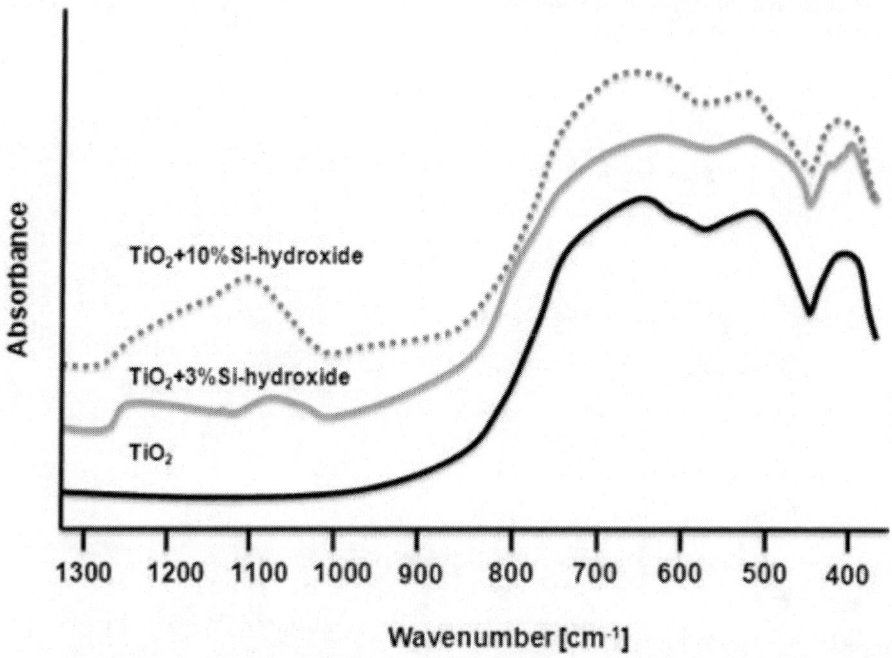

Figure 15. FTIR spectra of uncoated and SiO_2 surface treated pigmentary TiO_2 over the range of 1300-400 cm^{-1}.

The FTIR spectra of the sample obtained by hydrolysis of Na-aluminate show a band at 1114 cm^{-1}, which is attributed to Al–O stretching vibration [36] (Figure 16).

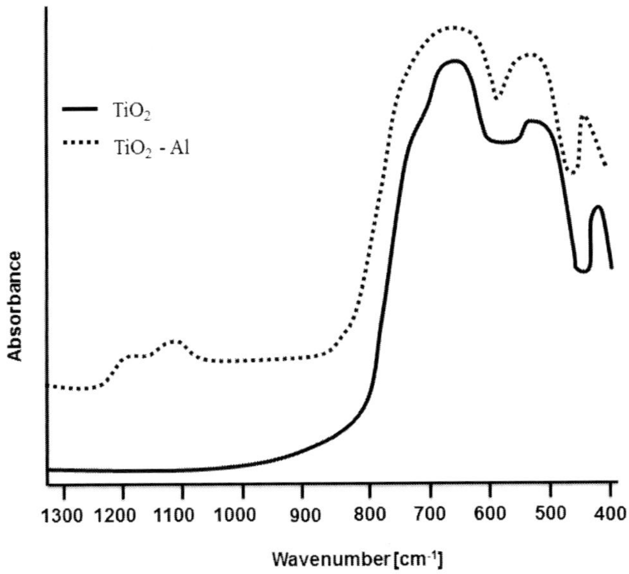

Figure 16. FTIR spectra of uncoated and alumina surface treated pigmentary TiO_2 over the range of 1300-400 cm^{-1}.

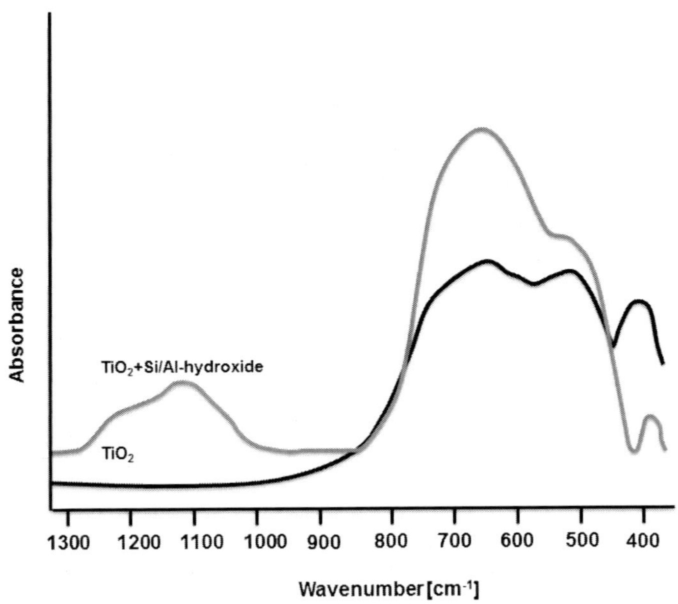

Figure 17. FTIR spectra of uncoated and Si-Al hydroxide surface treated pigmentary TiO_2 over the range of 1300-400 cm^{-1}.

When the binary silica/alumina-coated TiO_2 sample was prepared, the intensity of the band at 1077 cm^{-1} is increased and the band is blue shifted as compared with the Si–O–Si band of the SiO_2-coated TiO_2 sample (Figure 17). A broad shoulder band appears at 1185 cm^{-1}, which was probably ascribed to the stretching vibrations of Al–O and Si–O–Si bonds. The alumina coating layers are probably anchored at the surfaces of the silica-coated TiO_2

particles via Al–O–Si bonds because the absorption band of Al–O–Si bond is located around $1100 cm^{-1}$ [14]. The results are in agreement with Zhang et al. [15].

3.5 XRD Analysis

The amorphous structure of untreated and surface treated pigmentary TiO_2 was confirmed by XRD. Figure 18 shows the XRD patterns of the rutile TiO_2, sample, prepared by hydrolysis of Na-aluminate, the silica-coated TiO_2 sample, and the binary silica/alumina-coated TiO_2 samples. The XRD patterns show the diffraction peaks of rutile TiO_2, only. No diffraction peaks of alumina and silica can be found, indicating that silica and alumina present in coating layers are both in amorphous phase.

Amorphous alumina gel could be obtained only when the precipitation and aging are at low temperature and pH, and it could be changed to boehmite when the pH or temperature is high. However, the boehmite gel could not be converted to the amorphous form when the pH and temperature are low in the aging process. At low pH and temperature, in gel precipitation and aging, amorphous $Al(OH)_3$ gels formed [33]. Binary Al_2O_3/SiO_2 coating layers, both in amorphous phase were formed at TiO_2 surfaces.

The diffractogram, corresponding to surface treated samples, indicates that precipitated silica and alumina gels formed were amorphous.

Locations and relative intensities of the 2-theta peaks of the X-ray diffraction spectra of pigmentary TiO_2 in rutile crystalline structure are in accordance with the literature [1]. XRD patterns of the naked pigmentary TiO_2, the silica-coated TiO_2 pigmentary TiO_2 and silica/alumina-coated TiO_2 pigmentary TiO_2 are overlaping. An occurance of additional peaks can't be seen on the spectra after surface treatment with amorphous oxide layers.

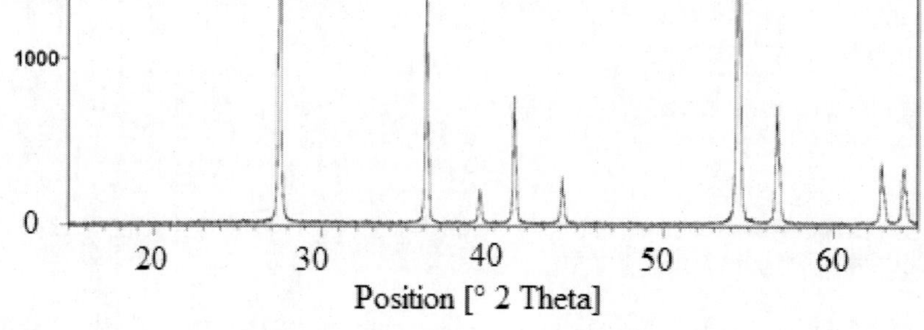

Figure 18. XRD patterns of the naked pigmentary TiO_2, the silica-coated pigmentary TiO_2 and alumina-coated pigmentary TiO_2.

Findings of various researchers about coatings properties are collected in Table II.

Table II. Findings of various researchers about coatings properties

Coatings morphology was studied by Zhang et al. [12], El-Toni et al. [23], Wu et al. [27, 33], Liu et al. [34]
Wu et al. reported about the effect of pH and temperature on the hydrolysis of Al^{3+}, the size and structure of the OH-Al species, and the coating morphology. Four kinds of coating morphologies were observed: dotted coating, film coating, floccule coating, and flake coating. Zhang et al. reports about formation of the continuous and dense binary Al_2O_3/SiO_2 coating layers when the pH value of reaction solution was 5. The thickness of the binary Al_2O_3/SiO_2 coating layer increased with the increase in the mole ratios of $NaAlO_2$ to TiO_2. Liu et al. reports about the evolution of island-like and uniform coating layers depending upon the ratio of Na_2SiO3 to TiO_2, reaction temperature, and pH. El-Toni et al. reports about the quality of the silica shell which improved with increasing coating time. Increasing coating time resulted in the formation of a denser silica layer. Silica shell thickness increases with increasing temperature. Promotion of silica coating with an increase of reaction temperature indicates that reaction temperature has a pronounced impact on hydrolysis of TEOS and subsequent condensation of silica molecules, which finally results in pronounced growth of the silica shell.
Surface microstructure was studied using FTIR by Lin et al. [9], Zhang et al. [12], Liu et al. [16], Kalele et al. [21], Pacewska et al. [26]
Findings of Zhang et al.: The FTIR spectra of the sample obtained by hydrolysis of $NaAlO_2$ show a band at $1130.9cm^{-1}$,which was attributed to Al–O stretching vibration. A broad shoulder band appeared at $1239.3cm^{-1}$, which was probably ascribed to the stretching vibrations of Al–O bonds. Findings of Kalele et al.: The peak at ~ 670 cm^{-1} is attributed to Ti-O bond and refers to symmetric O–Ti–O stretch while 505 cm^{-1} is due to the vibration of Ti–O bond. Findings of Liu et al.: The alumina coating layers anchored at the surfaces of TiO2 via Ti–O–Al bond. Findings of Pacewska et al.: A weak band at ~700 cm^{-1} probably results from combined absorptions of Al–O stretching vibrations and Al–OH rocking vibrational mode of molecular water. The band centered at 608 cm^{-1} reflects the combined absorptions of sulfate groups, Al–O stretching vibrations and Al–OH wagging librational mode of molecular water. Broad band at 750–900 cm^{-1} corresponds to Al–O modes in the lattice. Findings of Y.-L. Lin et al.: Compared with the spectrum of pure TiO_2, an absorption band in 1000–1300 cm^{-1} occurs in the spectrum of coated TiO_2 powders. It is known that the O–Si asymmetric flexible vibration of ordinary silicic acid leads to the absorption band in 800–1000 cm^{-1}, and that of amorphous silica leads to the absorption band in 1000–1200 cm^{-1}.
XRD method was used for coatings analysis by Wu et al. [27, 34]
Wu et al. reports that amorphous $Al\text{-}(OH)_3$ gel could be obtained only when the precipitation and aging of alumina coatings were at a low temperature and pH, and it could be changed to boehmite when the pH or temperature was high. In that case no diffraction peaks of alumina were found in the XRD pattern, indicating that alumina present in coating layers was in amorphous phase.
Electrokinetic properties of coated TiO_2 particles was studied by Li et al. [3], Lin et al. [9], Wu et al. [10]
Wu et al. reports about isoelectric point (IEP) of the pure TiO_2 particles to be about 4.1and the IEP of $Al(OH)_3$ species to be about 6.8. The ζ-potential-pH curve and isoelectric points (IEPs) for the TiO_2 particles coated with hydrous alumina are similar to those of pure $Al(OH)_3$ gel. Li et al. reports about (IEP) at 2.34 for silica nanoparticles and at 6.20 for titania particles. When silica loading increased to 9.33 wt%, though the IEP of coated nanoparticles isn't obtained due to the heavy aggregation at low pH value, the coated nanoparitcles altered the electrokinetic behavior similar to the silica nanoparticle. Lin et al. report that the coated TiO_2 has the similar electrokinetic behavior as pure SiO_2 particles. The IEP of both samples are roughly the same as 2.5. It shows that the surface characteristics of TiO_2 particle are significantly modified after coated with hydrous silica. The surface structure of hydrous silica-coated TiO_2 particles is –Si–OH. The surface electrokinectic behavior depends on the quality of surface coating. Dense film coating can seal the TiO_2 particles completely and improve the particle dispersion effectively. Whereas incomplete coating such as nucleation coating shows less modification ffect on TiO_2 particles.

Conclusion

TiO_2 particle surface was coated in a controlled manner to obtain uniform oxide layers which entirely encapsulate it. Inorganic surface modifiers are precipitated coatings of alumina and silica. When silica was deposited at acidic or neutral pH, it was deposited as submicroscopic particles joined together in a gel-like structure. Dense silica was deposited slowly at basic pH. The structure of the alumina coating on a TiO_2 pigment varied with the pH as well. Alumina, precipitated under acidic conditions, was amorphous, which was confirmed by the surface microstructure determination. Surface treatment influenced electrochemical properties of TiO_2 (nano) particles. Surface treated particles altered their electrokinetic behaviour in a way similar to oxide present in coating, which suggests that a complete oxide coating layer was obtained on the TiO_2 surface.

References

[1] Winkler, J. Titanium Dioxide, Vincentz, Hannover (2003).
[2] Chena, XD; Wanga, Z; Liaob, ZF; Maic, YL; Zhang, MQ. *Polymer Testing,* 2007, 26, 202–208.
[3] Li, QY; Chen, YF; Zeng, DD; Gao, WM; Wu, ZJ. *J. Nanopart. Res.*, 2005, 7, 295–299.
[4] Mori, T; Tanaka, K; Inomata, T; Takeda, A; Kogoma, M. *Thin Solid Films*, 1998, 316, 89–92.
[5] van Dyk, AC; Heyns, AM. *J. Colloid Interface Sci.*, 1998, 206, 381-391.
[6] Bałdyga, J; Jasińska, M; Jodko, K; Petelski, P. *Chem. Eng. Sci.,* 2012, 77, 207-216.
[7] Tyler, F. *Paint & Coatings Industry*, 2000, 16-32.
[8] Edward, CM. Titanium dioxide pigment and process making same. *United States Patent*, 4,022, 636, 1977.
[9] Lin, YL; Wang, TJ; Jin, Y. *Powder Technol.*, 2002, 123, 194-198.
[10] Wu, HX; Wang, TJ; Jin, Y. *Ind. Eng. Chem. Res.*, 2006, 45, 1337–1342.
[11] Lee, JW; Kong, SM; Kim, WS; Kim, JS. *Mater. Chem. Phys.;* 2007, 106, 39–44.
[12] Zhang, Y; Yin, H; Wang, A; Ren, M; Gu, Z; Liu, Y; Shen, Y; Yu, L; Jiang, T. *App. Surf. Sci.*, 2010, 257, 1351–1360.
[13] Fang, JX; Ding, BJ; Yang, ZM; Zhao, K; Gu, CQ. *J. Coll. Int. Sci.*, 2005, 283, 1–4.
[14] Liang, QL; Zhao, GY; Lu, JG. *Appl. Surf. Sci.*, 2007, 253, 5442–5446.
[15] Zhang, Y; Yin, H; Wang, A; Ren, M; Gu, Z; Liu, Y; Shen, Y; Yu, L; Jiang, T. *App. Surf. Sci.,* 2010, 257, 1351–1360.
[16] Liu, JM; Zhang, YS; Ge, C; Yin, HB; Wang, AL; Ren, M; Feng, H; Chen, J; Jiang, TS; Yu, LB. *Appl. Surf. Sci.;* 2009, 255, 7427–7433.
[17] Veronovski, N; 1, Verhovšek, D; Godnjavec, *J. Europ. Coat. J.*,2011, 01.
[18] Veronovski, N; Lešnik, M; Verhovšek, D. J. *Coat. Technol. Res.*, 2014, 11, 255-264.
[19] Veronovski, N; Verhovšek, D; Godnjavec, *J. Wood Sci. Technol.*, 2013, 47, 317-328.
[20] Viravathana, P; Marr, DWM. *J. Colloid Interface Sci.*, 2000, 221, 301–307.
[21] Kalele, S; Dey, R; Hebalkar, N; Urban, J; Gosavi; SW; Kulkarni, SK. Pramana, 2005, 65, 787-791.
[22] Bechger, L, Femius Koenderink, A; Vos, WL. *Langmuir,* 2002, 18, 2444-2447.

[23] El-Toni, AM; Yin, S; Sato,T. *J. Colloid Interface Sci.*, 2006, 300, 123–130.
[24] Park, OK; Kang, YS. *Colloid Surface A*, 2005, 257, 261-265.
[25] Nur H. *Mater. Sci. Eng.*, B, 2006, 133, 49–54.
[26] Pacewska, B; Kluk-Ploskonska, O; Szychowski, D. *J. Therm. Anal. Calorim.*, 87, 2007, 383–393.
[27] Wu, HX; Wang, TJ; Duan, JL; Jin, Y. *Ind. Eng. Chem. Res.,* 2007, 46, 3590-3594.
[28] Hayrapetyan, SS; Mangasaryan, LG; Tovmasyan, MR; Khachatryan, HG. *Acta Chromatogr.*, 2006, 16, 192-203.
[29] Ozimek, B; Grzechowiak, J; Radomyski, B. React. *Kinet. Catal. Lett.*, 1980, 13, 311-317.
[30] Gao, JC; Zou, J; Tan, XW; Wang, Y. Trans. *Nonferrous Met. Soc. China*, 2006, 16, 1252–1258.
[31] Determining the isoelectric point (i.e.p.), ZetaProbe Applications Colloidal Dynamics, Warwick, USA, 2.
[32] Wanga, C; Hana, Z; Wang, P; Laoa, NCT; Hong, P. Bull. *Chem. Soc. Ethiop.*, 2008, 22, 155.
[33] Wu, HX; Wang, TJ; Jin, Y. *Ind. Eng. Chem. Res.*, 2006, 45, 5274-5278.
[34] Liu, Y; Ge, C; Ren, M; Yin, H; Wang, A; Zhang, D; Liu, C; Chen, J; Feng, H; Yao, H; Jiang, T. *App. Surf. Sci.*, 2008, 254, 2809–2819.
[35] Shen, Z.; Zhao, ZG.; Wang, GT. Colloid and Surface Chemistry, Chemistry Industry Press, Beijing, China, 2004.
[36] Zhang, YS; Liu, YM; Ge, C; Yin, HB; Ren, M; Wang, AL; Feng, H; Chen, J; Jiang, TS; Yu, LB. *Powder Technol.;* 2009, 192, 171–177.
[37] Furlong, DN; Parfitt, GDJ. *Coll. Int. Sci.*, 1979, 69, 409.
[38] Kosmulski, M. *Coll. Int. Sci.,* 2009, 337, 439-448.
[39] Interpretation of reaction of Alum and Sodium Aluminate in Water Treatment http://www.business.u-net.com/~universal/reaction.htm Accesed on 31 July 2013.

In: Comprehensive Guide for Nanocoatings Technology. Vol. 1 ISBN: 978-1-63482-447-7
Editor: Mahmood Aliofkhazraei

Chapter 3

Ion Beam-Assisted Deposition

A. Guglya and I. Marchenko
National Science Center
"Kharkov Institute of Physics and Technology",
Kharkov, Ukraine

Abstract

This scientific paper describes the structure, phase composition and mechanical properties of nanocrystalline coatings obtained using the ion beam-assisted deposition technology. The results of computer simulation of IBAD processes for different energy values of bombarding ions and for relationships between the densities of ion fluxes and metal deposition rates is given. The data of electron-microscope analysis of the initial stages of film structure formation (of 5nm thick and more) is presented. And finally we demonstrate the influence produced by the ion energy and ion current density on the formation of nanocrystalline structure and its correlation with tribological properties of coatings.

1. Introduction

Perennial attempts made by the researchers in different countries of the world, who work in the field of production and application of thin-films, to develop technologies that would improve the structure-phase stability of such thin films resulted in the development of physical vapor deposition methods (PVD) [1-3]. The success achieved in this field can in many respects be attributed to the creation of different plasma plants and accelerated ion sources in the early sixties [4, 5]. It has been established that an increase in the ion component in the flux of deposited metal atoms and molecules results in a considerable change in the structure and properties of produced films [6, 7]. The use of nitrogen, carbon and boron ions as charged particles allowed for the formation of coatings made of nitrides, carbides and borides whose performance characteristics considerably surpassed those of metals [8-10].

Table 1. PVD, UMS and IBAD methods performances [11]

	PVD	UMS	IBAD
E, U_d, (V)	20-300	0-200	500-50000
Gas pressure, Pa	0,05 – 8 .0	0.05 – 1.0	10^{-2} – 10^{-3}
T, °C	200-600	100-400	20-200
Working gas	N_2	Ar + N_2	Ar or N_2
Damages, eV/at.	10 – 10^2	1,0 – 10	10^2 – 10^4
Depth of the damaged zone, nm	1-5	<1	10-100

An ever-increasing demand for the extension of the resource of articles operating under high mechanical loads, elevated temperatures and corrosion media promoted the development of such coating deposition technologies that allow for the formation of a coating structure by intensive bombardment with ions of >100eV. The purpose of ion bombardment is to provide controllable variation of the progress rate of diffusion processes, grain nucleation and growth conditions and also an increase in chemical activity of atoms. To produce such coatings nowadays solely no equilibrium technologies are used that include the stimulation bombardment of a deposited material with gas ions whose energy ranges from several tens of eV to tens of keV (Table 1, [11]). These methods include traditional PVD methods including unbalanced magnetron sputtering (UBM) and ion beam-assisted deposition (IBAD).

Physical phenomena that occur in the depth and on the surface of the material exposed to ion bombardment differ for different technologies. Figure 1 schematically shows most of those processes [11].

PVD and UMS methods usually deal with processes that occur on the film surface. As for the IBAD method the depth pass of gas ions reaches 100nm. Therefore their influence on the structure and phase composition is not limited by the film surface region; it captures approximately the same film thickness (Figure 1). The main processes that take place on the film surface in this connection are as follows: surface diffusion promotion and the creation of surface defects that are the major centers of grain nucleation and also maintenance of the high flow of radiation defects, which provides the formation of stoichiometric Me(N,O,C) compounds and highly dispersed structure stabilization.

While comparing the parameters of technological processes given in Table 1 the attention should be paid to the first two, in particular accelerating potential (ion energy) and working gas pressure in the chamber. It is seen that the IBAD method differs by high ion energy and low gas pressure in the process chamber. High energy provides good coating adhesion due to the ballistic mixing of condensed atoms with substrate atoms. PVD and UMS methods employ high process temperature for the mixing. The IBAD method requires no high temperature maintenance, which allows for the formation of the film structure in no equilibrium conditions. Particularly at such conditions the nucleation and growth of ultrafine grains of 5 to 10 nm (nano grains) take place. It should be noted that in the case of the IBAD method the structure formation is not ended at the nucleation stage.

The film material is exposed to the gas ion beam until the thicknesses of ~100nm is reached; this provides stabilization both of component composition and grain size. The advantage of low pressure inside the process chamber is that the formed coatings have low porosity and low pollution with gas impurities.

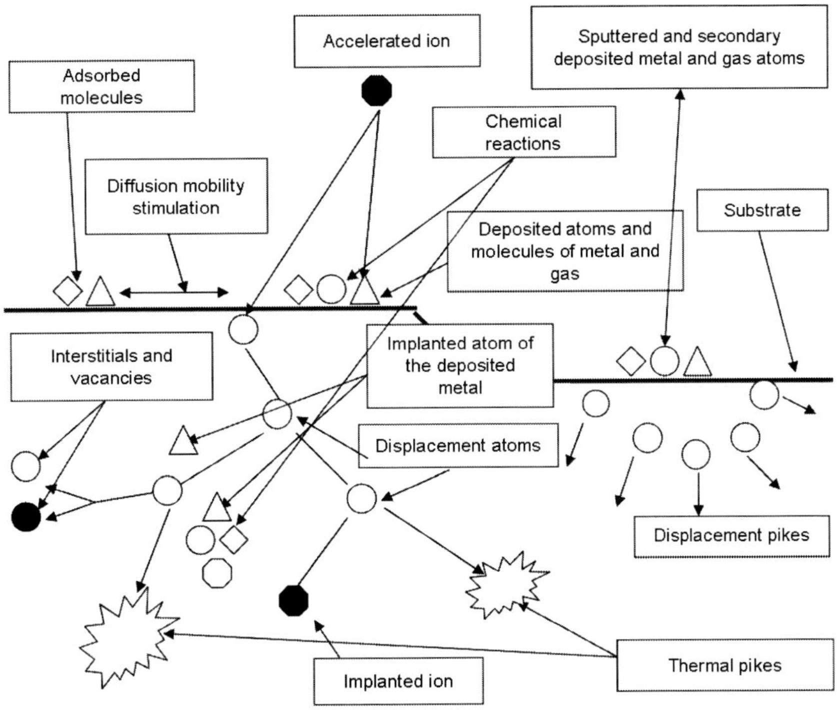

Figure 1. Surface and in-depth physical phenomena of the IBAD process [11].

And finally, in the case of the IBAD technology metal evaporators and gas ion sources are spaced and this provides an opportunity to independently control each parameter of the ion-assisted deposition process and have stable impact on the coating structure. Figure 2 gives diagrams of two the most spread configurations of the IBAD method.

The option given in Figure 2a has more opportunities for the rate control of metal evaporation and therefore it is employed by industrial ion beam-assisted deposition plants.

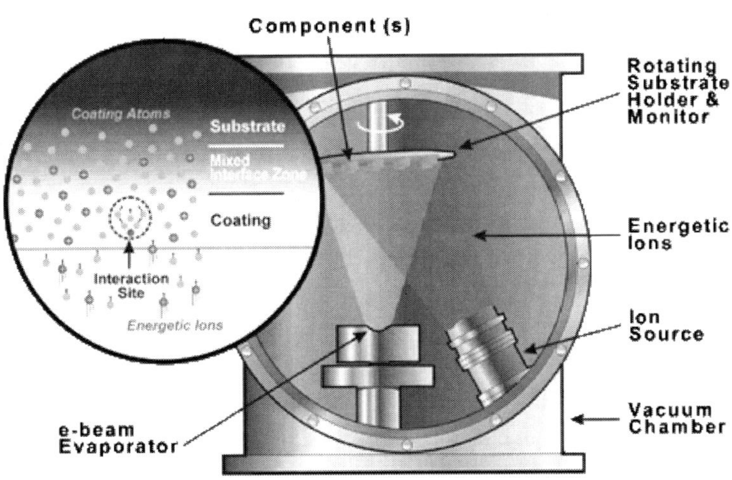

a

Figure 2. (Continued).

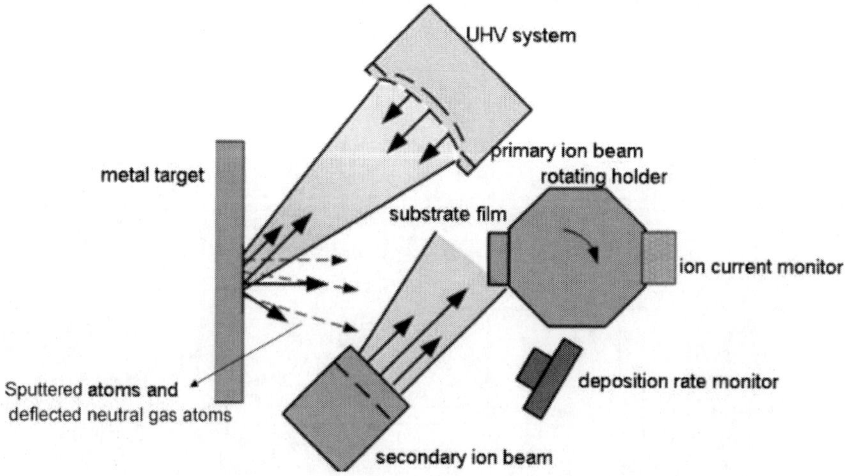

b

Figure 2. The main variants of IBAD technology realisation: a – e-beam metal evaporation + ion bombardment, b – ion-beam metal sputtering + ion bombardment.

2. Computer Simulation of the Ion Beam-Assisted Deposition

IBAD is a continuous process in which evaporated atoms and implanted ions (N, C, O) arrive at the surface of the target simultaneously. Figure 3 shows the physical process of IBAD.

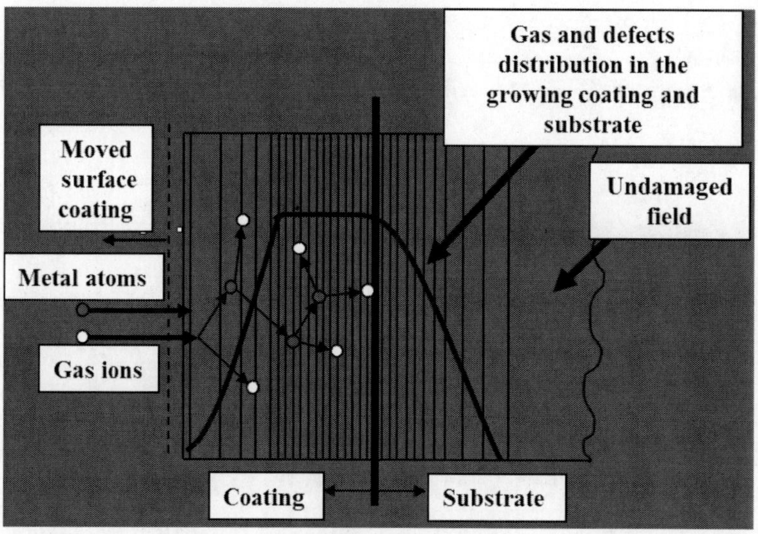

Figure 3. Schematic diagram of the synthesis of film by the IBAD process.

The compositional distributions of the synthesized films (coatings) are mainly determined by the ion energy, the atomic arrival ratio and the incident angle. In addition, ion sputtering, ion back scattering and diffusion processes might become important depending on the system under condensation.

To simulate the process of ion beam –assisted deposition we used a computer SRIM program of atom –atom collisions [12]. A peculiar feature of this method is that the portion of ions is reflected from the surface reducing thus total ion flux during the bombardment of deposited film with ions.

The interaction of ions with the surface results in the sputtering of deposited film and the rate of its growth is reduced. In the steady-state mode the film boundary moves at a rate V along the “x”-axis. In time t the profile of interstitial atoms $g(x,t)$ shifts together with the boundary of a growing film. During time t the concentration of interstitial atoms at a point x will change by a value of $\Delta c(x) = j_i \int_0^t g(x,t)dt$, where j_i is the flux density of incident ions. We consider the stationary process of film growth and therefore $g(x,t) = g(x - Vt,0)$. It means that the profile of an implanted dopant is simply shifted with time. The concentration at a point x can be found through the integration.

$$C(x) = \frac{j_i}{V} \int_{x-Vt}^{x} g(u)du \,, \tag{1}$$

On elapse of sometime t_{st} the stationary nitrogen distribution $C_{st}(x)$ should be established, which in the future shifts together with the surface. Its value can be derived if characteristics of atomic and ion flows are known.

$$C_{st} = \frac{j_i(1-\gamma)}{V} = \frac{j_i(1-\gamma)}{j_a - j_i S}\rho_s \,, \tag{2}$$

where γ is the ion reflection coefficient; S is the integral sputtering ratio; j_i is the ion current density, j_a is the flux density of deposited atoms, ρ_s is the near-surface atomic density of synthesized film material.

In the general case the film growth rate is a function of time and can be derived from the balance between the flow of deposited atoms j_a and sputtered near-surface atoms:

$$V = (j_a + j_i \cdot (1-S)) / \rho_s \,, \tag{3}$$

Figure 4 shows the dependence of nitrogen concentration in the substrate and deposited chrome film on time during the implantation of ions of 15 and 30 keV at room temperature, when the diffusion mobility of implanted dopant is low [13].

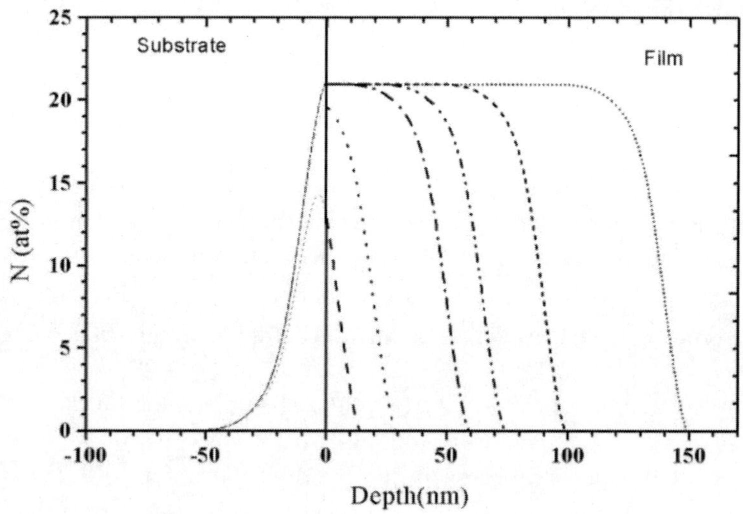

a

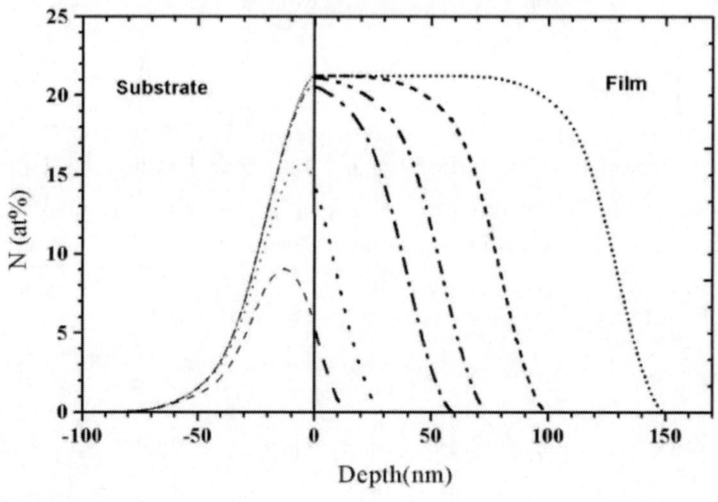

b

Figure 4. Distribution of nitrogen in the depth of growing Cr film under N_2^+ bombardment. The ion flux – 10^{14} ion/cm^2.s, the film growth rate – 8nm/min. a- E=15 keV and b-E=30 keV. Different curves correspond to the different thickness of the deposited film (- -15nm, · · · 30nm, -·-·- 60nm, -··-··- 75 nm, -- 100 nm, ... 150 nm) [13].

At the beginning of the deposition an abrupt increase in nitrogen concentration in the substrate is observed. Afterwards the dopant concentration reaches a maximum value c_{st}, which corresponds to the steady-state concentration in the depth of deposited film. The distance of nitrogen ion penetration is longer in case of the radiation with ions of 30keV. Accordingly, the substrate-film mixing zone is also larger.

A stationary stage of the IBAD process is also characterized by the availability of the region on the growing film surface whose nitrogen concentration varies from a maximum value to zero. It implies the existence of unformed zone. The higher the ion energy the wider this zone.

The width of mixing zone and that of under formed zone depends not only on the energy of bombarding ions but also on substrate and deposited film materials. Figure 5 shows the nitrogen distribution at the stationary stage of the IBAD process for the cases of chrome-on-chrome and aluminum-on-aluminum deposition [13]. It is seen that the penetration depth of nitrogen ions into the aluminum substrate exceeds that for the chrome substrate (a mixing zone is larger). At the same time the width of the under formed zone of chrome coating is smaller in comparison with that for aluminum coating.

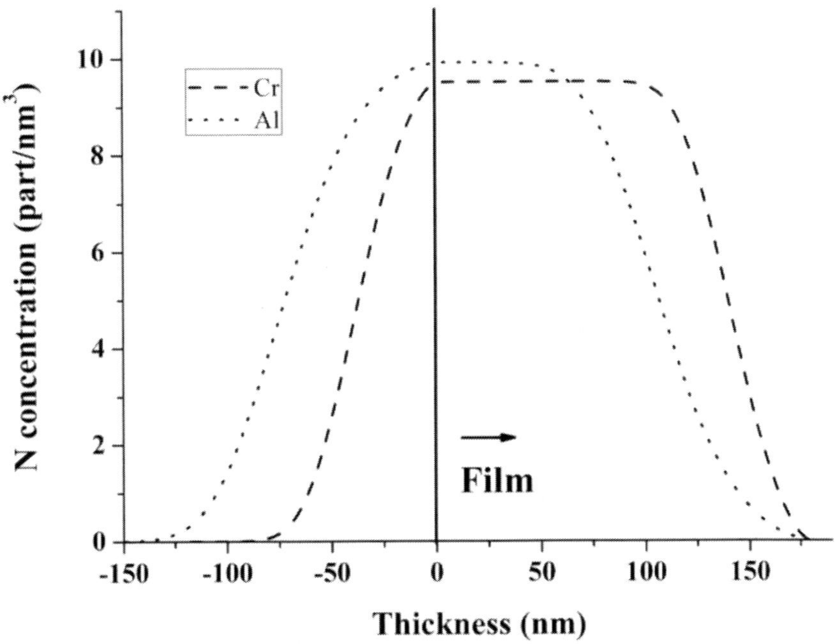

Figure 5. Sample thickness change of nitrogen concentration during deposition of aluminum and chrome onto substrates made of the same material at simultaneous bombardment with nitrogen ions of 30keV during 1800s, $j=10^{14}$ ion/cm^2/s [13].

To study the processes of a change in chemical composition of the near-surface region during the ion-assisted deposition taking into consideration diffusion processes the following diffusion equation can be used:

$$\frac{\partial C}{\partial t}=\frac{\partial}{\partial x}(D\frac{\partial}{\partial x}C)-V\frac{\partial}{\partial x}C+g(x), \tag{4}$$

where $C=C(x,t)$ is the nitrogen concentration, D is the diffusion coefficient, $g(x)$ is the concentration of nitrogen imbedded per time unit. The x coordinate is normal to the sample surface. The addend in the left side of equation is defined by the choice of coordinate system, which moves together with the solid body surface.

Figure 6 shows the sample depth change of the concentration of implanted nitrogen as the diffusion coefficient is increased [13]. Different curves correspond to different nitrogen diffusion coefficients in nm^2/s units.

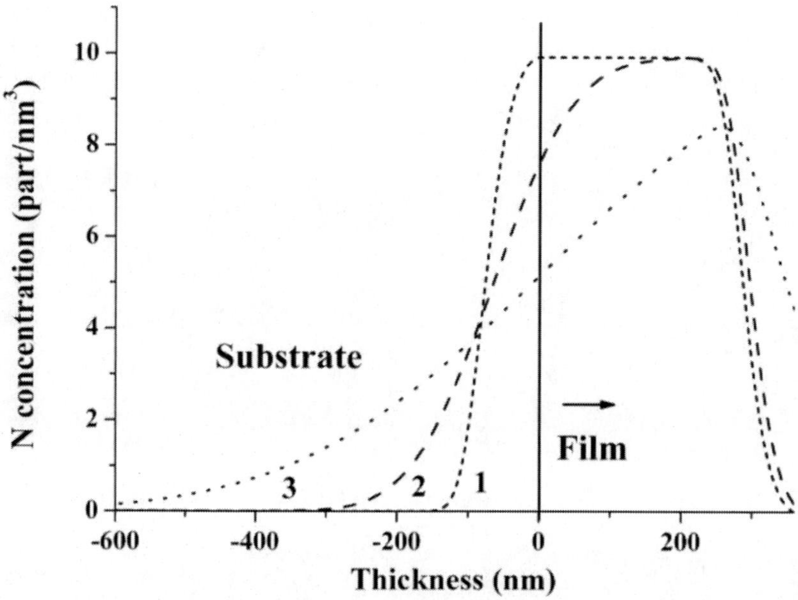

Figure 6. Sample thickness change in nitrogen concentration during the aluminum deposition onto the substrates made of the same metal at simultaneous bombardment with nitrogen ions of 30keV during 3600s 1- 0.01 nm^2/s, 2 – 1.0 nm^2/s, 3 – 10.0 nm^2/s [13].

Figure shows that the mixing zone thickness is increased with an increase in diffusion mobility. The nitrogen concentration at the growing film front is increased due to the migration of implanted nitrogen from the maximum position of occurrence to the film surface. It can also be seen that at high temperatures of the IBAD-process (D-10) the nitrogen concentration in the film depth fails to reach a stationary value during the estimated time (3600s) of film deposition.

To check the compliance of the results of computer simulation with experimental data using the secondary ion spectroscopy we investigated nitrogen distribution in the Cr-N-coating on the Al-substrate (Figure 7, [14]). Zero coordinate corresponded to the aluminum substrate surface. Figure shows that nitrogen concentration in a film reached constant stationary value, because the film thickness exceeded the thickness of the transition zone. Our attention is focused on an abrupt increase in nitrogen concentration at the film-substrate boundary and the concentration curve bent for aluminum at a depth of 25nm.

This Figure also gives computation data shown by the dashed line. Computation results fit well experimental data and reproduce special features of a nitrogen concentration curve. At the same time the mismatch of the design and experimental data is observed near the film surface. A higher nitrogen content revealed on the surface confirms the results that were published earlier on the investigation of the phase composition of Cr-N composite at the initial stage of its formation in the ion irradiation environment [15]. The electron microscopy showed that chrome nitrides CrN and Cr_2N are formed at process temperatures exceeding

200°C. The estimated nitrogen content in the film should not exceed 25%. Experimental investigations that were carried out testify that an increased value of nitrogen concentration at the growing film front can be related to the uptake of nitrogen from a residual atmosphere and formation of deposited metal nitrides within the chamber volume.

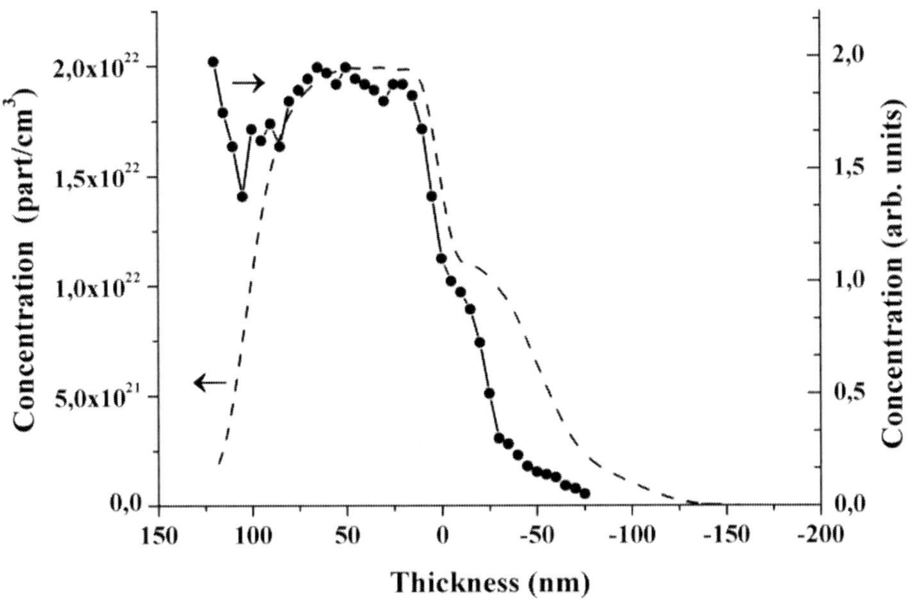

Figure 7. Data on a change in relative nitrogen concentration in the aluminum substrate and Cr-N film obtained using the SIMP method (marker-drawn), a dashed curve shows computer simulation data. The left ordinate axis is related to the design data. The ion current density is 25 μA/cm^2, the deposition rate is 0.1nm/s, and the temperature is 300°C [14].

3. Peculiarities of the Film Structure Formation at the Initial Stage of the IBAD Process

The results of computer simulation of a coating deposition process in the gas ion bombardment environment that are given in the previous chapter show that the concentration of implanted ions and the amount of defects created by them change at different process stages. Therefore the conditions at which the coating structure is nucleated and formed will be different. This becomes especially topical for the deposition of coatings using the ion beam-assisted technology because the control over coating deposition processes at the initial stage defines in the long run the formation of the substrate-coating transition zone structure and a coating adhesion. The previous papers that delved into the studies of the initial stage of coating formation [16,17] studied mainly the influence produced by the bombardment with inert gas ions on the nucleation of a coating structure. It has been shown that the ion bombardment results in the promotion of the coalescence of grain nuclei and formation of coatings with an explicit texture. While creating nitride coatings nitrogen ions should inevitably be used as bombarding particles since their presence must influence not only the structure of condensates but also their phase composition. The analysis of curves in

Figures 3 and 4 showed that the component composition and structure of films can continuously be changed until the thickness of 60 to 80nm is reached. This happens due to the implantation of nitrogen ions and an increase in the amount of radiation defects created by them. While studying the initial stages of the structure formation, for example, nitride coatings exposed to the ion bombardment attention should be paid to the nitrogen affinity to a definite material, i.e., to a value of Gibbs free energy of the given nitrides. The papers [18, 19] showed that the amount of nitrogen that can be adsorbed by the metal surface depends on the value of Gibbs free energy of appropriate nitrides.

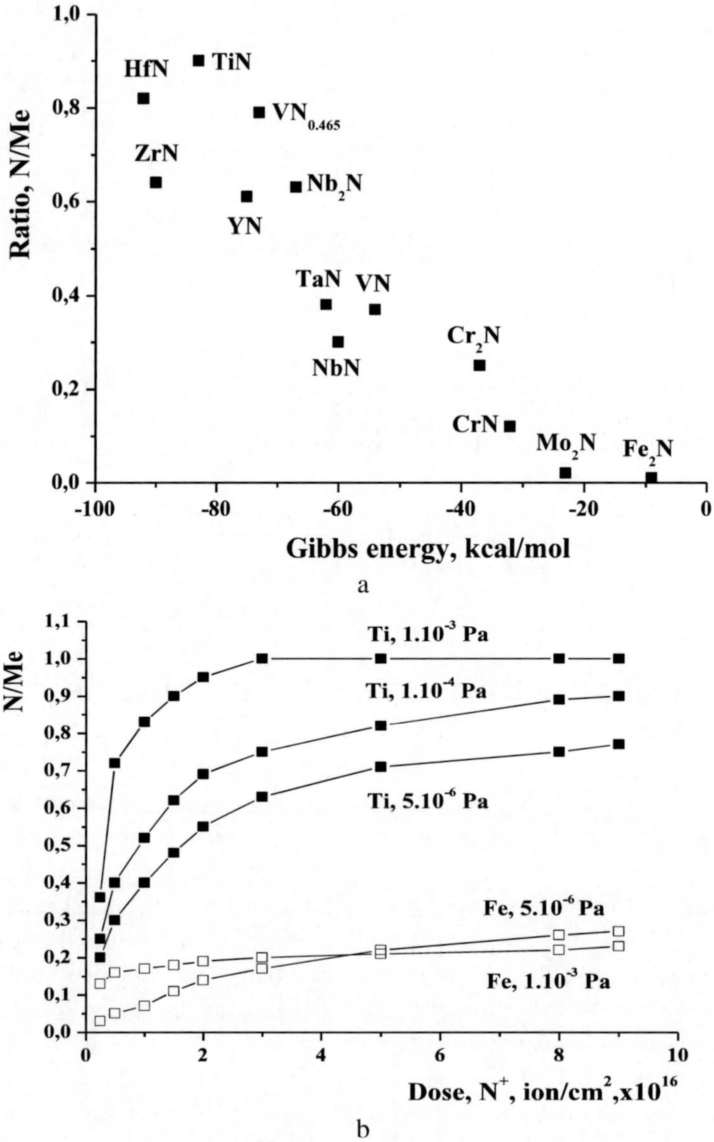

Figure 8. Relation between the saturation surface nitrogen-to-metal ratio and: a – Gibbs energy of nitride formation at 298K for various metals bombarded with 8 keV Ar^+ ions at 4 μA.cm^{-2} in nitrogen at 10^{-3} Pa [18]. b – dose nitrogen ion irradiation (E=6 keV) at different nitrogen pressure [20].

In particular Figure 8a gives theoretical relationships of N/Me ratio taken from the paper [18] for the series of nitrides exposed to the bombardment with argon ions in nitrogen environment as a function of their Gibbs energy. Figure 8b gives data taken from the paper [20], which show that using the same parameters for the irradiation of titanium whose nitride has low Gibbs energy with nitrogen ions its content in the material cardinally depends on nitrogen pressure.

The irradiation up to a maximum tested dose of 10^{17} ion/cm^2 at nitrogen pressure of $1\cdot10^{-4}$ Pa results in 90% nitrogen saturation and the pressure of $1\cdot10^{-3}$Pa provides 100 percent saturation and irradiation dose-independent saturation occurs at $2\cdot10^{16}$ ion/cm^2. For comparison the diagram gives data for iron whose free energy of nitride formation is much higher than that of titanium, in particular 8.5 kcal/mole. It is seen that a change in nitrogen pressure by two and a half orders actually results in no change of its content in iron.

We performed electron microscopy of initial stages of the formation of vanadium and chrome-nitride films that possess different Gibbs energy in conditions of ion-assisted deposition [21]. Figure 9, 10 give distribution histograms of film grains whose effective thickness is in the range of 5 to 20 nm.

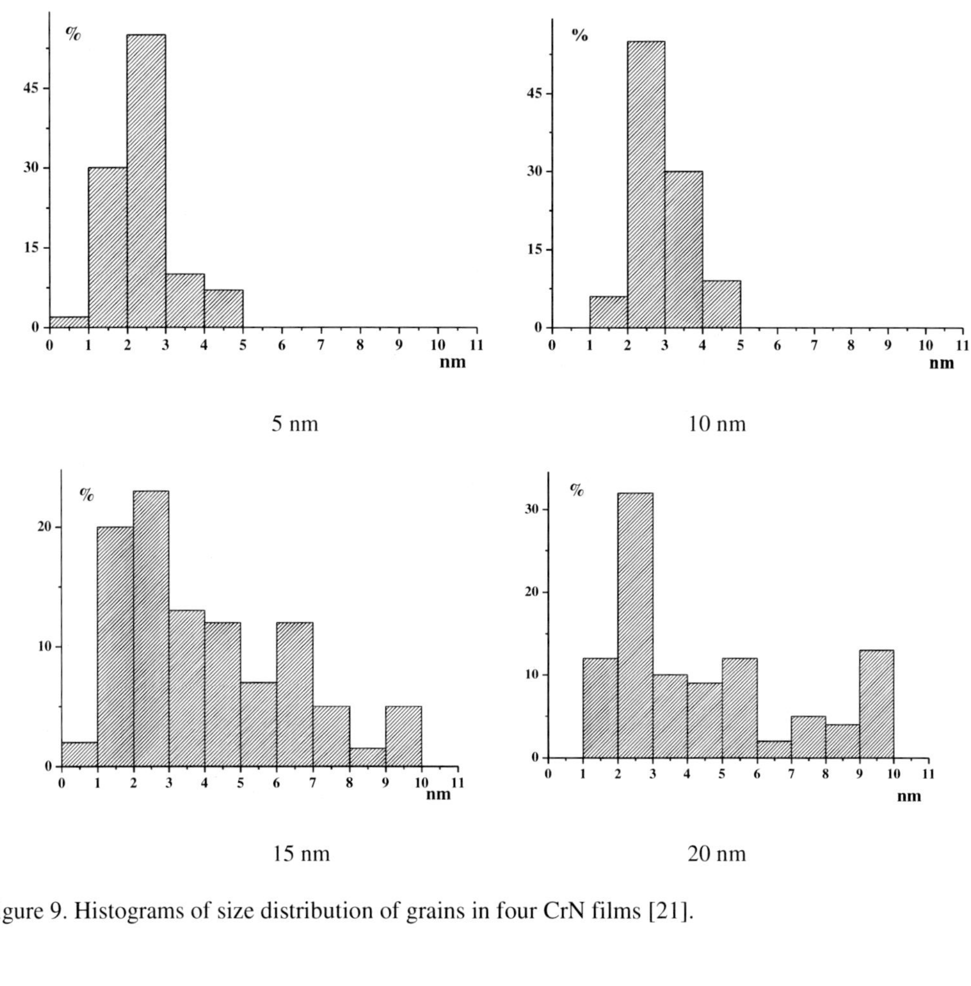

Figure 9. Histograms of size distribution of grains in four CrN films [21].

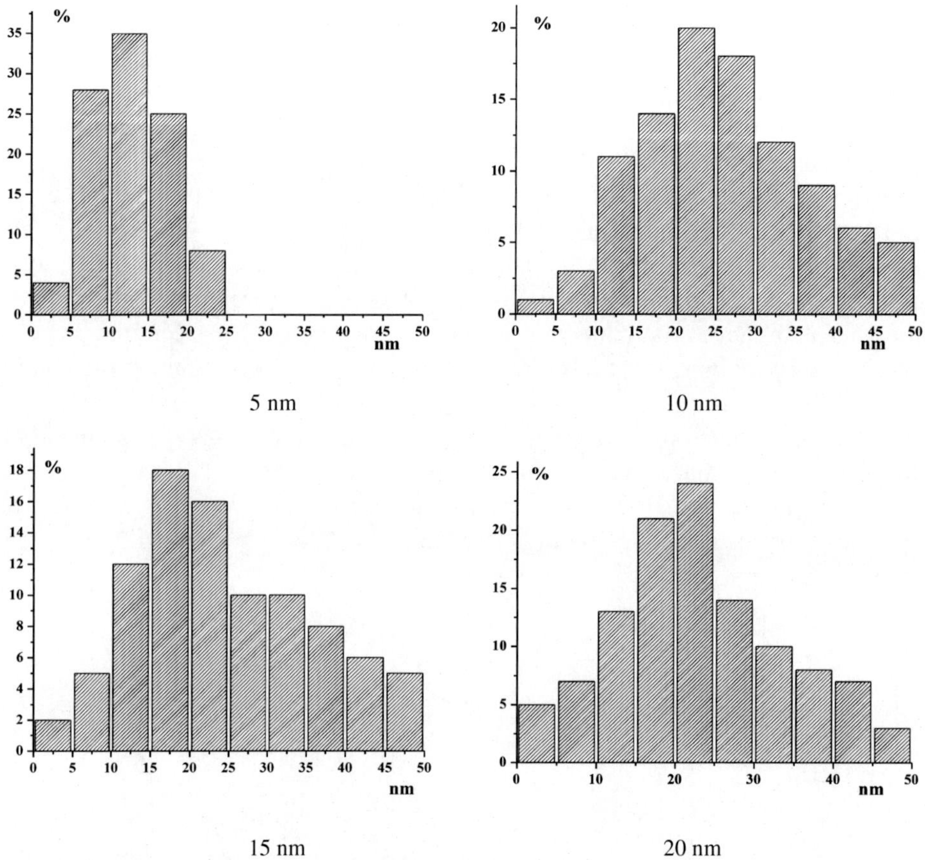

Figure 10. Histograms of size distribution of grains in four VN films [21].

The analysis of histograms of CrN grain distribution for different deposition time showed that the material structure at the given stage corresponds to the grain nuclei system whose distribution function changes with an increase in the deposition time and irradiation dose.

The first two films show uniform-sized distribution with sufficiently narrow peak for the size range of ≈ 2-3 nm, however starting from the third film the origination of nuclei of ≈ 5-8 nm is observed. This tendency is explicitly manifested in the fourth film.

It should be noted that an increase in the average size of nuclei occurs due to the appearance of the second peak in the region of large sizes, and not because of peak displacement on the distribution curve. The amount of small-size nuclei remains actually unchanged.

This is indicative of the progress of coalescence process with simultaneous nucleation of new grain population on vacant substrate sections. Vanadium grain size distribution curves have one maximum which shifts on the size scale from 15 to 25nm as the film thickness is increased. Further deposition results in no shift of the given peak toward large sizes.

In addition, an increase in the amount of small –size grains is observed, which occurs due to their origination on triple and quadruple grain junctions, not in the substrate plane. Due to the fact that the new grain population shades the space of intergranular junctions the formation process of a solid coating turns out to be incomplete. As a result the formed coating contains many pores of ~ 10 nm.

The diffraction analysis of chrome films showed that starting from the lowest thicknesses the formation of the crystalline structure of CrN occurs. Thereat, with an increase in the thickness of films their crystalline orientation is changed. The sequential orientation transition [211] → [211] + [100] → [100] is observed. A change in the structure of V films differs from that of Cr films. At low thicknesses a film consists of fully disoriented grains whose diffraction reflections form an annular electron-diffraction pattern.

As the coating deposition time increases the texture formation in the direction (001) occurs. The film structure corresponds to the fcc phase of vanadium nitride VN_{1-x}.

The attention is arrested by the fact that the vanadium nitride grain nuclei density does not exceed the value of $1.2\cdot 10^{11}$ cm^{-2}, while for chrome nitride the nuclei density varied in the range of $6\cdot10^{11} - 2\cdot10^{12}$ cm^{-2}.

Thus, we can state that in conditions of the irradiation with nitrogen ions during the chrome deposition a fine-crystalline dense structure is formed. The vanadium deposition in the same conditions results in the formation of a larger-grain messy structure. And finally, in spite of the fact that the estimated value of N^+/Cr,V ratio does not exceed 0.08 at the initial stage of coating formation (for thicknesses of 20 nm) the nitride phase is formed in both cases. The differences in formation processes of CrN_{1-x} and VN_{1-x} coatings can be explained within the framework of nitride formation thermodynamics. Figure 8a shows that for vanadium nitride the Gibbs energy is considerably lower in comparison with that of chrome nitride. Hence, nitrogen will react with vanadium more actively during the formation of nitride coatings by the ion-assisted deposition.

The deposition of vanadium and chrome occurs in conditions of continuous adsorption of nitrogen molecules by the substrate surface. The nitrogen dissociation energy is rather high and it reaches 205.8 kcal/mole, and therefore nitrogen fails to dissociate on the substrate at 200°C. Its dissociation seems to be more feasible in ion-assisted process conditions. This mechanism can take place if specificity of the statement of IBAD experiments is taken into consideration. For this technology a vacuum space right before the substrate is in the effective area of bombarding ions. At such a configuration the probability of the dissociation of molecular nitrogen in the chamber volume is rather high. This may result in the formation of chemical compounds both in the chamber and on the substrate surface. It means that this method of coating deposition allows for the formation of compounds at the initial stage in two ways, in particular on the substrate due to the surface diffusion of metal and nitrogen atoms and due to the formation of nitride molecules in space, their deposition and diffusion.

A criterion of applicability of either mechanism for ion-assisted technology can be a degree of nitrogen affinity to metal. The fact that nitrogen reacts easier with vanadium in comparison with chrome allows us to assume that major portion of stable VN molecules is formed at the earliest stage of the process, i.e., during spatial interaction of dissociated vanadium and nitrogen molecules. On the contrary the formation of chrome nitride mainly occurs after the deposition of metal onto the substrate surface.

In the paper [21] we substantiated that vanadium nitride molecules have a longer lifetime while depositing on the substrate surface and display higher diffusion mobility in comparison with individual chrome and nitrogen atoms. Therefore, while forming a coating by combining clusters into stable nuclei, the distribution density of those nuclei should be lower than the density of structures formed due to the atom merging. In the case of chrome nitride the nuclei density is defined by the diffusion mobility both of chrome atoms and nitrogen atoms and by the time of their existence on the substrate. Each of these parameters will be lower than those for VN clusters that were formed prior to their appearance on the substrate. Due to this fact the chrome nitride structure should be more fine-grained than that of vanadium nitride as the experiment showed.

4. Structure and Phase Characteristics of "Thick" Films and Coatings

It was mentioned above that the reactive gas used for the IBAD technology enters a formed coating by two channels, in particular in the form of accelerated ions and adsorbed molecules. In spite of low pressure of a working gas in the vacuum chamber and taking into account relatively low deposition rates used for this method that range from -0.1 to 1.0 nm/s the amount of gas absorbed during the deposition can considerably exceed the number of deposited metal atoms. Therefore, the phase composition and coating structure depend directly on the amount and final state of gas present in the metal lattice, to be more exact to what extent specific material reacts with gas being used.

Use of the IBAD technology allowed for more detailed analysis of a Cr-N compound. In particular, Ensinger et al. [22] showed that coatings with different phase composition (Figure 11) can be obtained by varying evaporation rate and ion current density. In order to form a stoichiometric CrN compound it is necessary to use low deposition rate of 0.1 to 0.3 nm/s and high ion current densities. However, all X-ray pictures show the reflection from the chrome lattice even at high densities.

In contrast to chrome transient metals of the fourth group, such as Ti, Hf, Zr have high affinity to reactive gases, therefore in conditions of ion-assisted deposition it is possible to obtain nitrides that contain no metal component (Figure 11b) [23].

As it was mentioned above the ion bombardment during metal deposition contributes to an increase in the concentration of grain nucleation places and diffusion mobility both of individual atoms and Me-(N,C,O) compounds. This results in a change of the formation mechanism of grain nuclei. The grains become equiaxed and the content of gas impurities is decreased. The diffusion mobility of atoms is defined not so much by the substrate temperature as by ion current density. The nucleating structure is characterized by high grain density, small size (in nano range) and narrow peak in the grain size distribution.

To stabilize the nanocrystalline structure and to improve consequently tribological characteristics the IBAD process is complemented with the evaporation or sputtering of the second metal, i.e., the three-component compounds are created [24]. A simultaneous deposition of two components requires the creation of appropriate conditions for the formation of nanocrystalline solid nitride phase surrounded by the second phase. In order to provide phase separation were selected metals that form no triple alloy Me_1Me_2N in their equilibrium state [25, 26]. The creation of nano composite structure is controlled by the process temperature and gas ion density. Nitrogen ions provide the formation of a solid nitride phase on the basis of one metal and a temperature provides diffusion of the second metal to the boundaries of nitride grains [26].

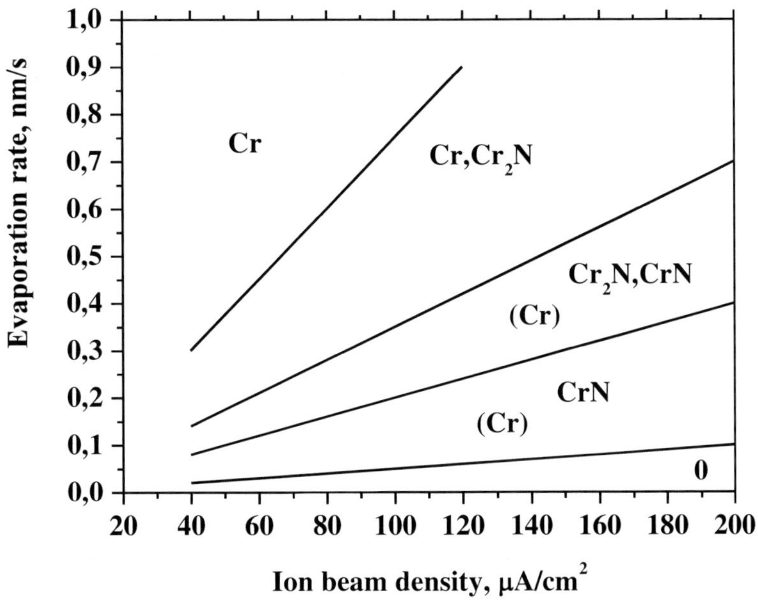

a

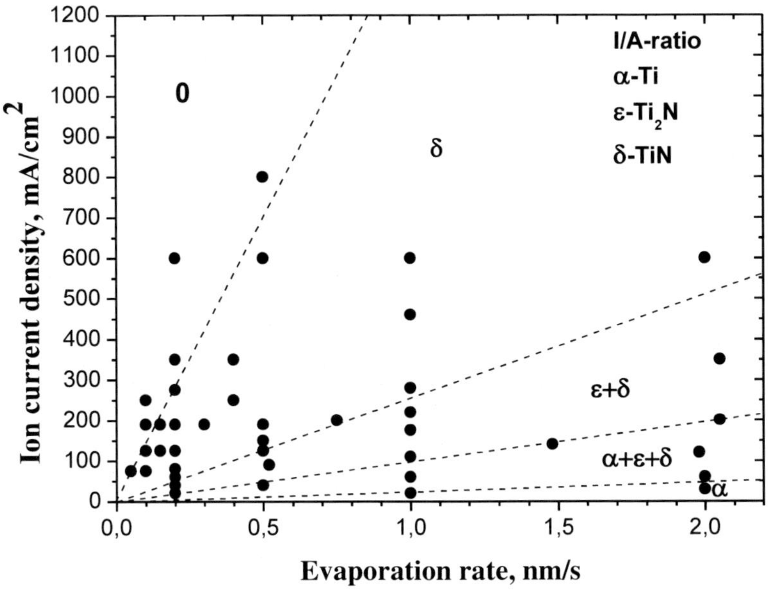

b

Figure 11. Regions of existence of different phases in the deposition rate/ion irradiation intensity diagram. a – Cr-N [22], b – Ti-N [23]. Ion energy - 30 keV.

This configuration allows for the creation of composite structures with different phase ratio by varying metal percentage, substrate temperature, and energy density imparted by the ion beam. As a rule, such materials demonstrate the highest hardness and acceptable tribological characteristics when the size of their grains approaches 5 to 10nm [24-26]. In such cases the intergranular space is filled very often with amorphous or crystalline boundary region.

Figure 12a gives the relationships of grain sizes of the Ti-Ni-N film as a function of Ni content [14]. It is seen that in the absence of nickel the grain size is higher for the deposition at 400 ° C. This means that the governing factor is the mobility of titanium and titanium nitride atoms. The addition of nickel results in its segregation at the nuclei boundaries of TiN grains and the blocking of their growth. As a result the grain size is decreased. The radiation-assisted nickel diffusion and not temperature becomes the basic factor that defines the grain size. As the nickel content is increased actually the same increase in thickness of amorphous nickel layer between the grains occurs at room temperature and 400°C (Figure 12b).

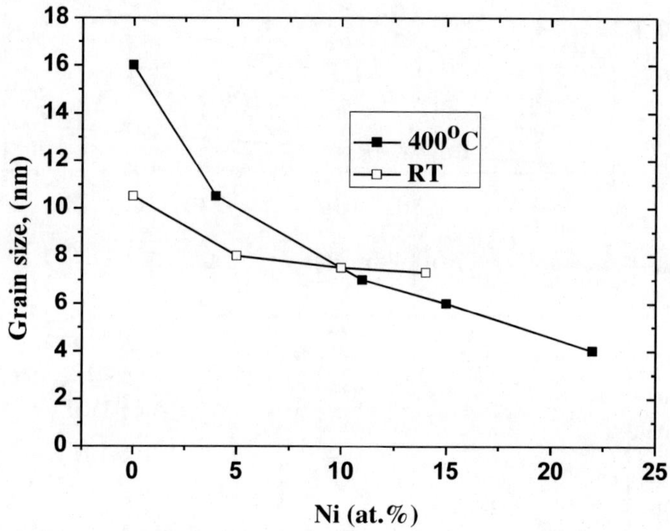

a

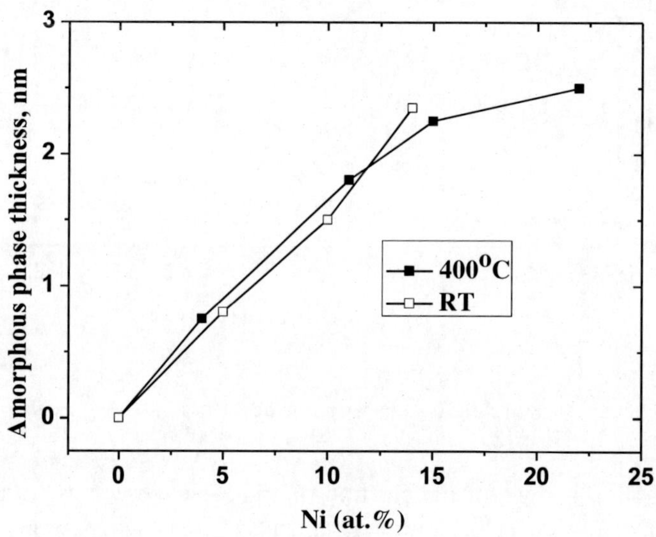

b

Figure 12. a – the grain size as a function of Ni content in the Ti-Ni-N coating. b – normalized average intergranular distance between TiN grains as a function of Ni content [26].

5. Tribological Characteristics

It is believed that coatings with high tribological characteristics should meet three basic requirements, in particular the portion, which is contiguous with the article, should possess a good article adhesion and the external portion of a coating should sustain loadings to which the article is exposed during the operation and an interim region must damp these loadings. In the classic case to solve this problem it is necessary to apply a three-layer coating in which each layer fulfills the required function. Naturally, the main parameter that defines the quality of transition zones between the layers will be a temperature of a coating deposition process. In the IBAD technology the function of temperature is fulfilled by the flow density of gas ions and their energies. For example, W. Ensinger et al. [27] analyzed the quality of TiN coatings applied onto steel AISI 52100 and aluminum alloy $AlCuMg_2$ using the IBAD method. It has been shown that structural and tribological characteristics of a coating are mainly defined by the ion beam energy and density. For the range of mean and high energies of 10 to 20keV the coatings that were obtained using the IBAD method demonstrate higher tribological characteristics, in particular adhesion, wear-resistance and hardness in comparison with those that were deposited using traditional PVD methods.

5.1. Microhardness

It is known that the hardness of polycrystalline structures is defined by their microstructure. In order to have an increased hardness the structure should be able to suppress the formation and motion of dislocations and micro cracks. This problem can be solved in many ways, in particular grain crushing, cold deformation, alloying and so on. However, these methods are not applicable to the case of nanocrystalline objects since the alloying ions leave the grain volume and segregate on its boundaries and the dislocations are not formed at all. Therefore, the microhardness of films obtained in ion bombardment conditions is defined by process parameters and a possibility to alloy the grain boundaries of a single-phase material by the second phase. The papers [28-31) studied the influence produced by the ion energy and ion current density on microhardness of Cr-N coatings (Figure 13). Lower values of microhardness were obtained in the paper [31] that used high densities of ion streams of 100 to 120 $\mu A/cm^2$; as a result the coating structure was a mixture of Cr + Cr_2N.The transition to lower intensities of ion current [29] results in the formation of CrN phase and an increase in microhardness. The increase was insignificant, possibly due to the large size of a coating grain, i.e., 200nm. The papers [28, 30] used higher energies that defined an increase in microhardness up to 18-30GPa. The highest increase was demonstrated by coatings with the smallest grain size of 5 to 8nm [28]. The data given above with regard to the influence produced by experimental parameters of the IBAD-process on microhardness of a Cr-N coating show that not only the phase composition influences this value but also the grain size. It was noted time and again that the Hall-Patch formula cannot be applied to nanocrystalline structures with the size grain of <10 to 20nm (for example [32, 33]) and the microhardness value can exceed the estimated value by a factor of 4 to 5. Today many models are available that could explain the observed effect. It is known, for example, that availability of a large amount of low-angle boundaries in a fine-grained structure creates conditions for the appearance of a stressed state in films [34]. As a rule, metal films deposited without ion stimulation experience tensile internal stresses that crave for the compensation of porosity that occurs at grain boundaries. The deposition of chrome in the ion-bombardment conditions cardinally changes the situation. As a rule an increase in the amount of energy introduced by the ion beam results in the change of the stress sign replacing tensile stresses by compressive ones [35-37].

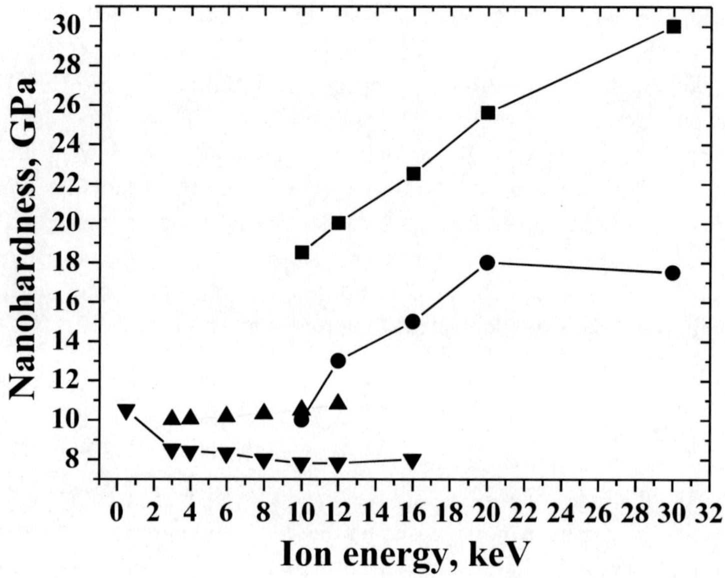

Figure 13. The microhardness dependence of CrN coatings on the ion bombardment energy. ● – [30], ▲ – [29], ▼ – [31], ■ – [28].

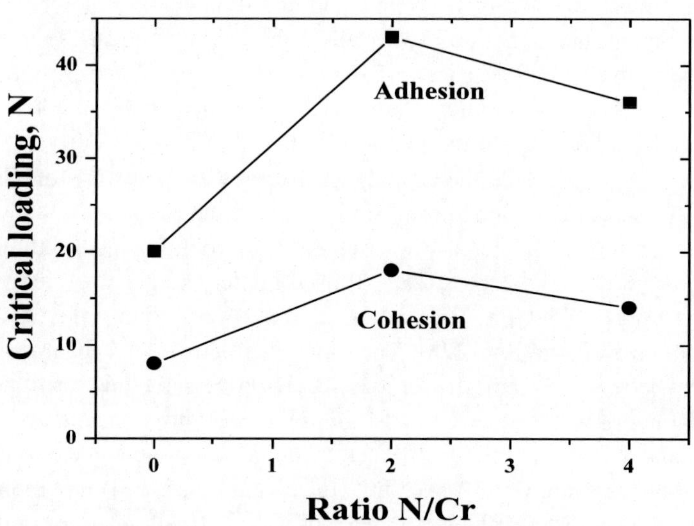

Figure 14. Dependence of a critical load of CrN coating separation on N/Cr ratio value [38].

Figure 14 shows the relationships of stress values in Cr-N coatings as a function of N/Cr ratio [38]. Such a trend of curves correlates with an increase in the CrN phase and decrease in the grain size revealed in the same work. That is, an increase in microhardness of CrN films with a decrease in the grain size is explained by an increase in compressive stresses. The grain value is just a parameter convenient for the comparison of different data [39].

5.2. Adhesion

The adhesion characteristics of coatings applied using ion bombardment technologies are defined by many factors, in particular careful preparation of the substrate surface, substrate-coating mixing zone depth, the formation efficiency of chemical compounds in the transition zone and also by the absence of both increased stresses and segregation of gas mixtures in this zone. The ion bombardment during the deposition ambiguously affects each factor mentioned above. The papers [38, 40] studied the influence of the ion energy on the adhesion value of the Cr-N coating. The authors did not manage to establish direct interrelation between the energy of nitrogen ions and the adhesion value [40].

On the contrary, the increase in energy from 20keV to 40keV resulted in a decreased adhesion. For additional improvement of adhesion characteristics of coatings we used the method of stimulation of mutual substrate-coating mixing through the preliminary implantation of nitrogen ions into the substrate with energy of 30keV to reach doses at which the substrate surface layer is saturated with nitrogen up to the concentration of Me/N=1.

The process temperature is not high and reaches 200°C. Then the irradiation is continued and the metal is deposited using the electron -beam evaporation technique, i.e., a duplex IBAD process [41].

The coating thickness is 130 nm. Figure 15 gives Cr and Nb distribution curves depending on the depth for the case of ordinary IBAD process (a) and for the case of duplex treatment, i.e., implantation +IBAD process (b).

It is seen that in spite of relatively low temperature of the experiment, the high flaw level peculiar for the IBAD technology contributes to the intensive progress of diffusion processes and formation of rather broad mixing zone. A preliminary ion bombardment stimulates this process even more.

5.3. Friction Coefficient and Wear

Comprehensive study of the friction coefficient and wear-resistance of CrN and TiN coatigs applied onto AlSI H13 steel was done in the paper [42]. It has been shown that the friction coefficient of CrN coatings applied on corundum at a relative humidity (RH) of 50 and 85% was about 0.3, slightly higher than that of TiN coatings. At RH<10% , it is a little bit higher but still has a low value before complete wear of CrN coatings, whereas TiN coatings exhibit the high coefficient of friction of about 0.8. In addition, a volumetric loss of CrN coatings during their friction is smaller in comparison with that of TiN coatings, especially at low RH.

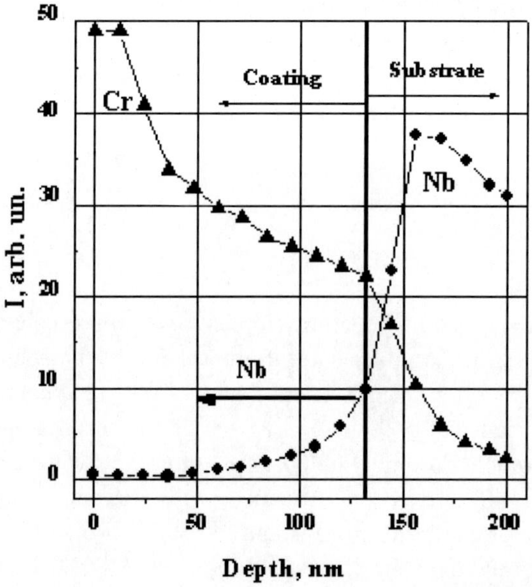

a

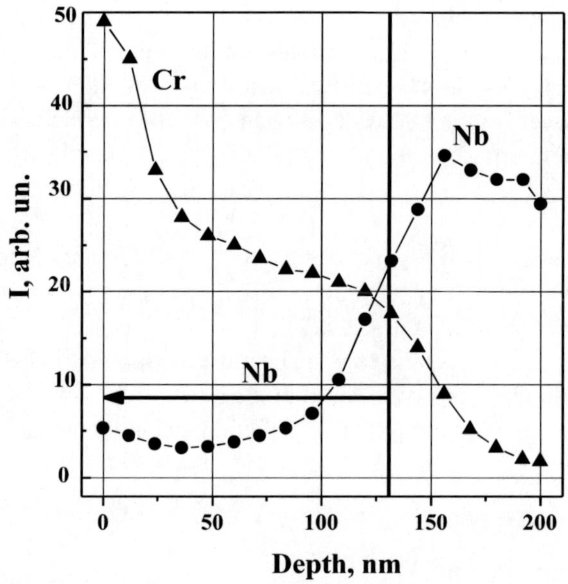

b

Figure 15. Cr and Nb distribution in a coating-substrate combination for the ordinary IBAD process (a) and for the duplex process, implantation 3.5x10^{17} cm^{-2} + IBAD (b) [41].

The influence of Cr/N ratio and the energy of nitrogen ions on the wear of CrN coating applied onto AISI 52100 steel were studied in the paper [28]. It has been shown that the fraction of Cr_2N phase is increased in the coating with an increase in Cr/N ratio and the wear is also cardinally increased (Figure 16). A three-fold increase in the ion energy produces actually no impact on the wear-resistance. Studying the influence produced by the basic parameters of the IBAD process (the ion energy and beam current density) on tribological characteristics of CrN coating showed that the friction coefficient slightly depends on both parameters [30]. The wear-resistance is significantly increased in a coated steel. The range of currents' densities embracing 8 to 32 μA/cm^2 and energies of 4 to 12 keV has been studied. The most optimal are energy values of 12 keV and minimal values of current density are 8 μA/cm^2.

The deposition of a three- component TiBN coating in conditions of low-energy radiation [43] showed that the change in the ion energy in the range of 0 to 800 eV cardinally changes hardness, friction coefficient and wear-resistance (Figure 17).

It is seen that the most acceptable values are reached at different energy values: a maximum hardness is reached at 400eV and a minimum wear and friction coefficients are observed at 150eV. The structure of TiBN composite consists of nano crystals TiB_2 and TiN surrounded by a soft h-BN phase. At such a configuration the best tribological characteristics are obtained at optimum mixing of nano grains in the soft phase.

An increase in hardness with an energy increase can be explained by that the rise in the ion energy up to 400eV and higher results in the preferred sputtering of lighter coating components B and N.

As a result the share of a harder component of TiN is increased and this contributes to an increase in hardness of TiN coatings at high energies. At lower energies a softer h-BN phase defines tribological properties to a large extent.

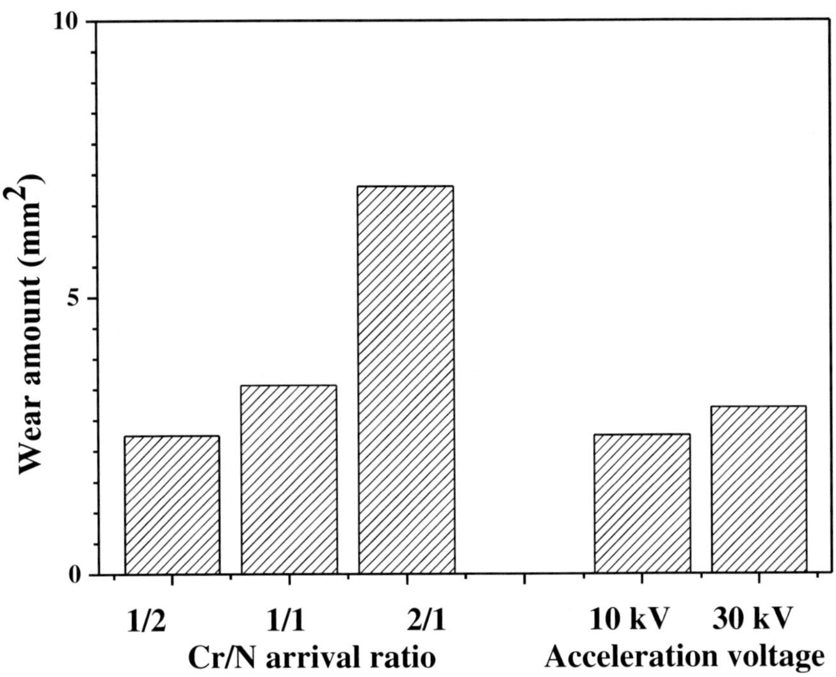

Figure 16. Wear of Cr-N coating after 200 test cycles as a function of Cr/N arrival ratio and ion energy [30].

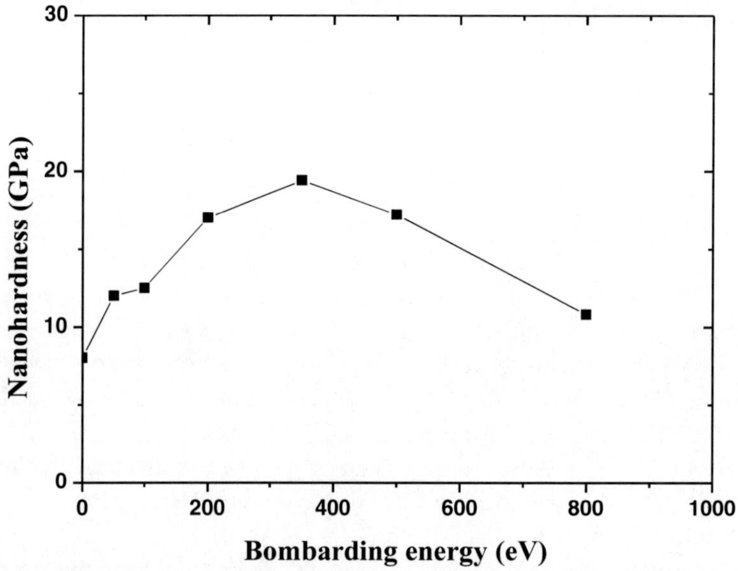

a

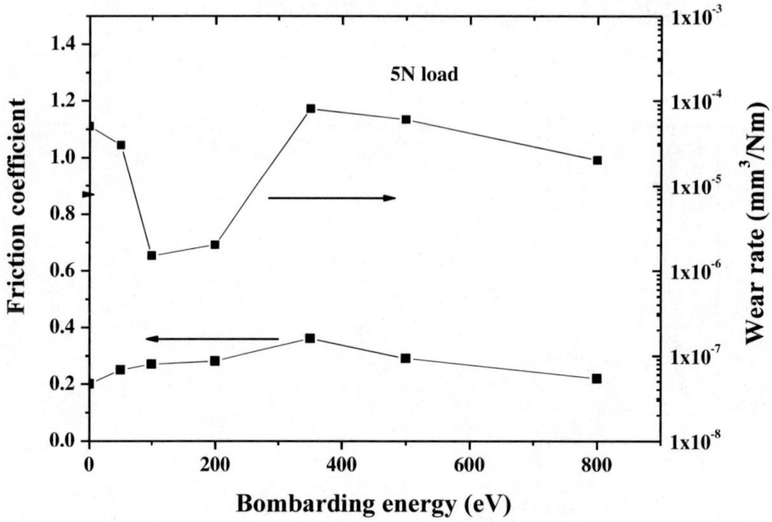

b

Figure 17. Nanohardness (a), friction coefficient and wear (b) of TiBN synthesized at different bombardment energy [43].

CONCLUSION

The research done in the field of nanostructured material science over the last years showed that materials with ultrafine grain have great prospects for the heavy-duty operation. The production of such materials is impossible using equilibrium technologies, Therefore the interest in nanocrystalline materials stimulated the development of technologies based on no equilibrium processes. In particular ion assisted deposition technologies are widely used for the formation of thin-film structures. High operating performances of the obtained coatings are defined by that the ion bombardment in combination with the deposition of atomic-&-molecular flows give an opportunity to design their structure on the atomic level. The intensive irradiation with heavy charged particles in combination with the deposition of overheated metal vapors onto the “cold” substrate creates conditions for the formation of structures with the grain size less than 10nm. By varying such parameters as ion current density, ion energy, nitrogen/boron/carbon/oxygen content and metal component concentration we can create structures whose tribological characteristics can vary in wide ranges.

The data given in the review allow for the formulation of key principles, which should be observed to provide the highest possible efficiency for the use of IBAD method:

- the ion beam -assisted deposition technology should be confined to the densities of ion fluxes within 100μA/cm^2 and the energy not lower than 5 to 10keV;
- the hardness of coatings is defined not so much by the phase composition as by the grain size and the level of internal stresses;
- the level of internal stresses is increased with an increase in the quantity of energy input into a coating per time unit and with a decrease in the grain size;
- the adhesion of coatings can be significantly improved due to the use of duplex technique;
- an increase in wear resistance is related to a decrease in the friction coefficient, the formation of coatings with low compression stresses, rather high hardness and good adhesion.

An additional advantage of the ion beam-assisted technology is that it can be used for the deposition of protective coatings onto the material with low melting temperature. Also is very important that the evaporation of metals with different affinity to active gases allows for the purposeful design of multicomponent structures that include both metal nanocrystalline grains and amorphous/ crystalline nitride (boride and carbide) phases.

REFERENCES

[1] Mattox D. *J. Vac. Sci. Tech.*, 1988, vol. A 7(3), 1105-1114.
[2] Weissmantel C. *Thin Solid Films*. 1982, vol, 92, 55-63.
[3] Cuomo J.; Rossnagel S. *Nucl. Instr.& Meth.in Phys. Res.*, 1987, vol.B19/20, 963-73.
[4] Ensinger W. *Rev. Sci. Instrum.*, 1992, vol. 63 (11), 5217-33.
[5] Andon Y.; Suzuki Y.; Matsuda K. et al. *Nucl. Instr. & Meth. Phys. Res*. 1993, vol, B6, 111-115.
[6] Petrov I.; Barna P.; Hultman L.; Greene J. *J. Vac. Sci. Tech.*, 2003, vol. A 21(5), S117-S128.
[7] Wolf G. *Vacuum*, 1989, vol. 39(11/12), 1105-1110.
[8] Vepřek S. *J. Vac. Sci. Tech.*, 1999, vol. A 17(5), 2401-2421.
[9] Musil J. *Surf. Coat. Tech.*, 2000, vol. 125, 322-330.
[10] Ensinger W. *Surf. Coat. Tech.*, 2000, vol. 128-129, 81-84.
[11] Guglya A.; Neklyudov I. Uspehi Fiziki Metallov, 2005, vol.6, 197 – 232 (in Russian).

[12] Ziegler J.F. www.srim.org

[13] Guglya A.; Marchenko I.; Neklyudov I. *Surf. Coat. Technol.* 2003, vol, 163-164, 286-292.

[14] Goncharov A.; Guglya A.; Marchenko I.; Neklyudov I. *Vacuum.* 2004, vol, 76, 299-302.

[15] Guglya A.; Litvinenko M.; Marchenko Yu.; Vasilenko R. *Voprocy atomnoy nauki i tehniki.* 2006. Vol.89 (4). 143-146 (in Russian).

[16] Marinov M. *Thin Solid Films.* 1977, vol,46, 267-274.

[17] Arnault, J.; Delafond, J.; Templier, C.; Chaumont, J.; Enea,O. *Nucl. Instr. & Meth. Phys. Res.* 1993, vol, B80/81, 1384-1387.

[18] Baba Y.; Sasaki T. *Mat. Sci. & Eng.* 1989, vol, A 115, 203-207.

[19] Barin I.; Knacke O. Thermodynamical properties of inorganic substances. Springer: Berlin. 1973; 126.

[20] Takano I.; Isobe S.; Sasaki T.; Baba Y. Applied Surface Science, 1989, vol, 37, 25-29.

[21] Guglya A. Fizika Metalov & Metalovedeniye. 2010, vol.109, 72-82 (in Russian).

[22] Ensinger W.; Kiuchi M. Surf.& Coat. Tech. 1997, vol.94-95, 433-436.

[23] Ensinger, W.; Volz K.; Kiuchi M. *Surf.& Coat. Tech.* 2000. Vol.128-129. 81-84.

[24] Akbari A., Riviere J.P., Templier C., Le E. *Rev. Adv. Mater. Sci.* 2007, vol. 15, 111-117.

[25] Zhang C., Luo J., Li W., Chen D. *J. Tribol.* 2003, vol. 125, № 2, 445-447.

[26] Akbari A., Riviere J., Templier C., Bourhis E. *Surf. & Coat. Tech.* 2006, vol. 200, 6298-6302.

[27] Ensinger,W.; Schröer,A.; Wolf,G. *Nucl. Instr. & Methods in Phys. Res.* 1993, vol, B80/81, 445-454.

[28] Vanzha, O.; Guglya, A. *Metalophysika i noveyshie tehnologii,* 2007, vol, 29. №9, 1185-1199 (in Russian).

[29] Fu, Y.; Zhu, X.; Tang, B.et al. *Wear.* 1998, vol, 217, 159-64.

[30] Sugiyama, K.; Hayashi, K.; Sasaki, J.; Ichiko, O.; Hashiguchi, Y. Nucl. Instr.&Meth. in *Phys. Res.* 1993, vol, B80/81, 1376-82.

[31] Engel, P.; Schwarz, G.; Wolf, G. *Surf. & Coat. Tech.* 1999, vol, 112, 286-290.

[32] K. Lu. *Mater. Sci. Eng.* 1996, vol, R16, 161-221.

[33] Andrievski, R. *J. Mater. Sci. Technol.* 1998, vol, 4, 97-103.

[34] Palatnik, L.; Fuks, M.; Kosevich, V. Mekhanism obpasovaniya i substruktura kondensirovanyh plenok. Nauka: Moscow, 1972; 183.

[35] Mayrhofer, P.; Tischler, G.; Mitterer, C. *Surf. & Coat. Tech.* 2001, vol, 142-144, 78-84.

[36] Oden, M.; Almer, J.; Hakansson, G.; Olsson, M. *Thin Solid Films.* 2000, vol, 377-378, 407-12.

[37] Ehrlich, A.; Kühn, M.; Richter, F.; Hoyer, W. *Surf. & Coat. Tech.* 1995, vol,76-77, 280-286.

[38] Demaree, J.; Fountzoulas, C.; Hirvonen, J. *Surf. & Coat. Tech.* 1996, vol, 86-87, 309-315.

[39] Chuang, C.; Chao, C.; Chang, R.; Chu, K. *J. Mat. Proc. Tech.* 2008, vol, 201, 770-774.

[40] Tian, L.; Tang, B.; Liu, D. et al. *Surf. & Coat. Tech.* 2005, vol, 191, 149-54.

[41] Guglya, A.; Neklyudov, I.; Shkuropatenko, V.; Vanzha, A.; Virich, V., *Problems of Atomic Science & Technology.* 2005, vol, 86. №3, 171-175 (In Russian).

[42] Chen, H.; Wu, P.; Quaeyhaegens, C. et al. *Wear*. 2002, vol, 253, iss.5-6, 527-532.
[43] He, J.; Miyake,S.; Setsuhara,Y. et al. *Wear*. 2001, vol, 249, 498-502.

In: Comprehensive Guide for Nanocoatings Technology. Vol. 1 ISBN: 978-1-63482-447-7
Editor: Mahmood Aliofkhazraei

Chapter 4

PARTICLE COATING BY ROTATING PCVD REACTOR

Anna Nasonova[1] and Kyo-Seon Kim*

Department of Chemical Engineering, Kangwon National University
Chuncheon, Kangwon-Do, Republic of Korea

ABSTRACT

The particle coating by rotating Plasma Chemical Vapor Deposition (PCVD) reactor will be discussed in this chapter. Rotating PCVD technique is used to synthesize nano-sized thin films with high uniformity. The rotating PCVD reactor allows to produce a single-layer, as well as double-layer thin films coated on small-size substrates of various shapes (spherical, cylindrical, oval, etc.) and materials (glass, polymer, zeolite, etc.) with using liquid precursors (TTIP (titanium tetra-isopropoxide), TEOS (tetraethyl orthosilicate), etc.) and/or gaseous precursors. This chapter will explain the principles of rotating PCVD process, discuss simulation and experimental results and show possible applications of this technique. Being flexible in changing of process parameters, rotating PCVD is the promising method for synthesis of various thin films with several desirable properties.

1. INTRODUCTION

Recently, there is growing interest in coating thin films with the desirable electrical, magnetic, mechanical and optical properties onto bulk materials or particles, and the thin film with excellent physical properties can be applied to various fields which require high-performance materials. For example, the hydrogenated amorphous silicon (a-Si:H) thin films are used for the solar cell, image sensor, electrophotographic drum, thin film transistor, etc. There are wet and dry (gas phase) coating processes. The dry coating process has several

---

* Corresponding author. Tel: +82-33-250-6334, E-mail: kkyoseon@kangwon.ac.kr;
[1] Present address : Department of Chemical and Biological Engineering, Korea University, Seoul 136-713, Republic of Korea.

advantages such as easy scale-up, high throughput, comparable low costs, and short processing time, compared with the wet coating process [1-6].

To improve stability and dispersion characteristics of the particles and also to enhance the mechanical properties of them, several researchers have studied on coating thin films on the particles by using the gas phase processes. Hung et al. [7] used a counter flow flame reactor to prepare the SiO_2-coated TiO_2 particles and observed the dependence of precursor concentrations and temperature on the surface morphology of the TiO_2 particles. Powell et al. [8,9] prepared the TiO_2 particles coated with SiO_2, Al_2O_3 and Al_2O_3/SiO_2 by using a hot-wall, continuous flow tubular reactor, and suggested that the precursor concentrations of coating materials and the reactor temperature can be important variables to control the surface characteristics of coated particles. Fotou et al. [1] also used the continuous flow tubular reactor for preparing the TiO_2 particles coated with metal oxides (ZrO_2, Al_2O_3/ZrO_2, and SiO_2/ZrO_2), and proposed different film formation mechanisms for those metal oxides with different morphologies. Zhang et al. [6] obtained uniformly coated Ag particles with organic materials at a low cooling rate in a jet flow aerosol reactor. Nakagawa et al. [10] prepared the fine Cu particles coated with a thin film of TiO_2 by using H_2–Ar arc-plasmas, and reported the role of H_2 in preparing fine composite particles and maintaining a steady state of arc-plasmas. The circulating fluidized bed plasma chemical vapor deposition (PCVD) reactor was used to coat the amorphous TiO_2 film on the glass microbeads as photocatalysts for water treatment [11] and also to prepare the hard SiC particles coated with the aluminum oxide film to increase the chemical stability of the SiC particles [12]. Researches mentioned above are summarized in Table 1.

Table 1. Particle coating methods

Method	Substrate material	Particle size	Coating material	Coating thickness	Precursor	Ref.
Counterflow diffusion flame	TiO_2	<200 nm	SiO_2	10-30 nm	$TiCl_4$, $SiCl_4$	7
Hot-wall tubular aerosol reactor	TiO_2	~1 μm	SiO_2	10-80 nm	$TiCl_4$	8,9
Hot-wall tubular reactor	TiO_2	<1 μm	ZrO_2, Al_2O_3/ZrO_2, SiO_2/ZrO_2	5-30 nm	$TiCl_4$, $ZrCl_4$, $AlCl_3$, $SiCl_4$,	1
Gas phase aerosol method	Ag	<400 nm	Organic coating	1-7 nm	Carnauba wax, stearic acid, melamine, $AgNO_3$	6
Arc plasma irradiation technique	Cu	10-30 μm	TiO_2	20-100 nm	Cu, TiO_2	10
Circulating fluidized bed plasma – CVD reactor	Glass microbeads	124 μm	TiO_2	7-120 nm	$Ti(OtBu)_4$	11
Circulating fluidized bed plasma – CVD reactor	SiC	150 μm	Al_2O_3	0.3-1.3 μm	$Al(OC_4H_9)_3$	12
Rotating PCVD	Glass or polypropylene particles	1-3 mm	SiO_2, TiO_2, SiO_2/TiO_2	50 nm – 1.5 μm	$Ti\{OCH(CH_3)_2\}_4$, $Si(OC_2H_5)_4$	42-48

The PCVD involves the dissociation and/or chemical reactions of gaseous reactants in an activated by plasma environment, followed by the formation of a stable solid product. The deposition involves homogeneous gas phase reactions, which occur in the gas phase, and/or heterogeneous chemical reactions which occur on/near the vicinity of a substrate surface leading to the formation of powders or films, respectively. PCVD uses electron energy (plasma) as the activation method to enable deposition to occur at a low temperature and at a reasonable rate. Supplying electrical power at a sufficiently high voltage to a gas at reduced pressures (<1.3 kPa), results in the breaking down of the gas and generates a glow discharge plasma consisting of electrons, ions and electronically excited species. The vapor reactants are ionized and dissociated by electron impact, and hence generating chemically active ions and radicals that undergo the heterogeneous chemical reaction at or near the heated substrate surface and deposit the thin film. The temperature of the electron can be on the order of 20,000 K or higher, while the temperature of the vapor reactants may remain near room temperature, depending on the pressure at which the discharge is operated.

The main advantage of PCVD over other CVD methods is that the deposition can occur at relatively low temperatures on large areas. It also offers the flexibility for the microstructure of the film and deposition tobe controlled separately. The ion bombardment can be substituted for deposition temperature to obtain the required film density. Such low temperature deposition is important for applications that involve the use of temperature sensitive substrates. The deposition temperature of PCVD can be lowered further by the use of metalorganic precursors (MOPECVD). For example, Rie et al. [13,14] have explored the use of metalorganic as precursor in the PECVD process to lower the deposition temperatures from 450–550 °C to about 140 °C. This deposition temperature is considerably lower than the conventional thermal CVD (900–1100 °C). This has widened the scope of substrates that can be used. They have deposited TiN, TiC, TiCN, ZrCN, ZrBCN, BCN single and multilayer systems on steel substrates, hard metals, cermet substrates and age hardening aluminium alloys to improve the wear and corrosion resistance.

The PCVD processes have widely been applied to make thin films in the semiconductor manufacturing processes and also can be used for particle coating, but they are quite notorious from the point of view of particle contamination because a lot of particles in the size ranges from nanometers to microns are easily generated in the plasma processes [15-22]. The growth of nano-sized particles in the plasma reactor has been found out to be suppressed efficiently by using the pulse-modulated plasma technique. There are several experimental researches to show that the pulsed plasma processes can be used to prepare the thin films of amorphous silicon [23-28], polycrystalline silicon [27,28], titanium nitride [29], amorphous hydrogenated silicon carbide [30], diamond [31], silicon dioxide [32,33], and fluorocarbon [34-37].

Rotating PCVD processes have been widely used to prepare uniform thin films for semiconductor fabrication. Kim et al. [38-44] extended the application of the PCVD process for uniform particle coating and proposed a rotating cylindrical PCVD reactor for particle coating based on a numerical analysis. They suggested that the rotating PCVD reactor could be used to provide thin films on particles with higher quality, as compared with other particle-coating processes. In the rotating cylindrical PCVD reactor, the particles will rotate together with the cylindrical PCVD reactor, and these particles can be divided into two classes: particles falling in the gas phase and particles located on the cylinder wall, as shown in Figure 1.

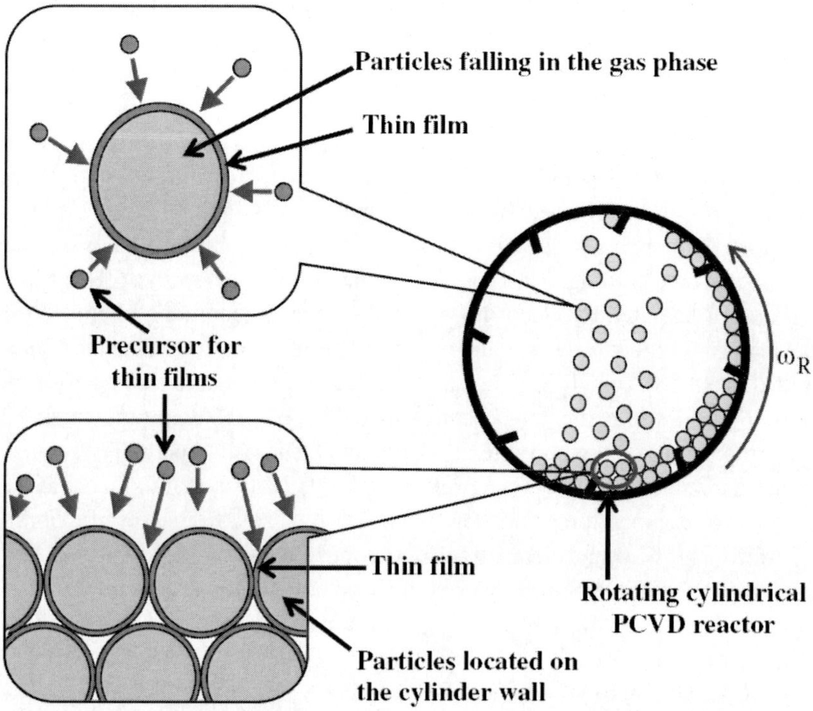

Figure 1. Cross-sectional view of rotating cylindrical PCVD reactor with particles [44].

For the particles falling in the bulk plasma of the gas phase, their total surface area is available for deposition of precursors for thin films and the particles will be coated uniformly with the precursors generated by the plasma reactions. On the other hand, for the particles located on the cylinder wall, the surface area of the cylinder wall exposed to plasmas is available for deposition of precursors, and the upper section of the particles will be coated with the precursors while they stay on the cylindrical tube wall. Kim's group [38-44] made theoretical analysis of particle coating in rotating cylindrical PCVD as well as an experimental analysis; detailed explanation of this analysis will be given later in this chapter.

2. Theoretical Analysis of Particle Growth

The thickness of thin films and growth rate of TiO_2 thin films on the glass beads in a rotating PCVD reactor were calculated by solving the model equations numerically by Kim's group [38-41]. Calculations were made for TiO_2 thin film and Si thin film deposited onto particles with using TTIP and SiH_4 as precursor.

A schematic of the rotating cylindrical PCVD reactor is shown in Figure 2. All gases are supplied into a cylindrical quartz tube whose diameter and length are 6.5 and 35 cm, respectively. Low-temperature plasmas are generated by an inductively coupled rf plasma process with a spiral-shape electrode installed around the reactor. The plasma reactor rotates, with a rotation speed ω_R, to deposit the chemical species uniformly onto the surface of glass particles. The particles in the rotating PCVD reactor can be classified into two types: particles falling in the gas phase and particles on the cylinder wall during the plasma process. The particles are coated with active chemical species generated inside the plasma reactor. For the falling particles, the total surface area of the particles is available for deposition of the active chemical species. On the other hand, for the particles located on the cylinder wall, the surface area of particles exposed to the plasmas is just available for particle coating.

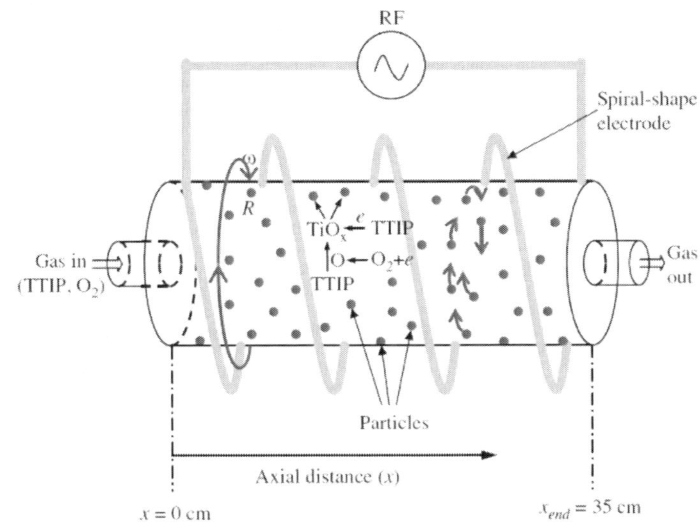

Figure 2. Schematic of the inductively coupled plasma reactor for particle coating [39].

In the TTIP + O_2 plasmas, highly energetic electrons collide with O_2gas molecules to generate reactive radicals such as O and O('D).TTIP can be decomposed to produce TiO_x, $Ti(OH)_x$, or $Ti(OH)_x(OR)_{4-x}$, by the electron-impact dissociation reaction and also by reaction with reactive chemical species, mainly O radicals. The TiO_x concentration will depend on the concentrations of TTIP and O, and 9 important plasma reactions can be considered for the evolution of TTIP, TiO_x, and O radicals inside the reactor [39].

The number concentrations for chemical species, *i*, in the plasma reactor can be calculated from Eq. (1) and (2) as follows:

$$\frac{dN_i}{dt} = \sum_j^9 \alpha_{ij}(RXN)_j - \nabla(u_s N_i - D\nabla N_i) \qquad (i = 1\text{-}5) \qquad (1)$$

$$\frac{dN_{TiO_x}}{dt} = (k_8 N_e + k_9 N_o)N_{TTIP} - \nabla\left(u_s N_{TiO_x} - D_{TiO_x}\nabla N_{TiO_x}\right) - J_{TiOx}\left(\frac{2}{R}\right) - J_{TiOx}\left(\frac{D_P^2}{R^2}\right)N_T W_{FP} \qquad (2)$$

In Eqs. (1) and (2), the first, second, and third terms on the right-hand side (RHS) are the effects of plasma chemical reactions, convection, and diffusion, respectively. The fourth and fifth terms in Eq. (2) show the disappearance rates of TiO_x by deposition on the reactor wall and on the particles falling in the gas phase, respectively. N_i, N_{TiOx}, N_e, N_o, and N_{TTIP} are the number concentrations of the *i*th chemical species, TiO_x, electron, O radicals, and TTIP, respectively (#/cm^3). $(RXN)_j$ is the rate of the jth reaction, and is the stoichiometric coefficient of the ith species in the jth reaction. u_s is the gas velocity inside the reactor (cm/s). R, D_P, N_T, D_i and W_{FP} are the reactor radius (cm), particle diameter (cm), number of particles per unit length of the reactor (#/cm), diffusion coefficients of the chemical species, and fraction of particles falling in the gas phase, respectively. J_{TiOx} is the flux of TiO_x onto the reactor wall and also onto the falling particles based on the gas kinetic theory, and is expressed as follows (cm^{-1}s^{-1}):

$$J_{TiO_x} = N_{TiO_x}\left(\frac{s_{TiO_x}}{1-\frac{s_{TiO_x}}{2}}\right)\sqrt{\frac{k_B T_g}{2\pi m_{TiO_x}}} \tag{3}$$

In Eq. (3), the sticking coefficient of TiO_x, s_{TiOx}, is assumed to be 1, and k_B, T_g, and m_{TiOx} are the Boltzmann constant, gas temperature (K), and molecular weight of TiO_x (g), respectively.

By assuming no slippage of the particles at the cylinder wall, we can expect uniform thickness of the particles at the cylinder wall where the particles exist. The particle size increases with the deposition of TiO_x and the change of particle volume, V_P is expressed in Eq. (4):

$$\frac{dV_P}{dt} = V_M\pi\left[\left(\frac{2RJ_{TiO_x}W_P}{N_T}\right) + J_{TiO_x}D_P^2W_{FP}\right] \tag{4}$$

In Eq. (4), V_M is the molecular volume of chemical species depositing on the particles (cm^3) and $W_P$is the fraction of the cylinder wall covered by the particles. The first and second terms in the RHS of Eq. (4) show the effects of deposition of TiO_x onto the particles located on the cylinder wall and onto the particles falling in the gas phase, respectively.

The rotation speed of the cylinder (ω_R) can affect the film growth rate on the particles because W_{FP} depends on ω_R and the relationship between W_{FP} and ω_R can be formulated as follows:

$$W_{FP} = \frac{\tau_{FP}\omega_R}{W_P + \tau_{FP}\omega_R} \tag{5}$$

where τ_{FP} is the residence time of particles falling in the gas phase (s). τ_{FP} depends on the particle falling velocity in the gas phase of the plasma reactor. To obtain τ_{FP}, the particle falling velocity was calculated from the particle motion in the gas phase, considering gravity and the drag force acting on the particles. W_{FP} depends on ω_R in Eq. (5). If ω_R becomes infinite, W_{FP} approaches 1 and most of the particles are in the process of falling down in the gas phase, but, if ω_R approaches 0, W_{FP} becomes proportional to ω_R.

Using model equations shown above, Kim et al. [39] calculated TiO_x concentration and film thickness on the particles, along the axial distance. Figure 3 shows the evolution of (a) TiO_x concentration profile and (b) film thickness on the particles, along the axial distance for various times. As shown in Figure 3(a), the TiO_x concentration increases at the start of the reactor, due to the fast generation rate of TiO_x, but, for $3.6 < x < 35$, decreases with an increase of reactor length as the TiO_x generation rate decreases because of the decrease in the TTIP concentration. As the particles grow with time due to the deposition of TiO_x, the surface area of the particles available for deposition increases. The TiO_x concentration profile decreases with time, because of the faster deposition rate of TiO_x withincreasing film thickness on the particles. In Figure 3(b), the film thickness on the particles depends on the TiO_x concentration profiles, because TiO_x deposits onto the particles. The particles near the inlet grow faster because of the high concentration of TiO_x. The film thickness becomes a maximum at around x = 3.6 cm, where we can see the highest TiO_x concentration, and decreases with an increase in the reactor length because the TiO_x concentration decreases.

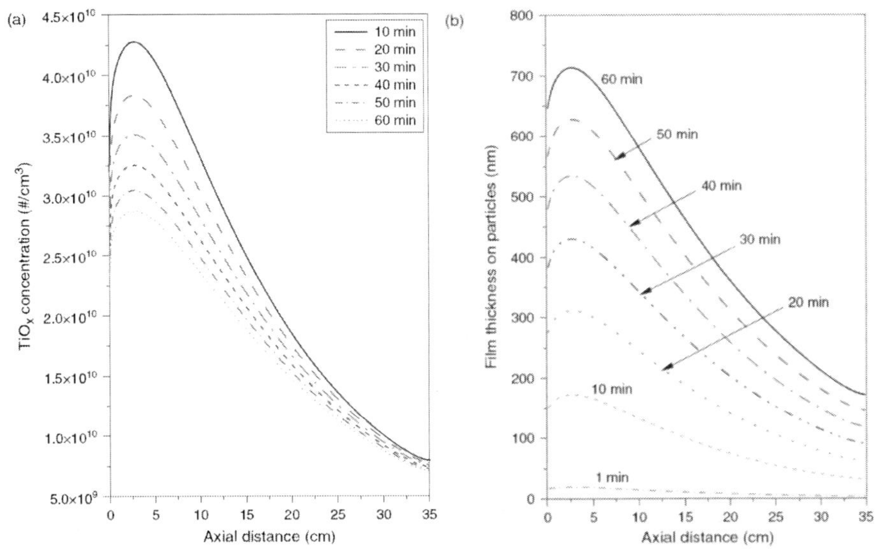

Figure 3.(a) Evolution of TiO_x concentration profile along the axial distance for various times, (b) Evolution of film thickness on the particles along the axial distance for various times [39].

Similar numerical analysis was presented by Kim et al. for the a-Si thin-film growth on particles in the rotating pulsed SiH_4 plasma process [41]. The evolutions of chemical concentrations (SiH_4, SiH_x, $SiH^+_x$and polymerized negative ions) in the pulsed plasmas have been shown during the plasma-on and -off. During plasma-on, SiH_4 is consumed by the electron impact dissociative reactions, but, during plasma-off, the disappearance reaction of SiH_4 stops because the electrons disappear in the plasma reactor. During plasma-on, $SiH_x$and SiH^+_x are generated quickly by a fast dissociative reaction of SiH_4, but, during plasma-off, SiH_x disappears rapidly by a reaction with hydrogen and also by the deposition onto the reactor wall and particles, and SiH^+_x is consumed quickly by fast neutralization reactions with the negative ions. The negative ions are polymerized by the reactions with SiH_4 during plasma-on, but, disappear by neutralization reactions during plasma-off. The growth rate of the film thickness profile depends on the SiH_xconcentration because the particles grow with the SiH_xdeposition. As the plasma-on time increases or as the plasma-off time decreases, the thin film thickness on the particles increases more quickly with faster SiH_xdeposition onto them. Figs. 4(a) and (b) illustrate the changes of the film thickness on the particles at the reactor center as a function of time for various t_{on}'s and t_{off}'s, respectively. During plasma-on, the film thickness on the particles increases quickly with the deposition of highly concentrated SiH_x, but, during plasma-off, it cannot increase quickly because the SiH_x concentration becomes quite low. As t_{on} increases or as t_{off} decreases, SiH_x deposits onto the particles for a longer time and the film growth rate becomes higher.

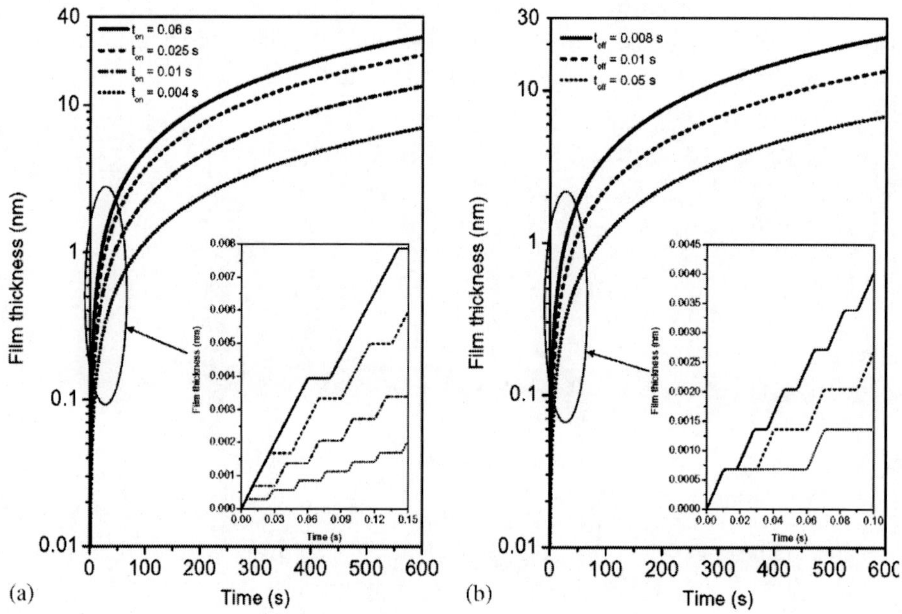

Figure 4. Changes of the film thickness on the particles at the reactor center as a function of time for various: (a) t_{on}'s and (b) t_{off}'s [41].

3. Coating Methods by Rotating PCVD

Rotating cylindrical PCVD reactor can be used for various kinds of thin films on particle-shaped substrates of different sizes and materials. Figure 5 shows the schematic of experimental setup for coating thin films on particles by rotating PCVD process. The inner diameter and the length of the cylindrical quartz reactor are 55 mm and 280 mm, respectively. A mechanical rotary vane pump is used for evacuation of the gas stream from the reactor. An ultrasonic nebulizer is used to supply precursor as a source of thin film material into the plasma reactor. Precursor mists are made by the ultrasonic nebulizer at 40 °C and are supplied into the plasma reactor with N_2 carrier gas during deposition onto the particles. The flow rate of precursor is controlled by changing the power of the ultrasonic nebulizer. The feed line for precursor is heated to help the evaporation of precursor droplets and also to prevent condensation of precursor. O_2 is supplied to the reactor separately from precursor to prevent reaction between O_2 and precursor in the feed line. The flow rates of all gases are controlled by mass flow controllers (MFC). A water-cooled spiral-shape coil electrode is located outside the cylindrical reactor to rotate the plasma reactor independently. RF power of 13.56 MHz frequency is applied to the electrode to generate inductively coupled plasmas in the cylindrical reactor. The volume of plasmas in the cylindrical plasma reactor is 2661 cm^3 and, for the maximum applied power of 30 W, the power density in the reactor are calculated to be 0.01 W/cm^3. The rotation speed of the cylindrical quartz reactor is controlled by a DC motor. The precursors for the TiO_2 thin films are generated from TTIP by plasma reactions in the gas phase. The precursors of the thin films deposited on the substrate contain some carbon and hydrogen compounds. After coating the substrate particles with thin films, the coated particles should be heated at 500 °C for 1 h to remove the carbon and hydrogen compounds [42-48].

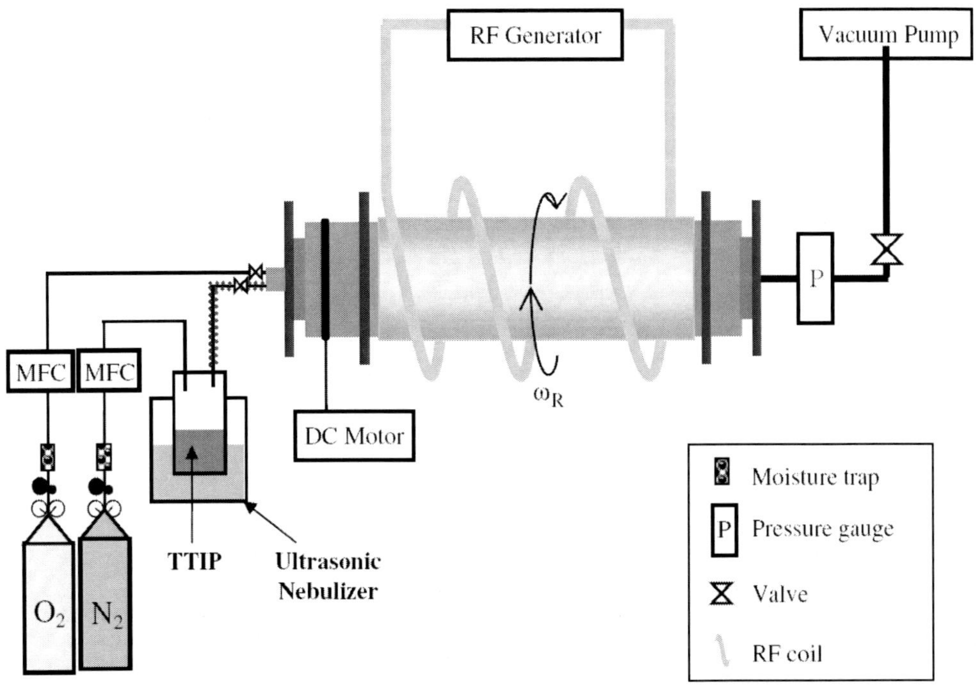

Figure 5. Schematic diagram of experimental setup for coating of TiO_2 thin films on glass beads by PCVD process [44].

This rotating PCVD reactor can be used to prepare a single-layer or multi-layer thin films on substrates of various materials with different shapes using various gas/liquid precursors.

4. Experimental Results

4.1. TiO_2 Single Layer on Glass Particles

Kim's group [42, 44] performed experiment for glass particles coating with TiO_2 thin films by using rotating cylindrical PCVD reactor. The coating process is similar with the one explained above. The morphologies and cross-sectional views of the TiO_2 films were analyzed by SEM before and after heat treatment. They took several SEM images of thin films coated on the glass beads for each process condition and obtained the film thicknesses on glass beads from SEM images to calculate the average film thicknesses for all process conditions. Two of those SEM images are shown in Figure 6 (a,b). They obtained the net weight increases of the TiO_2 thin film by measuring the weights of particles before and after the coating process, and also after heat treatment. They did the particle coating experiments several times under the same process conditions and measured the net weight increases of the TiO_2 thin film and confirmed that the net weight increases were almost same under the same process conditions. They calculated the densities of the thin films from the measured film thickness and net weight increase before and after heat treatment. The average film densities before heat treatment and after heat treatment are 0.93 g/cm^3 and 1.04 g/cm^3, respectively.

(a)

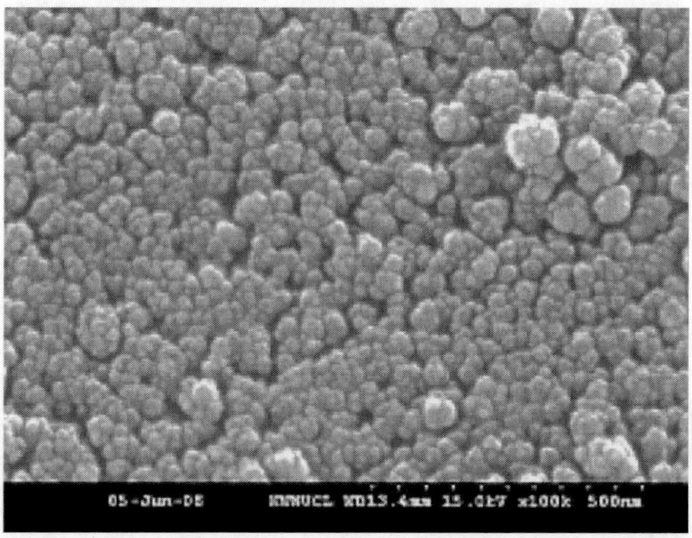

(b)

Figure 6. TiO_2 coatings on glass particles: (a) cross-section view; (b) surface of TiO_2 thin film coated on glass particle [42, 44].

Figure 7 shows the thickness of the TiO_2 thin films before and after heat treatment for various mass flow rates of TTIP. For TTIP mass flow rates of 1.333, 2.667, and 3.333 mg/min, the film thicknesses after heat treatment are about 99, 216, and 281 nm, respectively. The particles grow by the deposition of precursors for the thin films and the increase of the particle volume can be expressed by the fluxes of precursors onto the particles located on the cylinder wall and also onto the particles falling in the gas phase. The change of particle volume (V_P) by the deposition is expressed in Eq. (4), assuming that the particles keep a spherical shape during the deposition process [39]. The concentration of precursors for deposition is almost proportional to the TTIP concentration and, in Figure 7, the thickness of the TiO_2 thin film on the glass beads increases proportionally to the mass flow rates of TTIP.

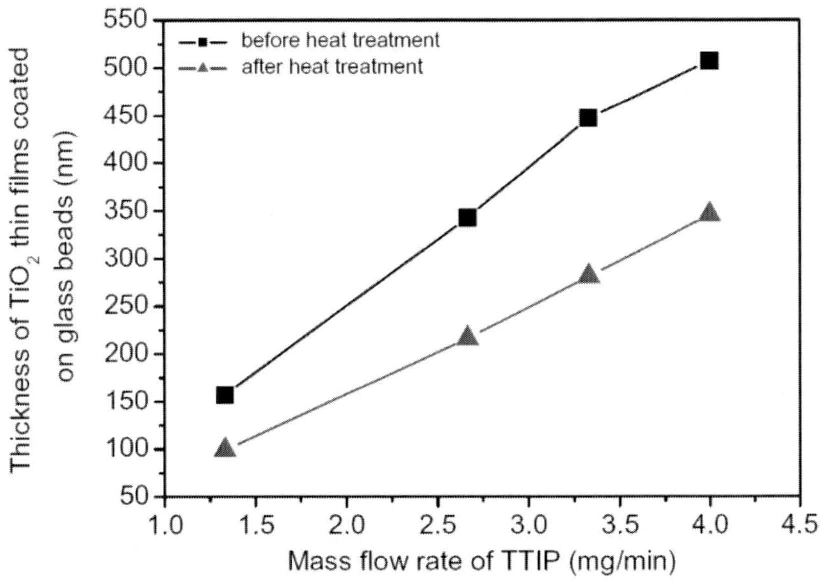

Figure 7. Thickness of TiO_2 thin films before and after heat treatment for various mass flow rates of TTIP (P_w = 10W, P = 600 mtorr, ω_R = 10 rpm) [44].

4.2. TiO_2 and SiO_x Single Layer on PP Particles

Kim et al. [43, 48] coated PP (polypropylene) particles with TiO_2 and SiO_x single-layer films via rotating cylindrical plasma PCVD. The coating process is similar with the one explained above. The morphologies and cross-sectional views of the TiO_2 films were analyzed by SEM.

Figure 8 shows the SEM images of the cross-section of the SiO_xsingle-layer films that were used to coat the PP beads. The SiO_xlayer on PP was found to be quite uniform, without particle contamination. As the deposition time increased from 5 to 30 min, the thickness of the SiO_x single-layer films increased from 150 to 600 nm. Figure 9 shows the SEM images of the cross-section of the TiO_y single-layer films on the PP beads. The thickness of the TiO_y single layer increased with the deposition time (for the 20 and 40 min deposition times, the thickness of the layer became 400 and 700 nm, respectively).

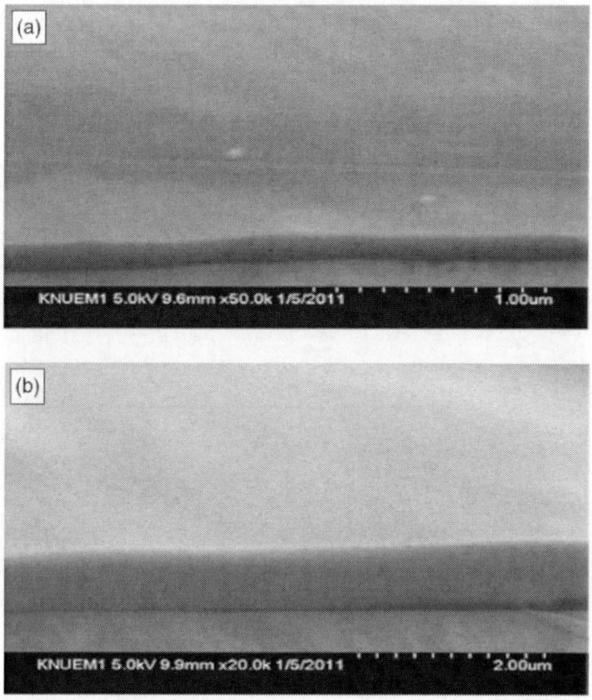

Figure 8. SEM images of the cross-section of the SiO_x single-layer films that were used to coat PP beads for various deposition times (a: 10 min; b: 20 min) [43].

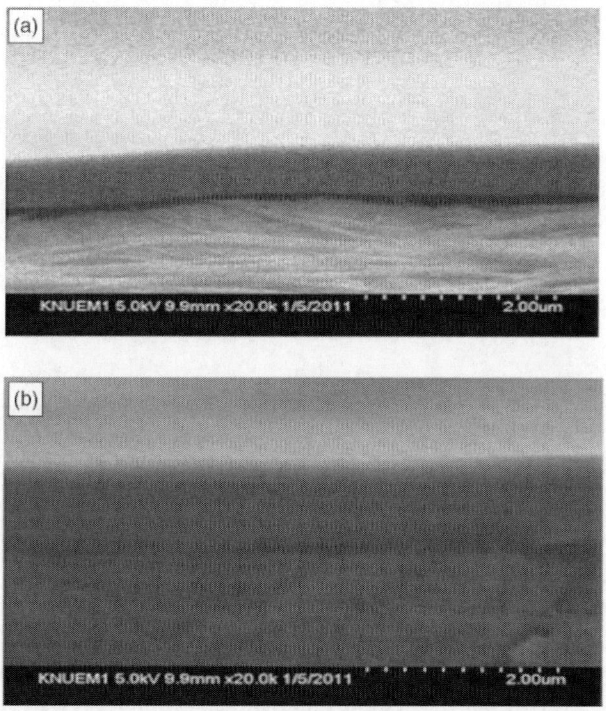

Figure 9. SEM images of the cross-section of the TiO_y single-layer films that were used to coat PP beads for various deposition times (a: 15 min; b: 40 min) [43].

4.3. TiO_2/SiO_x Double Layer Coated on PP Particles

The photocatalytic activity of TiO_2/SiO_x double-layer films can be enhanced in the double-layer structure compared with the TiO_2 single layer due to the presence of a SiO_xbottom layer. SiO_xthin films may act as a trap for the electrons generated in the TiO_2 layer, thus preventing electron–hole recombinations.

Kim's group [43] coated polypropylene (PP) particles with TiO_2/SiO_x double layer files via rotating cylindrical PCVD. They controlled thickness of the SiO_x and TiO_2 layers by changing the deposition time. To coat the PP particles with TiO_2/SiO_x double-layer films, the SiO_xbottom layer was first deposited on the PP particles by supplying TEOS as a precursor into the reactor. Before the deposition of the TiO_y top layer, the SiO_x single-layer films on the PP particles were annealed in a N_2 stream at 120 °C for 2 h. The TiO_y top layer was deposited on the SiO_x bottom layer by supplying TTIP and N_2 carrier gas to the ultrasonic nebulizer. The TiO_y/SiO_x double-layer films on the PP particles were annealed in an O_2 atmosphere at 120 °C for 2 h. The cross-sectional views of the SiO_x and TiO_y single-layer films and the TiO_2/SiO_x double-layer films on the PP beads were measured via SEM.

Figure 10 shows the SEM images of the cross-section of the TiO_2/SiO_xdouble-layer films that were used to coat PP beads for various deposition times, to prepare the TiO_2 top layer. For the deposition times of 10, 20, and 40 min, the thicknesses of the TiO_2 top layer were about 50, 300, and 600 nm, respectively. The TiO_2 molecule formed in the gas phase could not be quickly deposited on the SiO_x layer. Further, the TiO_2 concentration in the gas phase became high, and the TiO_2 particles were nucleated at a high TiO_2 concentration. As a result, some of the TiO_2 molecules were deposited to form a thin film, but some of the TiO_2 particles were also deposited onto the SiO_x layer. The film quality prepared via PCVD largely depends on the substrates. The TiO_2 thin films prepared on a SiO_x layer had many particle contaminants on their surfaces, as shown in Figure 10, while the SiO_x and TiO_2 single layers on the PP beads had none.

5. Applications of Paricles Coated by Rotating PCVD Reactor

Various substrates coated with different kinds of thin films can be applied for various purposes. The application of TiO_2-based photocatalysts is quite potential for the air and water pollutant removals [49-52], VOCs decomposition [53-55], water splitting [56, 57], etc. because of their high activity, stability and availability for photodegradation. The photodegradation efficiency of TiO_2photocatalyst can be improved by using small particles coated with TiO_2 thin films, because the surface area of TiO_2photocatalyst becomes far larger compared with that by the plate substrate coated with TiO_2 thin films. The rotating PCVD process is efficient method to coat the TiO_2 thin films uniformly on the surface of particles for obtaining the photodegradation of high efficiency. For example, glass particles and zeolite particles coated with TiO_2 can be used for air pollutant removal in dielectric barrier discharge process, and PP beads coated with TiO_2 can be used for photodegradation of phenol.

Figure 10. SEM images of the cross-section of the TiO_2/SiO_xdouble-layer films that were used to coat PP beads for various deposition times ((a) 10 min; (b) 20 min; (c) 40 min) [43].

5.1. Application of TiO_2-Coated Particles by Rotating PCVD Reactor to Air Pollutant Removal

Non-thermal plasma is one of the methods applied for air pollutants removal. The removal of pollutants such as NO and SO_2 from industrial processes and vehicle exhausts is important for environmental pollution control. The removal efficiencies of those pollutants can be improved by the combination of low-temperature plasma and photocatalyst process.

In this example, TiO_2photocatalyst was deposited as a very uniform thin film on a surface of glass particles. Figure 11 shows the experimental setup to remove NO and SO_2 by dielectric barrier discharge-photocatalyst hybrid (DBD-PH) reactor. The glass particles coated with TiO_2 thin films by this rotating PCVD process are packed in the glass tube which has the inner and outer diameters of 27 mm and 30 mm, respectively. The copper rod wire is located at the center of this glass tube and the outside wall of the reactor is wrapped over by a stainless steel mesh which acts as a ground electrode. The nonthermal plasma discharge is generated in this cylinder-wire type DBD-PH reactor by applying high voltage power into the reactor. The UV light is generated from the plasma discharge and activates the TiO_2photocatalyst. The reactive radicals generated from the plasma discharge and photocatalyst induce the chemical reactions necessary to remove NO and SO_2. The initial concentrations of NO and SO_2 were $[NO]_0$ =300 ppm and $[SO_2]_0$ = 260 ppm, respectively, pulse frequency (f) was 900 Hz and residence time of gas stream (τ_r) was 1s as the standard conditions. The initial O_2 concentration was fixed to be 21%. The total gas flow rate was adjusted to 5 l/min by balancing the N_2 gas flow rate. All experiments were done at 1 atm and 298 K [58,59].

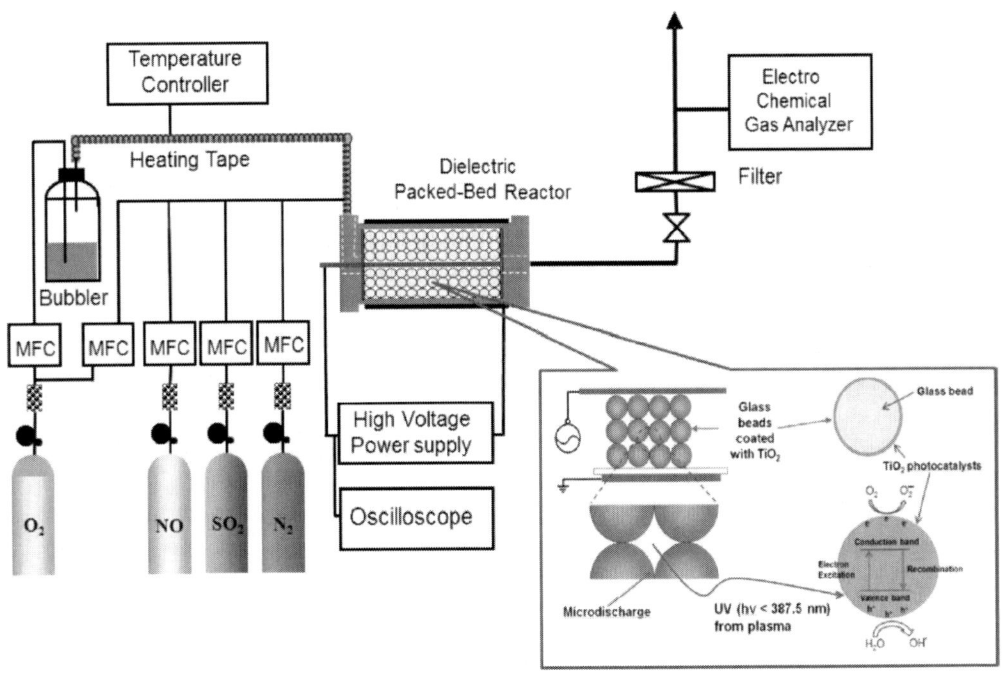

Figure 11. Experimental setup for NO and SO_2 removals by a dielectric barrier discharge-TiO_2 photocatalyst hybrid process [58].

Figure 12 (a) and (b) show the removal efficiencies of NO and SO_2, respectively, for various initial concentrations of NO and SO_2 and for the TiO_2 thin film thicknesses of 600 nm and 440 nm as a function of the applied peak voltage. As the initial NO and SO_2 concentrations increase, the removal efficiencies of NO and SO_2 decrease because more energy is required to remove the NO and SO_2 of higher concentrations by plasma discharge. Also the TiO_2 thin film thickness of 600 nm shows the higher removal efficiencies of NO and SO_2 than those of 440 nm. These experimental results show that the complete removals of NO and SO_2 are possible by controlling the process conditions in DBD-PH process.

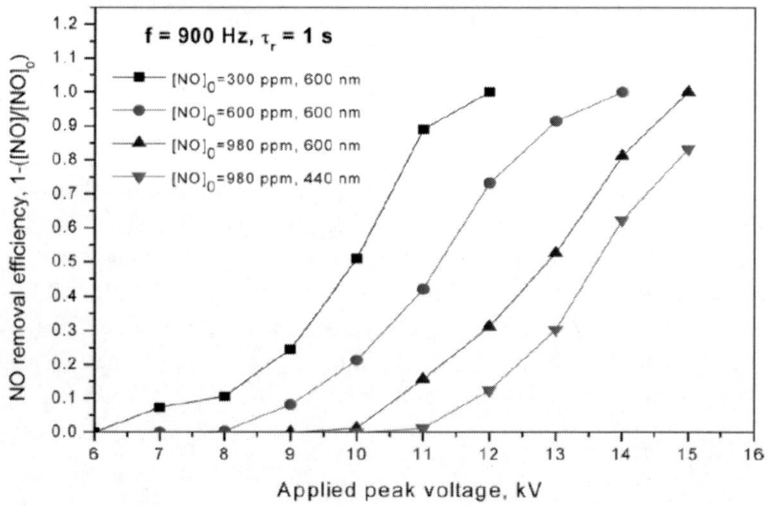

(a)

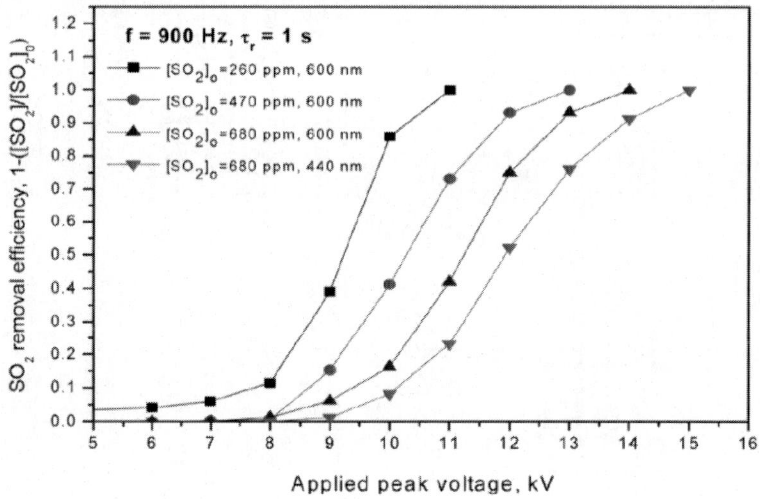

(b)

Figure 12 (a) NO and (b) SO_2 removal efficiencies for various initial concentrations of NO and SO_2, respectively, as a function of applied peak voltage [58].

Zeolite is a catalyst and also another dielectric material, which can be used in DBD reactor instead of glass beads. Zeolite particles also can be coated with TiO_2 photocatalyst by this rotating PCVD [46].

Figs. 13(a,b) show SEM images of the cross section of the zeolite particles (a) before TiO_2 coating and (b) after TiO_2 coating by rotating PCVD for 30 min. Zeolite is a porous material with large surface area. Figure 13(b) shows that TiO_2 is deposited on the surface of the zeolite particles, but the TiO_2 only partially covers the surface of the zeolite, thereby not blocking all the zeolite pores and still preserving the large surface area of the zeolite. The amount of TiO_2 deposited onto the zeolite surface can be easily controlled by changing the deposition time in the PCVD process.

(b)

Figure 13. SEM images of cross section of zeolite particles: (a) before coating and after TiO_2 coating (b) deposited for 30 min [46].

Figure 14 shows comparisons of NO and SO_2 removal efficiencies, respectively, by the dielectric barrier discharge-catalyst (DBD-C) and dielectric barrier discharge catalyst-photocatalyst (DBD-C-P) hybrid processes for various initial concentrations of NO and SO_2 as a function of the applied peak voltage. A stronger applied voltage enhances the generation of reactive radicals (O, N, OH, H, etc.) because of the faster electron-impact dissociation reaction to generate those radicals and results in higher removal efficiencies of NO and SO_2 because of their faster oxidation reactions with the generated reactive radicals. The maximum removal efficiencies of NO and SO_2 at $[NO]_0$ = 200 ppm and $[SO_2]_0$ = 200 ppm reach 77% and 100%, respectively. As the initial NO and SO_2 concentrations increase to 600 ppm, removal efficiencies decrease because more energy is required to remove the NO or SO_2 with higher initial concentrations. For the DBD-C-P hybrid process, the UV light generated inside the plasmas can be used additionally to remove NO and SO_2 by the TiO_2photocatalyst and the removal efficiencies of NO and SO_2 both become higher as compared with the DBD-C hybrid process. The differences in NO and SO_2 removal efficiencies between the DBD-C and DBD-C-P hybrid processes become larger at medium applied voltage (10–13 kV), where the applied voltage is not high enough for complete removal of NO and SO_2 but some UV light generated from the DBD expedites the removal of NO and SO_2. The maximum differences in NO and SO_2 removal efficiencies between the DBD-C and DBD-C-P hybrid processes are about 15% and 10%, respectively.

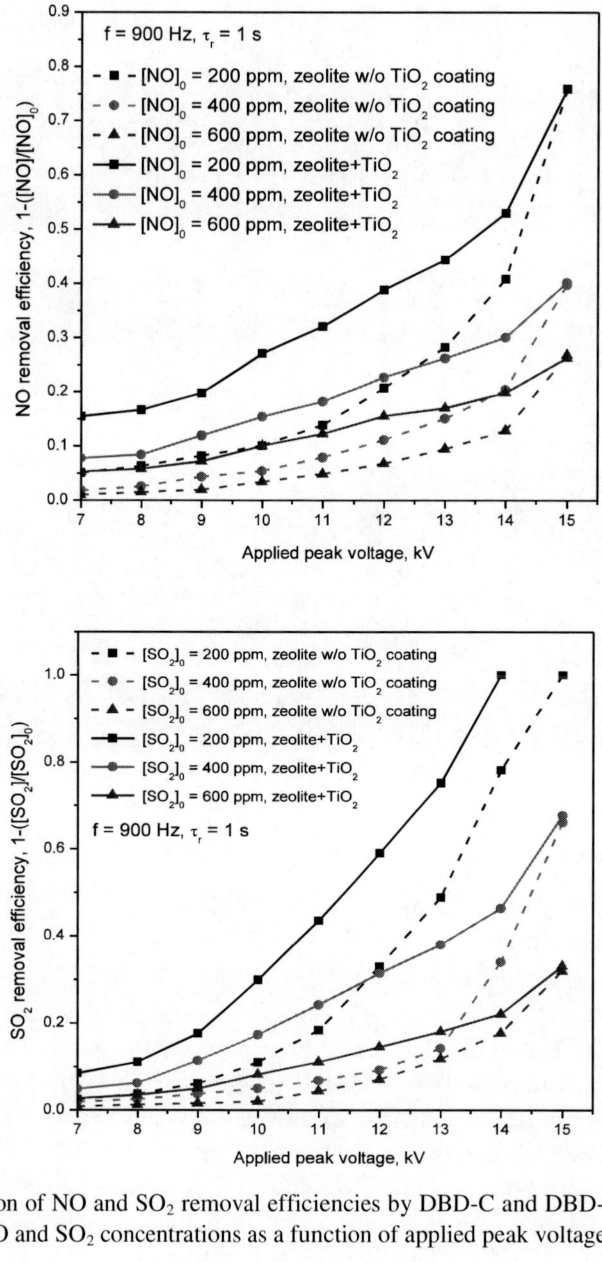

Figure 14.Comparison of NO and SO_2 removal efficiencies by DBD-C and DBD-C-P hybrid processes for various initial NO and SO_2 concentrations as a function of applied peak voltage [46].

5.2. Application of TiO_2-Coated Particles by Rotating PCVD Reactor to Water Pollutant Removal

There is another example of using TiO_2-coated particles for photodegradation of water pollutants. Kim et al. [60] applied TiO_2-coated PP particles coated by this rotating cylindrical PCVD for the photodegradation of phenol in aqueous solution.

Photodegradation experiments of phenol with the PP particles coated with TiO_2 thin films were done in a slurry type batch reactor with 300 ml phenol aqueous solution as shown in Figure 15. The PP particles were suspended with TiO_2 thin films in aqueous solution which was stirred the solution continuously by a magnetic stirrer for enough mixing inside the reactor. O_2 (50 ml/min) was continuously bubbled into the reactor for all experiments. The UV lamp (wavelength < 365 nm, 1100 $\mu W/cm^2$) was placed in the center of the batch reactor to illuminate the phenol solution. The phenol concentration in the solution was measured by UV–vis spectrophotometry. During their experiments, no flake of TiO_2 films was found in the solution and the thin films seem to be deposited on the PP beads strongly.

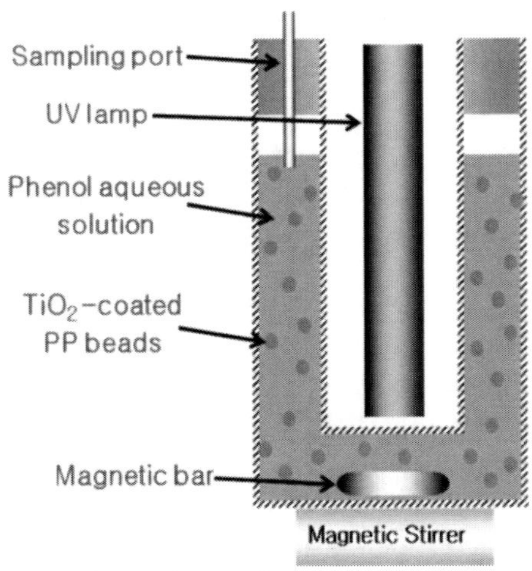

Figure 15. Schematic diagram of experimental setup for photodegradation of phenol by TiO_2 thin films coated on polypropylene beads [60].

Figure 16(a) shows the changes of phenol concentration for various initial phenol concentrations. As the initial phenol concentration increases, the degradation rate of phenol becomes faster (the slope of phenol concentration profiles becomes steeper) and, for the initial phenol concentration of 100 ppm, more than 90% of phenol was decomposed by TiO_2 thin films coated on the PP beads for 60 min of illumination. Figure 16(b) shows the changes of phenol concentration for various numbers of PP beads suspended in the aqueous solution. The amount of TiO_2photocatalysts is proportional to the number of PP beads in this study. The phenol is decomposed more quickly by increasing the number of PP beads, because the total surface area of photocatalysts available for photodegradation becomes larger.

If there are too high concentrations of PP beads, some particles cannot be illuminated enough for photodegradation because of the shading effects by the PP beads, but, in these experimental conditions, we could not observe the decrease in photocatalytic activities for phenol removal even with the highest concentration of 400 PP beads inside the reactor. The uniform and compact coating of TiO_2 thin films on the PP beads was achieved by the rotating cylindrical PCVD reactor. There was no void on the thin films, and O_2 and H_2O cannot penetrate the thin films to reach the PP beads and there was no serious photodegradation of PP beads at the interface between TiO_2 thin film and PP beads for the long-term use of these photocatalysts.

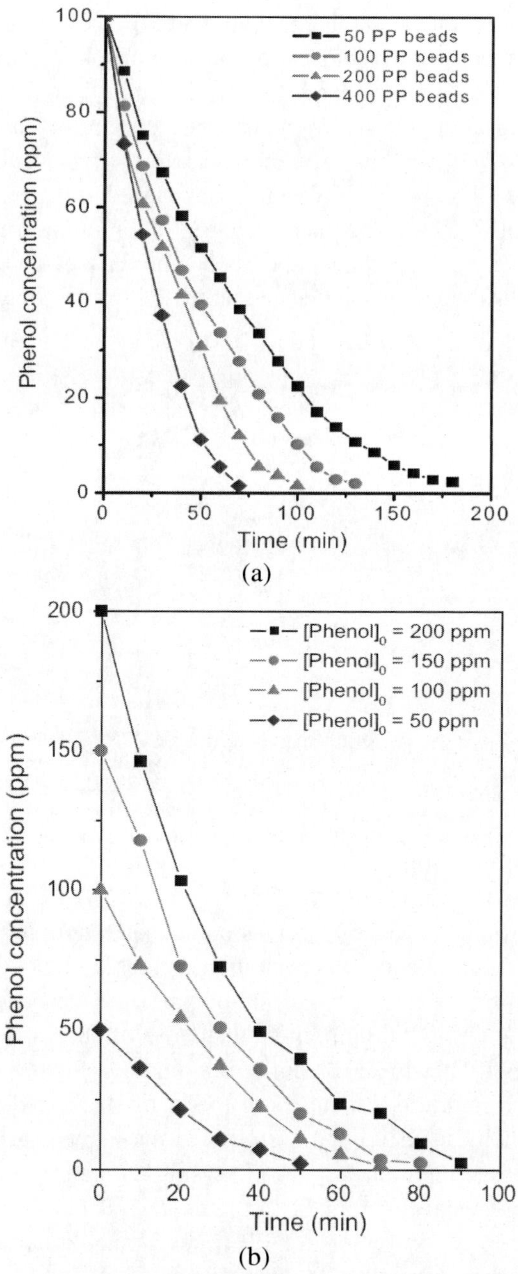

Figure 16. Changes of phenol concentration for various initial phenol concentrations (a) and for various numbers of polypropylene beads coated with TiO_2 thin films (b) [60].

CONCLUSION

This chapter has presented a review of current development of rotating PCVD process and its applications. It has revealed that rotating PCVD is an important technique for the deposition of highly uniform thin films on a various substrates. Rotating PCVD process allows to synthesize thin films of various thicknesses. The desirable thickness of thin films can be easily achieved by controlling various parameters, such as deposition time, precursor feed rate, number of particles, reactor pressure, applied power and rotation speed of reactor. Particles coated with thin films can be applied for water and air pollutant removals, VOCs decomposition, water splitting, etc. Data demonstrated in this chapter shows that a rotating cylindrical PCVD reactor can be applied to coat the particles with high-quality thin films. The rotating PCVD can be used for preparation of the particles coated with metal or organic thin films for the enhanced stability and dispersion characteristics, and mechanical properties of the particles.

ACKNOWLEDGMENTS

This work was supported by the Regional Innovation Center for Environmental Technology of Thermal Plasma (ETTP) at Inha University, designated by the Ministry of Trade, Industry and Energy, Republic of Korea.

REFERENCES

[1] Fotou, G.P.; Kodas, T.T.; Anderson, B. Coating Titania Aerosol Particles with ZrO_2, Al_2O_3/ZrO_2, and SiO_2/ZrO_2 in a Gas-phase Process. *Aerosol Sci. Tech.* 2000, 33, 557-571.

[2] Koga, K.; Kai, M.; Shiratani, M.; Watanabe, Y.; Shikatani, N. Cluster Suppressed Plasma Chemical Vapor Deposition Method for High Quality Hydrogenated Amorphous Silicon Films. *Jpn. J. Appl. Phys.* 2002, 41(2), 168-170.

[3] Koga, K.; Kaguchi, N.; Shiratani, M.; Watanabe, Y. Correlation Between Volume Fraction of Clusters Incorporated into a-Si:H Films and Hydrogen Content Associated with Si–H_2 Bonds in the Films. *J. Vac. Sci. Technol., A* 2004, 22(4), 1536-1539.

[4] Shiratani, M.; Fukuzawa, T.; Watanabe, Y. Particle Growth Kinetics in Silane RF Discharges. *Jpn. J. Appl. Phys.* 1999, 38, 4542-4549.

[5] Taga, Y. Recent Progress of Nanotechnologies of Thin Films for Industrial Applications. *Mater. Sci. Eng., C* 2001, 15, 231–235.

[6] Zhang, L.; Ranade, M.B.; Gentrya, J.W. Formation of Organic Coating on Ultrafine Silver Particles Using a Gas-phase Process. *J. Aerosol Sci.* 2004, 35, 457-471.

[7] Hung, C.-H.; Miquel, P.F.; Katz, J.L.; Formation of Mixed Oxide Powders in Flames: Part II. SiO_2–GeO_2. *J. Mater. Res.* 1992, 7, 1870-1875.

[8] Powell, Q.H.; Fotou, G.P.; Kodas, T.T.; Anderson, B.M.; Guo, Y. Gas Phase Coating of TiO_2 with SiO_2 in a Continuous Flow Hot-Wall Aerosol Reactor. *J. Mater. Res.* 1997, 12, 552-559.

[9] Powell, Q.H.; Fotou, G.P.; Kodas, T.T.; Anderson, B.M. Synthesis of Alumina- and Alumina/Silica-Coated Titania Particles in an Aerosol Flow Reactor. *Chem. Mater.* 1997, 9, 685-693.

[10] Nakagawa, T.; Hieda, M.; Shindo, D.; Saito, F.; Yubuta, K.; Nishimura, K.; Fujii, T. Preparation of a Fine Powder Composed of Oxide–Metal Composite by an Arc-Plasma Method. *Powder Technol.* 2003, 130, 456-461.

[11] Karches, M.; Morstein, M.; Rudolf von Rohr, P.; Pozzo, R.L.; Giombi, J.L.; Baltanás, M.A. Plasma-CVD-Coated Glass Beads as Photocatalyst for Water Decontamination. *Catal. Today* 2002, 72, 267-279.

[12] Karches, M.; Morstein, M.; Rudolf von Rohr, P.; Encapsulation of Abrasive Particles by Plasma CVD. *Surf. Coat. Technol.* 2003, 169–170, 544-548.

[13] Michalski, A.; Olszyna, A. Electron-Assisted Hot-Filament Chemical Vapour Deposition of E-BN Layers. *Surf. Coat. Tech.* 1993, 60, 498-501.

[14] Rie, K.T, Gebauer, A., Prohl, C. Deposition of Boron Containing Coatings Using MO-PACVD Processes to Protect Aluminum Casting Tools. *Journal de Physique IV* 1995, 5, 637-645.

[15] Fukuzawa, T.; Kushima, S.; Matsuoka, Y.; Shiratani, M.; Watanabe, Y. Growth of Particles in Cluster-Size Range in Low Pressure and Low Power SiH_4 RF Discharges. *J. Appl. Phys.* 1999, 86(7), 3543-3549.

[16] Kim, K.-S.; Ikegawa, M. Particle Growth and Transport in Silane Plasma Chemical Vapor Deposition. *Plasma Sources Sci. Technol.* 1996, 5, 311-322.

[17] Kim, K.-S.; Kim, D.-J. Modeling of Rapid Particle Growth by Coagulation in Silane Plasma *Reactor. J. Appl. Phys.* 2000, 87(6), 2691-2699.

[18] Kim, D.-J.; Kim, K.-S. The Factors Affecting the Particle Distributions inside the Silane PCVD Reactor for Semiconductor Processing. *Aerosol Sci. Technol.* 2000, 32, 293-308.

[19] Kim, D.-J.; Kim, K.-S. Analysis on Nanoparticle Growth by Coagulation in Silane Plasma Reactor. *A.I.Ch.E. J.* 2002, 48(11), 2499–2509.

[20] Kim, K.-S.; Kim, D.-J.; Yoom, J.H.; Park, J.Y.; Watanabe, Y.; Shiratani, M. The Changes in Particle Charge Distribution During Rapid Growth of Particles in the Plasma Reactor. J. *Colloid Interface Sci.* 2003, 257, 195–207.

[21] Kim, D.-J. Kim, K.-S.; Zhao, Q.-Q. Production of Monodisperse Nanoparticles and Application of Discrete-Monodisperse Model in Plasma Reactors. *J. Nanopart. Res.* 2003, 5, 221–223.

[22] Koga, K.; Matsuoka, Y.; Tanaka, K.; Shiratani, M.; Watanabe, Y. In situ Observation of Nucleation and Subsequent Growth of Clusters in Silane Radio Frequency Discharges. *Appl. Phys. Lett.* 2000, 77 (2), 196-198.

[23] Hollenstein, C.; Dorier, J.-L.; Dutta, J.; Sansonnens, L.; Howling, A.A. Diagnostics of Particle Genesis and Growth in RF Silane Plasmas by Ion Mass Spectrometry and Light Scattering. *Plasma Sources Sci. Technol.* 1994. 3, 278-286.

[24] Howling, A.A.; Courteille, C.; Dorier, J.-L.; Sansonnens, L.; Hollenstein, C. From Molecules to Particles in Silane Plasmas. *Pure Appl. Chem.* 1996, 68 (5), 1017-1022.

[25] Itagaki, N.; Fukuda, A.; Yoshizawa, T.; Shindo, M.; Ueda, Y.; Kawai, Y. Plasma Parameter Measurements and Deposition of a-Si:H Thin Films in Pulsed ECR Plasma. *Surf. Coat. Technol.* 2000, 131, 54-57.

[26] Kirimura, H.; Maeda, H.; Murakami, H.; Nakahigashi, T.; Ohtani, S.; Tabata, T.; Hayashi, T.; Kobayashi, M.; Mitsuda, Y.; Nakamura, N.; Kuwahara, H.; Doi, A. Study of Deposition Process in Modulated RF Silane Plasma. *Jpn.J. Appl. Phys.* 1994, 33, 4389-4394.

[27] Madan, A.; Morrison, S. High Deposition Rate Amorphous and Polycrystalline Silicon Materials Using the Pulsed Plasma and Hot-Wire CVD Technique. *Solar Energy Materials and Solar Cells* 1998, 55, 127.

[28] Madan, A.; Morrison, S.; Kuwahara, H. Progress in High Deposition Rate Amorphous and Polycrystalline Silicon Materials Using the Pulsed Plasma and Hot Wire CVD Deposition Technique. *Sol. Energy Mater. Sol. Cells* 1999, 59, 51-58.

[29] Hiramatsu, K.; Ohnishi, H.; Takahama, T.; Yamanishi, K. Formation of TiN Films with Low Cl Concentration by Pulsed Plasma Chemical Vapor Deposition. *J. Vac. Sci. Technol., A* 1996, 14, 1037-1040.

[30] McCurdy, P.R.; Truitt, J.M.; Fisher, E.R. Pulsed and Continuous Wave Plasma Deposition of Amorphous, Hydrogenated Silicon Carbide from SiH_4/CH_4 Plasmas. *J. Vac. Sci. Technol., A* 1999, 17, 2475-2484.

[31] Khachan, J.; Gardner, D. The Effect of Frequency and Duty Cycle of a Pulsed Microwave Plasma on the Chemical Vapor Deposition of Diamond. *J. Appl. Phys.* 1999, 86, 6576-6579.

[32] Bapin, E.; Rudolf von Rohr, P. 2001. Deposition of SiO_2 Films from Different Organosilicon/O_2 Plasmas under Continuous Wave and Pulsed Modes. *Surf. Coat. Technol.* 2001, 142–144, 649-654.

[33] Qi, Y.; Mantei, T.D. Pulse-Modulated Plasma-Enhanced Chemical Vapor Deposition of SiO_2 Coatings from Octamethylcyclotetrasiloxane. *Surf. Coat. Technol.* 2004, 177–178, 394-398.

[34] Hynes, A.M.; Shenton, M.J.; Badyal, J.P.S. Pulsed Plasma Polymerization of Perfluorocyclohexane. *Macromolecules* 1996, 29, 4220-4225.

[35] Licheng, M.H.; Rajeshwar, K.; Timmons, R.B. Film Chemistry Control and Electrochemical Properties of Pulsed plasma Polymerized Ferrocene and Vinylferrocene. *Langmuir* 1997, 13, 5941-5950.

[36] Licheng, M.H.; Timmons, R.B.; Lee, W.W.; Chen, Y.; Hu, Z. Pulsed Plasma Polymerization of Pentafluorostyrene: Synthesis of Low Dielectric Constant Films. *J. Appl. Phys.* 1998, 84(1), 439-444.

[37] Savage, C.R.; Timmons, R.B.; Lin, J.W. Spectroscopic Characterization of Films Obtained in Pulsed Radio-Frequency Plasma Discharges of Fluorocarbon Monomers, *ACS Adv. Chem.*1993, 236, 745-768.

[38] Kim, K.S.; Kim D.J. Analysis on Uniform Particle Coating by the Rotating Cylindrical PCVD Reactor, *J. Aerosol Sci.* 2006, 37(11), 1532-1544.

[39] Kim D.-J.; Baeg J.-O.; Moon S.-J.; Kim K.-S. Uniform Coating of TiO_2 Thin Films on Particles by Rotating Cylindrical PCVD Reactor, *J. Nanosci. Nanotechnol.* 2009, 9(7), 4285-4292.

[40] Kim D.-J.; Kim K.-S. Prediction of TiO_2 Thin Film Growth on the Glass Beads in a Rotating PCVD Reactor, *J. Nanosci. Nanotechnol* 2010, 10(5), 3211-3215.

[41] Kim K.-S.; Kim D.-J.; Zhao Q.Q. Numerical Analysis on Particle Coating by the Pulsed Plasma Process, *Chem. Eng. Sci.* 2006, 61, 3278-3289.

[42] Kim D.-J.; Kang J.-Y.; Kim K.-S. Preparation of TiO_2 Thin Films by a Rotating Plasma Chemical Vapor Deposition Reactor, *J. Ind. Eng. Chem.* 2010, 16(6), 997-1000.

[43] Pham H.C.; Kim K.-S. Thin Film Coating of Particles through the plasma process, *J. Nanoelectron. Optoe.* 2011, 6(3), 264-267.

[44] Kim D.-J.; Kang J.-Y., Kim K.-S. Coating of TiO_2 Thin Film on Particles by Plasma Chemical Vapor Deposition Process, *Adv. Powder Technol.* 2010, 21, 136-140.

[45] Kim D.-J.; Pham H.C.; Kim K.-S. Surface Review and Letters. *Application of PCVD Process to Uniform Coating of TiO_2 Thin Films on Polypropylene Beads*, 17(3), 2010, 329-335.

[46] Nasonova A.; Kim K.-S. Effects of TiO_2 Coating on Zeolite Particles for NO and SO_2 Removal by Dielectric Barrier Discharge Process, *Catal. Today*, 2013, 211, 90-95.

[47] Nasonova A.; Pham H.C.; Kim D.-J.; Kim W.-S.; Charinpanitkul T.; Kim K.-S. Application of TiO_2-Coated Alumina Beads to Dielectric Barrier Discharge-Photocatalyst Hybrid Process for NO and SO_2 Removals, *J. Nanosci. Nanotechnol.* 2011, 11(2), 1323-1327.

[48] Pham H.C.; Kim K.-S. Preparation of TiO_2/SiO_xDouble-Layer Films on Glass Beads and Its Application to NO and SO_2 Removal, *J. Nanosci. Nanotechnol.* 2012, 13(8), 5568-5571.

[49] Chang M.B.; Balbach J.H.; Rood M.J.; Kushner M.J. Removal of SO_2 from Gas Streams Using a Dielectric Barrier Discharge and Combined Plasma Photolysis. *J. Appl. Phys.* 1991, 69(8), 4409-4417.

[50] Hoffmann M.R.; Martin S.T.; Choi W.; Bahnemann D.W. Environmental Applications of Semiconductor Photocatalysis. *Chem. Rev.* 1995, 95, 69–96.

[51] Mills A.; Davies R.H.; Worsley D. Water Durification by Semiconductor Photocatalysis. *Chem. Soc. Rev.* 1993, 22, 417-425.

[52] Matthews R.W. Photooxidation of Organic Impurities in Water Using Thin Films of Titanium Dioxide. *J. Phys. Chem.* 1987, 91, 3328–3333.

[53] Lee B.-Y.; Park S.-H.; Lee S.-C.; Kang M.; Choung S.-J. Decomposition of Benzene by Using a Discharge Plasma–Photocatalyst Hybrid System, *Catal. Today*2004, 93–95, 769–776.

[54] Stafford U.; Gray K.A.; Kamat P.V. Photocatalytic Degradation of Organic Contaminants: Halophenols and Related Compounds. *Heterogen. Chem. Rev.* 1996, 3, 77–104.

[55] Ogata A.; Einaga H.; Kabashima H.; Futamura S.; Kushiyama S.; Kim H.H. Effective Combination of Nonthermal Plasma and Catalysts for Decomposition of Benzene in Air. *Appl. Catal. B: Environ.*, 2003, 46, 87-95.

[56] Kudo A., Photocatalyst Materials for Water Splitting, *Catalysis Surveys from Asia*, 2003 7(1), 31-38.

[57] Meng N.; Michael K.H.; Leung, D.; Leung, Y.C.; Sumathy K., A Review and Recent Developments in Photocatalytic Water-Splitting Using TiO_2 for Hydrogen Production, *Renew. Sust. Energ. Rev.* 2007, 11, 401–425.

[58] Pham H.C.; Kim K.-S. Effect of TiO_2 Thin Film Thickness on NO and SO_2 Removals by Dielectric Barrier Discharge-Photocatalyst Hybrid Process, *Ind. Eng. Chem. Res.* 2013, 52(15), 5296-5301.

[59] Nasonova A.; Cuong P.H.; Kim D.-J.; Kim K.-S. NO and SO_2 Removal in Non-Thermal Plasma Reactor Packed with Glass Beads-TiO_2 Thin Film Coated by PCVD Process, *Chem. Eng. J.* 2010, 156(3), 557-561.

[60] Kim D.-J.; Pham H.C.; Park D.-W.; Kim K.-S. Preparation of TiO_2 Thin Films on Polypropylene Particles by a Rotating PCVD Process and Its Application to Organic Pollutant Removal, *Chem. Eng. J.* 2011, 167(1), 308-313.

In: Comprehensive Guide for Nanocoatings Technology. Vol. 1 ISBN: 978-1-63482-447-7
Editor: Mahmood Aliofkhazraei

Chapter 5

SYNTHESIS, PROCESSING AND APPLICATION OF NANOSTRUCTURED COATINGS

Kal Renganathan Sharma*
Department of Physics, College of Science and Technology,
Texas Southern University, Houston, TX, US

ABSTRACT

Nanocoatings are used in cell phones, solar power plants and electric cars. Nanocoatings are used in order to improve the light-to-electricity conversion efficiency in solar cells. Flake production, functionalization, particulation, sacrificial coating, specifications are critical in order to prepare lubricous, high performance, nanocoatings Nanocoatings are used in GMRs, giant magneto resistive thin film heads that go into the preparation of hard drives in desktop computers. They are also used in thermal barrier coating and corrosion-resistant ceramic coating applications. resistance. SSA, specific surface area can be used to characterize nanocoating such as the degree of nanostructuring. TBC can be used to prevent fire spread. For materials with relaxation times greater than ($R^2/15.33\alpha_{nr}$) the temperature in the naval warhead can be expected to undergo subcritical damped oscillations.

X-ray lithography and electroplating techniques were used in order to create a zone plate of thickness of 500 nm with zone dimensions less than 50 nm. PU coating can be grafted with nano-alumina in order to improve the performance of the coating by increasing the adhesion between nanoalumina and organic substrate. Coating machines and rollers can be used to manufacture solar cells at lower cost. A PV, photovoltaic device was created by casting a solution of conducting polymer and fullerenes into a thin film.

Nanostructured alumina/titania powders can be used to improve abrasive wear resistance of ceramic coatings by 400%. Process comprises of dispersion of nanopowders into a colloid. Binders and granules are spray dried using a Metro 9 MB plasma gun. Thermal spray robots are used to obtain nanocoating. Nanostructures can result from control of morphology of phase separated polymer blends. Nanostructuring can be

* E-mail: jyoti_kalpika@yahoo.com; Tel.: (713) 313 1994.

controlled during preparation of metal chalcogenide nanostructured films by electrodeposition. These films can be formed within mesopores of silica.

The use of graphene layers as anti-reflection coatings in solar cells can result in the reduction in the cost of the solar cells. Light to electricity conversion was found to increase to 14.6%. Graphene sheets with less than 24,000 atoms/25 nm are metastable. Nanocoatings are used in the design and application of nanorobots for drug delivery. Principles of photodynamic therapy, fullerene chemistry, nanostructuring, x-rays, computers, pharmacokinetic and robotics are applied to design a nanorobot in order to treat Alzheimer's disease. Drug is inactive in its caged form. It is activated by irradiation at desired site in the human anatomy. The irradiation causes breakage of the fullerenes and the drug becomes available for action. Graphene is used in naval applications in order to demonstrate an infrared-transparent, electromagnetic shielding coating that can be applied to electro optic sensor windows and domes. Transmission greater than 90% in the 3-5 micron wavelength region can be achieved.

Protein folding, secondary structure of proteins, nucleic acid structure, ribosomes, chromosomes and phospholipid membranes are examples where self-assembly is critical.

Fluidized bed can be used to obtain the nanocoating. The precursor material is calcined in ammonia at 1000°C for 48 hours in a fluidized bed reactor. A red tri-tantallumpenta nitride, Ta_3N_5 coated mica pigment is produced. Bubbles and slugs and other instabilities formed during the operation of the fluidized bed are reduced by use of vibrators/agitators. The substrate particle size range is between 1 – 500 μm.

1.0. INTRODUCTION

According to Arthur C. Clarke:

> "Any sufficient advanced technology is virtually indistinguishable from magic".

Nanocoatings offer performance advantages over traditional coatings that customers are willing to pay for. It is a by-product of the miniaturization drive triggered by the computer chip revolution. The material and manufacturing techniques that are used in solar power plants, cell phone industry have evolved in the past 50 years. Some of the critical areas in effecting nanocoatings are as follows [1]:

(i) process for manufacturing optical flakes that go into inks that are used for dollar bills and other currencies;
(ii) functionalization of nanoparticles for in the polymer coating;
(iii) particulation for an outcome of hard, lubricious razor blade coatings;
(iv) materials and methods for barrier, clear and sacrificial coats;
(v) meet necessary qualifications, such as aerospace vehicles.

Alloys are designed using computer models with increased sophistication. DoD, Department of Defense solicits alternates for Cadmium, Cd plating with the sole purpose of protection of aircraft components from corrosion. Side—side-by evaluations of whether the coatings is absolutely needed and whether the same set of properties can be met using a steel of a different alloy composition have been conducted. When the reports from operating experience of aircrafts poured in often times it was found that the real issue for maintenance is service failure and that also because of stress corrosion cracking. Questek came up with a steel grade made up of nanophase precipitated hardened variety with hardened equal to 300 M. This can be used to address the concerns of service failure and cadmium concerns. It is currently in the stage of patenting and qualification. Nanocoatings can offer an alternate to this critical node technology. Reduction in cost and time to market of an order of magnitude can be realized. At a glance survey of the different nanocoatings discussed in this chapter is presented in Table 1.0.

Table 1.0. At a Glance Survey of Different Investigations in Nanocoatings

Ref #	Brief Description	Author	Source	Year	Vol.
[1]	Intellectual Property Landscape	K. Legg	Rowan Technology Group, Chicago, IL	2003	June
[2]	Thermal Barrier Coating to Protect Super alloys that Turbine Blades or Made out of	Siemens Corp.	US Patent 7,413,798	2008	
[3,4]	Critical Thickness below which Oscillations in Temperature can be Expected	K. R. Sharma	231st ACS National Meeting CHEMCON	2006 2004	Atlanta, GA Mumbai, India
[5]	E-beam and X-ray Lithography for Fabrication of Zone Plates for Hard X-ray	T. N. Lo et. al.	AIP Conference Proceedings	2007	1466-1469
[6]	High Performance Waterborne Nano-Alumina-Grafted Polyurethane Coating	B. K. Jha et.al.	Int. J of Nanoscience	2011	Vol 10, 4,5, 1119-1124
[7]	39 Different Nanostructuring Methods	K. R. Sharma	Nanostructuring Operations in Nanoscale Science and Engineering, McGraw Hill Professional	2010	NY
[8]	Primitive Solar Cell Prepared by Thin film Cast from Solution blend of Conducting Plastic and Bucky balls	Heeger	http://nobel.se	2000	
[9]	Progress in Solar Cell Efficiency using Coatings	P. Fairley	Technology Review	2004	July/ August, 35-40
[10]	Plasma Sprayed Nanostructured Alumina/Titania Coatings with Abrasive Wear Characteristics	Y. Wang et. al.	Wear	2000	Vol, 237, 176-185
[11]	Sub-100 nm Structure Formation in Polymer Films	U. Steiner	Nanostructure Science and Technology, Springer	2005	NY
[12]	Polymer Phase Behavior, Partial Miscibility and Nanostructure Formation	K. R. Sharma	Polymer Thermodynamics: Blends, Copolymers and Reversible Polymerization, CRC Press	2012	Boca Raton, FL
[13]	Preparation of Metal Containing Nanostructured Films	Tulane Educational Fund	US Patent 7,001,669	2006	
[14]	AR Coating in Solar Cells using Colloidal Nanostructures	E. Shi et. al.	Nanoletters	2013	Vol. 13, 4, 1776-1781
	IR Transparent Electromagnetic Shielding	K. R. Sharma	Graphene Nanomaterials, Momentum Press	2014	NY
[15]	Roll-to-Roll Production of Graphene Film for Transparent Electrode	S. Bae et. al.	Nature Nanotechnology	2010	Vol. 5, 574-578
[16]	Nanocoatings for Drug Delivery	B. Gromley	Wall Street Journal	2007	Sep' 12th
[17]	Curcuma Longa/Fullerene Nanorobots for Therapy of Alzheimer's	K. R. Sharma	Journal of Scientific Research and Reports	2013	Vol. 2, 1, 206-227

There are three phases in the development nanocoating application. The initial phase is *conception* of the idea, production in bench unit, scale-up to pilot and plant, obtaining licenses. The second stage is the *definition of what critical nodes* the nanocoating occupies. What is the developed product for? Which aspects of the IP, intellectual property needs patent protection? The third stage is *marketing* – exploration of market potential and finding out what customers need, identification of outside the box markets, spelling out unique features of the new product. Coatings are the interface between the product and the environment. The external appearance of the product as the user sees is determined by the coating used. The interaction of the product with the environment is through the coating. Coatings are requisite in some products. Coating is the product in small scale.

1.1. GMR Thin Film Read Heads

Nanocoatings are used in GMR thin film heads, MEMS, microelectromechanical sensors, fuel cells. GMR, giant magneto resistive thin film heads are used in hard drives of desktop computers. The seven layers used in preparation of GMR thin film read head is shown in Figure 1.0.

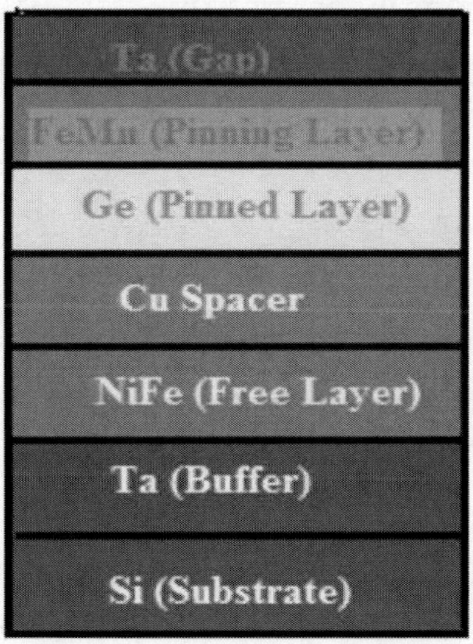

Figure 1.0. Seven Layers of Giant Magneto resistive Thin Film Heads.

The seven layers are as follows;

(i) a Silicon substrate at the bottom
(ii) Tantalum buffer layer
(iii) Nickel-Iron free layer
(iv) Copper space
(v) Germanium pinned layer
(vi) Iron-Manganese pinning layer and
(vii) Tantalum gap layer on top.

Prototypical magneto resistive structures include a pair of ferromagnetic layers that are separated by nonmagnetic spacer layer. The direction of magnetization can change according to the magnetic field. The electrical resistance of GMR structure depends on the relative orientations of magnetization directions of the free and pinned layers. Data stored in form of small magnetic fields recorded on a magnetic disk can be detected using GMR structure. The interleaving insulating layer is a nano oxide layer.

1.2. TBC, Thermal Barrier Coatings

In gas turbine environments, ceramic nanocoatings can be used to prevent corrosion and oxidation of the blades. The blades are usually made of composite or metal alloy materials. Blades made of Nickel and cobolt based super alloys are coated with a bond coat. The coat is made of metal such as nickel, cobolt or iron, chromium, aluminum and yttrium stabilized zirconia. The process of coating is by EB-PVD, electron beam physical vapor deposition. Columnar structures are obtained by this method. Planar structures are obtained by APS, air plasma spraying method. Nanostructured coatings can result in improved thermal cycle life, reduced tendency for sintering and reduced thermal conductivity. Presence of nanostructures can be characterized by using a quantity called SSA, specific surface area. This is the ratio of the surface area to the volume. SSA values can be obtained using SANS, small angle neutron scattering mapping [2]. Other methods for obtained the stated objectives are SEM, surface electron microscope, TEM, transmission electron microscope and x-ray diffraction techniques.

The thermal barrier coating is used to protect the material of construction from the corrosive, high temperature environment of the gas turbine. An oxide layer is coated on top of the bond coat. Over this is deposited an insulating layer. Other than the bond coat the TBC also consists of mixed oxide particles such as zirconium and yttrium dispersed in alumina matrix. The average particle size of the mixed oxide particles is in the range of 10 – 100 nm. The thickness of the oxide layer is about 17 – 33% of the bond coat layer. The particle size of the mixed oxide layer can be controlled by varying the parameters of the EB-PVD process. The parameters of interest are partial pressure of oxygen, partial pressure of inert gases, ingot feed rate, substrate temperature, substrate rotation rate and flux density of vapor species.

Another nanofeature is the aluminum projections extending from the mixed oxide layer and into the insulating layer. Columnar grains with cross-sectional widths of 1- 5 nm in the different layers form another nanofeature. The growth of such splats can be facilitated by change of deposition parameters, size of the powder used, cooling rate of the deposited coating and heat treatment finishing steps. Traverse and lateral grains with radius of curvature less than 1 A and lengths of 5-80 nm can be attributable to less sintering. These feather like structures were identified in the material after 1000 hours of exposure of temperatures of 1400°C. Mass flow is directed during sintering into individual columns during the sintering process due to increased surface area per column. Cross-column necking and bridging is prevented by directed mass flow. Separated columns result and the strain tolerance of the material is retained as was in the original coated system. Column separation is effected by control of deposition parameters. The elastic modulus of the coat decreases linearly with increase in specific surface area, SSA. Lower elastic modulus can lead to desirable performance of the TBC. The degree of nanostructuring in the coating can be gauged using SSA values. The SSA reported [2] was as high as 20,000 cm^2/cm^3. The SSA of the coating was found to decrease with increase in temperature. With increased temperature sintering was found simultaneously in the granular and intersplat boundaries. The composition of the coating can be modified in order to increase the resistance to change in SSA of the coat with increase in temperature.

Magnesium oxychloride sorrel cement coatings can be used to control and delay the spread of flames in naval warheads. Mathematical model [3, 4] was developed for estimation of the time taken for the transient temperature of the naval warhead to reach alarm temperatures. Thermal barrier coatings, TBC, are advanced material systems applied to metallic surfaces operated at elevated temperatures. These coatings serve as insulators of heat. TBC can be used in naval warheads. In the event of fire, metallic bulkheads and decks spread fire by heat transfer which increases the temperature on the back side above the ignition temperature of common combustibles in adjacent and overhead compartments. In order to keep temperatures on the back side below the ignition temperature of the common combustibles, fire insulation such as mineral wool of Structogard is used by the Navy. The role of the TBC is to prevent fire spread. Coated onto a combustible substrate they can prevent the ignition and spread of flame. These can be used in applications with high fire potential. Magnesium oxychloride sorrel cement coatings have been developed by Delphi Research to be used as a barrier system. The ceramic coating can be prepared using chemical vapor deposition process or spray coating process. The thermal conductivity of the coating may be reduced by increasing the porosity of the coat. Magnesium oxychloride cement coating co-bonded with high alumina calcium aluminate cement colloidal silica was found to be effective. One of the critical parameters in the design of such systems is the resistance offered by the material to temperature rise fires. ASTM standards discuss this parameter in the UL 1709 test. UL 1709 test lays down the performance criteria for insulation material. The backside average temperature rise should not exceed 250 ^{0}F above the ambient when tested under the conditions prescribed in the test. UL-1709 fire is specified at 2000^0 F and 200 kw/m^2 for 30 minutes or more. The thrust of the research in development of TBC is looking to advance fire insulation technologies and bring in new materials that reduce weight and cost. The nature of heat transfer during the fire spread is transient. The insulation materials are usually heterogeneous in nature. Under short timescales the Fourier's law of heat conduction does not accurately depict the real events. Mathematical models that can predict the time of rise of insulation temperature specified in UL 1709, that use Fourier parabolic law of heat conduction cannot be relied upon. This is because of the presence of "blow-up" or singularities in the expression for heat flux in the analytical solution obtained using Fourier's law. Damped wave conduction and relaxation may become significant during the short time scales expected when TBC is used to delay onset of fire. It cannot be neglected. The analytical solution for temperature profile in a naval warhead coated with a thermal barrier coating, TBC is derived for the both the naval warhead and the coating by the method of separation of variables. The external fire temperature, T_{fire} drives the problem. The final condition in time is used as the fourth constraint and second time constraint to obtain bounded and meaningful solutions. These solutions no not disobey the Clausius's Inequality, The solutions were obtained by the method of separation of variables. A critical coating thickness was derived, below which the transient temperature in the coating will undergo subcritical damped oscillations. This shall drive oscillations in temperature within the naval warhead.

In the literature, coatings are usually by design made “thin” as possible. The results of a theoretical study on transient temperatures can be used in order to show that they cannot be made too thin. A desirable mode of operation is when the transient temperature does not oscillate and hence the coating thickness has to be greater than the critical thickness of coating as predicted. In a similar manner, the naval warhead cannot be made of materials with relaxation time greater than, $R^2/15.33\alpha_{nr}$ the temperature in the naval warhead will undergo subcritical damped oscillations in time [3]. The exact solutions for transient temperatures within the naval warhead and in the TBC are seen to be bifurcated. Both solutions are given by an infinite Bessel series of the first kind and zeroth order and decaying exponential in time. Under some conditions the time domain portion of the solution becomes cosinous.

1.3. Electroplating

The technical hurdles in nanomanufacturing are the large fabrication precision and high aspect ratio. State of the art x-ray lithography and electroplating techniques were used [5] in order to create a zone plate with thickness of 500 nm and with zone dimensions less than 50 nm. Selection of photoresist and the capability to electroplate thick Au layer into small trenches will be critical parameters in performance of the device. The substrate used in fabrication of the x-ray mask is prepared by deposition process of Si_3N_4 film on Si wafer. The desired zone plate pattern is created using e-beam lithography with capability to write less than 10 nm on photoresists. This step is followed by electrodeposition of gold layer onto the nanopatterned photoresist structure of sufficient thickness. The resist is then removed. Potassium hydroxide, KOH, is used in order to etch the Si wafer down to the Si_3N_4 film. The desired z-ray hard mask is produced. The zone plate pattern with the x-ray lithography method is defined by hard mask. Precisions of up to 30 nm outermost zone in the mask can be obtained.

High penetration into the material is one reason for choice of x-rays with wave lengths of less than 1 nm. The resolution of the device cannot be any better than the smallest zone. For an image resolution of less than 50 nm, the smallest zone has to be less than 50 nm. E-beam lithography is the one feasible approach for structures less than 500 nm. X-ray lithography can be used for structures thicker than 50 nm. A typical zone plate has concentric circles with alternate empty and filled zones. The width of the zone decreases when the diameter of the outermost zone increased.

1.4. Nanografting

The properties of polyurethane, PU coating such as mechanical and corrosion resistance can be significantly improved by grafting techniques. The interfacial adhesion between nano alumina and organic substrate is increased by grafting. Nanostructured alumina is incorporated in the PU conventional coating and applied on mild steel. Bifunctional polymers [6] such as MTMS, methyl-trimethoxysilane and GPTMS, 3-glycidoxy-propylmethyldiethoxysilane were grafted onto the surface of the nanostructured alumina. This prevents agglomeration of the dispersed particles. Dispersion of particles in coating was achieved by ultrasonication. AFM , atomic force microscopy analysis was used to confirm that grafting nano alumina particles results in reduction of agglomeration of the dispersed phase. The RMS surface roughness parameter was also found to decrease. The coating formulation comprises of four steps: (i) Aqueous uranedio and hexamethoxymethyl-melamine are mixed in molar ratio of 3:1 and ultrasonicated for 10 minutes at room temperature; (ii) 10% each of the mixture, alumina and MTMS and GPTMS were dispersed separately; (iii) 0.4 wt% of blocks acid catalyst para-toluenesulphonic acid, p-TSA were added and the resulting solution ultrasonicated for 20 minutes at room temperature and; (iv) resulting solution applied on cleaned mild steel surface by brush method at room temperature and placed in furnace at 150 – 160°C for 30 min for final curing. The average dry coating thickness was found in the range of 15-20 μm for brush coating.

1.5. Printing Solar Cells

Electrical power can be produced using solar cells. Solar cells can be made using production line of coating machines and rollers. The processes used are similar to that found in a modern printing press. Konarka has developed technology that can cut the cost in half of silicon based panels that are lighter in weight and offers flexibility. Flexible films can be laminated onto laptops and cell phones can trigger flow of electric current. Such cells are targeted for mixing with automotive paint that can be used to tap into sun power to charge batteries of HEV, hybrid electric vehicles. The printable solar revolution was spearheaded by start-ups around 2005 such as Konarka, General Electric, Siemens, and STMicroelectronics. C_{60}, fullerenes can be used as ingredient in solar cells. Several methods to prepare fullerenes was discussed [7].

Siemens has combined two of the important advances in materials science, i.e., electrically conducting polymers and characterization of fullerenes as the third allotrope of carbon.

In early 1990s physicists Sariciftci and A. Heeger [8] created a primitive PV, photovoltaic device by pouring a solution of conducting plastic and Bucky balls onto a glass plate and then by spinning the plate in order to spread the solution into a film and forming a sandwich of the film between the two electrodes. The conducting polymers are used to absorb photons and then generating electrons and then attracted by Bucky balls and routed to an electrode.

The light to electricity conversion was small at 1%. Sariciftci headed up a team of research scientists at Kepler University and then at Siemens and identified a mixture of solvents, temperatures and drying conditions and obtained better blends of fullerenes and plastic. The power output was found to double. This was attributable to more jumps of electrons from plastic to fullerenes.

Further improvements in light to electricity conversion was achieved by manipulation of nanomaterials and change to nanocoating methods. The identification of the causative factor in jump of electricity is not conclusive. Attributable mechanism are more regular structuring of polymers and fullerenes that go into the making of the solar cell. 10% of the light can be converted to electricity. Konarka flexible film is used in cell phones and lap tops. TiO_2, titanium dioxide coated with light-absorbing dyes can be used in place of fullerenes. These particles are bathed in an electrolyte and embedded in a plastic film. Nanosolar's approach is by spraying. Cocktail of alcohol, surfactants and titanium compounds are sprayed on a metal foil.

A high efficient solar cell is made by evaporation of alcohol from the metal foil, self-assembly of surfactant molecules into elongated tubes. A molecular scaffold is formed and fused. Nanoholes are bored in 30 seconds. These nanoholes are filled with conductive polymers. Electrodes are erected and entire block is covered using plastic and the high efficient solar cell is ready to use.

The materials used are sensitive to oxygen, blowing sand, intense sun light, extreme temperature shifts [9].

1.6. Plasma Spray Coating

Nanostructured alumina/titania powders can be used to improve the abrasive wear resistance of ceramic coatings. 400% improvement in the performance of the coatings were reported [10]. Alumina is an effective coating used in the tool industry. Alumina coated carbide tools have a speed capability 250% higher than uncoated carbide tools. Nanoparticulated powders can be reformed into spherical granules. This is achieved by dispersion of the nanopowders into a colloid. A binder is subsequently added and granules are formed by spray drying. Sprayable granules can be form metals, ceramics and composite powders. A Metco 9MB plasma gun was used [10] in the thermal spraying process. Coating thickness achieved were 250-600 μm. Mild steel coupons were used as substrate and was blasted in order to remove dust. A GM-Fanuc six axis thermal spray robot was used and controlled using a computer. The plasma deposition parameters were: (i) primary argon gas pressure (690 kPA); (ii) secondary hydrogen gas pressure (380 kPa); (iii) argon gas flow rate (120); (iv) powder carrier gas flow rate (40-70); (v) powder feed rate (400-900 g/h); (vi) current (600 amps); (vii) voltage (65V); (viii) moving speed of gun (500 mm/s); (ix) spray distance (100 mm); (x) thermal spraying 10 passes after one cycle.

1.7. Nanostructuring in Thin Films

In some applications in industrial electronics polymer structures are required on length scales down to individual molecules. A "bottom-up" approach is preferred to a "top-down" approach in order to achieve this. A lateral resolution less than 100 nm can be created by allowing for surface instabilities and pattern formation in polymer films. Steiner [11] discussed the methodology for demixing of polymer blend and pattern formation. Nanostructure formation is by capillary instabilities.

A compatible blend is one when some improvement of property of blends is shown in the application when two polymers are mixed. Different from this is the molecular mixing achieved in miscible blends. Nanostructures can result from control of the morphology of phase separated polymer blends. Polymer blends can also be found to be partially miscible [12]. Miscible polymer blends a single glass transition temperature much like a homopolymer made out of a single monomer. Immiscible polymer blends will exhibit the same two glass transition temperatures as their homopolymers that went into the make of the blend. The partially miscible blends can be expected to exhibit glass transition temperatures distinctly different from the homopolymer glass transition temperatures.

When polymers undergo phase separation in thin films the kinetic and thermodynamic effects are expected to be pronounced. The phase separation processes can be controlled to effect desired morphologies. Under suitable conditions a film deposition process can lead to pattern replication. Demixing of polymer blends can lead to structure formation. The phase separation process can be characterized by the binodal and spinodal curves. UCST is the upper critical solution temperature, which is the temperature above which the blend constituents are completely miscible in each other in all proportions. Not found that often in other systems other than among polymers is the LCST behaviour. The LCST is the lower critical solution temperature. This is the temperature above which the polymers that were miscible below this temperature now exhibit immiscibility. The binodal and spinodal curves can be calculated by the stability criteria of Gibbs free energy.

$$\Delta G = \Delta H - T\Delta S$$

$$\Delta G \leq 0$$

$$\frac{\partial^2 G}{\partial \phi^2} > 0$$

where G is the Gibbs free energy, H is the enthalpy and S is the entropy and ϕ is the phase composition. When the free energy change is negative the polymer blend can be expected to be miscible. The immiscible and miscible region are separated by the binodal curve. The spinodal curve can be found within the binodal where the curvature of the free energy curve becomes negative. When the material falls on the spinodal region of the phase diagram spinodal decomposition can be expected to occur.

Phase *morphology* with a *single characteristic length scale* can be synthesized by quenching a partially miscible polymer blend below the critical temperature of demixing. A well-defined spinodal pattern is formed and becomes larger with passage of time. Polymer films can be made by spin coating solvent casting method. The polymers and solvent form a homogeneous mixture at first. Solvent evaporation during spin coating causes an increase in polymer phase volume resulting in traversal to the spinodal region of the phase diagram. This can be expected to lead to polymer-polymer demixing. Characteristic phase morphology can be found in the polymer film The polystyrene and poly vinyl pyridine are mixed in the THF, Tetrahydrafuran solvent.

Different phase separated morphologies can be found in different polymer solvent systems. The pattern formation consists of several stages. In the initial stage phase separation results in a layered morphology of the two solvent swollen phases. As more solvent evaporated this double layer is destabilized in two ways: 1) capillary instability of the interface and ii) surface instability. Each of the mechanisms result in different morphological length scales. Core shell spherical domains in phase separated ternary systems have also been found. The shell thickness can be a few nanometers.

1.8. Electrochemical Deposition

Electrodeposition can be used in order to prepare metal chalcogenide nanostructured films. Thin films of metals and semiconductors are needed in advanced technologies. Precise control of the nanostructuring can lead to improvement in performance of thin film based devices such as nanosensors, magnetic storage media, and nanoscale optical devices. Wet methods for nanostruturing include chemical bath deposition, CBD, electrochemical deposition, ED. Template based synthesis is a more efficient method of preparation of nanostructured thin film nanowire arrays. A variety of nanostructures have been electrochemically deposited within porous alumina and polycarbonate templates providing arrays of nanoscale wires. Nanowires possess dimensions of about 20-250 nm and length of 1-10 mm. Materials used in preparation of nanowires include metals such as Gold, Silver, Cobalt, Copper, Nickel, Palladium and Platinum and compound semiconductors such as Cadmium Sulfide, CdS, Cadmium Selenide, CsSe, IndiumPhospide, InP, GaAs, Gallium Arsenide, and conductive polymers such as polyacetylene.

An electrodeposition method was patented by University of Tulane [13] to make nanostructured films within the pores of mesoporous silica by formation of a metal silica nanocomposite. Nanostructured film need possess high structural integrity and mechanical strength. The nanocomposite is obtained by annealment in order to strengthen the deposited metallic composition. The deposition temperature is at a fraction of the melting temperature of deposited metal. Removal of silica from the nanocomposite is by dissolution in a suitable etching solvent such as HF, hydrofluoric acid.

The dimensions of nanoscale wires of metal chalcogenide thin films assumed are determined by the dimensions of the pores in the template used to prepare the film. The pore structure is mesoporous. Templates may be hexagonal with arrays of straight pores. The pore structure is interconnected, continuous and forms a nanomesh. Segments with nanoscale cross-sectional dimension are present. Cubic mesoporous silica templates having a network of three dimensional interconnected pores can be use.

A mesoporous silica film is prepared by deposition of a surfactant micelle-templated silica sol onto an electrically conductive substrate. Next step is calcination of the deposit in order to provide a film that is surfactant-free. The pore structure of the film is a function of the choice of the surfactant and the concentration of surfactant. The film is deposited on a conductive metal surface such as Cu/Au/Silver in order to provide an electrode for electrodeposition of metal chalcogenide within the pores of the film. The film with its conductive substrate is immersed in a solution of metal ion complex and elemental chalcogen such as sulphur, selenium or tellurium and a voltage potential is applied between the conductive substrate in contact with the mesoporous silica template and a counter electrode in the solution. The nanowires grow in a single direction growing outward through the pores of the silica forming a nanocomposite of silica and metal chalcogenide. The nanocomposite is later annealed and the structural integrity of the film is imparted in this step. The silica is removed by solvent etching techniques.

These nanostructured films can be used in solar cells catalytic membranes and information storage media. A self-supporting metal chalcogenidenanomesh film can be used as a photo electronic component in photovoltaic solar cells. The solar cell comprises of CdSe, cadmium selenide nanomesh film and a layer of CdSe nanomesh in contact with an optically transparent indium-oxide, ITO substrate. The pores of the nanomesh are filled with a hole transport agent such as polythiopene, a conductive polymer. A reflective metallic layer is coated over the nanomesh and hole transport agent. Semi conductive CdSe nanomesh is excited by the light entering through the ITO substrate. Electron is generated that is injected into the hole transport agent and migrates to the conductive metal layer generating an electric current. This is expected to increase the photovoltaic efficiency of solar cells from 14% to the order of 40-50% by use of nanostructured components.

1.9. Graphene Based Solar Cells

The cost of solar cells can be reduced by addition of graphene layer to silicon solar cells.

The light to electricity conversion efficiency is about 15%. Cells are expensive. One reason for the higher costs is the thin charge-carrying silicon layer on top. This requirement needs process at high temperatures. Thin silver wires serve as an electrode. Cao [14] of Peking University, China recently showed that graphene film can replace this charge-carrying layer. It can also be used as an electrode. This is less expensive to fabricate. But the light to electricity conversion efficiency is about 8.6 – 10.0%. The scientists at the National Center for Nanoscience and Technology at Beijing, China coated a 65 nm thick layer TiO_2 atop silicon graphene solar cell (Figure 2.0).

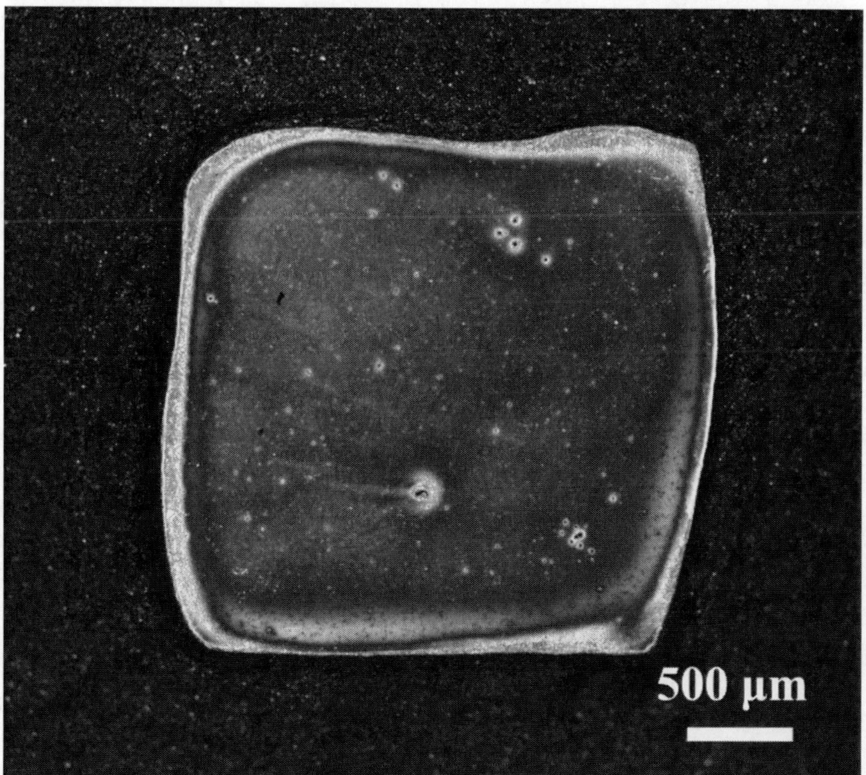

Figure 2.0. 65 nm Thick TiO_2 Layer Coating on Graphene Silicon Solar Cell.

The amount of visible light reflected from the silicon surface is decreased from 30% to less than 10% on account of the coating. Less reflected light would mean more light to electricity conversion efficiency. The efficiency was found to increase to 14.6%. Graphene layers are grown on copper foil using chemical vapor deposition and transferred to silicon wafers.

Unscrolled CNT is graphene. 2D nanosheets cannot be generated without an epitaxial substrate. The substrate is used to provide atomic bonding in the third dimension (Landau-Pearls argument). This is an important stability consideration when attempting to synthesize material with thickness of an atom.

Raman spectra is used to confirm monolayer structure of graphene. Another important thermodynamic stability consideration is the metastability. Gibbs free energy change of formation of graphene structures has to be negative for stable structures.

Unstable structures may be expected when the second derivative of the Gibbs free energy with respect to phase volume is negative. Metastable structures can result when the Gibbs free energy change is negative and the phase stability criterion is not met.

Graphene sheets with less than 24,000 atoms/25 nm are metastable. The honey comb lattice structure has been confirmed for graphenes.

Graphene made by CVD, chemical vapor deposition methods can be rolled into thin films by a transfer process. The process [15] comprises of three steps: (i) Adhesion of polymer supports to the graphene on copper foil. Two rollers are used to get the graphene film grown on a copper foil to be attached to a polymer film coated with adhesive film as it passes through; b; (ii) Etching of copper layers. Graphene is synthesized in an annular plug flow reactor, APFR. The annular reactor space is generated by an 8 inch outer quartz tube and an inner 7.5 inch copper foil, wrapped quart tube. The use of annular reactor in place of the tubular reactor is to minimize radial temperature gradients. This was found to cause inhomogeneity in the film formation. The inner tube is heated to 1000 ^{0}C. Hydrogen, H_2 is allowed to flow at 8 s.c.c.m and 90 mtorr. Annealing process comes next. Annealment for 30 minutes allows for increase in grain sized in copper foil from a few micron to 100 microns. This has been found to increase the graphene growth. Methane, CH_4 is allowed to mix with the flowing hydrogen at a flow rate of 24s.c.c.m at 460 m torr. The sample is rapidly cooled at about $10^{\circ}C.s^{-1}$. The graphene film grown on copper foil is attached to a thermal release tape by applying pressure on the rollers at 0.2 MPa. Copper foil is etched in a plastic bath filled with etchant.

The etched film is washed with deionized water to remove and unused etchant. The graphene film is ready for transfer to a target substrate such as a curved surface. 150- 200 $mm.min^{-1}$ transfer rates by thermal treatment can be achieved by letting the graphene film pass through the rollers at mild heat of 90-120 ^{0}C. Multilayered graphene can be made by repetition of this process.

The product that comes as a result of this would be different from the bilayer or multilayer material formed during the reaction by other methods. This can be viewed as physical stacking of the formed graphene layers. Screen printing can be used to generate 4 wire touch panels.

Continuous production of graphene on large scale is possible. Scalability of the process is high. Processability is good. Carbon has limited solubility in Copper even at 1000 ^{0}C. The copper may have a catalytic effect on the graphene formation reaction. has limited Transparent electrodes can be made using graphene in large scale and replace the currently used indium tin oxide, ITO, electrodes. Monolayer graphene was confirmed using Raman spectra. Bilayer and multilayer islands were found from atomic force microscope, AFM and transmission electron microscope, TEM. Stacked layers reduces the optical transmittance by 2.2-2.3% a layer and the conductivity also decreases. Dopants can be added as desired. p junction formation by doping can be achieved by addition of nitric acid, HNO_3. Sheet resistance can be increased by chemical doping. Poly methyl methacrylate, PMMA can be used as polymer supports.

Some of the challenges in using this method is the formation of polycrystalline graphene due to occurrence of nucleation again to form a second layer. Oxidation of copper has to be avoided. High rates of evaporation of copper from the foil can hinder graphene growth. Copper is not as effective as Ni, nickel to lower the energy barrier to form graphene.

Further reductions in fabrication costs for graphene single layer is by painting or spraying a graphene solution onto the silicon wafers in order to produce the graphene film. These devices have to make more stable. Solar device efficiency slips after 20 days of outdoor performance.

1.10. Nanorobot Drug Delivery

Nanocoatings can be used in design and application of nanorobots for drug delivery. In March 2004 a nanoelevator was built at UCLA, University of California at Los Angeles. This molecular scale device is 2.5 nm high and can move up and down by 1 nm. This can be used in drug delivery systems. Drugs are packed with nanoagents and are used to knock-out cancer cells. Per the General Manager of Flagship Ventures, Cambridge, MA there is a swarm of important therapeutics arriving that uses nanotechnology. Nanoparticles are allowed to act as agents. In 1995, Doxil was approved for use in the United States [16]. The chemotherapy doxorubicin is encapsulated into a fat bubble/liposome and another coating of hair like strands made of rubbery material is applied. The agent is 100 nm in diameter. Doxil liposomes are small and are able to circulate for long periods of time. They possess the capability of penetration of vasculature or arrangement of blood vessels or tumors. Upon arrival at the tumor tissue the drug is slowly released. Johnson and Johnson sells doxil in this country United States and can be used for treatment of patients with ovarian cancer.

Principles from photodynamic therapy, fullerene chemistry, nanostructuring, x-rays, computers, pharmacokinetics and robotics are applied in developing a strategy for nanorobot, treatment of Alzheimer's disease. The curcuma longa that has shown curative effects in rats' brain with Alzheimer's is coated on fullerenes. The drug is inactive when in caged or coating form. It is infused intrathecally into the cerebrospinal system. Irradiation of the hypothalamus and other areas of the brain where Alzheimer's disease is prevalent lead to breakage of fullerenes and availability of the drug with the diseased cells.

Due to better mass transfer better cure is affected. The other plausible reactions such as addition polymerization of fullerene, polycurcumin formation and other hydrolysis reactions are modeled along with the drug action under the Denbigh scheme of reactions. The fractional yield of drug curcumin interaction is a function of intensity of radiation, frequency of radiation, patient demographics, age, gender, other disorders etc. Chromophore in curcumin is used as a sensor and computer imaging and feedback control design can result in more bioavailability for curcumin therapeutic action to cure Alzheimer's disease. The principles used in the design, the strategy of the design of the nanorobot drug delivery system with a specific target and pharamacokinetic formulation of the associated competing parallel reactions were studied [17].

1.11. Self-Assembly

Science underlying nanocoatings is being developed. One item to watch in this area is self-assembly. P. J. Stang was awarded the Priestly Medal by the American Chemical Society this year, 2013. Priestly medal is the highest honor bestowed by the ACS. Among other things he was recognized for his research contributions in molecular self-assembly to make supersized Lego type molecules. Self-assembly according to Prof. P. J. Stang is the spontaneous reaction and organization of simple building units into well-defined more complex ensembles based upon the recognition elements embedded in the building units. A number of non-covalent interactions can be found in natural systems.

These interactions include hydrophobic/hydrophilic, van der Waals, Pi orbital stacking, dipole interactions, electrostatic, London forces, Keesome forces, hydrogen bonding. Molecular self-assembly can be found in living organisms. Protein folding, secondary structure of proteins, nucleic acid structure, ribosomes, chromosomes and phospholipid membranes are examples where self-assembly is critical. The protein coats of viruses, encasing their genetic material, comprise of self-assembled capsids resembling polyhedrons such as dodecahedrons and icosahedrons.

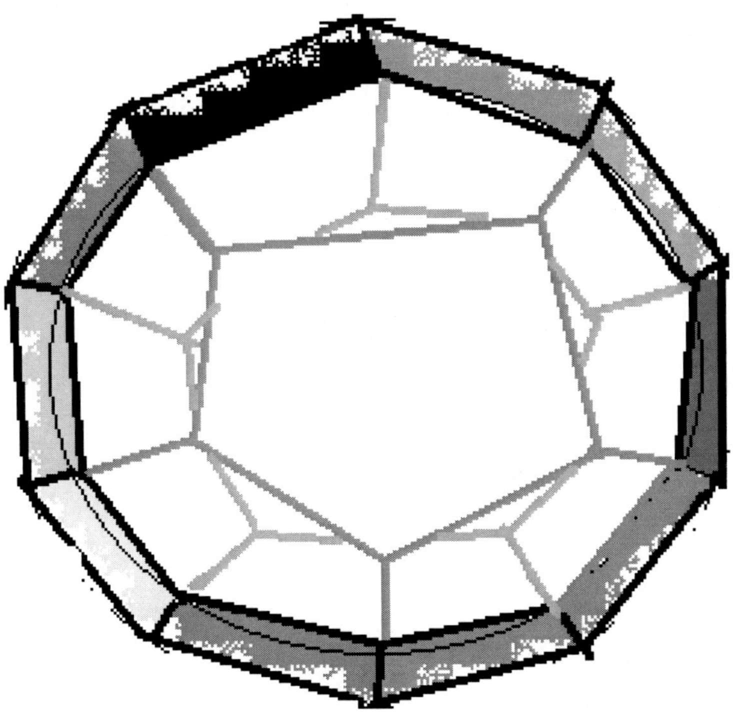

Figure 3.0. Dodecahedron Nanoscale Architecture.

Attempts to mimic natural systems in the lab were a long road. Use of coordination chemistry and use of dative metal ligand bonds obviate the need for directionality of weak interactions and precise positioning of these interactions in order to obtain functional assemblies. Due to the d orbitals' geometry high directionality is provided by use of metal ligand interactions. Metal-ligand interactions have bond energies of about 15-25 kcal. These values lie between that of covalent bonds (60 -120 kcal.mol-1) and weak interactions of 0.5 – 12 kcal. mol-1. Kinetics of coordination bond formation may be controlled. Reversible, self-repairing superstructures can be formed by thermodynamic control.

Several hydrogen bonds can be replaced with one dative meta-ligand bond in the self-assembly process. Over two decades Prof. P. J. Stang, Distinguished professor at University of Utah, has developed by exploiting coordination chemistry and metal-ligand interactions, wide variety of large, nanoscale, complex 2-dimensional and 3-dimensional assemblies with well-defined shapes and sizes.

Some of these shapes are trigonal bipyramid, tetrahedron, cube, anamantanoid, double square, dodecahedron (Figure 3.0), truncated tetrahedron, cuboctahedron.

1.12. Fluidized Bed Nanocoating for Interferance Pigment

Nitride and oxynitride nanocoatings with thicknesses in 5 – 500 nm range are used in pearlescent pigment formulation [18]. Optimal interferences are seen at 50 – 350 nm. A liquid-solid suspension in water is made from 100 gm of mica muscovite for 2 liters of water. The particle size is 10- 50 μm diameter. With the suspension maintained at 75°C a tantalum pentachloride, $TaCl_5$ solution in hydro chloric acid, HCl is slowly added to the reactor. The pH is maintained at 9 in the kettle. Preparation is filtered off, water washed and dried at 110°C for 12 hours. For every 100 gm of mica, 5 gm of $TaCl_5$ is precipitated onto the solid.

The precursor material is calcined in ammonia at 1000°C for 48 hours in a fluidized bed reactor. A red tri-tantalum penta nitride, Ta_3N_5 coated mica pigment is produced. These pigments are semi-transparent with an angle-dependent optical effect. Because the calcination is performed in a fluidized bed a product with less agglomerates is formed at reduced reaction times. The process can be performed either in the batch mode or in the continuous mode. Ammonia is an interesting choice for nitration. The conversion from oxides to nitrides/oxynitrides depends on a number of parameters such as gas flow rates, reaction times, temperature profiles. Higher nitride to oxide ratio is obtained at higher reaction times. The structure of the product obtained depends on the reaction time. The color of the compound and the color strength of the compound is found to be dependent on the structure. Bubbles and slugs and other instabilities are reduced by use of vibrators/agitators. The substrate particle size range is between 1 – 500 μm.

Post processing of the pigments upon attaining desired thinness surface treatment steps can be used to increase the weather stability and chemical stability of the pigment. It can be used to facilitate the application of pigment in different media. Pearlescent pigments can be used for coloring different materials: plastics; glasses; ceramic products; agricultural foils; cosmetic formulations; automotive coatings; inks and; printing inks. Printing inks for intaglio printing, bronzing, flexographic printing can be made out of nanocoated pigments. They can also be used for laser making of papers, plastics. National security applications of the pigment includes banknotes, ID cards, concert tickets and plastic films.

1.13. Infra Red Transparent Electromagnetic Shielding

Graphene is an interesting material of choice in naval applications in order to demonstrate an infrared-transparent, electromagnetic shielding coating that can be applied to electro optic sensor windows and domes [19]. Transmission greater than 90% in the 3-5 micron wavelength region can be achieved. Sheet resistance can be made less than 10 Ω per square. The coating can be found to be chemically stable in the atmosphere and in daylight and exhibit good resistance to rain and solid particles. The coating use temperature can be as high as 600°C.

Electromagnetic shielding of electro optic sensor electronics in military systems is currently provided by metal grids that are electrically conductive applied to the dome or sensor window. However, the optical system is compromised via geometric blockage, diffraction and unwanted reflection of light. Grids provide electrical shielding but are difficult to be deposited on curved shapes. Damage is caused by erosion from rain and particle impact on the external surface of the dome or window. Graphene thin film coating with high electrical conductivity and optical transparency can provide adequate electromagnetic shielding and superior optical performance. With its hardness erosion resistance can be expected to be high. The graphene layer can be part of a stack of layers in order to minimize reflection losses and provide high erosion resistance. The coating can be deposited on an infrared-transparent substrate such as sapphire or silicon. This allows for infrared optical properties to be measured. Silicon can be used only if it is chemically inert under the film deposition conditions.

CONCLUSION

Miniaturization drive towards production of cell phones, solar power plants has led to progress towards preparation of nanocoatings. Flake production, functionalization, particulation, sacrificial coating, specifications are critical in order to prepare lubricous, high performance, nanocoatings. Questek developed a steel grade made of nanophase precipitated dispersion. The three phases during the development of nanocoating are; (i) conception, production in bench scale, scale-up to pilot an plant; (ii) obtaining license, purpose of development of product and; (iii) marketing. Coating is the interface between product and environment.

GMR, giant magneto resistive thin film heads used in hard drives of desktop computers comprises of seven layers. The seven layers are silicon substrate, tantalum buffer layer, nickel-iron free layer, copper spacer, germanium pinned layer, iron-manganese pinning layer and tantalum gap layer. In addition to GMR, nanocoatings are also used as thermal barrier coatings. Magnesium oxychloride sorrel cement coatings are used, for example. Ceramic coatings can be used to prevent corrosion and oxidation of blades. Electron beam physical deposition process is used to obtain nanocoating. SSA, specific surface area can be used to characterize nanocoating such as the degree of nanostructuring. TBC comprises of oxide layer, bond coat layer, insulating layer, mixed oxide particles with particle size in the range of 10-100 nm. Aluminum projections with width of 1-5 nm and lateral grains with radius of curvature less than 1 A and length 5-8 nm are other nanofeatures. TBC can be used to prevent fire spread.

Transient temperature in TBC was modelled including the effects of damped wave conduction and relaxation. There is a critical thickness of coating below which subcritical damped oscillations in temperature can be expected. For materials with relaxation times greater than $R^2/15.33\alpha_{nr}$ the temperature in the naval warhead can be expected to undergo subcritical damped oscillations.

X-ray lithography and electroplating techniques were used in order to create a zone plate of thickness of 500 nm with zone dimensions less than 50 nm. PU coating can be grafted with nano-alumina in order to improve the performance of the coating by increasing the adhesion between nanoalumina and organic substrate. Coating machines and rollers can be used to manufacture solar cells at lower cost. Flexible film are laminated on top of cell phones. A PV, photovoltaic device was created by casting a solution of conducting polymer and fullerenes into a thin film. Nanocoatings has been shown to improve the light to electricity conversion efficiency in solar cells.

Nanostructured alumina/titania powders can be used to improve abrasive wear resistance of ceramic coatings by 400%. Process comprises of dispersion of nanopowders into a colloid, binded and granules spray dried using a Metro 9 MB plasma gun. Thermal spray robots are used to obtain nanocoating. Nanostructures can result from control of morphology of phase separated polymer blends. Kinetics and thermodynamics of phase separation can be modeled. Polymer morphology with a single characteristic length scale can be made by quenching partially miscible polymer blend below the critical temperature of demixing. Spinodal patterns are formed that evolve with passage of time. Nanostructuring can be controlled during preparation of metal chalcogenide nanostructured films by electrodeposition. These films can be formed within mesopores of silica.

Cost of solar cells can be reduced by use of graphene layers as AR coatings. Light to electricity conversion was found to increase to 14.6%. Graphene sheets with less than 24,000 atoms/25 nm are metastable. This can result when Gibbs free energy of mixing is negative and the phase stability criterion is not met. Large single layer graphene sheets that are made by roll transfer are stable. Annular Plug Flow Reactor is used to decompose acetylene autocatalytic decomposition and carbon is grown by deposition on rolls.

Nanocoatings are used in the design and application of nanorobots for drug delivery. Principles of photodynamic therapy, fullerene chemistry, nanostructuring, x-rays, computers, pharmacokinetic and robotics are applied to design a nanorobot in order to treat Alzheimer's disease. Drug is inactive in caged form and activated by breakage of fullerenes by irradiation at desired site.

Protein folding, secondary structure of proteins, nucleic acid structure, ribosomes, chromosomes and phospholipid membranes are examples where self-assembly is critical. Graphene is an interesting material of choice in Naval applications in order to demonstrate an infrared-transparent, electromagnetic shielding coating that can be applied to electro optic sensor windows and domes. Transmission greater than 90% in the 3-5 micron wavelength region can be achieved.

Fluidized bed can be used to obtain the nanocoating. The precursor material is calcined in ammonia at 1000°C for 48 hours in a fluidized bed reactor. A red tri-tantallumpenta nitride, Ta_3N_5 coated mica pigment is produced. Bubbles and slugs and other instabilities formed during the operation of the fluidized bed are reduced by use of vibrators/agitators. The substrate particle size range is between 1 – 500 μm.After the pigments of the desired thinness is obtained a surface treatment step can be used to increase the weather stability and chemical stability of the pigment.

REFERENCES

[1] K. Legg, "Nanocoatings IP Landscape", Rowan Technology Group, Chicago, IL, June 2003, http://www.rowantechnology.com.

[2] A. J. Burns and R. Subramanian, "Thermal Barrier Coating having Nano Scale Features", US Patent 7,413,798, 2008, Siemens Power Generation, Inc., Orlando, FL.

[3] K. R. Sharma, "Critical Thickness of Coating and Onset of Subcritical Oscillations", 231st ACS National Meeting, (2006), Atlanta, GA.

[4] K. R. Sharma, "Some Issues in Thermal Barrier Coating", CHEMCON 2004, (2004), Mumbai, India.

[5] T. N. Lo, Y. T. Chen, C. J. Liu, W. D. Chang, T. Y. Lai, H. J. Wu, I. K. Lin, C. I. Su, B. Y. Shew, J. H. Je, G. Margaritondo and Y. Hwu, "Using E-Beam and X-ray Lithography Techniques to Fabricate Zone Plates for Hard X-ray", AIP Conference Proceedings, Vol. 879, 2007, 1466-1469.

[6] B. K. Jha, M. K. Bhadu and A. S. Khanna, "High-Performance Nano-Alumina-Grafted Waterborne Polyurethane Coating", *Int. J. of Nanoscience*, Vol. 10, (2011), 4, 5, 1119-1124.

[7] K. R. Sharma, "Nanostructuring Operations in Nanoscale Science and Engineering, McGraw Hill Professional, 2010, New York.

[8] A. Heeger, "Semiconductors and Metallic Polymers: The Fourth Generation of Polymeric Materials, 2000, http://nobel.se

[9] P. Fairley, "Solar-Cell Rollout", Technology Review, July/August, 2004, 35-40.

[10] Y. Wang, S. Jiang, M. Wang, S. Wang, T. D. Xiao, P. R. Strutt, "Abrasive Wear Characteristic of Plasma Sprayed Nanostructured Alumina/Titania Coatings", *Wear,* Vol. 237, 2000, 176-185.

[11] U. Steiner, Structure Formation in Polymer Films: From μm to sub 100 nm Length Scales, in Nanoscale Assembly – Chemical Techniques (Nanostructure Science and Technology Eds. W. T. S. Huck), Springer, 2005, New York, NY.

[12] K. R. Sharma, Polymer Thermodynamics: Blends, Copolymers and Reversible Polymerization, CRC Press, 2012, Boca Raton, FL.

[13] Y. Lu and D. Wang, Process for the Preparation of Metal-Containing Nanostructured Films, US Patent 7,001,669, 2006, Tulane Educational Fund, New Orleans, LA.

[14] E. Shi, H. Li, L. Yang, L. Zhang, Z. Li, P. Li, Y. Shang, S. Wu, X. Li, J. Wei, K.Wang, H. Zhu, D. Wu, Y. Fang and A. Cao, "Colloidal Antireflection Coating Improves Graphene-Silicon Solar Cells", Nano *Letters,* Vol. 13, 4, (2013), 1776-1781.

[15] S. Bae, H. Kim, Y. Lee, X. Xu, J. S. Park, Y. Zheng, J. Balakrishnan, T. Lei, H. R. Kim, Y. I. Song, Y. J. Kim, K. S. Kim, B. Ozyilmaz, J. H. Ayn, B. H. Hong and S. Iijima, "Roll-to-Roll Production of 30 inch Graphene Films for Transparent Electrodes", *Nature Nanotechnology,* vVol. 5, 574-578, 2010.

[16] B. Gromley, "Betting on Nanotech Therapies", *Wall Street Journal,* Sep.' 12th, 2007.

[17] K. R. Sharma, "On Photodynamic Therapy of Alzheimer's Disease Using Intrathecal Nanorobot Drug Delivery of Curcuma Longa for Enhanced Bioavailability", *Journal of Scientitic Research and Reports,* vol. 2, 1, (2013), 206-227.
[18] S. Bertaux, P. Reynders, J. U. Wichmann and E. Schweda, "Pearlescent Pigments", US Patent 6,719,837, Merck Patent Gesellschaft mit beschrankter Haftung, Darmstadt, DE.
[19] K. R. Sharma, Graphene Nanomaterials, Momentum Press, (2014), New York.

In: Comprehensive Guide for Nanocoatings Technology. Vol. 1 ISBN: 978-1-63482-447-7
Editor: Mahmood Aliofkhazraei

Chapter 6

INKJET PRINTING OF GROUP-11 METAL STRUCTURES

Christian Gäbler [1]***, Christian Schliebe*** [1]***, David Adner*** [1]***,***
Thomas Blaudeck [2]***, Alexander Hildebrandt*** [1]
and Heinrich Lang [1,*]

[1]Faculty of Natural Sciences, Institute of Chemistry, Department of Inorganic Chemistry, Technische Universität Chemnitz, Chemnitz, Germany
[2]Faculty of Electrical Engineering and Information Technology, Center for Microtechnologies, Technische Universität Chemnitz, Chemnitz, Germany

ABSTRACT

The inkjet printing of copper, silver and gold patterns using nanoparticle- and molecule-based inks will be discussed. Diverse applications including low-cost conductive structures for passive wiring, photovoltaic and near-field communication applications as well as microelectronic components are discussed.

1. INKJET PRINTING TECHNOLOGY

1.1. Introduction

During the last decade, the field of printed electronics has matured from a vision to a serious field of research in microelectronic engineering. In its self-conception, printing and coating techniques replace vacuum processing and lithographic structuring to achieve electronic components of increased complexity even on non-silicon substrates. Passive circuitry and wiring will soon be followed by complex logics such as RFID tags integrable into stretchable and bendable materials including textiles. Recently, inkjet printing has

---

* Corresponding author: E-mail: heinrich.lang@chemie.tu-chemnitz.de; Phone: +49(0)371 531 21210; Fax: +49(0)371 531 21219.

become recognized by a number of striking advantages over other printing techniques, because it allows the nearly loss-free deposition of small ink volumes (picoliter range) to obtain thin films (sub-micrometer thickness) in a patterned manner (feature size: several micrometer). Also, it is a contact-free technique that is suitable for diverse substrates, both rigid and flexible and for a wide range of production scales, from lab scale to prototyping to large-scale industrial production. [1] Foremost, inkjet printing exceeds all other printing methodologies with respect to its efficient handling of the appropriate inks and hence allows the deposition not only of graphical ink, but of fluids containing costly materials such as precious metals. The ignition for this conceptual idea can be assigned to the late 1980s, when Teng and Vest suggested a home-built inkjet printing system including a mechanical stage for the deposition of silver neodecanoate ink. Their aim was the patterning of silicon solar cells with current collector grids [2–4] and microcircuits. [5] Apart from inkjet printing, liquid precious metal inks have already been known since the 17th century for the decoration of ceramics such as bone china and porcelain. The first particle-based gold inks consisted of gold powder dispersed into natural resins. [6] The onset of metal-organic decomposition (MOD) inks dates back to the 19th century, when gold halides were used together with sulphurized terpenes or balsams. [6] Exposed to high temperatures, they turn into bright films on the croppery with eye-catching effects.

In this article, we describe the use of nanoparticle-based and MOD inks for inkjet-based generation of coinage metals copper, silver, and gold patterns on solid substrates.

1.2. The Workflow of Inkjet-Based Metallization

1.2.1. Technology Map of Inkjet Printing Technologies

Inkjet-based techniques are generally described as "the digitally controlled ejection of drops of a fluid from a print head onto a substrate".[7] Herein, the term "digitally controlled ejection" refers not only to the generation of drops inside the print head but also to their placement relative on the substrate usually by a precision-controlled mechanical stage. A typical drop formation process, an exemplary voltage pulse ("wave form") for piezo drop-on-demand inkjet printing and a general technology map of inkjet printing is shown in Figure 1. [1,7,8]

In continuous inkjet (CIJ), drops are formed at constant frequency without intermittency. Only drops that are part of the image pattern are directed electrostatically to the substrate, the others are collected for recirculation. Due to specific requirements concerning the electrostatic properties of the carrier fluid and the significant deterioration of the ink, due to evaporation and agglomeration, CIJ has not proven suitable for printed electronics and its costly ink concepts. In contrast, drop-on-demand (DOD) techniques have in common that they form and eject drops only when required by the creation of a pressure pulse. The particular method for this generation can be used to categorize DOD inkjet further: thermal expansion (TIJ), piezoelectric transduction (PIJ) and electrostatic acceleration (EIJ). In TIJ, drops are generated when a resistive element within the fluid chamber is digitally heated to 400 °C, vaporizing an adjacent small fluid volume. This process generates a bubble and it leads to a pressure wave that is able to create a drop at the nozzle orifice. Although there are certain advantages of TIJ including high integration density of the print head and low costs of associated hardware, the drawbacks for the usage of metal inks prevail as they typically do not withstand the local high temperatures.

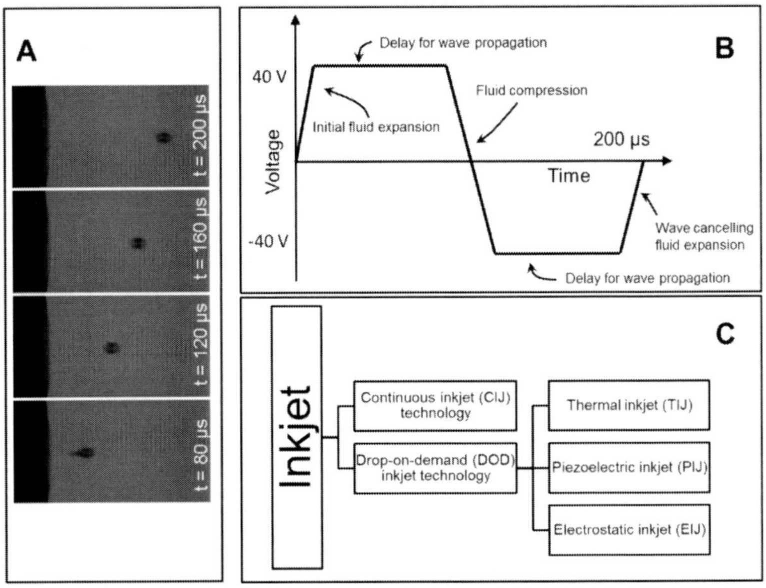

Figure 1. (A) Typical drop formation process of an aqueous silver(I)-2-[2-(2-methoxyethoxy)-ethoxy]acetate precursor ink [12] from the nozzle of a piezo drop-on-demand inkjet system. (B) Qualitative scheme of a bipolar wave function for the actuation of a piezo drop-on-demand print head. [1] The values indicate typical orders of magnitude. (C) Technology map of inkjet printing. [7,8]

In PIJ, a pressure pulse is generated in the fluid reservoir by a piezoelectric transducer the voltage signal of which is designed to create and eject a drop from the reservoir. Compared to TIJ, the print heads represent a microelectromechanical system (MEMS) with higher complexity and hence costs due to lithographic patterning. [7] In EIJ (or electrohydrodynamic inkjet, EHIJ), drops are generated and accelerated in an electrical field between the nozzle orifice and the substrate. [9] Although the minimal feature size is two orders of magnitude lower than the nozzle orifice, the integration of these printers in scalable production is limited. PIJ gives the highest degree of freedom for ink developers and is the method of choice in functional printing.

1.2.2. Functional Inkjet Printing

In graphical printing, color reproduction is achieved by a subtractive mixing of incident and remitted light. This process takes place in adjacently deposited dots, and the color quality is present right after the deposition, before the drying step. In contrast, the patterns of metallic ink remain not functional, that is conductive, at all right after their deposition. In order to describe functional printing, a workflow diagram for metallization by metal-organic decomposition (MOD) inks is shown in Figure 2. [10] Following drying of printed ink dots on the substrate, a number of post-printing techniques to transform the printed ink into conductive layers was investigated. The particular treatment of pre-functional ink layers after their deposition by printing is hence a crucial step as only doing so provides the printed products with the desired electrical, chemical, or physical properties. [11]

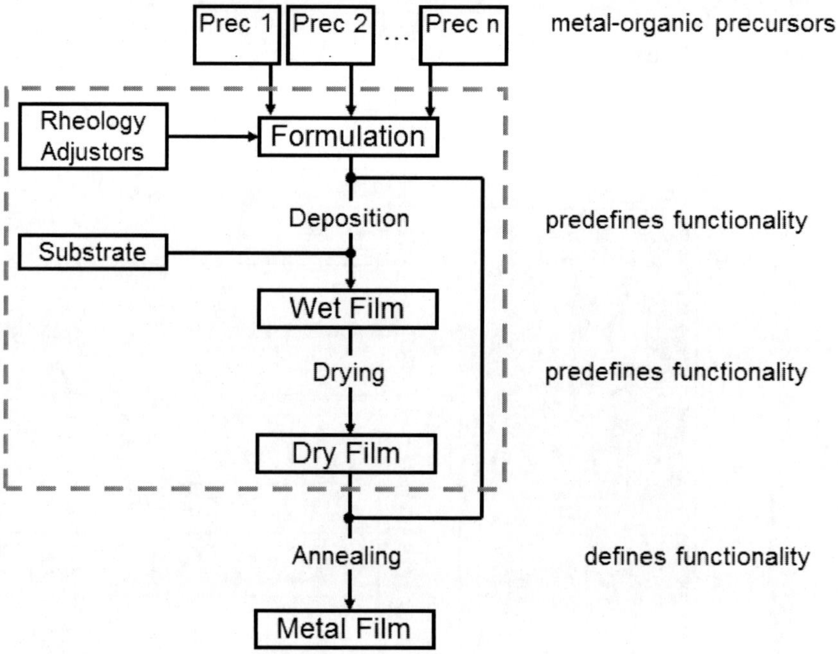

Figure 2. Workflow diagram of functional printing for the example of metallization by MOD inks. [10] Process steps inside the dashed box are related to graphical printing. [11]

In PIJ, a pressure pulse is generated in the fluid reservoir by a piezoelectric transducer the voltage signal of which is designed to create and eject a drop from the reservoir. Compared to TIJ, the print heads represent a microelectromechanical system (MEMS) with higher complexity and hence costs due to lithographic patterning. [7] In EIJ (or electrohydrodynamic inkjet, EHIJ), drops are generated and accelerated in an electrical field between the nozzle orifice and the substrate. [9] Although the minimal feature size is two orders of magnitude lower than the nozzle orifice, the integration of these printers in scalable production is limited. PIJ gives the highest degree of freedom for ink developers and is the method of choice in functional printing.

1.3. Ink and Substrate Requirements for Inkjet-Based Patterning

1.3.1. Ink Requirements

The processability of an ink through the print head is determined by its aptitude to form spherical droplets at the nozzle orifice. To this end, maximum values for the viscosity (0.04 Pa·s) and minimum values for the surface tension (0.02 N/m) of the ink are reported [8], minimizing creation of detrimental satellite drops or filaments. Fluids with at least nearly Newtonian behavior are the preferred carrier liquids. Another criterion for the ink is its stability against evaporation at the fluid meniscus that has contact with the atmosphere at the nozzle orifice. [12] The evaporation rate has to be sufficiently low, which is sometimes in contradiction to the requirement of fast drying droplets on the substrate. For particle inks, the size of the dispersant should not exceed 5% of the nozzle orifice diameter as aggregation of the constituents may lead to instability of the drop generation. [8]

1.3.2. Pattern Formation on the Substrate

The dynamics of film formation is governed by physical properties like the interface energies between involved players: substrate, ink, surrounding atmosphere, and temperature distribution within the droplet affect the evaporation dynamics of the solvent. Upon their impingement on the substrate, the spherical drops originating from the inkjet nozzle will turn into sessile droplets with the shape of a distorted spherical cap. [13,14] Features that may occur during droplet drying are coffee staining (spatial gradient in the evaporation rate) and Marangoni flows (spatial gradient in the surface tension or temperature profile). To get some degree of control, binary or ternary mixtures of solvents with different boiling points can be employed.

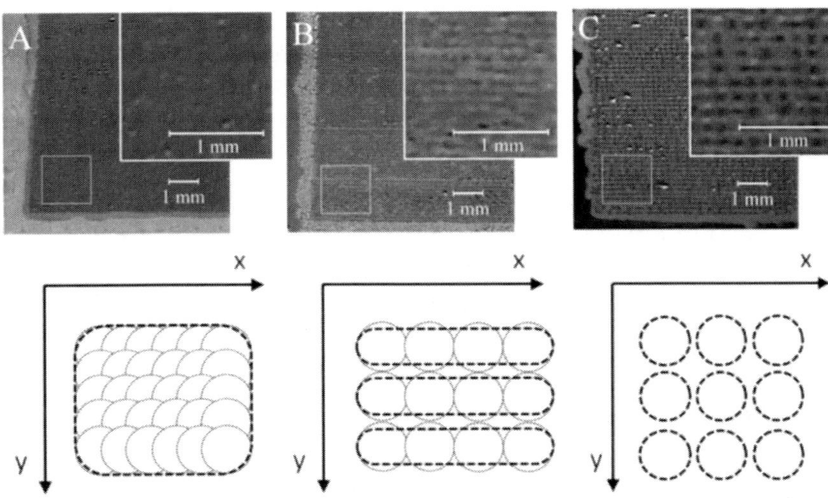

Figure 3. Top: Images of three different coalescence regimes for deposited mesitylene ink droplets on SiO_2 scaling with the chosen drop spacing: (A) d = 80 μm (2D film), (B) d = 120 μm (line), and (C) d = 160 μm (isolated droplets). Bottom: Graphical explanation of the three coalescence regimes (A-C). The origin of the x- and y-axes represents the printing origin. The deposition starts with the first droplet in the uppermost line in x-direction, proceeding after a line feed in y-direction with the subsequent lines. Due to the difference in delay times in x- and y-direction, the choice of the drop spacing d is critical with respect to d_{cx} and d_{cy} that are different in both directions. (A) A 2D film is formed for $d < d_{cx}$, (B) linear liquid beads are formed for $d_{cy} < d < d_{cx}$, (C) isolated droplets are formed for $d > d_{cx}$. Reproduced from Belgardt, C.; Sowade, E.; Blaudeck, T.; Baumgärtel, T.; Graaf, H.; von Borczyskowski, C.; Baumann, R. R. Phys. Chem. Chem. Phys, 2013, *15* (20), 7494-7504 with permission of the RSC.

Depending on the time delay between the impingements of neighboring drops (as given by the firing frequency of the print head) and their center-to-center distance (related to the distance between the respective inkjet nozzles), the result will be individual droplets or a film of droplets coalesced with one or more neighboring counterparts. For inkjet-based metallization, the formation of 1D ('beads', 'lines' for wiring) and 2D structures ('films' for electrodes, conductive pads etc.) is of primary interest. As introduced by Soltman and Subramanian for polymeric ink materials, printed lines can be 'scalloped', 'uniform', 'bulged', or like 'stacked coins', depending on the chosen drop spacing, firing frequency, and drying times. [15] For industrial inkjet printing, the choice of the print direction affects the quality of a printed feature. Figure 3 explains the procedure of forming zero-, one- and two-dimensional structures for a single-nozzle printer. The concept can be generalized for print heads that consist of an array of nozzles in x- and y-direction.

1.4. Post-Printing Techniques

1.4.1. Drying

Although often underestimated in its significance, the process of solvent drying should be conceptually separated from the functionality forming steps. Evaporative drying is related to a conversion of the printed wet layer from liquid to solid and release of the solvent to atmosphere. The drying process is determined by the heat and material transport in the boundary film on the liquid-gas interface. The main parameters for the drying speed are the surface temperature, the speed of the gas molecules parallel to the substrate surface and the partial pressure difference. Evaporative drying is accelerated by an enforced convection. Following the solvent drying, there can be steps to annihilate other additives such as surfactants, dispersants, humectants, and adhesion enhancers present in the ink.

1.4.2. Annealing and Sintering

'Annealing' or 'sintering' are processes of traditional metallurgy, related to the manufacture of solid-state matter as steel or ceramics. Whilst 'annealing' relates to a general process that includes heating and subsequent cooling invoking desired phase transitions in the material to make it harder or less brittle, 'sintering' is particularly related to the coalescence of powders under the influence of heat, forming strong solids. [16] In printed electronics, both terms are used synonymously. In analogy to the metallurgical sintering concept, across a printed film-like feature of nanoparticle ink, conductivity will be given only if the particles form electrical contacts along a continuous percolating path. Printed electronic has put forward several post-printing concepts, the most common of which are to date thermal, flash, and chemical sintering.

Thermal sintering. Whilst the metallization of inorganic solids, *i.e.,* porcelain, from the liquid phase is traditionally followed by a high-temperature curing step, polymeric substrate materials in printed electronics (polyethylene terephthalate, polycarbonate, or polypropylene) limit thermal sintering to 120-170°C, [17] while polyimides (e.g., Kapton) allow conversion temperatures of 200°C. [18] For nanoparticle inks, thermal treatment removes the organic surfactants and establishes electrical contact between the nanoparticles. [19] In the case of MOD inks, thermal treatment triggers the decomposition of the metal-organic compounds, releases the organic surfactants, initiates the growth of metal nanoparticles by Ostwald ripening, and has them coalesce. [20,21] As a dynamic counterpart, rapid thermal annealing (RTA) using temperature bursts of a duration of seconds [22] has been successfully applied to inorganic materials like chalcogenides for the fabrication of low-cost silicon solar cells. [23] It benefits from shorter cycle times, reduced thermal exposures, and higher temperature uniformities.

Flash sintering. Flash sintering, also known as photonic curing, uses intense light pulses of microseconds durations which are selectively absorbed by sub-micrometer surfacial layers of deposited ink, preventing detrimental thermal impact on the substrate. The respective technology was brought to market by NovaCentrix. [24a] Kang et al. report about inkjet-printed curing of CuO layers on porous PET in less than 1 s.[24b] Occasionally, flash sintering turned out to be effective in combination with other sintering techniques. Perelaer et al. combined flash sintering with microwave techniques and reported a conductivity of 40% of bulk silver within 15 seconds of exposure for silver nanoparticle inks. [25] Recently, Galagan et al. compared sintering of inkjet-printed current collector grids for organic photovoltaic cells and reached similar conductivities after 5 s of flash sintering and 6 h of thermal sintering. [26]

Chemical sintering. Apart from physical methods, there are techniques with which the final functionality of a printed precursor layer is achieved through a chemical reaction. Primarily applied to MOD inks, aqueous solutions of salts and reducing agents are printed consecutively on top of each other to initiate an immediate chemical reduction of the metal cations to small metallic particles. [27] The first application of this procedure was claimed in 1996 by Agfa-Gevaert, who produced a lithographic printing plate by inkjet printing using a silver MOD ink and a reducing agent on aluminum. [28] Recently, the usage of ethylene glycol vapor to a printed silver nitrate solution was reported, reaching ca. 1.9% of bulk silver conductivity. [29] Apart from these techniques, laser sintering, [30–33] microwave sintering, [34–37] plasma sintering, [38,39] and electric sintering have been reported. [40,41] Table 1 summarizes all annealing techniques discussed here with respect to their performance and integration aspects.

Table 1. Summary of annealing techniques for printed thin films of metal at the example of silver

Annealing technique	Refs.	Electric performance a)	Scalability to large volume b)	Scalability to large area	Thermal impact on substrate	Technical effort	Safety & Environmental issues
Thermal	17-21	very high 60	very low 1000	high	high	low	low
RTA	22,23	no report --	high 1	high	medium	high	low
Flash	24-26	high 40	very high 0.01	high	low	medium	low
Chemical	27-29	very low < 2	medium 10	medium	low	low	very high
Laser	30-33	medium 20	high 1	low	high	high	medium
Micro-wave	34-37	low < 10	low 100	low	high	very high	medium
Plasma	38, 39	high 40	medium 10	medium	medium	high	Low
Electric	40, 41	very high 60	very high 0.01	medium	medium	medium	medium

Legend: a) reported final conductivities for printed Ag inks (in% Ag bulk silver); b) approximate duration per cm^2 printed ink (order of magnitude, in sec)

The complexity and importance of the interplay between different ink concepts, their printability and film formation properties as well as the final electrical performance on the substrate is documented in Table 2. An increase in the silver content in a metal-organic decomposition (MOD) ink may lead to a lower final conductivity due to an insufficient film formation on the substrate. However, a binary ink composition adding a surfactant to an aqueous solution of the silver precursor restores (and even outnumbers) the initial electric performance.

Table 2. Processability, morphological and electrical properties of inkjet-printed layers at the example of silver(I)-2-[2-(2-methoxyethoxy)-ethoxy]acetate as precursor of an aqueous metal-organic decomposition ink concept

Ink composition	Refs.	Printability on piezo drop-on-demand inkjet printer	Contour definition after thermal sintering	Layer homogeneity after thermal sintering	Electric performance a)
23 wt% silver precursor in water	20, 21	good	good	good 300 ±100	high 43 ± 7
40 wt% silver precursor in water	12	good	insufficient	good 300 ± 80	medium 28 ± 13
40 wt% silver precursor in 99 wt% water and 1 wt% sodium dodecyl sulfate	12	good	good	good 300 ± 50	low 53 ± 28

Legend: a) after thermal sintering; b) reported approximate thickness and roughness (in nm); c) reported final conductivities for printed Ag inks (in% Ag bulk silver)

2. Conductive Inks of Cu, Ag and Au

2.1. Introduction

Ideal metal-containing inks are cheap, stable, easy to handle and result in highly conductive metal structures and hence copper, silver and gold inks are best suited for many applications. In this respect, two different main concepts of "conductive metal inks" are used today: a suspension of metal nanoparticles in a suitable solvent and a solution of metal-organic compounds. Whereas each type of inks has different advantages and disadvantages, both inks require a further post-printing technology to convert the preliminary functional ink layer into a conductive metal structure. In the following, both types of group-11 (Cu, Ag, Au) inks will be discussed with main focus on MOD inks.

2.2. Nanoparticle Inks

In nanoparticle containing inks, metal particles are suspended either in water or in an organic solvent, mainly toluene or ethylene glycol. These solvents evaporate at comparatively low temperature after the printing process, but prevent to dry out at the nozzle. The advantage of nanoparticle-containing inks over MOD inks is their high metal content (Chapter 2.3.).

This was one driving force for their development towards commercial ability to date (suppliers: e.g., InkTec, Sun Chemical, American Elements) and wide usage in printed electronics.

Since nanoparticles tend to agglomerate, they have to be stabilized by adding, for example, polyethylene glycols, amines or amides. However, these stabilizing components must be removed after printing to establish high conductivities. Furthermore, the organic stabilizers have influence on the sintering temperature and viscosity of the particles. [42,43] To prevent agglomeration and thereby the clogging of the nozzle, a particle diameter less than 1% of the nozzle diameter is recommended. [44] Thus the nozzle diameter and therefore the minimal feature size of the printed pattern are limited.

2.2.1. Silver Nanoparticles Inks

Presently, the most reported metal inks are silver inks, due to their better oxidation stability compared to copper and their lower cost regarding gold nanoparticles. Also the high conductivity of the bulk material favors silver as the perfect metal for printing conductive structures. The enormous interest in silver nanoparticles also created a vast diversity in protocols for their generation and stabilization. In majority, they were synthesized by reacting, for example, [$AgNO_3$] with a reducing agent in presence of a stabilizer in an appropriate solvent. [45] Therefore, a wide range of reductive reagents were investigated. For example, sodium borohydride is used extensively for the preparation of silver nanoparticles. Environment-friendly reducing agents are polysaccharides. [46] In addition, the stabilizing component is of importance for a stable colloidal suspension. Two kinds of common used stabilizers exist. Polymeric stabilizers such as poly(vinylpyrolidon) (PVP) and poly(sodium styrenesulphonate) (PSS), which allow the generation of nanoprisms [47] and hydrophobic alkyl chains bearing polar endgroups like dodecyl amine or oleic acid, allowing a large scale procedure ($100\ g \cdot L^{-1}$ of reaction volume) for synthesizing monodisperse silver nanoparticles in oleic acid. [48] Both methodologies are used equally. Huang et al. reported in 2005 the preparation of a silver nanoparticle ink by stabilizing the particles with PVP and formaldehyde (Figure 4), with a silver content of 25 wt%. [49] Sintering temperatures below 150 °C increase the time for formation of low resistive structures, while at 260 °C the lowest resistivity (16 μΩcm, Bulk silver: 1.51 μΩcm) was achieved after 3 minutes.

A different approach to silver colloid inks was chosen by Moon et al. [50] using the polyol method for particle generation. [51] They also investigated the influence of a high boiling solvent incorporated in the ink formulation. By addition of ethylene glycol a decrease of resistivity was observed (3.5 μΩcm). In contrast to the PVP-stabilized colloids, [49,50] Zhou et *al.* investigated the influence of a long hydrocarbon chain with different linking groups [52]. Dodecylamine (DDA) and dodecanthiol (DDT) were incorporated during a reduction agent-driven synthetic methodology.

After printing these particles on a polyimide substrate the sheet resistivity was measured (DDA, 2.9 μΩcm, DDT, 18 μΩcm). This allows quantifying the influence of the amino or thiol linking group, which is bonded to the metal surface, explaining these behaviors with a stronger silver-sulfur interaction. Therefore, higher energies would be needed to cleave the Ag-S bond, resulting in higher annealing temperatures (Figure 5). A different approach was chosen by Cross and coworkers. [53] They decreased the amount of carbon atoms needed for stabilization using acids with 6, 8 or 10 carbons and mixtures thereof. The hexanoic and decanoic acid mixtures showed a rather high conductivity of 57% pertaining of bulk silver, while only decanoic acid reaches 46%. At an annealing temperature of 250 °C a conductivity of 87% regarding to bulk silver was obtained.

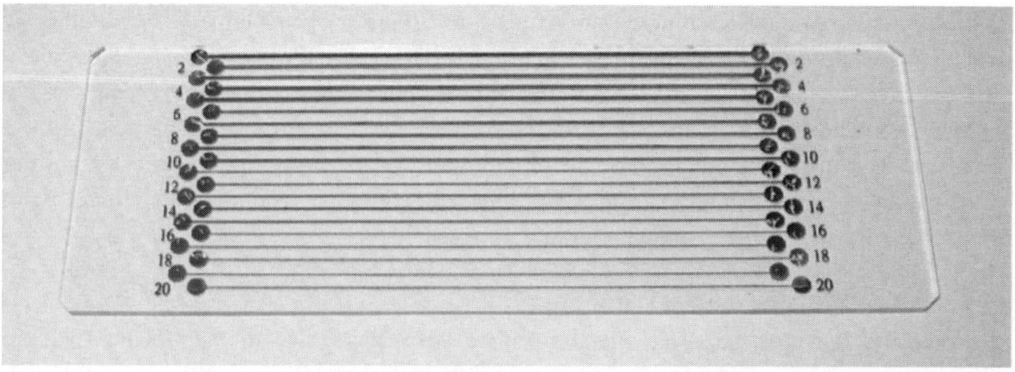

Figure 4. Test patterns having various line thicknesses on glass printed with a Ag-NP ink. Reproduced from Lee, H.; Chou, K. Huang, K. *Nanotechnology*, 2005, *16,* 2436-2441, with permission from IOP Publishing.

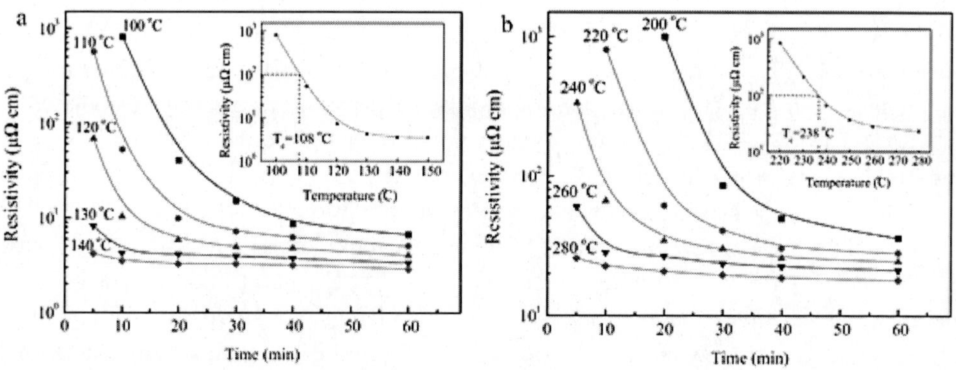

Figure 5. Electrical resistivities of inkjet-printed Ag-DDA (a) and Ag-DDT (b) nanoparticle inks after heating at different temperatures for various times. Reproduced from. Mo, L.; Liu, D.; Li, W.; Li, L.;Wang, L.; Zhou, X. *Appl. Surf. Sci.* 2011, *257*, 5746-5753, with permission from Elsevier.

2.2.2. Gold Nanoparticle Inks

The combination of high conductivity with oxidation stability and its biocompatibility, makes gold a prospective material for, for example, printed electronics, nanomedicine [54–56] and sensing applications [57].

Typical precursors for synthesis of gold nanoparticles are $H[AuCl_4]$ or its hydrate $H[AuCl_4] * 3\ H_2O$ which can be converted to gold nanoparticles by treatment with a reducing agent. A very common method is the Turkevich methodology [58] using citrate as reducing and stabilizing reagent (Scheme 1).

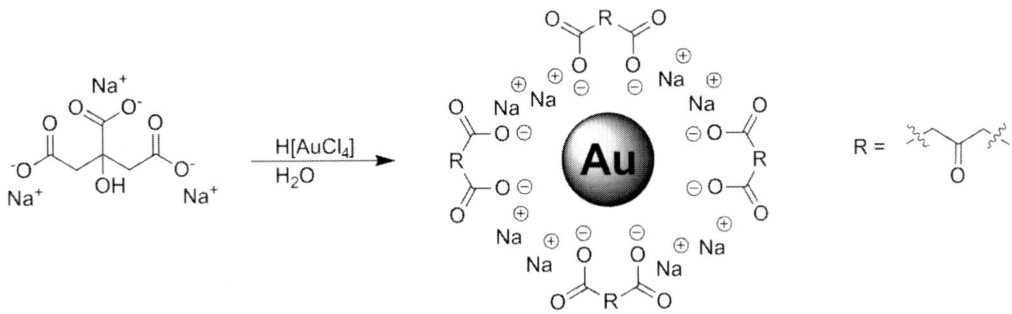

Scheme 1. Au-NP formation and stabilization using the Turkevich method. According to Zhao, P.; Li, N.; Astruc, D. *Coord. Chem. Rev.* 2013, *257*, 638-665, with permission from Elsevier.

A further synthetic route is the two-phase Brust-Schiffrin [59] method allowing the preparation of thiol-stabilized nanoparticles, which are the most used particles for inkjet printing processes. Furthermore, water soluble and biocompatible thiol-capped Au-NP could be prepared (Figure 6). [60] Another interesting method to gold nanoparticle formation and stabilization is the usage of single-source (= all-in-one) precursors. They allow the generation of Au-NPs without the addition of a reducing agent or a stabilizing component by simply heating the precursor molecule. In this respect, several phosphane gold ethylene glycol carboxylates were recently reported by Tuchscherer and Lang. [61,62] Thermal decomposition of these precursors in various solvents led to the formation of 3 – 7 nm nanoparticles (Scheme 2). These precursors allow the formulation of MOD inks (Section 2.3.3). In addition, several other synthetic routes for generation of Au-NPs exist. [63]

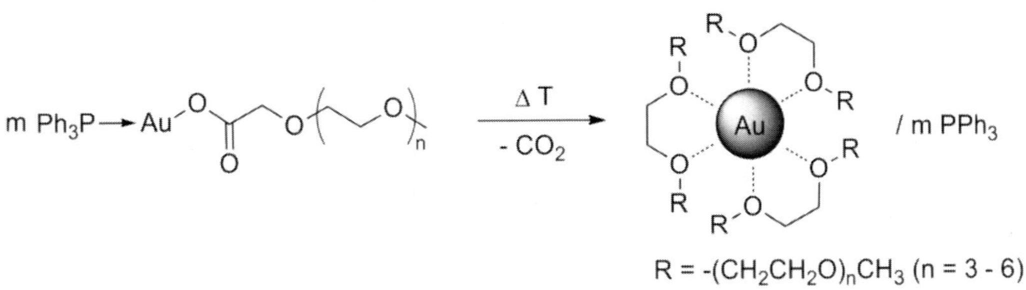

Scheme 2. Thermal induced generation of Au-NP from $[Au(PPh_3)(O_2CCH_2(OCH_2\ CH_2)_nCH_3)]$ and schematic view of stabilized Au-NP by the *in situ* generated ethylene glycol matrix. Reproduced from Tuchscherer, A.; Schaarschmidt, D.; Schulze, S.; Hietschold, M.; Lang, H. *Dalton Trans.* 2012, *41*, 2738-2746 with permission of *RSC*.

Toluene as solvent for Au-NP ink formulations allows high particle concentrations of 15 to 35 wt%. [64–66] The particle size vary from 2 to 5 nm, [49,66–68] but also printable inks with particle sizes up to 20 nm [69,70] are known. Furthermore, inkjet printing of commercially available inks with particle sizes of 100 nm is possible. [71] The organic shell commonly consists of thiols (e.g., 1-dodecanthiol, 1-butanthiol) or citrate and conversion of the printed structures to electrically conductive layers by elimination of these stabilizers requires typical temperatures of 100 - 300°C. [69,72] In contrast, treatment of a 1-butanethiol-stabilized gold nanoparticle ink with an oxidizing agent affords conductive films at ambient temperature. [73] A water-based ink using mercaptosuccinc acid as stabilizer was presented by Yamauchi et al. [69] The ink was sonicated and conductive films with resistivities of 45 μΩcm were obtained after drying at 25°C without any post-treatment. In addition to the citrate or thiol-stabilized nanoparticle inks, a double layer polymer stabilized ink was reported. [65] Several inkjet-printed patterns of 1 to 50 layers were sintered at 500 °C and highest conductivity was found for the 50-layer pattern (Figure 7).

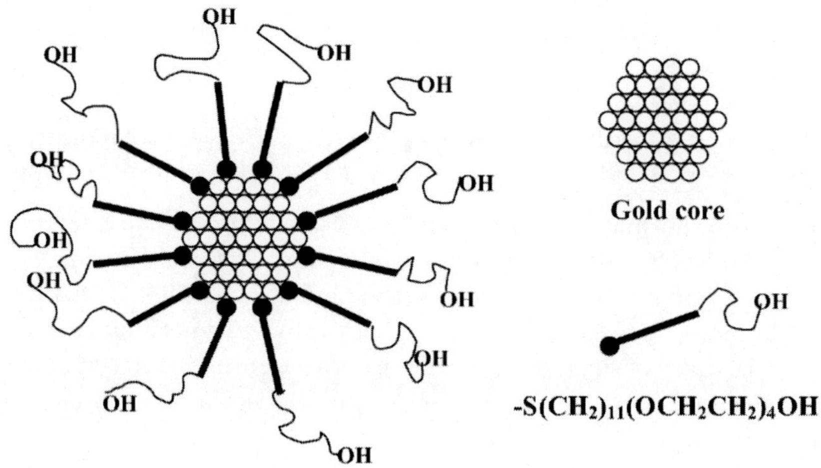

Figure 6. Schematic representation of Au-NP protected by a $-S(CH_2)_{11}(OCH_2CH_2)_4OH$ monolayer. The hydrophobic C11-chain confers stability to the cluster, while the hydrophilic tetraethylene glycol unit ensures solubility in water. Reproduced from Kanaras, A. G.; Kamounah, F. S., Schaumburg, K.; Kiely, C. J.; Brust, M. *Chem. Comm.* 2002, 2294-2295 with permission of the RSC.

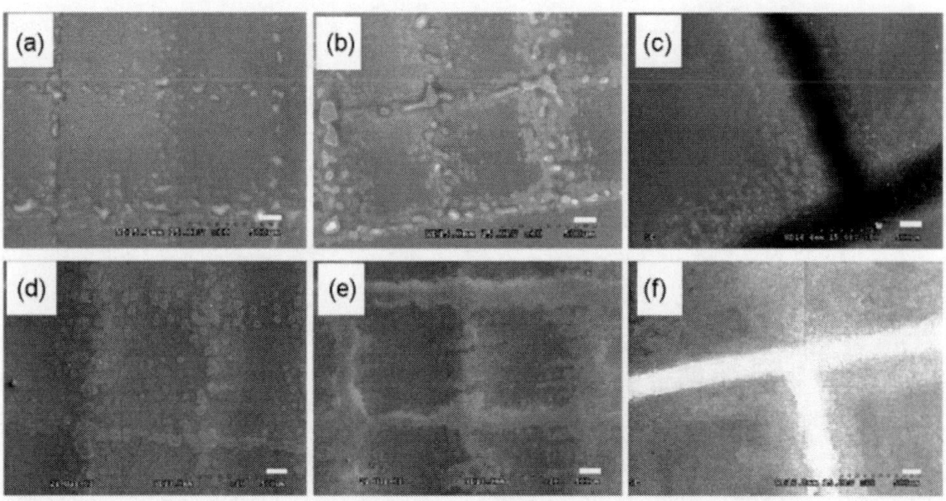

Figure 7. (a), (b), (c) show SEM-images of 1-, 5- and 50-layer patterns at 25 °C; (d), (e), (f) after sintering at 500 °C for 3 h. (white bars = 100 μm). Reproduced from Cui, W.; Lu, W.; Zhang, Y.; Lin, G.; Wei, T.; Jiang, L. *Colloids Surf. A*, 2010, *358*, 35-41 with permission from Elsevier.

2.2.3. Copper Nanoparticle Inks

Because of its high electrical conductivity ($1.72 \cdot 10^{-8}$ Ωm) in combination with its low price, copper is a worthy alternative to silver and gold. [74] However, in contrast to silver and gold, its low standard potential ($Cu \rightleftharpoons Cu^{2+} + 2e^-$, $\varepsilon_0 = 0.34$ V) allows surface oxidation of the metal in air (e.g., by the half-reaction $O_2 + 2\ H_2O + 2e^- \rightleftharpoons 4\ OH^-$, $\varepsilon_0 = 0.401$ V). The resulting oxidation sensitivity seriously limits the application of Cu–NP inks, especially since structures produced from inks with surface-oxidized nanoparticles are known to show only inferior conductivities and demand high sintering temperatures. [75]

Commercial available copper nanoparticles [76–78] with sizes between 5 and 20 nm have been produced by plasma methods, [78] flame synthesis, [79] or chemical reduction [76]. Depending on the fabrication process, the particles can be coated with carbon [78], aliphatic chains [76] or polymers. In lab-synthesis, Cu-NPs are usually prepared by reduction of a copper salt (*e.g.*, copper(II) acetate) with a reducing agent, such as hydrazine, sodium borohydride or polyols in the presence of a stabilizing component. Other preparation methods include spray pyrolysis [80], sputtering approaches [81], sonoelectrochemical processes [82] or thermal decomposition of precursor complexes. [83,84] A review concerning different stabilizers has recently been published in which polyvinylpyrrolidone with high molecular weights is most recommended for printing applications. [85] The preparation of samples consisting of monodisperse Cu-NP with size distribution maxima ranging from 5 nm [86] to 65 nm [87] is possible, but also polydisperse samples can be prepared and have been used for inkjet printing. [88,89] The application of Cu-NPs ink formulations often facilitates a high metal content, e.g., 50 wt%. [89] For the printed structures, resistivities of 15 – 40 μΩcm [77,90] (12% or 4.3% of bulk copper conductivity) are characteristic, but also values as low as 5 μΩcm (35%) have been found. [91]

The mentioned oxidation sensitivity of the Cu-NPs is considered to be the critical factor in obtaining structures with high electrical conductivity. Furthermore, it limits the possibility to store the inks for a longer time. On the other hand, many copper(I) and copper(II) complexes are known to be stable species, which can be stored for years. Thus, especially for copper, the use of MOD inks can be very advantageous.

2.3. Metal-Organic Decomposition (MOD) Inks

Instead of dispersed nanoparticles it is also possible to apply metal-organic solutions based on carboxylates, amines or acetylacetonates. Because of the absence of any particles, the risk of agglomeration and nozzle clogging is negligible.

Similar to nanoparticle inks, MOD inks require a thermal sintering to convert the appropriate metal-organic species into nanoparticles, resulting in conductive metal thin patterns or films. Also laser-assisted and microwave sintering, electrical sintering, UV radiation and chemical conversion can be employed as metallization processes for the printed metal-organic structures and layers (Chapter 1.4).

2.3.1. Copper MOD Inks

In contrast to Cu-NP inks, many copper MOD inks are very stable and can be stored under ambient conditions without the risk of oxidation or decomposition. Among various compounds which have been investigated as precursors for MOD inkjet printing, two classes of substances are dominant, which are copper(II) carboxylates and copper(I) β-diketonates. Copper(II) formate (tetrahydrate) is the most investigated precursor for inkjet printing, [92–

94] and other printing techniques [95–99]. Its decomposition temperature is with 150 °C exceptional low, and the formation of solely volatiles during decomposition (e.g., CO_2, CO, H_2O) (Scheme 3) makes the compound an ideal precursor for ink formulation. However, the limited solubility of the compound (in water up to 3.1 wt.% Cu) usually requires the addition of ligands, such as ammonia or short-chained aliphatic amines, in order to obtain an ink with sufficient metal content (10 - 25 wt%). [92–94] The mentioned additives are known to react chemically with copper(II) ions, however, the reaction products do not distinguish from copper formate regarding their decomposition process.

$$[Cu^{II}(H_2O)_4(O_2CH)_2] \xrightarrow{\Delta T} Cu^0 + CO_2 + CO + H_2O + ...$$

Scheme 3. Thermal decomposition of copper(II) formate.

Copper(II) formate is commercially available at low cost, and can be prepared by electrochemical means [92–94] or from Cu_2O and H_2COH [97]. Water is typically used as solvent, which helps to reduce the environmental impact and allows a safe printing process. Solubility-mediating additives like ethylene glycols can be added to adjust the inks viscosity. [100]

In a typical printing process, e.g., following the procedure reported by Choa et al. [92], an aqueous ink with *ca.* 15% metal content was printed on polyimide. Heating to 150 – 200°C resulted in the evaporation of the solvent, followed by reduction of the Cu(II) ions to metallic copper by decarboxylation and the vaporization of the decomposition byproducts. Metal formation can be supported by a reducing atmosphere, e.g., hydrogen. [93,101] The resulting structures consisted of oxide-free copper. Cross-sectional SEM images showed that the metal is well sintered without voids (Figure 8). The electrical resistance was measured to be in the order of 45 μΩ cm (3.8% of bulk copper conductivity). In further investigations the authors could show that plasma treatment of the polymer afforded a better wetability due to the more hydrophilic surface, [93] and a further optimization was possible by adding 3-aminopropyltriethoxysilane as adhesion promoter [94]. When comparing copper(II) formate with copper(II) citrate, it was found that the citrate decomposes at about 180 °C, yielding pure copper along with volatile species. The printing of an aqueous precursor solution was followed by reductive annealing at 150 – 250 °C, yielding dense Cu layers consisting of fine grains. However, no sufficient neck-formation was present and the resulting structures had high resistivities of 6.5 mΩ cm. Recently, a mixture of copper(II) formate and -citrate was used as an aqueous ink by Oh et al., who investigated sintering by intensive pulsed light (IPL).[100] Following the printing process, the patterns were exposed to light pulses with energies of 40 – 60 $J \cdot cm^{-2}$. Electrical conductivities close to that of bulk copper were obtained, which suggests that sintering by IPL can be superior to thermal methods.

An solution of copper(II) hexanoate in 2-propanol was successfully applied by Wagner et al. for the source/drain metallization of thin film transistors. [101] Following the printing process, the ink decomposed at 200°C *in vacuo*, which led to a mixture of copper and copper oxide. Subsequent annealing in a reducing atmosphere (H_2) led to the formation of metallic copper. Even though significant impurities were found (16 at.% C, 7 at.% O), the resistivity of *ca.* 10 μΩ cm (17.2% of bulk copper conductivity) is exceptional low for copper MOD inks.

Copper(II) carboxylates such as copper(II) 2-[2-(2-methoxyethoxy)ethoxy]acetate have recently been investigated since the ethylene glycol chains induce a high solubility both in water and organic solvents and allow to generate MOD inks of high metal content without the need of solubility-enhancing additives. By introduction of methyl and phenyl groups at the carbon atom in α-position to the carboxylato group the decomposition temperature of the complexes could be lowered to 174°C. An aqueous ink prepared from $[Cu[O_2CCH_2(OC_2H_4)_2OCH_3]_2]$ was printed on glass. By thermal decomposition at 300 °C, copper structures with only minor carbon and oxygen impurities and a conductivity of $8.1 \cdot 10^6$ S · m^{-1} (14% of bulk copper conductivity) were obtained without the need of reductive sintering. [102]

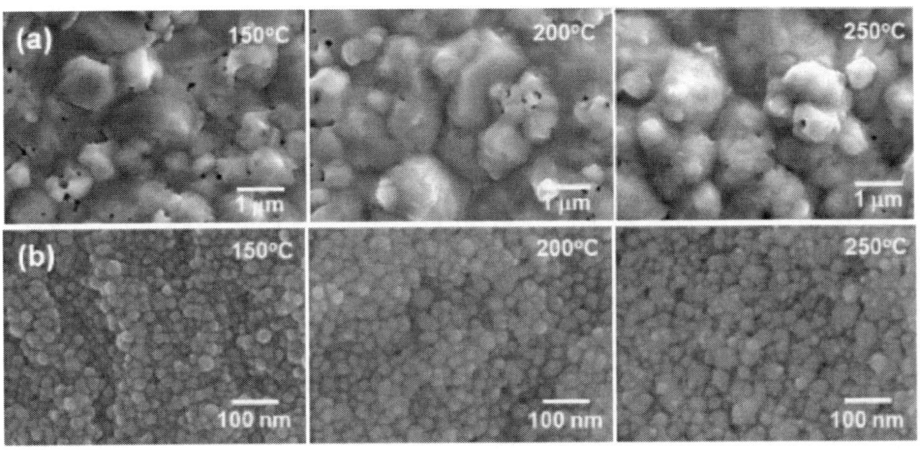

Figure 8. FE-SEM images of patterns prepared from inkjet-printed (a) copper(II) formate and (b) copper(II) citrate ink as function of heat-treatment temperature. Reproduced from Lee, Y.-I.; Lee, K.-J.; Goo, Y.-S.; Kim, N.-W.; Byun, Y.; Kim, J.-D.; Yoo, B.; Choa, Y.-H. Jap. *J. Appl. Phys*. 2010, *49*, 086501 with permission of the JSAP.

Another important representative for MOD inks is copper(I) (vinyltrimethylsilane)-(hexafluoroacetylacetonate), [Cu(VTMS)(hfac)]. [103] Its straightforward decomposition at low temperature, its stability and high solubility in many solvents made it a worthy precursor for diverse copper deposition processes. Its successful application for inkjet printing was demonstrated in 2000 by Homma et al. [104] In this study, printing was performed by dropping a precursor solution on a heated surface. Copper formation was observed at temperatures as low as 100°C, even though higher conductivities were obtained at higher temperatures. The resistivity could be lowered to 45 µΩcm, which corresponds to 3.8% of the metals conductivity. The low conductivity was attributed to accumulation of the reaction byproducts and to the formation of discrete copper particles.

Since [Cu(hfac)(VTMS)] is a liquid, printing is possible without using a solvent. [105] Also complexes featuring two or four [Cu(hfac)] fragments, coordinating to one tetravinylsilane, e.g., tetraallylsilane triscopper(I) hexafluoroacetylacetonate [106] or polymeric copper(I) hexafluoroacetylacetonate poly(vinylsiloxane), were prepared. [107] However, their use as precursors for MOD inks produced only structures with high resistivities (200 µΩ cm, 0.8% of bulk copper conductivity), mainly due to impurities.

Reactive inkjet printing can be a noteworthy alternative for temperature sensitive substrates. No thermal decomposition is necessary, since the fellow-up byproducts can be removed by washing. Calvert et al. produced conductive copper patterns by contemporarily printing two separate aqueous inks, i. e., copper citrate and sodium borohydride, on substrates such as paper or poly(vinylidenfluoride). A conductivity of 3% of bulk copper could be achieved.

2.3.2. Silver MOD Inks

Due to its high conductivity, it becomes obvious why silver is so valuable for modern electronic applications. Therefore, the demand of convenient silver precursors, suitable for inkjet printing arises. Recently, Lewis et al. developed a colorless and transparent silver ink with low viscosity which makes it favorable for inkjet printing (Figure 9) using the modified Tollens process. [108,109] During the reaction of silver acetate, hydroxyl amine and formic acid a soluble silver(I) diamine complex is formed, which is stabilized by one molecule of formic acid. This ink contains about 22 wt% of silver. After printing the desired patterns on vinyl acetate, the substrates were annealed at 90°C. During this step the volatile amine ligands evaporate and the formic acid reduces silver(I) to silver(0). In addition to vinyl acetate, also other substrates like poly(ethylene terephthalate)-, polyimide-, glass- and cellulose-based materials can be applied as substrates in MOD inkjet printing. The finest pattern obtainable with this ink are of size 5 μm and regarding the nozzle diameter, it is magnitudes smaller then compared with nozzles need for particle containing inks, due to the clogging effect caused by particle sedimentation. The printed structures exhibit a rather high conductivity, which is as high as for bulk silver (6.25×10^5 S/cm). Lewis *et. al.* compared their ink with literature known particle-containing systems and found that the obtained structures derived from the nanoparticle ink have a lower conductivity as structures printed with the described MOD ink. [110] This effect was caused by the stabilizers, which were needed in particle inks, despite their higher silver content (55 – 75 wt%).

A modified approach was recently reported by adding 2-methoxyethanol, ethylene glycol and polyvinyl pyrrolidone influencing the printing abilities of the MOD ink. [111] Clear solutions were obtained which were printed on silicon nitride. After sintering these samples at 50°C, however, no silver could be detected, additional heat treatment at 350°C led to a successful decomposition of the ink and the respective silver patterns obtained were characterized by XRPD and EDX (Figure 10). The resistivity of these structures was determined to 3.3 μΩcm, which can be explained by voids resulting of an incomplete sintering of the silver structures. Therefore, an additional heat treatment at 500°C was performed resulting in a resistivity of 3.1 μΩcm, for comparison bulk silver 1.6 μΩcm. [112]

In contrast to solution stable silver(I) diamine complexes (vide supra), Vest and Teng [2,4] concentrated on silver(I) carboxylates as printable MOD inks. Due to the polymeric nature of silver(I) carboxylates these complexes are insoluble in most common organic solvents. Out of this, silver(I) neodecanoate was prepared. This carboxylate is highly soluble, due to steric effects of the ramose alkyl groups. The silver content of a xylene solution contains ca. 20 wt% of silver. In contrast to the reactive inks discussed above silver(I) carboxylates need higher temperatures for decomposition, i.e., silver(I) neodecanote decomposes at 175°C. Highest decomposition rates are observed at 230 °C as determined by thermogravimetric analysis.

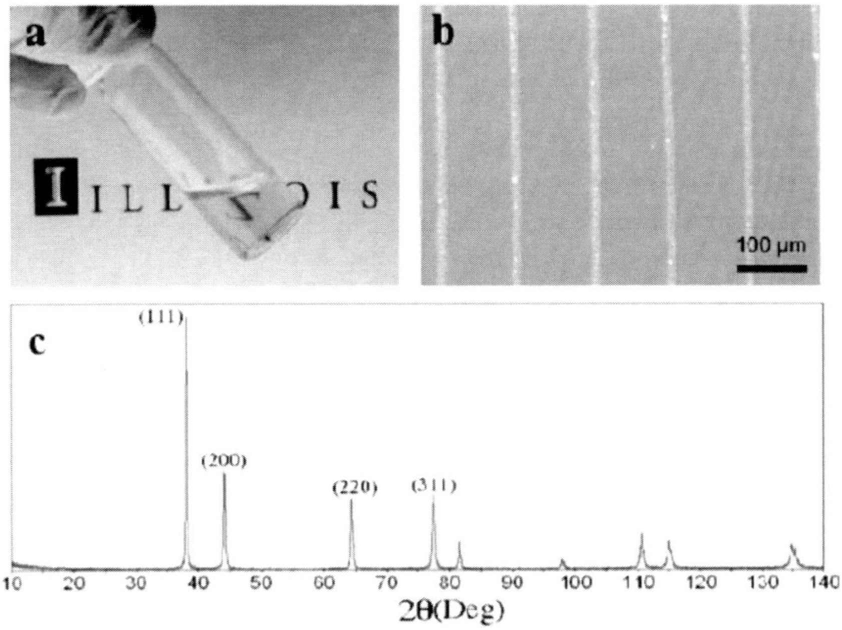

Figure 9. (a) Optical image of the reactive silver ink in a scintillation vial. (b) 1D array of conductive silver lines (~5 μm wide) printed on a silicon substrate via direct-write assembly using a 100 nm nozzle. (c) XRD patterns of films patterned from the reactive silver ink annealed at 90°C for 15 min. Asterisks denote peaks corresponding to [AgO_2CMe]. Reproduced from Walker, S. B.; Lewis, J. *J. Am. Chem. Soc.*, 2012, *134*, 1419-1421, with permission of the ACS.

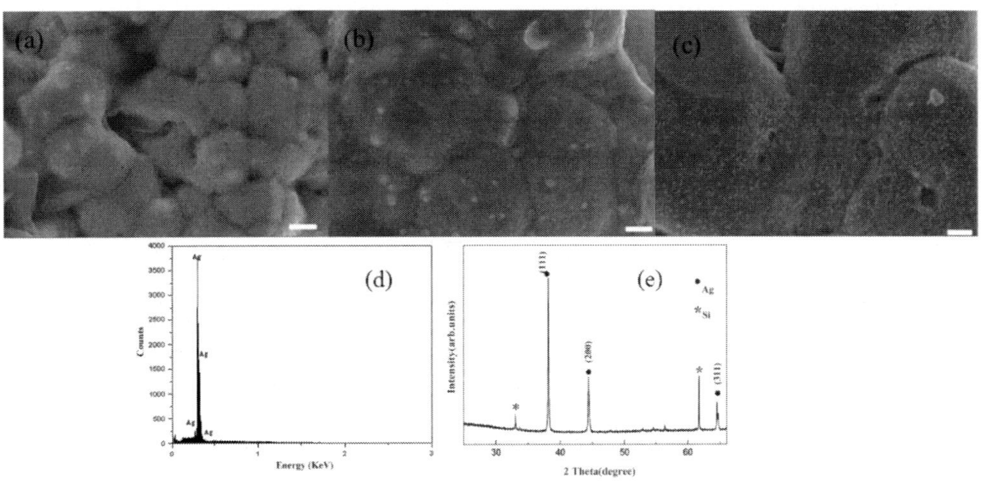

Figure 10. SEM images of inkjet-printed Ag films sintered at 500°C (white bar is 200 nm, 20 printed cycles): (a) 10 min (b) 30 min and (c) 60 min. (d) EDX of a Ag film sintered at 500°C for 60 min. (e) XRPD analysis of printed Ag patterns (8 printed cycles) after heating at 350°C for 5 min. Reproduced from Huang, Q.; Shen, W.; Song, W. *Appl. Surf. Sci.* 2012, *258*, 7384-7388 with permission from Elsevier.

A moderate resistivity was obtained for the printed silver lines. It also could be shown that the resistivity decreases by applying more silver layers. A short thermal treatment at 800°C also decreased the resistivity. Vest and Teng focused on the metallization of a silicon substrate, while O'Brien et al. printed on glass substrates. [113] They were able to show that an optimal decomposition temperature was 150°C and is best suited for the silver(I) decanoate-glass system. These silver tracks exhibit a resistivity of 3.8 x 10^{-8} Ωm. Increasing the temperature resulted in only small changes of this value, decreasing this temperature led to an enormous increase in resistivity, due to incomplete conversion of the carboxylate.

Despite silver(I) neodecante, which is only soluble in aromatic solvents, an example of a water-soluble and hence enviromentaly-friendly silver(I) carboxylate, $[AgO_2C(CH_2O CH_2)_3H]$, was prepared by Lang and Jakob. [20,21] The synthesis of this transition metal complex was achieved by mixing an ethanol/acetonitrile solution of $[AgNO_3]$ with triethylamine and 2-[2-(2-methoxyethoxy)ethoxy]acetic acid. The precipitating colorless silver(I) carboxylate was collected, washed, dried and redissolved in deionized water to obtain the appropriate ink, ready for printing. The prepared solution contains 9.1 wt% of silver and exhibits a slightly lower surface tension then pure water. Thermal decomposition of the silver(I) carboxylate was measured by thermogravimetric analysis showing two decomposition steps between 200 and 300°C (Figure 11). Printed structures on glass or silicon were annealed at 250°C exhibiting 43% of the bulk conductivity of silver. [20] The authors also applied this ink on polymeric substrates like PET and PI. For metallization a thermally assisted UV sintering technique combining photoinduced and thermal decomposition was employed. Decomposition of the precursor layer to a metal was achieved at a threshold temperature of 100°C which had to be maintained throughout the post-treatment process. The conductivity of those samples is comparable with those ones prepared on glass or silicon (2.7 x 10^7 S/m *vs.* 1.6 x 10^7 S/m). The addition of 1% sodium lauryl sulfate to the aqueous silver ink further improved the final conductivity to 3.2 x 10^7 S/m and yielded a better contour definition of the printed structures after sintering purely aqueous inks. [12]

All these methods do have in common that they carry one mol of silver per mol of compound. Increasing the silver ratio would increase the economic benefit of further printing processes. Therefore, Zou et al. investigated silver(I) citrate as a potential precursor. [114] The advantage of this system is that per mol compound three mol of silver can be generated. This complex was prepared according to a convenient ion exchange methodology starting from trisodium citrate and silver nitrate, respectively, the resulting silver(I) citrate was then resolved in methanol, isopropanol and 1,2-diaminopropane. Optimal decomposition temperatures were first determined by thermogravimetric analysis for silver(I) citrate and the ink thereof. While silver(I) citrate degrades at 190°C, the respective ink decomposes at 170°C. According to this analysis, a thermal treatment of the printed metal layers on glass and PET substrates were set to 150°C. At this temperature good sheet conductivities (20 μΩcm) were obtained of the silver films. Higher curing temperatures decreased the resistivity down to 5 μΩcm.

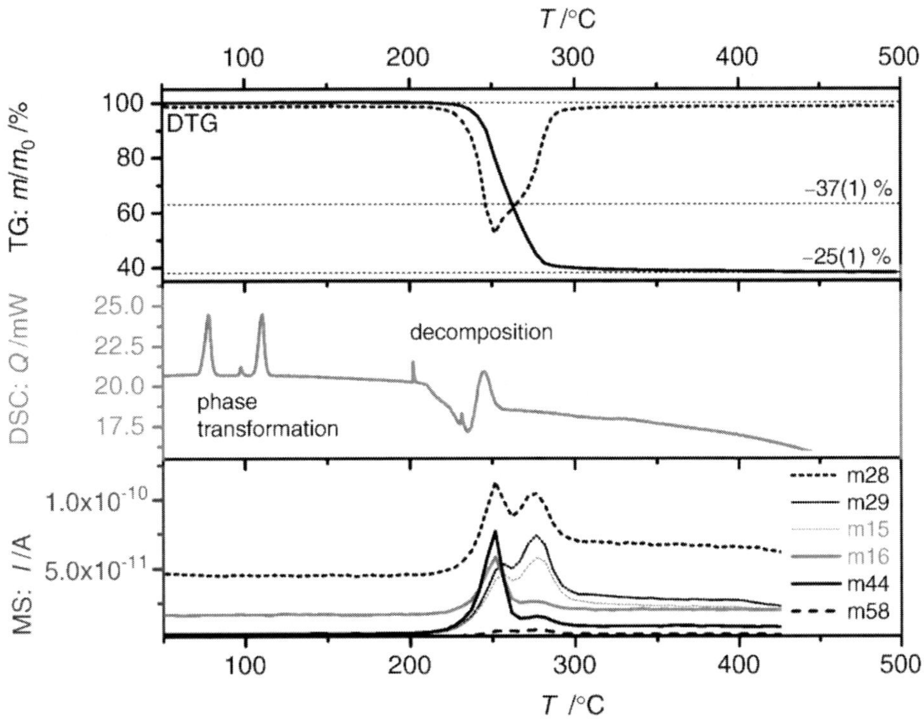

Figure 11. TG/DTG (black) and DSC (grey) traces of $[AgO_2C(CH_2OCH_2)_3H]$ in argon (flow rate 100 mL/min) with a temperature increase of 10 K/min complemented with mass spectrometric scans of the thermal induced decomposition of the precursor (m/z = 15 (CH_3^+); m/z = 16 (CH_4^+); m/z = 28 (CO^+); m/z = 29 (CHO^+); m/z = 44 (CO_2^+); m/z = 58 ($C_2H_2O_2^+$)). Reproduced from Jahn, S. F.; Jakob, A.; Blaudeck, T.; Schmidt, P.; Lang, H.; Baumann, R. R. *Thin Solid Film*, 2010, *518*, 3218-3222 with permission from Elsevier.

2.3.3. Gold MOD Inks

In contrast to copper and silver, the development of gold-based MOD inks is still in an early stage and today only a few publications exist. [2,115,116] Soluble gold-containing metal-organic precursors act as the metal source and therefore, real solutions can be printed. Potential compounds for this application are precursors which were used in chemical-vapor deposition processes, e.g., gold β-diketonates like $[Au(CH_3)_2R^1]$ (R^1 = acac (= $CH_3C(O)CHC(O)CH_3$), tfacac (= $CF_3C(O)CHC(O)CH_3$), hfac (= $CF_3C(O)CHC(O)CF_3$)) and carboxylates such as $[Au(CH_3)_2(CO_2R^2)]$ (R^2 = CH_3, ${}^tC_4H_9$). [117] Unfortunately, these compounds are difficult to handle due to their high reactivity. A different approach was followed by Nur et al. by usage of gold mercaptopropionyl glycine. [115] Because of its thiole group, mercaptopropionyl glycine can bind to heavy metals and is used as drug, e.g., for the treatment of cystinuria. [118] Furthermore, tiopronin was applied as stabilizing component for gold nanoparticles. [119] Nur et al. developed a printable MOD ink by stirring gold mercaptopropionyl glycine (31 wt%) with triethylamine in methanol. [116] The ink was printed on alumina and after drying heated in air to 580°C. Decomposition of the organo-gold ink starts at 200°C and is completed at 580°C. After the first step (200°C), sulfur was not present anymore in the sample as indicated by EDX analysis. The authors conclude that the second decomposition step (520°C) is probably associated with oxidative loss of carbon. However, electrically conductive samples were produced by heating to 420°C, i.e., already after the first decomposition step. Conductivities up to 10% of bulk gold could be reached. An alternative route for development of printable MOD gold inks is the usage of single-source (= all-in-one) precursors. These complexes were primary developed for application in nanoparticle synthesis (Section 2.2.2.). Known examples are gold(I) thiolate complexes [120,121] and phosphane-stabilized gold(I) alkyl-[122] as well as ethylene glycol carboxylates. [61,62] In case of the carboxylates the particles were formed by a thermally induced decarboxylation step. An organic matrix consisting of phosphane and ethylene glycol or alkyl chains stabilizes the formed particles. For inkjet printing of electrically conductive patterns, the elimination of this stabilizing organic shell is necessary. To investigate the decomposition temperature of the organic materials, thermogravimetric analysis was performed. Coupling this analysis with mass spectrometry facilitate the study of decomposition processes. A MOD ink was recently reported by Lang and Baumann and coworkers [116] In this study, the "all-in-one" precursor was synthesized by the reaction of [AuCl(P^nBu_3)] with silver(I) ethylene glycol carboxylate in toluene (Scheme 4). Filtration and evaporation of the solvent afforded [Au(P^nBu_3)($O_2C(CH_2OCH_2)_3H$)]. This precursor contains 34.2 wt% of gold. [115]

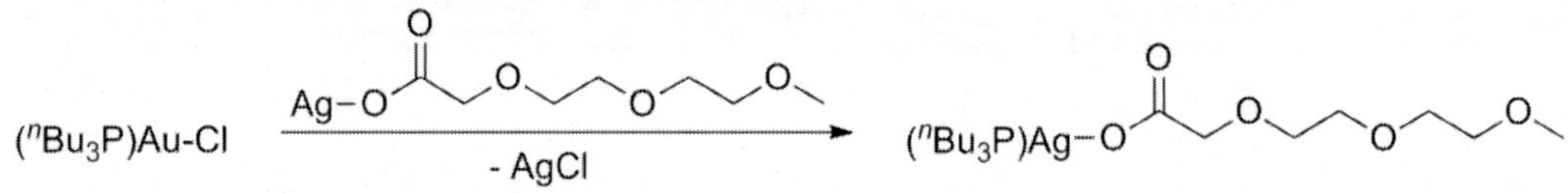

Scheme 4. Synthesis of [Au(P^nBu_3) ($O_2C(CH_2OCH_2)_3H$]. Reproduced from Schoner, C.; Tuchscherer, A.; Blaudeck, T.; Jahn, S. F.; Baumann, R. R.; Lang, H. *Thin Solid Films*, 2013, *531*, 147-151, with permission from Elsevier.

Decomposition of [Au(P^nBu_3) ($O_2C(CH_2OCH_2)_3H$] starts at 228 °C and is completed at 252°C. At the beginning, CO_2 is released followed by the elimination of P^nBu_3. In the final step the ethylene glycol chain is decomposed via Au-C, C-C and C-O bond cleavage. [116] The precursor was dissolved in toluene to produce a 20 wt% MOD ink with a gold content of 6.9 wt%. The resulting ink was used for printing double-layer structures. To achieve maximum conductivity a sintering step at 280°C for 10 min was applied. Conductivities of 43% of gold were measured (Figure 12).

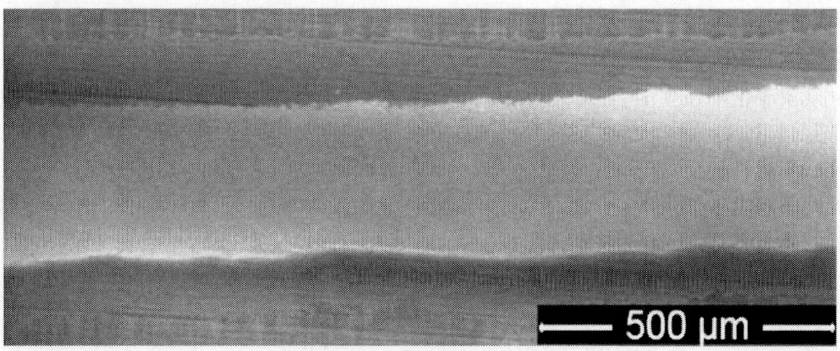

Figure 12.SEM image of an inkjet-printed and sintered (280 °C, 10 min) line of a 20 wt% gold MOD ink on glass substrate. Reproduced from Schoner, C.; Tuchscherer, A.; Blaudeck, T.; Jahn, S. F.; Baumann, R. R.; Lang, H. *Thin Solid Films*, 2013, *531*, 147-151, with permission from Elsevier.

3. Applications of Inkjet-Printed Cu, Ag and Au Structures

Since the pioneering works by Teng and Vest in the late 1980s, inkjet-based metallization has led to industrial manufacturing of low-cost passive electrical components with 'macroscopic' dimensions in the range of 20 – 1000 μm, e.g., coils for magnet resonance imaging (MRI) [123], antennae in radio-frequency identification devices (RFID) [38,124–126] and current collector grids for roll-to-roll printed OPV devices. [26,127,128]

Inkjet-printed gold structures can be used in chemical and biological sensing, due to their chemical inertness. The affinity of gold to thiols allows simple modifications on the surface of printed gold electrodes to immobilize capture molecules. A chemiresistor sensor device using inkjet-printed 1-hexanethiol capped gold nanoparticles was presented by Raguse et al. [129] Due to an increase in the impedance of the sensor, detection of toluene, dichloromethane and ethanol in aqueous KCl solution was possible with detections limits of 0.1, 10 and 3000 ppm, respectively. Skotadis et al. developed a gas sensor for humidity and ethanol, based on printing of commercial available gold nanoparticles. [71] Concentrations of water or ethanol down to 2500 ppm could be detected. Another interesting application is biosensing. Jensen et al. presented inkjet-printed Au-NP arrays for immunosensing of a cancer biomarker protein with good selectivity and low detection limits (Figure 13). [68] Furthermore, the determination of triglycerides with an amperometric biosensor based on a inkjet-printed Au/PEDOT-PSS nanocomposite was recently reported. [130] It was found that increasing the number of printed Au/PEDOT-PSS layers from 2 to 7, increases the response current but decreases sensitivity and linear dynamic range. Thus, the 5 layer electrode was preferred due to the wide dynamic range and the low detection limit of triglyceride.

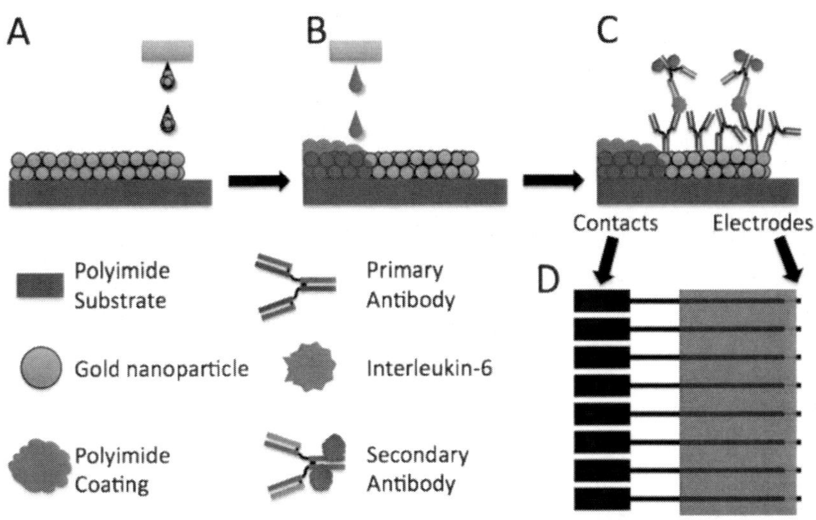

Figure 13. Fabrication of Au-NP arrays: (A) inkjet printing of an Au-NP ink onto PI. (B) inkjet printing of a PAA ink to insulate the electrode leads: (C) building an immuno-assay for electrochemical detection of human IL-6. (D) the Au-NP pattern (black) shown overlaid with the PAA printed pattern (grey). Reproduced from Jensen, G. C.; Krause, C. E.; Sotzing, G.A., Rusling, J. F. *Phys. Chem. Chem. Phys.* 2011, *13*, 4888-4894 with permission of the RSC.

In contrast to gold, the main applications of inkjet-printed copper and silver patterns can be found in the field of electrical engineering. Fox example, printing silver(I) neodecanoate on a Ti/Pd grid and heat treament thereafter, produced solar cells with efficiencies of 8%. [4] Applications regarding microelectromechanical systems (MEMS) and flexible electronics were published by Jacobson et al. [131] Using a custom silver nanoparticle ink and inkjet printing techniques only, they were able to print electromechanical structures like electrostatic driven motors reaching angular speeds up to 100° per second. Next to this, electrostatic devices even heatuators were printed by applying up to 400 layers of ink. Once powered up these MEMS a tip motion from 200 μm was reached. Also resonant inductive coils suitable for RFID tags are feasible. Another example for a completely flexible RFID antenna printed on paper was given by Song et al. (Figure 14), [132] using a custom made ink, accessible in a large scale process. After printing the antenna and bonding the chip on the structure the device fulfills all requirements known from classical production methods. Moon fabricated an organic thin film transistor (OTFT) with printed source and drain electrodes covered by α,ω-dihexylquaterthiophene. [133,134] Comparing the printable device with a vacuum deposited OTFTs, there were no major difference in both devices. They were also able to involve the inkjet printing techniques into a TFT production process and could show the economic benefits of printing such devices.

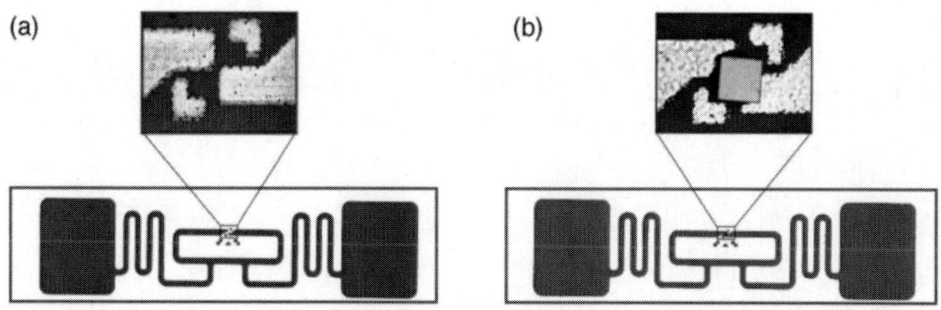

Figure 14. (a) An Ag-NP-based ink-jet printed flexible RFID antenna on paper, chip location close-up view (top). (b) a chip bonded RFID antenna, chip close-up view (top). Reproduced from Zhang, Z.; Zhang, X. *Nanotechnology*, 2011, *22*, 425601, with permission from IOP Publishing.

Printing conductive copper patterns can be superior to other techniques, especially where high numbers of devices with low to moderate complexity have to be produced, e.g., in the fabrication of wearable electronics, flexible RFID tags etc. Basic investigations regarding conductive copper lines have been performed. [76,77] Different substrates, such as glass, flexible polymers or paper [135] have been investigated and performed equally well. Electrical devices as LEDs could be operated by the printed circuits, however, a certain voltage drop has to be taken into account. [76] RFID tags have been printed on paper by Magdassi et al. [88], whereas Moon et al. [75] succeded in printing complex patterns with feature sizes of ~130 μm. Wagner et al. [101] as well as the Moon group [136] used copper as source/drain material in thin film transistors (Figure 15). Moon printed a copper particle ink on zinc tin oxide acting as semiconducting layer, whereas Wagner applied copper(II) hexanoate on α-Si:H. Both of them were able to fabricate the desired structures at moderate temperatures, for which electromigration of copper is still slow.

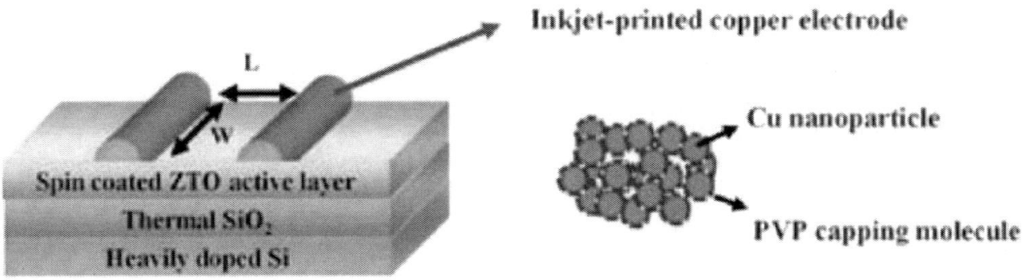

Figure 15. Schematic illustration of a TFT structure with inkjet-printed Cu-NP source/drain on the spin-coated active layer (zinc tin oxide annealed at 500°C). Reproduced from Woo, K.; Bae, C.; Jeong, Y.; Kim, D.; Moon, J. *J. Mater. Chem.*, 2010, *20*, 3877-3882 with permission of the RSC.

CONCLUSION

This article describes different concepts for inkjet printing of group-11 (Cu, Ag, Au) metal patterns and areas based on nanoparticle-containing and MOD inks, respectively. Applying these methodologies, high conductive inkjet-printed patterns and layers are accessible after a post-printing sintering step. Whereas the nanoparticle ink approach is well studied and inkjet printable inks are commercial available, the development of MOD inks is still in an early stage but opens the possibilities of increased ink stability, high conductivities and decreased minimal feature size of the printed patterns. It will be of interest to see the further development of nanoparticle-containing and especially of MOD inks in inkjet-printing processes and their usage in chemical and biosensing, nanomedicine, and printed electronics with applications such as OLEDs, RFID tags, and logics based on thin film transistor and microelectrodes.

ACKNOWLEDGMENT

We gratefully acknowledge the Fonds der Chemischen Industrie for financial support. We thank Dr. P. J. Smith, Dr. S. F. Jahn, M. Hösel, C. Schoner, and J. Mammitzsch for fruitful discussion on this topic.

REFERENCES

[1] Sridhar, A.; Blaudeck, T.; Baumann, R. R. *Mater. Matters* 2011, *6*, 12–15.

[2] Teng, K. F.; Vest, R. W. *IEEE Trans. Components, Hybrids, Manuf. Technol.* 1987, *10*, 545–549.

[3] Teng, K. F.; Vest, R. W. *IEEE Electron Device Lett.* 1988, *9*, 591–593.

[4] Teng, K. F.; Vest, R. W. *IEEE Trans. Components, Hybrids, Manuf. Technol.* 1988, *11*, 291–297.

[5] Teng, K. F.; Vest, R. W. *IEEE Trans. Ind. Electron.* 1988, *35*, 407–412.

[6] Bishop, P. T. *Gold Bull.* 2002, *35*, 89–98.

[7] Hudd, A. In *The Chemistry of Inkjet Inks*; Magdassi, S., Ed.; World Scientific: New Jersey, 2010.

[8] Wallace, D. In *Inkjet-Based Micromanufacturing*; Korvink, J. G.; Smith, P. J.; Shin, D.-Y., Eds.; Wiley-VCH: Weinheim (Germany), 2012.

[9] Choi, K.-H.; Rahman, K.; Muhammad, N. M.; Khan, A.; Kwon, K.-R.; Doh, Y.-H.; Kim, H.-C. In *Recent Advances in Nanofabrication Techniques and Applications*; Cui, B., Ed.; InTech: Rijeka (Croatia), 2011.

[10] Vest, R. W. In *Ceramic Films and Coatings*; Wachtmann, J. B.; Haber, R. A., Eds.; Noyes Publications: Westwood, 1993.

[11] Baumann, R. R.; Willert, A.; Blaudeck, T. *Adv. Print. Media Technol.* 2010, *37*, 327–334.

[12] Schoner, C.; Jahn, S. F.; Jakob, A.; Blaudeck, T.; Lang, H.; Baumann, R. R. *MRS Proc.* 2010, *1285*, 13–18.

[13] Watanabe, M.; Eiwa, Y.; Tokunaga, S. *Sensors Actuators B Chem.* 2011, *157*, 727–734.

[14] Belgardt, C.; Sowade, E.; Blaudeck, T.; Baumgärtel, T.; Graaf, H.; von Borczyskowski, C.; Baumann, R. R. *Phys. Chem. Chem. Phys.* 2013, *15*, 7494–7504.

[15] Soltman, D.; Subramanian, V. *Langmuir* 2008, *24*, 2224–2231.

[16] Pan, J. Z.; Huang, R. Y. *Key Eng. Mater.* 2008, *368-372*, 1668–1672.

[17] Ko, S. H.; Pan, H.; Grigoropoulos, C. P.; Luscombe, C. K.; Fréchet, J. M. J.; Poulikakos, D. *Nanotechnology* 2007, *18*, 345202.

[18] Aminuzzaman, M.; Watanabe, A.; Miyashita, T. *Thin Solid Films* 2009, *517*, 5935–5939.

[19] Greer, J. R.; Street, R. A. *Acta Mater.* 2007, *55*, 6345–6349.

[20] Jahn, S. F.; Jakob, A.; Blaudeck, T.; Schmidt, P.; Lang, H.; Baumann, R. R. *Thin Solid Films* 2010, *518*, 3218–3222.

[21] Jahn, S. F.; Blaudeck, T.; Baumann, R. R.; Jakob, A.; Ecorchard, P.; Rüffer, T.; Lang, H.; Schmidt, P. *Chem. Mater.* 2010, *22*, 3067–3071.

[22] Eftekhari, G. *Jpn. J. Appl. Phys.* 1992, *31*, 832–834.

[23] Wang, X.; Li, S. S.; Kim, W. K.; Yoon, S.; Craciun, V.; Howard, J. M.; Easwaran, S.; Manasreh, O.; Crisalle, O. D.; Anderson, T. J. *Sol. Energy Mater. Sol. Cells* 2006, *90*, 2855–2866.

[24] a) Schroder, K. A.; McCool, S.; Hamill, D.; Wilson, D.; Furlan, W.; Walter, K.; Willauer, D.; Martin, K. US Patent. US7820097, 2010. b) Kang, H., Sowade, E., Baumann, R.R. *ACS Appl. Mater. Interfaces*, 2014, *6*, 1682–1687.

[25] Perelaer, J.; Abbel, R.; Wünscher, S.; Jani, R.; van Lammeren, T.; Schubert, U. S. *Adv. Mater.* 2012, *24*, 2620–2625.

[26] Galagan, Y.; Coenen, E. W. C.; Abbel, R.; van Lammeren, T. J.; Sabik, S.; Barink, M.; Meinders, E. R.; Andriessen, R.; Blom, P. W. M. *Org. Electron.* 2013, *14*, 38–46.

[27] Bidoki, S. M.; Lewis, D. M.; Clark, M.; Vakorov, A.; Millner, P. A.; McGorman, D. *J. Micromechanics Microengineering* 2007, *17*, 967–974.

[28] Leenders, L.; Oelbrandt, L.; van den Bogaert, J.; Desie, G. EU Patent. EP0697282, 1996.

[29] Wu, C.-C.; Hsu, S. L.-C.; Chiu, C.-W.; Wu, J.-T. *J. Micromechanics Microengineering* 2013, *23*, 085026.

[30] Shtyrkov, E. I.; Khaibullin, I. B.; Zaripov, M. M.; Galyatudinov, M. F.; Bayzazitov, R. M. *Sov. Phys. Semicond.* 1975, *9*, 1309.

[31] Yung, K. C.; Wu, S. P.; Liem, H. *J. Mater. Sci.* 2008, *44*, 154–159.

[32] Lungenschmied, C.; Dennler, G.; Neugebauer, H.; Sariciftci, S. N.; Glatthaar, M.; Meyer, T.; Meyer, A. *Sol. Energy Mater. Sol. Cells* 2007, *91*, 379–384.

[33] Kim, T.-Y.; Hwang, J.-Y.; Moon, S.-J. *Jpn. J. Appl. Phys.* 2010, *49*, 05EA09.

[34] Ramírez, M. A.; Bueno, P. R.; Longo, E.; Varela, J. A. *J. Phys. D. Appl. Phys.* 2008, *41*, 152004.

[35] Perelaer, J.; de Gans, B.-J.; Schubert, U. S. *Adv. Mater.* 2006, *18*, 2101–2104.

[36] Perelaer, J.; Klokkenburg, M.; Hendriks, C. E.; Schubert, U. S. *Adv. Mater.* 2009, *21*, 4830–4834.

[37] Valeton, J. J. P.; Hermans, K.; Bastiaansen, C. W. M.; Broer, D. J.; Perelaer, J.; Schubert, U. S.; Crawford, G. P.; Smith, P. J. *J. Mater. Chem.* 2010, *20*, 543.

[38] Wolf, F. M.; Perelaer, J.; Stumpf, S.; Bollen, D.; Kriebel, F.; Schubert, U. S. *J. Mater. Res.* 2013, *28*, 1254–1261.

[39] a) Wünscher, S.; Stumpf, S.; Perelaer, J.; Schubert, U. S. *J. Mater. Chem.* 2014, *2*, 1642–1649. b) Reinhold, I.; Hendriks, C. E.; Eckardt, R.; Kranenburg, J. M.; Perelaer, J.; Baumann, R. R.; Schubert, U. S. *J. Mater. Chem.* 2009, *19*, 3384.

[40] Allen, M.; Leppäniemi, J.; Vilkman, M.; Alastalo, A.; Mattila, T. *Nanotechnology* 2010, *21*, 475204.

[41] Allen, M. L.; Aronniemi, M.; Mattila, T.; Alastalo, A.; Ojanperä, K.; Suhonen, M.; Seppä, H. *Nanotechnology* 2008, *19*, 175201.

[42] Cummins, G. *Circuit World* 2012, *38*, 193–213.

[43] Liang, L.; Shen, C.; Du, S.; Liu, W.; Xie, X.; Gao, H. *Phys. Rev. B* 2004, *70*, 205419.

[44] Romano, F. J. *Inkjet!: History, Technology, Markets, and Applications*; Digital printing Council, PIA/GATFPress: Pittsburgh, PA, 2008.

[45] Rycenga, M.; Cobley, C. M.; Zeng, J.; Li, W.; Moran, C. H.; Zhang, Q.; Qin, D.; Xia, Y. *Chem. Rev.* 2011, *111*, 3669–3712.

[46] Sharma, V. K.; Yngard, R. a; Lin, Y. *Adv. Colloid Interface Sci.* 2009, *145*, 83–96.

[47] Aherne, D.; Ledwith, D. M.; Gara, M.; Kelly, J. M. *Adv. Funct. Mater.* 2008, *18*, 2005–2016.

[48] Lee, K. J.; Jun, B. H.; Choi, J.; Lee, Y. Il; Joung, J.; Oh, Y. S. *Nanotechnology* 2007, *18*, 335601.

[49] Lee, H.-H.; Chou, K.-S.; Huang, K.-C. *Nanotechnology* 2005, *16*, 2436–2441.

[50] Kim, D.; Jeong, S.; Park, B. K.; Moon, J. *Appl. Phys. Lett.* 2006, *89*, 264101.

[51] Kim, D.; Jeong, S.; Moon, J. *Nanotechnology* 2006, *17*, 4019–4024.

[52] Mo, L.; Liu, D.; Li, W.; Li, L.; Wang, L.; Zhou, X. *Appl. Surf. Sci.* 2011, *257*, 5746–5753.

[53] Ankireddy, K.; Vunnam, S.; Kellar, J.; Cross, W. *J. Mater. Chem. C* 2013, *1*, 572.

[54] Dreaden, E. C.; Alkilany, A. M.; Huang, X.; Murphy, C. J.; El-Sayed, M. a. *Chem. Soc. Rev.* 2012, *41*, 2740–2779.

[55] Kannan, R.; Katti, K. V. In *Biomedical Applications of Nanotechnology*; Labhasetwar, V.; Leslie-Pelecky, D. L., Eds.; John Wiley & Sons, Inc.: Hoboken, New Jersey, 2007; pp. 173–189.

[56] Kim, D.; Jon, S. *Inorganica Chim. Acta* 2012, *393*, 154–164.

[57] Saha, K.; Agasti, S. S.; Kim, C.; Li, X.; Rotello, V. M. *Chem. Rev.* 2012, *112*, 2739–2779.

[58] Turkevich, J.; Stevenson, P. C.; Hillier, J. *Discuss. Faraday Soc.* 1951, *11*, 55.

[59] Brust, M.; Walker, M.; Bethell, D.; Schiffrin, D. J.; Whyman, R. *J. Chem. Soc. Chem. Commun.* 1994, 801.
[60] Kanaras, A. G.; Kamounah, F. S.; Schaumburg, K.; Kiely, C. J.; Brust, M. *Chem. Commun.* 2002, 2294–2295.
[61] Tuchscherer, A.; Schaarschmidt, D.; Schulze, S.; Hietschold, M.; Lang, H. *Inorg. Chem. Commun.* 2011, *14*, 676–678.
[62] Tuchscherer, A.; Schaarschmidt, D.; Schulze, S.; Hietschold, M.; Lang, H. *Dalton Trans.* 2012, *41*, 2738–2746.
[63] Zhao, P.; Li, N.; Astruc, D. *Coord. Chem. Rev.* 2013, *257*, 638–665.
[64] Määttänen, A.; Ihalainen, P.; Pulkkinen, P.; Wang, S.; Tenhu, H.; Peltonen, J. *ACS Appl. Mater. Interfaces* 2012, *4*, 955–964.
[65] Cui, W.; Lu, W.; Zhang, Y.; Lin, G.; Wei, T.; Jiang, L. *Colloids Surfaces A Physicochem. Eng. Asp.* 2010, *358*, 35–41.
[66] Chung, J.; Ko, S.; Bieri, N. R.; Grigoropoulos, C. P.; Poulikakos, D. *Appl. Phys. Lett.* 2004, *84*, 801.
[67] Cortie, M. B.; Coutts, M. J.; Ton-That, C.; Dowd, A.; Keast, V. J.; McDonagh, A. M. *J. Phys. Chem. C* 2013, *117*, 11377–11384.
[68] Jensen, G. C.; Krause, C. E.; Sotzing, G. A.; Rusling, J. F. *Phys. Chem. Chem. Phys.* 2011, *13*, 4888–4894.
[69] Yamauchi, H.; Dobashi, T.; Sato, S. *Chem. Lett.* 2012, *41*, 1154–1156.
[70] Szczech, J.; Megaridis, C.; Zhang, J.; Gamota, D. *Microscale Thermophys. Eng.* 2004, *8*, 327–339.
[71] Skotadis, E.; Tang, J.; Tsouti, V.; Tsoukalas, D. *Microelectron. Eng.* 2010, *87*, 2258–2263.
[72] Huang, D.; Liao, F.; Molesa, S.; Redinger, D.; Subramanian, V. *J. Electrochem. Soc.* 2003, *150*, G412.
[73] Coutts, M. J.; Cortie, M. B.; Ford, M. J.; McDonagh, A. M. *J. Phys. Chem. C* 2009, *113*, 1325–1328.
[74] Perelaer, J.; Smith, P. J.; Mager, D.; Soltman, D.; Volkman, S. K.; Subramanian, V.; Korvink, J. G.; Schubert, U. S. *J. Mater. Chem.* 2010, *20*, 8446.
[75] Jeong, S.; Woo, K.; Kim, D.; Lim, S.; Kim, J. S.; Shin, H.; Xia, Y.; Moon, J. *Adv. Funct. Mater.* 2008, *18*, 679–686.
[76] Jang, S.; Seo, Y.; Choi, J.; Kim, T.; Cho, J.; Kim, S.; Kim, D. *Scr. Mater.* 2010, *62*, 258–261.
[77] Kang, J. S.; Kim, H. S.; Ryu, J.; Thomas Hahn, H.; Jang, S.; Joung, J. W. *J. Mater. Sci. Mater. Electron.* 2010, *21*, 1213–1220.
[78] Yabuki, A.; Arriffin, N. *Thin Solid Films* 2010, *518*, 7033–7037.
[79] Luechinger, N. A.; Athanassiou, E. K.; Stark, W. J. *Nanotechnology* 2008, *19*, 445201.
[80] Kim, J. H.; Germer, T. A.; Mulholland, G. W.; Ehrman, S. H. *Adv. Mater.* 2002, *14*, 518–521.
[81] Chan, G. H.; Zhao, J.; Hicks, E. M.; Schatz, G. C.; Van Duyne, R. P. *Nano Lett.* 2007, *7*, 1947–1952.
[82] Salkar, R. A.; Jeevanandam, P.; Kataby, G.; Aruna, S. T.; Koltypin, Y.; Palchik, O.; Gedanken, A. *J. Phys. Chem. B* 2000, *104*, 893–897.
[83] Hambrock, J.; Becker, R.; Birkner, A.; Weiß, J.; Fischer, R. A. *Chem. Commun.* 2002, 68–69.

[84] Adner, D.; Korb, M.; Schulze, S.; Hietschold, M.; Lang, H. *Chem. Commun. (Camb).* 2013, *49*, 6855–6857.
[85] Magdassi, S.; Grouchko, M.; Kamyshny, A. *Materials (Basel).* 2010, *3*, 4626–4638.
[86] Deng, D.; Jin, Y.; Cheng, Y.; Qi, T.; Xiao, F. *ACS Appl. Mater. Interfaces* 2013, *5*, 3839–3846.
[87] Lee, Y.; Choi, J.-R.; Lee, K. J.; Stott, N. E.; Kim, D. *Nanotechnology* 2008, *19*, 415604.
[88] Grouchko, M.; Kamyshny, A.; Magdassi, S. *J. Mater. Chem.* 2009, *19*, 3057.
[89] Cuya Huaman, J. L.; Sato, K.; Kurita, S.; Matsumoto, T.; Jeyadevan, B. *J. Mater. Chem.* 2011, *21*, 7062.
[90] Kim, D.; Jeong, S.; Lee, S.; Park, B. K.; Moon, J.; Kim, J. S. *Thin Solid Films* 2007, *515*, 7706–7711.
[91] Ida, K.; Tomonari, M.; Sugiyama, Y.; Chujyo, Y.; Tokunaga, T.; Yonezawa, T.; Kuroda, K.; Sasaki, K. *Thin Solid Films* 2012, *520*, 2789–2793.
[92] Lee, Y.-I.; Lee, K.-J.; Goo, Y.-S.; Kim, N.-W.; Byun, Y.; Kim, J.-D.; Yoo, B.; Choa, Y.-H. *Jpn. J. Appl. Phys.* 2010, *49*, 086501.
[93] Goo, Y.-S.; Lee, Y.-I.; Kim, N.; Lee, K.-J.; Yoo, B.; Hong, S.-J.; Kim, J.-D.; Choa, Y.-H. *Surf. Coatings Technol.* 2010, *205*, S369–S372.
[94] Lee, Y.-I.; Choa, Y.-H. *J. Mater. Chem.* 2012, *22*, 12517.
[95] Kim, S. J.; Lee, J.; Choi, Y.-H.; Yeon, D.; Byun, Y. *Thin Solid Films* 2012, *520*, 2731–2734.
[96] Choi, Y.-H.; Lee, J.; Kim, S. J.; Yeon, D.-H.; Byun, Y. *J. Mater. Chem.* 2012, *22*, 3624.
[97] Joo, M.; Lee, B.; Jeong, S.; Lee, M. *Thin Solid Films* 2012, *520*, 2878–2883.
[98] Yabuki, A.; Arriffin, N.; Yanase, M. *Thin Solid Films* 2011, *519*, 6530–6533.
[99] Yabuki, A.; Tanaka, S. *Mater. Res. Bull.* 2012, *47*, 4107–4111.
[100] Wang, B.-Y.; Yoo, T.-H.; Song, Y.-W.; Lim, D.-S.; Oh, Y.-J. *ACS Appl. Mater. Interfaces* 2013, *5*, 4113–4119.
[101] Wagner, S. *IEEE Electron Device Lett.* 2000, *21*, 384–386.
[102] Adner, D.; Wolf, F. M.; Möckel, S.; Perelaer, J.; Schubert, U. S.; Lang, H. *Thin Solid Films* 2014, *565*, 143-148.
[103] Hampden-Smith, M. J.; Kodas, T. T. *Polyhedron* 1995, *14*, 699–732.
[104] Homma, T.; Takasaki, A.; Yamaguchi, M.; Kokubun, H.; Machida, H. *J. Electrochem. Soc.* 2000, *147*, 580–585.
[105] Rozenberg, G. G.; Bresler, E.; Speakman, S. P.; Jeynes, C.; Steinke, J. H. G. *Appl. Phys. Lett.* 2002, *81*, 5249–5251.
[106] Rickerby, J.; Steinke, J. H. G. *Chem. Rev.* 2002, *102*, 1525–1550.
[107] Rickerby, J.; Simon, A.; Jeynes, C.; Morgan, T. J.; Steinke, J. H. G. *Chem. Mater.* 2006, *18*, 2489–2498.
[108] Walker, S. B.; Lewis, J. A. *J. Am. Chem. Soc.* 2012, *134*, 1419–1421.
[109] Yin, Y.; Li, Z.-Y.; Zhong, Z.; Gates, B.; Xia, Y.; Venkateswaran, S. *J. Mater. Chem.* 2002, *12*, 522–527.
[110] Russo, A.; Ahn, B. Y.; Adams, J. J.; Duoss, E. B.; Bernhard, J. T.; Lewis, J. a. *Adv. Mater.* 2011, *23*, 3426–3430.
[111] Huang, Q.; Shen, W.; Song, W. *Appl. Surf. Sci.* 2012, *258*, 7384–7388.
[112] *Handbook of Chemistry and Physics*; Lide, D. R., Ed.; 77th ed.; CRC Press: Boca Raton, 1996; pp. 12–41.

[113] Dearden, A. L.; Smith, P. J.; Shin, D.-Y.; Reis, N.; Derby, B.; O'Brien, P. *Macromol. Rapid Commun.* 2005, *26*, 315–318.

[114] Nie, X.; Wang, H.; Zou, J. *Appl. Surf. Sci.* 2012, *261*, 554–560.

[115] Nur, H. M.; Song, J. H.; Evans, J. R. G.; Edirisinghe, M. J. *J. Mater. Sci. Mater. Electron.* 2002, *3*, 213–219.

[116] Schoner, C.; Tuchscherer, A.; Blaudeck, T.; Jahn, S. F.; Baumann, R. R.; Lang, H. *Thin Solid Films* 2013, *531*, 147–151.

[117] Grodzicki, a.; Łakomska, I.; Piszczek, P.; Szymańska, I.; Szłyk, E. *Coord. Chem. Rev.* 2005, *249*, 2232–2258.

[118] Claes, D. J.; Jackson, E. *Pediatr. Nephrol.* 2012, *27*, 2031–2038.

[119] Templeton, A. C.; Chen, S.; Gross, S. M.; Murray, R. W. *Langmuir* 1999, *15*, 66–76.

[120] Nakamoto, M.; Yamamoto, M.; Fukusumi, M. *Chem. Commun. (Camb).* 2002, *29*, 1622–1623.

[121] Nakamoto, M.; Kashiwagi, Y.; Yamamoto, M. *Inorganica Chim. Acta* 2005, *358*, 4229–4236.

[122] Yamamoto, M.; Nakamoto, M. *Chem. Lett.* 2003, *32*, 452–453.

[123] Mager, D.; Peter, A.; Tin, L. Del; Fischer, E.; Smith, P. J.; Hennig, J.; Korvink, J. G. *IEEE Trans. Med. Imaging* 2010, *29*, 482–487.

[124] Virtanen, J.; Björninen, T.; Ukkonen, L.; Sydänheimo, L. *IEEE Antennas Wirel. Propag. Lett.* 2010, *9*, 440–443.

[125] Koski, K.; Koski, E.; Virtanen, J.; Björninen, T.; Sydänheimo, L.; Ukkonen, L.; Elsherbeni, A. Z. *Int. J. Adv. Manuf. Technol.* 2011, *62*, 167–182.

[126] Virtanen, J.; Virkki, J.; Sydanheimo, L.; Tentzeris, M.; Ukkonen, L. *IEEE Trans. Autom. Sci. Eng.* 2013, *10*, 796–806.

[127] Yu, J.-S.; Kim, I.; Kim, J.-S.; Jo, J.; Larsen-Olsen, T. T.; Søndergaard, R. R.; Hösel, M.; Angmo, D.; Jørgensen, M.; Krebs, F. C. *Nanoscale* 2012, *4*, 6032–6040.

[128] Abbel, R.; van den Boomen, J.; van Lammeren, T.; de Koning, T.; Valeton, J. J. P.; Meinders, E. R. *MRS Proc.* 2011, *1323*, 87–92.

[129] Raguse, B.; Chow, E.; Barton, C. S.; Wieczorek, L. *Anal. Chem.* 2007, *79*, 7333–7339.

[130] Phongphut, A.; Sriprachuabwong, C.; Wisitsoraat, A.; Tuantranont, A.; Prichanont, S.; Sritongkham, P. *Sensors Actuators B Chem.* 2013, *178*, 501–507.

[131] Fuller, S. B.; Wilhelm, E. J.; Jacobson, J. M. *J. Microelectromechanical Syst.* 2002, *11*, 54–60.

[132] Zhang, Z.; Zhang, X.; Xin, Z.; Deng, M.; Wen, Y.; Song, Y. *Nanotechnology* 2011, *22*, 425601.

[133] Kim, D.; Jeong, S.; Lee, S.; Park, B. K.; Moon, J. *Thin Solid Films* 2007, *515*, 7692–7696.

[134] Kim, D.; Jeong, S.; Shin, H.; Xia, Y.; Moon, J. *Adv. Mater.* 2008, *20*, 3084–3089.

[135] Li, D.; Sutton, D.; Burgess, A.; Graham, D.; Calvert, P. D. *J. Mater. Chem.* 2009, *19*, 3719–3724.

[136] Woo, K.; Bae, C.; Jeong, Y.; Kim, D.; Moon, J. *J. Mater. Chem.* 2010, *20*, 3877–3882.

In: Comprehensive Guide for Nanocoatings Technology. Vol. 1 ISBN: 978-1-63482-447-7
Editor: Mahmood Aliofkhazraei

Chapter 7

ELECTROLESS NICKEL-PHOSPHOROUS VS ELECTROLESS NICKEL-BORON: COMPARISON OF HARDNESS, ABRASION RESISTANCE, SCRATCH TEST RESPONSE AND CORROSION BEHAVIOR

V. Vitry and F. Delaunois
Metallurgy Lab, UMONS, Belgium

ABSTRACT

Electroless nickel coatings are often used because they are resistant to scratch, abrasion and corrosion.

However, it is difficult to choose between electroless nickel-phosphorous and electroless nickel- boron coatings for any given application because their properties are not often compared.

Both types of coatings were completely compared: their structure, morphology, hardness, abrasion resistance, scratch test response, and corrosion resistance were investigated and the best candidate for each feature was determined.

Heat treated coatings were submitted to the same comparison.

INTRODUCTION

Electroless nickel plating is a process based on the reduction of nickel salts in aqueous solution by a chemical reactive. The main particularity of the method is that it does not need any external current source to operate.

The development of electroless nickel plating is the result of numerous prior discoveries (see figure 1), the first one being that of the hypophosphite compound, in the early 19th century [1]. A few decades later, Wurtz discovered in 1844 the possibility of using sodium hypophosphite to reduce nickel ions [2] and a few of his contemporaries studied the process [3-5] but only Bretean [6] was able to obtain anything else than a black powdery deposit at the bottom of the vessel. In 1916, Roux patented a process that allowed depositing nickel by a chemical process but the process plated all surfaces in contact with the solution [7]. It was only a century after Wurtz's discovery that Brenner and Riddel were able to develop a catalytic electroless nickel plating process that allowed plating selected surfaces only [8-10]. History remembers them as the inventors of the process because they rediscovered it by accident while trying to electroplate the internal surface of tubes [11-13]. At this time, only nickel-phosphorous was available because the alkali metal borohydrides that are the most popular reducing agent for the synthesis of nickel-boron were only discovered in 1949 [14]. However, the first electroless nickel-boron plating process was patented as soon as 1955 [15].

Electroless nickel plating is a direct concurrent of nickel electroplating but there are several differences between the processes that make each one suited for distinct applications (Figure 2).

Firstly, as electroless plating is a purely chemical process, it can be applied to non-conducting substrates such as polymers. Secondly, there are no edge effects due to current repartition (contrary to electroplating) and the coating has a uniform thickness, with uniform properties. Moreover, the electroless technology is suitable for complex shapes and blind holes. Nevertheless due to the uniform thickness, the leveling ability of electroless nickel is poor. Another difference is that electroless coatings are rarely pure nickel: the most popular electroless nickel coatings being nickel-phosphorous and nickel-boron alloys. Lastly, electroless nickel operations are a bit more delicate than electroplating: the plating process is spontaneous and catalytic. For it to work efficiently, the bath has to be close to instability. However, it has to stay stable to be operating. The fine tuning between stable enough to be handled and instable enough for quick plating is the main difficulty of electroless plating operation [1, 16-17].

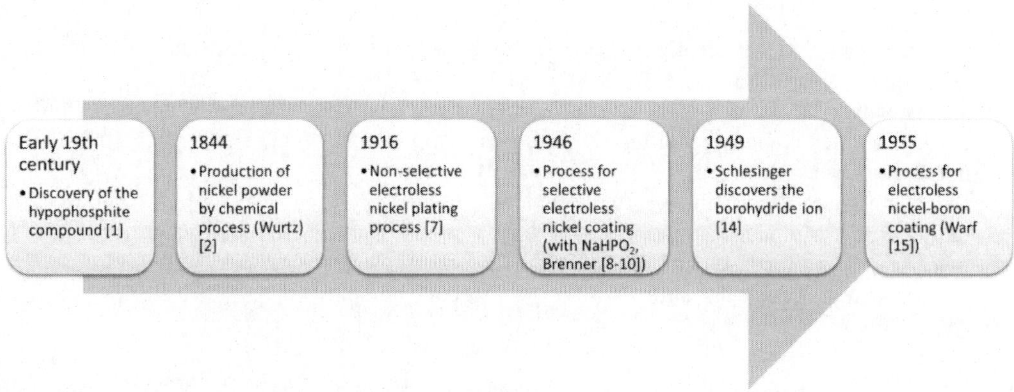

Figure 1. Historical overview of electroless nickel processes.

Electroless Nickel Plating

The principle of electroless nickel plating is relatively simple: nickel salts are placed in aqueous solution. They are then reduced by the action of a chemical agent. However, in practice, the process is relatively more complex, due to the need for stability, selectivity of plating and high plating rate.

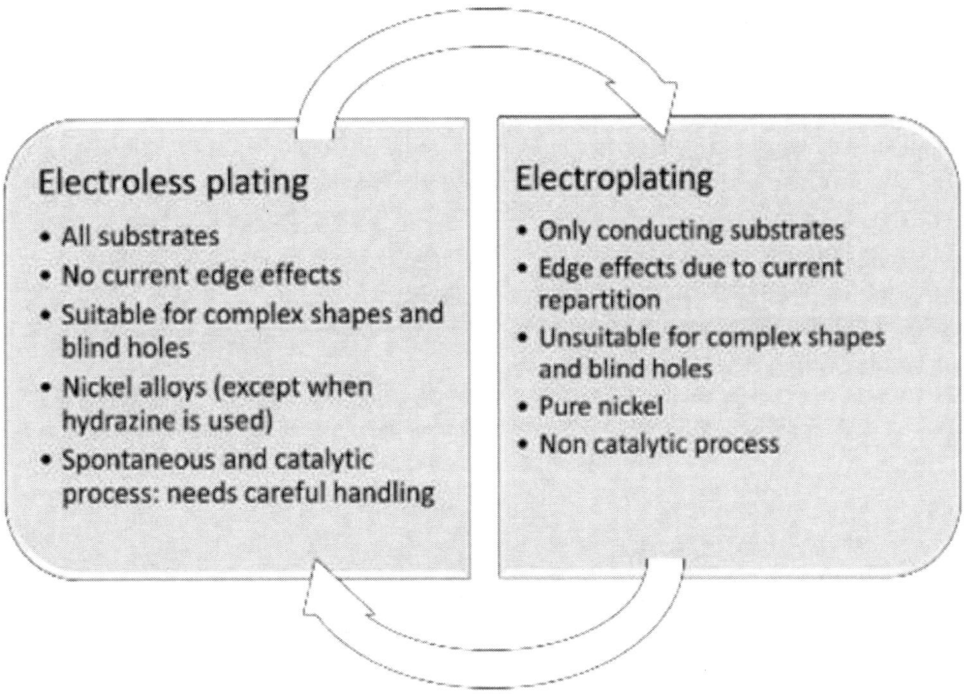

Figure 2. Comparison of electroless plating and electroplating.

To obtain an operational electroless nickel plating bath, several conditions have to be respected; they are described in the following sections.

Thermodynamic and Kinetic Conditions

Electroless plating is a spontaneous process that occurs without external current source. It is thus indispensable that the reducing agent is adequately chosen, with a lower redox potential than that of the metal that has to be reduced. This is the thermodynamic condition [16-19].

It is important to note that, even when this condition is respected, there is no guarantee that deposition will take place in a useful manner: this condition gives no information on the plating rate. There is thus a kinetic condition, linked to the overpotential for the oxidation of the reducing agent on the surface that has to be plated: this overpotential has to be as small as possible to ensure that the deposition current (and thus the plating rate) is high [16-19]. This kinetic condition has to be respected for the substrate during the initial stages of plating but also for the metal deposited itself at later stages. This is the reason why the process is sometimes named ''autocatalytic''.

The need for both of the above conditions to be respected at all times limits the choice of reducing agents and the substrates on which electroless nickel plating can take place without previous surface modification. Figure 3 shows the overpotential values for the oxidation of several reducing agents on various substrates. It shows that 4 of the 5 investigated reducing agents are able to produce electroless nickel: sodium hypophosphite, sodium borohydride, dimethylamineborane (and the other amineborane compounds) and hydrazine. However, formaldehyde is not suitable for this process. Most investigated substrates are adequate but, electroless nickel plating on copper and platinum with sodium hypophosphite requires specific preparation because the overpotential for the oxidation of the reducing agent on those materials is too high [16, 18-19].

According to the chosen reducing agent, diverse deposits can be obtained (figure 4): hydrazine is used when pure nickel is needed. This reducing agent is not very popular because it does not allow high plating rates. It is thus used only when pure nickel is absolutely needed [17].

When sodium hypophosphite is used, phosphorous is incorporated in the coating, up to 12%, depending on plating bath chemistry and plating conditions. The resulting coatings are called electroless nickel-phosphorous (EN-P) and represent the majority of the commercialized products [1, 17].

When amineborane compounds or sodium borohydride are used, boron is incorporated in the coating. The amount of boron is typically 0.5 to 1.5% with an amineborane bath and 1 to 8% with a borohydride bath. Those coatings are called electroless nickel-boron (EN-B).

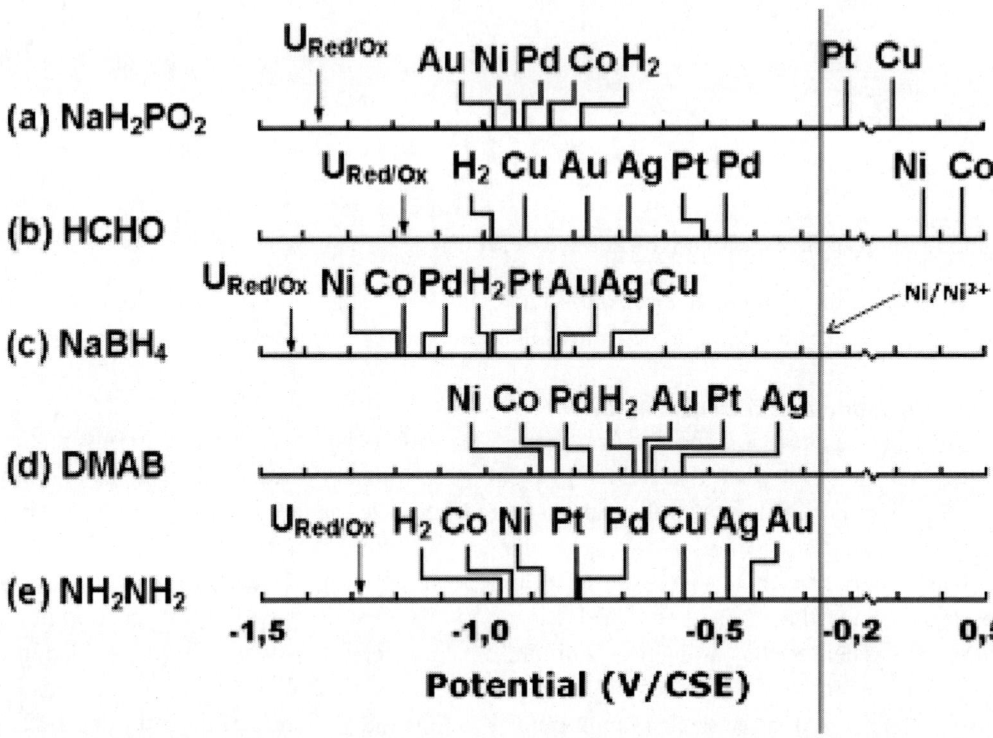

Figure 3. Overpotential values for the oxidation of reducing agents on various substrates at 25°C. Catalytic activity of various metals at 25°C, under an anodic current of 10^{-4} A.cm^2, for diverse reducing agents (adapted from Ohno (1991) [19], with permission).

They are less often used than EN-P but they present very interesting properties and have been the object of many studies in recent years [1, 16-18, 20-40].

The most powerful of those reducing agents is the borohydride ion, that can provide up to 8 electrons for the reduction of nickel (against 2 for hypophosphite), and that is also very light (much lighter that hypophosphite) so that only 600g of sodium borohydride are needed to reduce 1kg of nickel [41]. However, this reducing agent is only stable in alkaline to very alkaline solutions, which makes the baths very corrosive [42].

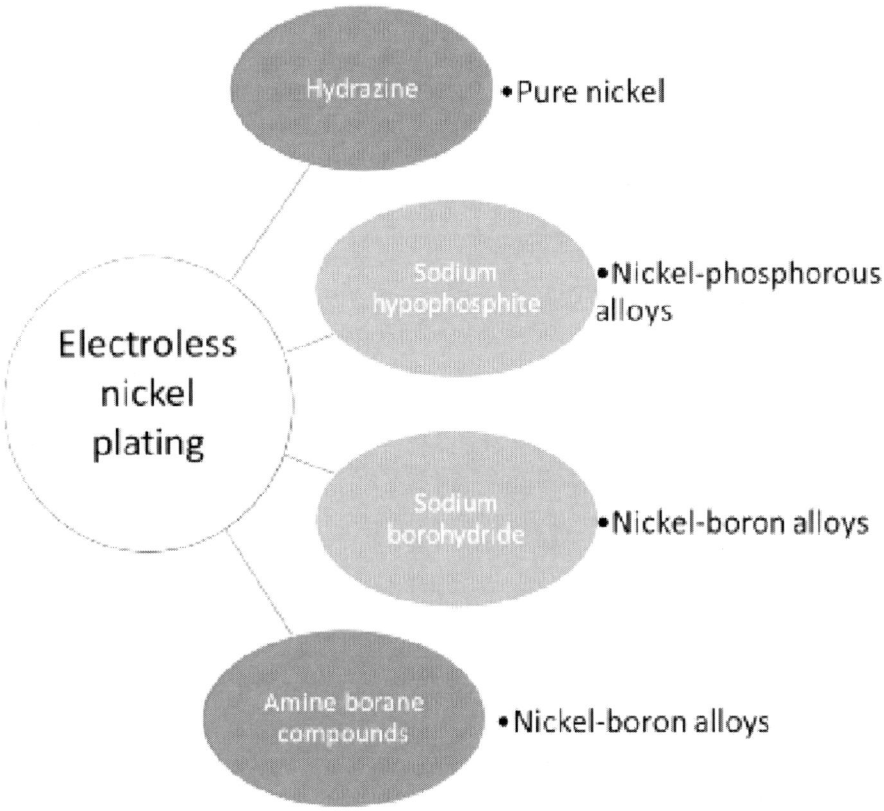

Figure 4. Reducing agents suitable for electroless nickel plating and the nature of the deposits obtained with those.

Bath Stability

Electroless deposition is a spontaneous process that involves the use of very strong reducing agents that are close to instability in the conditions they are used in. As a result, to ensure preferential deposition on the substrate and to avoid plating other surfaces, the bath needs to be stabilized. The role of the stabilizer is to block part of the primary catalytic sites (i.e., of the surface of the sample) to regulate the plating rate and to fully block the secondary catalytic sites (the surface of the cell and all the particles in suspension in the solution) to avoid unwanted nickel deposition and nickel powder formation in the solution. There are 2 favorite kinds of stabilizers: heavy metal salts (Pb^{2+}, Cd^{2+}, Tl^{3+}, …) that are less and less popular due to increasing restrictions on their use, and organic compounds such as thiourea. The stabilizer usually acts by adsorption on the sites and is needed only in very small amounts but its role is essential for smooth plating operations [16-19, 43-51].

High Nickel Solubility

Electroless nickel plating requires that nickel ions are available in important quantity throughout the plating process and that the nickel ion concentration stays stable during operations, in order to keep the plating potential (and thus the plating rate) as constant as possible. The availability of a constant amount of nickel during long plating operations can only be obtained by complexing the nickel. Several kinds of complexants are used, such as citrate, glycolate or ethylene diamine [16-19].

In alkaline plating baths, the complexant is particularly important because the precipitation of insoluble nickel hydroxide must be avoided at all costs.

pH Stability

The pH of the plating bath must be kept as constant as possible, to avoid modifications of the plating conditions and, in the case of sodium borohydride, the degradation of the reducing agent. For this reason, most baths are buffered [1, 16-18].

Practical Classes of Electroless Nickel

As was said previously, the chemistry of electroless deposits is influenced by the reducing agent used and by bath chemistry and operating parameters. In turn, the chemistry of the coating (the nature and amount of alloying elements incorporated into the nickel coating) influences the properties of the deposit. Electroless nickel coatings are thus divided in 4 classes, according to chemistry [16, 18, 52-53], each class possessing roughly homogeneous properties. Those 4 classes are:

- Low phosphorous electroless nickel (lowP EN), which contain 1 to 5% P.
- Mid phosphorous electroless nickel (midP EN), which contain 5 to 9% P. This class constitutes the majority of electroless nickel used in industrial applications.
- High phosphorous electroless nickel (highP EN), which contain 10 to 12% P.
- Electroless nickel-boron (EN-B), which contain 1 to 8% B. There are 2 variants in this class: coatings that contain boron and thallium and other that contain boron and lead.

As midP EN is much more used than the low and high phosphorous variations, it will be used as base for the comparisons. Information about the other nickel-phosphorous grades will be included when available.

Coating Morphology

In 1990, Riedel said (in his book 'electroless nickel' that stays to this date the most complete work on the subject) that the morphologies of electroless nickel-boron and nickel-phosphorous were very different [17]. He described for nickel-phosphorous an homogeneous structure that was laminated in the case of bath containing heavy metal salts and for nickel-boron a very heterogeneous structure with a cauliflower-like surface texture (said to be 'less uniform' than that of EN-P). As Riedel's book is the 'bible' of electroless nickel, this has been passed on for a long time.

However, looking at recent results, the morphological differences between EN-P and EN-B are not as big as expected: great progress has been made in nickel-boron plating technologies and the newer coatings are homogeneous in their unetched state and reveal a columnar morphology after etching, as shown on figure 5a [21, 25-27, 30, 39, 54-55]. This feature is not observed on SEM or optical micrographs of EN-P. However, columnar features have been observed for some electroless nickel-phosphorous coatings by TEM on cross sections (figure 6) [56]. The morphologies observed by Tsai [56] were very similar to some observed by Vitry on EN-B by SEM [21]. As for the surface texture, the cauliflower-like texture typically associated with nickel-boron (figure 5b) [21, 25-27, 30, 39, 54-55, 57-58] can also be observed, but at a smaller scale, on electroless nickel-phosphorous coatings (figure 7) [56, 59-64].

Those results are not unexpected because the synthesis scheme of both kinds of coatings is very similar, with an expected columnar growth. As for the laminated or layered structure sometimes observed on EN-P [17, 65] with heavy metals salts stabilizers, it is probably linked to local variations of P content in the coating, due to changes in the plating conditions[17, 53]. A similarly laminated structure has been observed for electroless nickel-boron deposits when periodic bath replenishment has been implemented [21]. It is truly possible that the observed difference between laminated EN-P and columnar EN-B are more linked to parameters other than the difference in coating chemistry: most EN-P coatings studied and observed are thick replenished coatings obtained by industrial processes with the use of bath replenishments, contrary to EN-B which are mainly lab-made (and thus not replenished). Moreover, metallographic etching is not systematically used in the study of electroless coating. Nickel-boron, which is synthesized in an alkaline bath, etches easily with nital or other classical metallographic (acidic) etchants, which may not be the case of nickel-phosphorous, which is synthesized in an acidic bath. However, even when replenished EN-P coatings are etched, the morphological features linked to replenishment are very prominent and may mask other features (see figure 8).

Structure

It is important to differentiate the morphology (or microstructure) of the coating from its structure: a columnar coating can be amorphous just as a seemingly homogeneous coating can be fully crystalline.

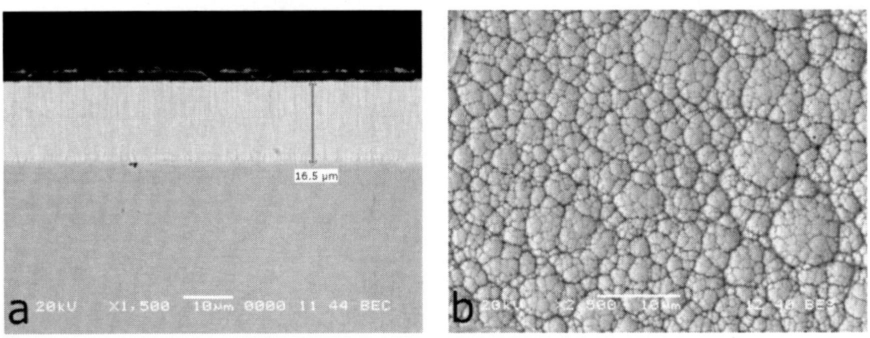

Figure 5. SEM micrographies of an as-deposited electroless nickel-boron coating a) cross section b) surface texture [26], with permission.

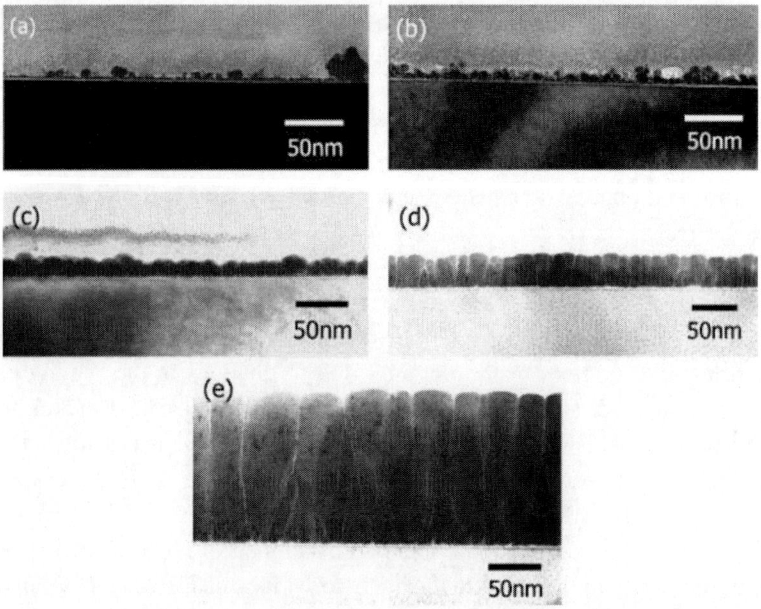

Figure 6. TEM cross-section images of the electroless NiP alloy plating for various deposition times deposited at pH 5.2: (a) 2 s; (b) 4 s; (c) 6 s; (d) 15 s; and (e) 60 s [56], with permission.

The structure of electroless nickel coatings is obviously influenced by their chemistry. The equilibrium phase diagrams for nickel and phosphorous and nickel and boron predict a very low solubility of those alloying element with nickel and the formation of a series of intermetallic phases: Ni_3P, Ni_5P_2, $Ni_{12}P_5$, Ni_2P, Ni_5P_4 and NiP_2 for the phosphide phases, and Ni_3B, Ni_2B, Ni_4B_3 and NiB for the boride phases [67]. In the composition range of electroless coatings, the expected phases are Ni, Ni_3P, Ni_5P_2 and Ni_2P for nickel-phosphorous alloys, and Ni, Ni_3B and Ni_2B for nickel-boron.

Figure 7. Surface and cross-sectionalmorphology of nickel coatingdeposited on chemical conversion coating for AZ91D alloy: (a) low magnification, (b) higher magnification and (c) cross-section [60], with permission.

In the as-deposited state, structure of electroless nickel coatings is very different from that predicted by phase diagrams due to their synthesis. The boride and phosphide phases are not formed (because the energy level in the plating bath is not high enough) and phosphorous and boron form highly supersaturated solid solutions in nickel [16]. This causes a decrease of the size of crystallites with increasing P or B content, leading, for very high alloying element contents, to structures that appear fully amorphous [16, 53, 59, 68-74].

Farber [69] studied the repartition of phosphorous in electroless nickel coatings. He was able, from his work and grain sizes measured by other authors, to establish a model for the evolution of grain size in as-deposited (or heat treated at low temperature) electroless nickel-phosphorous, as a function of the phosphorous content.

a

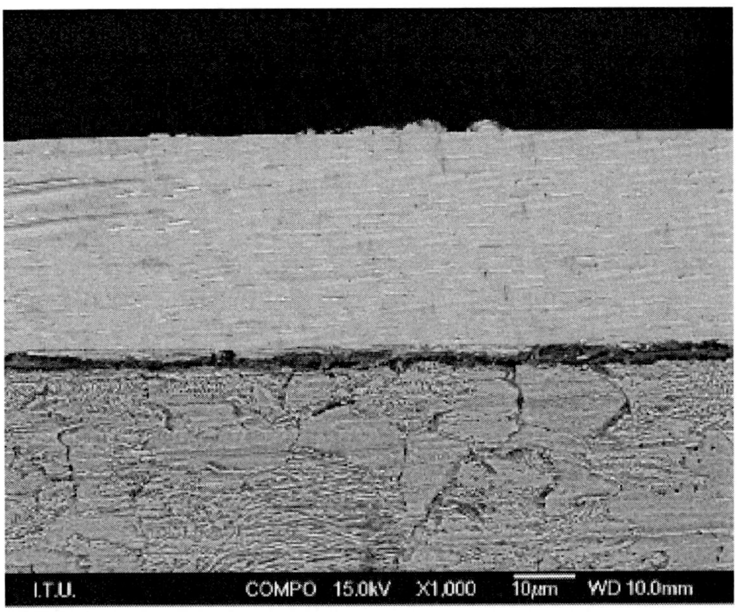

Figure 8. SEM cross-section images of (a) Ni–P [66], with permission.

This model indicates that electroless nickel-phosphorous becomes completely amorphous for P contents higher than 15 wt.% (25 at.%). Jiaquiang, using slightly different values for the modeling, obtained a value of 20 wt.% [75]. Vitry [16] re-established this model for electroless nickel-boron but she did not dispose of some information that Farber obtained by tomographic atom probe. She observed a nanocrystalline-amorphous transition for 4.5 wt.% B (20 at.%) and the results of her modeling were confirmed by measurements carried out by Gaevaskaya on electrodeposited nickel-boron [76].

However, structural studies on electroless nickel-boron showed that, even for coatings with boron content over 5 wt.%, short range order was conserved and the coating was not fully amorphous [30].

There are no equivalent results for EN-P because the transition between nanocrystalline and amorphous state happens for a phosphorous content higher than what is used in electroless plating. Moreover, TEM studies of electroless coatings are relatively infrequent in the literature and most structural results are obtained by X-ray diffraction without confirmation by TEM [53, 59, 64 77-78] (Some didn't even check by XRD before affirming their coatings were amorphous [75]). This causes a surevaluation of the amorphous character of electroless nickel coatings because nanocrystalline coatings with very small grain sizes can easily be mistaken for fully amorphous coatings, as Park and Lee said in their paper as soon as 1988 [63].

The structural state of coatings after heat treatment has been extensively studied for both nickel-boron and nickel-phosphorous, mainly by X-ray diffraction [30, 53, 64, 68, 70-73, 77-78], but a few authors did carry out TEM analysis on EN-P [63] or EN-B [30, 70-71]. Heat treatment of electroless nickel coatings brings the energy necessary to return to equilibrium and leads to the crystallization of various phase (Nickel, nickel phosphides, nickel borides) depending on the initial chemistry. The standard heat treatment conditions are 1h at 400°C, preferably under protective atmosphere to avoid oxidation of the coating (but this is not mandatory) [35]. If the heat treatment is carried out in those conditions, it brings sufficient energy for complete crystallization, but not for grain growth, leading to nanocrystalline coatings with relatively small grain size (rarely over 100nm) [30, 77-80]. The phases most often found in heat treated electroless nickel are Ni and Ni_3P for EN-P and Ni and Ni_3B for EN-B [35]. In some cases, coexistence of phases that are not adjacent in the phase diagram (like Ni and Ni_2B) is observed [39]. This is due to heterogeneities in the coating linked to the plating process that cannot be suppressed by diffusion during the heat treatment (that is realized at relatively low temperatures).

Several authors studied the crystallization temperature of electroless nickel-phosphorous [53, 63, 75, 77, 81-84] and electroless nickel-boron [31, 39, 57, 85-88]. They observed, in the case of electroless nickel-phosphorous, a maximum temperature of 350°C for 19 at.%, while Pal obtained, for EN-B, a decrease of crystallization temperature from 350 °C to 300 ° with increasing boron content up to 25 at.%, followed by a sudden increase at higher boron contents.

Hardness

The hardness of electroless nickel-phosphorous coatings is in the 500-700 hv_{100} range for mid-phosphorous coatings. It is influenced by the P content of the deposit: hardness decreases with P content, with higher hardness for lowP-EN and lower values (close to 500 hv_{100}) for highP-EN [1, 35, 89-90]. Some authors have reported an increase of hardness up to 8 at.% (which stays in the low P range) followed by a decrease for higher concentrations [91].Taheri [65] reported similarly that the maximum hardness was obtained for 4 wt.%, when P was in solid solution. All those results are compatible with the following hypothesis: low P content in the coating brings hardening by solid solution and Hall-Petch effect (due to the decrease of grain size), while the mechanism for hardness decrease at higher P content is still not well elucidated but could be linked to inverse Hall-Petch effect. The high hardness at low P content is also sometimes attributed to residual stress [53].

The reason why electroless nickel-boron coatings are harder that EN-P is not known, but the resulting effect has been proved extensively and the hardness of as-deposited EN-B can be as high as 900 hv_{100}[25-26, 31, 34, 39, 57, 88, 92-96]. Das reported that, contrary to EN-P, the hardness of EN-B increased with B content[35]. However, the low values obtained by some authors for coatings with boron content over 6 wt.% [57, 94] suggest that higher boron content could bring a decrease of hardness similar to what is observed for EN-P. It is nevertheless interesting to note that the hardening effect of boron is more important and is conserved for smaller grain sizes. This may be due to the fact that the solid solution formed by boron in nickel is interstitial contrary to phosphorous that is included in substitution.

Effect of Heat Treatment on Hardness

The effect of heat treatment on the hardness of electroless nickel is similar for EN-P and EN-B: heat treating at a suitable temperature (usually 400°C) for an adequate time (1h) brings an increase of hardness up to 1000 hv_{100} for EN-P [53, 79-80, 84, 89] and 1400 hv_{100} for EN-B [25-26, 31, 39, 57, 87-88, 92-96].

The evolution of hardness according to heat treatment temperature and time is similar for both kinds of coatings, as shown on figures 9 and 10: when treatment time is kept at one hour (or close to that), hardness increases up to 400°C then decreases sharply. When temperature is kept around 400°C, hardening is observed up to 1 hour before the hardness decreases [31, 39, 53, 79-80, 89]. The reduction of hardness observed for high temperatures and long times is obviously linked to coarsening of the coating structure. However, the explanation of the initial hardening is no so simple: it is often attributed to the creation of grain boundaries in an amorphous material [17, 53, 72-73 85, 87-88, 97-98]. This hypothesis, while seducing, may not be used as sole reason for hardening because, as was shown hereover, electroless nickel coatings are far less amorphous than most people think. How would creating grain boundaries bring such an increase in hardness if grains are already present in the coating?

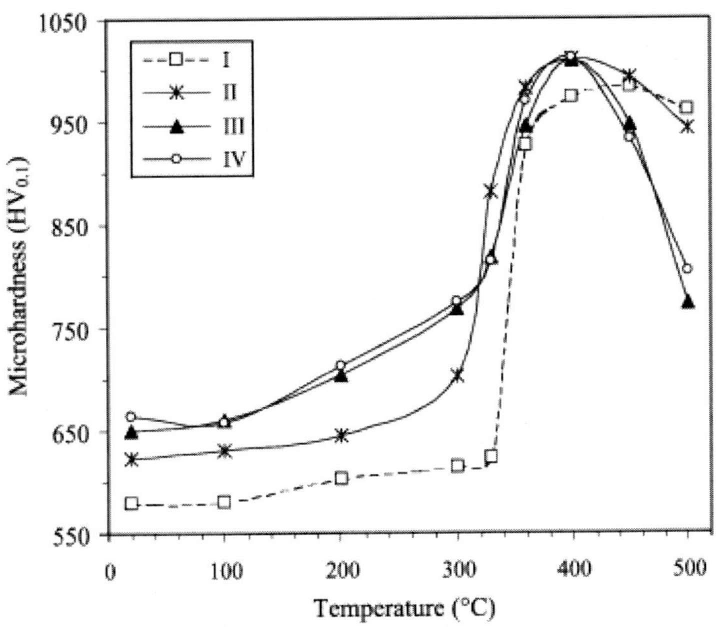

Figure 9. Vickers hardness measured along the centre line of cross sections for Samples I–IV after heating at different temperatures for 1 h (isothermal). Sample composition (wt.% P). I: 12%; II: 8.8%; III: 6.7%; IV: 6.3% [53], with permission.

Inverse Hall-Petch effect could play a role in this (initial grains are usually very small and could be in a range were it is in action) but there is another fact that is nearly always overlooked by researchers: the nickel-phosphide and nickel-boride phases that are precipitated during heat treatment of electroless coatings are hard phases, much more hard than nickel. For example, Ni_3B presents an orthorhombic structure similar to that of a notoriously hard phase, cementite (Fe_3C). The study of peak (maximal) hardness obtained after heat treatment for various EN-P coatings showed an increase for higher P content [80], which is in agreement with an hardening linked to the presence of hard intermetallic phases (in this case phosphides).

Wear Resistance

Electroless coatings are often used in applications where hardness is not the only required feature: good resistance to wear and/or to corrosion is also often needed. For this reason, a lot of research was carried out on the friction behavior and wear resistance of electroless nickel coatings with and without heat treatments. The wear resistance of materials is linked to their hardness, but also to other features, like surface morphology [65].

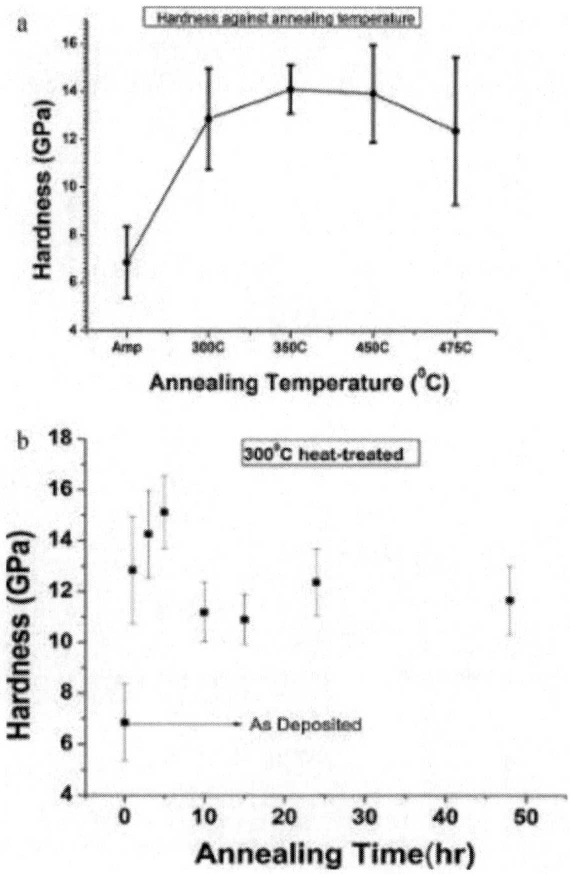

Figure 10. The variation of hardness with annealing time and temperature, respectively [39], with permission.

There are several ways of measuring wear. Some of them measure resistance to erosion, others to abrasion and other kinds of wear. It is thus difficult and not always useful to compare results obtained from different tests: the wear mode depends mainly of the applications intended for the parts. However, it is hugely important, during wear testing of coatings, to use a method allowing measuring the wear of the coating separately from the substrate. The wear testing method must also be reproducible and the wear must be quantifiable [99].In his review, Sahoo [35] took inventory of wear tests carried out on electroless nickel. The most popular are by far Pin on disc, Ball on disc and Block on ring but some authors used original or unusual wear test methods. Others did not completely describe their experiments, in most cases because their work was already a comparison between different coatings (often composites) or between coatings with different heat treatments [87]. This makes their work meaningful, but this work is as a result very difficult to use in a comparative study.

The first important parameter in wear is the coefficient of friction. It has been measured on electroless nickel-phosphorous in dry conditions by several researchers, sometimes without information about the phosphorous content of the coating, as summarized on figure 11. Values from 0.6 to 0.9 were routinely obtained [65, 100-104]. The lowest value (0.5) was obtained by Novak [105]. Liu [106] measured the friction coefficient with a microtribometer with a cemented carbide counterpart and a load of 5N. He obtained friction coefficients from 0.15-0.2 to 0.3-0.4 depending on the roughness of the substrate. In lubricated conditions, Ramalho [107] measured values in the 0.05-0.09 range in sliding wear.

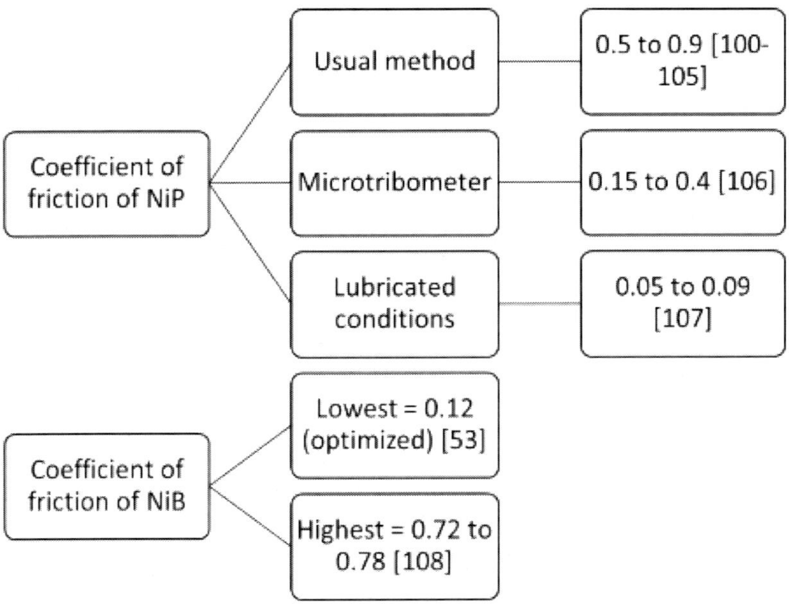

Figure 11. Summary of friction coefficient values for electroless nickel-phosphorous.

Studies measuring the tribological properties of different types of EN-P are relatively rare. Taheri [104] measured the friction coefficient of Low, Mid and High Phosphorous electroless nickel coatings in dry conditions. He obtained values between 0.725 and 0.850 depending on the P content, on the heat treatment and on the roughness of the test surface. The effect of P content was less clear than that of test surface roughness (a decrease of roughness causes a decrease of friction coefficient) and that of heat treatment (that brings a slight decrease of the friction coefficient).

As a point of comparison, Etcheverry [52] studied both lowP EN and hard chrome using the pin-on-disc test. He obtained lower and more stable coefficient of friction for EN-P and the damage was similar for both types of coatings.

Like for EN-P, several researchers measured the friction coefficient of electroless nickel-boron. The usual values are relatively lower than on nickel-phosphorous: 0.72 to 0.78 for Krishnaveni [108], an astonishing 0.12 for Sahoo (after optimization of the plating parameters) [53], 0.3 to 0.4 for Bulbul [38] and Correa [36].

The values of coefficient of friction cited hereover were obtained by different methods and different research groups. It is thus relatively difficult to judge if the differences are really significant, moreover so because there is no comprehensive study available in which a same group measured the friction coefficient of both EN-P and EN-B, and because the friction coefficient is influenced by the roughness of the coating [65, 104, 106].

For the comparison of actual wear of electroless nickel coatings, let's begin with results obtained by the pin-on-disc test. Those may be difficult to compare because of variations in test conditions, but also because not all researchers measured wear in the same way: some measured weight loss while others measured wear depth. In both groups, some calculated a specific wear rate while others kept the absolute value.

Amongst the people that used a specific wear rate (see Table 1), Palaniappa [100] obtained, on EN-P a specific wear rate in the range of $3*10^{-10}$ kg N^{-1} m^{-1}, while Novak [105] reported a value of $572*10^{-6}$ mm^3 N^{-1} m^{-1} in the as-plated state for a load of 5N for EN-P. Using 8.2 as the density of electroless nickel, it is possible to compare those results: Palaniappa's value becomes then $36*10^{-6}$ mm^3 N^{-1} m^{-1}.

Some people also used specific wear rates for the measure of the wear of EN-B: Krishnaveni [108] measured, on as-plated EN-B, with a load of 40N, a value of $2.46*10^{-10}$ kg N^{-1} m^{-1} ($30*10^{-6}$ mm^3 N^{-1} m^{-1}), and Bulbul [38] measured, on a EN-B coating with a blackberry-like surface texture, a value of $15*10^{-6}$ mm^3 N^{-1} m^{-1}. Both of those values are slightly lower that Palaniappa's [100] but much lower than that obtained by Novak [105].

Table 1. Summary of specific wear rate data for electroless nickel-phosphorous and nickel-boron

Author	*Coating*	*Specificwear rate*	
		kg N^{-1} m^{-1}	mm^3 N^{-1} m^{-1}
Palaniappa [100]	NiP	$3\ 10^{-10}$	$36\ 10^{-6}$
Novak [105]	NiP		$572\ 10^{-6}$
Krishnaveni [108]	NiB	$2.46\ 10^{-10}$	$30\ 10^{-6}$
Bulbul [38]	NiB		$15\ 10^{-6}$
Sankara Narayanan [94]	NiP (9 wt.%)	$4.6\ 10^{-10}$	$56\ 10^{-6}$
Sankara Narayanan [94]	NiB	$2.45\ 10^{-10}$	$30\ 10^{-6}$

In one of the rare works comparing the wear properties of both EN-P and EN-B,Sankara Narayanan [94] obtained a value of $4.6*10^{-10}$ kg N^{-1} m^{-1} ($56*10^{-6}$ mm^3 N^{-1} m^{-1}) for a midP EN (9 wt.% P) and one of $2.45*10^{-10}$ kg N^{-1} m^{-1} ($30*10^{-6}$ mm^3 N^{-1} m^{-1}) for EN-B. This confirms the trend that electroless nickel-boron has a better resistance to the pin-on-disc wear test than nickel-phosphorous.

The effect of heat treatment on wear of electroless nickel phosphorous is not very clear: Taheri [65] observed improved behavior after heat treatment for all types of EN-P coatings (lowP EN, midP EN, highP EN), Sankara Narayanan [94] obtained a value of $1.62*10^{-10}$ kg N^{-1} m^{-1} ($19.8*10^{-6}$ mm^3 N^{-1} m^{-1}) and Novak [105] reported a value of $4.52*10^{-6}$ mm^3 N^{-1} m^{-1} after heat treatment at 400°C. He also investigated higher heat treatment temperatures but observed a decrease of the wear resistance compared to the sample treated at 400°C. However, according to Palaniappa, wear increases slightly after heat treatment [100].

On EN-B, Krishnaveni [108] observed an improvement of wear resistance after heat treatment, with a value of $1.04*10^{-10}$ kg N^{-1} m^{-1} ($12.7*10^{-6}$ mm^3 N^{-1} m^{-1}) and Sankara Narayanan [94] reported a value of $0.88*10^{-10}$ kg N^{-1} m^{-1} ($10.7*10^{-6}$ mm^3 N^{-1} m^{-1}), confirming the previous results.

Like was said before, some researchers did not convert their results into specific wear rates. Sharma [102] measured the wear resistance of EN-P by the pin-on-disc method with various loads up to 30 N and obtained a weight loss of 8mg for 500m, with a prominence of adhesive wear for the maximal load. For smaller loads, he measured much lower weight loss (<1mg for 500m). By observation of the wear tracks, Staia [101] also noticed that the wear of as plated samples of MidP EN (6.7 wt.% P) was mainly adhesive but he noticed that the wear of heat-treated EN-P was abrasive.

Different configurations than pin-on-disc were also used for the study of the wear of electroless nickel coatings: On nickel-phosphorous, Yu [103] used a pin-on-block configuration for reciprocating wear test, on highP EN coatings (9.4 and 14.4 wt.% P). He observed lower wear on the 9.4 wt.% P coating in the as-deposited state and the reverse behavior after heat treatment, together with a global decrease of wear due to heat treatment; Trzaska [109] used the three-rollers-cone method, with quick-cutting steel as a counterpart, with oil lubrication and a load of 50 MPa. He obtained a wear depth of 0.08 mm after 30 minutes of testing but only 0.02 mm after 20 minutes (before the coating was completely removed). Gawne [89] used different wear methods on midP EN (8.5 wt.% P) and highP EN (12 wt.% P), including Falex test for metal-to-metal adhesive wear (under a load of 22.5 kgf). He observed a decrease of wear after heat treatment. Ramalho [107] obtained very low wear rates in sliding wear under lubricated conditions.

On nickel-boron, Correa [110] used a dry sliding ball-on-disc test with an alumina counterpart and observed a specific wear rate of $7.34*10^{-11}$ kg N^{-1} m^{-1} ($8.9*10^{-6}$ mm^3 N^{-1} m^{-1}), with barely visible wear tracks on the EN-B. SEM observation showed plastic deformation of some nodules in the coating. This was confirmed by the work of Das [35] who used a block on roller test in dry conditions, with a load of 25 N. He managed, modifying the plating parameters, to obtain an optimized coating for wear, with a wear depth (for test duration of 300s) of 4.4 μm and, similarly to Correa [110], observed that some nodules of the worn surface were deformed while others remained unaffected by the wear test (Figure 12). Kaya [111] measured wear in the pin-on-block configuration. He did not calculate specific wear rate but had much lower wear than on 1040 steel.

Abrasive wear of both electroless nickel-phophorous and nickel-boron has also been measured, mainly by the Taber Abrasion Test, but also sometimes by other means. On EN-P, Brooman [112] obtained values from 15 (6 wt.% P) to 11 (8.5 wt.%) for the Taber Wear Index of electroless nickel-phosphorous with CS-10 wheels. It is a pity that those results cannot be compared with those obtained on nickel-boron by Vitry and Kanta [16. 26. 28] who used CS-17 wheels (which are harsher than CS-10 wheels). They reported Taber Wear Index of 28 in the as-deposited state and 13 after heat treatment at 400°C.

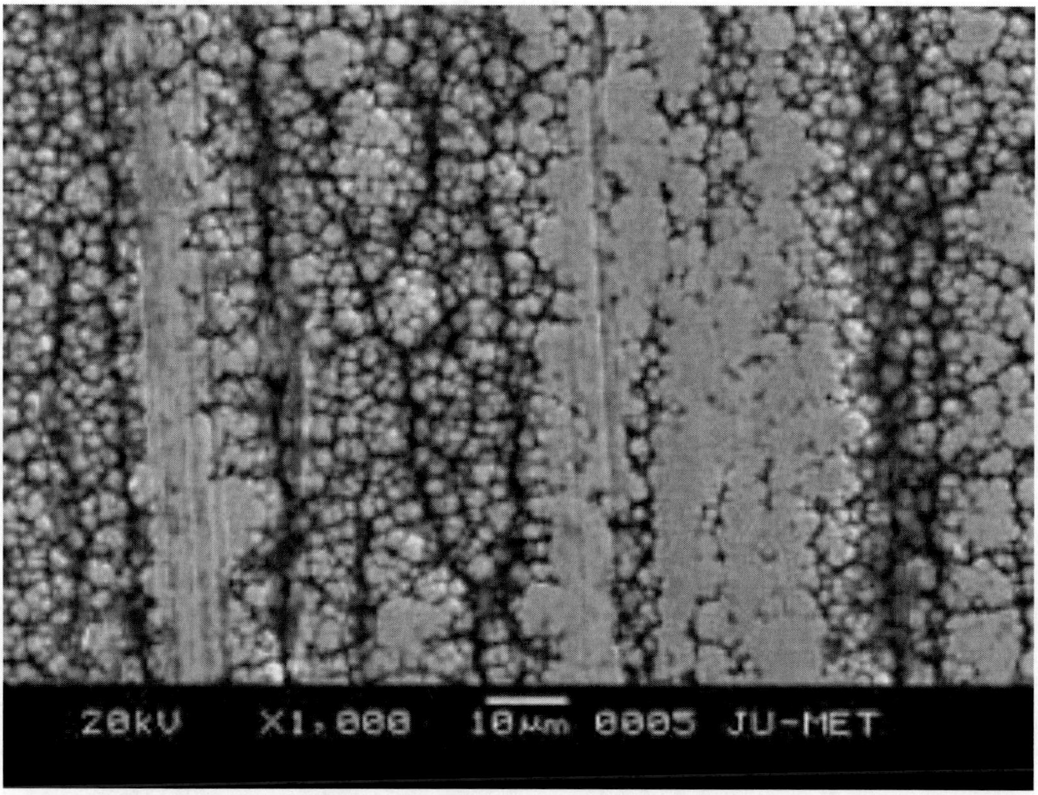

Figure 12. SEM micrograph of the worn surface of electroless nickel-phosphorous [35], with permission.

In his review, Hari Krishnan [1] reported values of 17 for EN-P and 9 for EN-B, for CS-10 wheels but he did not mention where nor how he obtained these result. He even reports values of 9.4 for EN-P and 1.4 to 2 for EN-B but gives no indication about the chemistry of the samples or heat treatments.

Alternative studies of abrasive wear include that of Apachitei [113] who used a reciprocal movement abrasive test with 320 mesh SiC paper and observed an increase of wear loss with phosphorous content from 3.9 to 10.2 wt.%. Heat treatment improved the wear loss for low and midP EN, but not really for highP EN.

Abrasive wear of various electroless coatings has been studied by Allen [114] on AISI 4140 steel. He measured wear of 3 EN-P and one EN-B coating, using a sand abrasion test and obtained wear coefficients of 0.0334, 0.0275 and 0.0292 $mm^3\ kg^{-1}\ m^{-1}$ for the nickel-phosphorous coatings and 0.0240 for the nickel-boron. As a comparison, he obtained values lower than 0.01 for hard chrome.

Similarly, using the dry sand and rubber wheel test, as described in ASTM G65 procedure B, Riddle [115] obtained an adjusted mass loss of 0.387 for EN-B against 1.025 for lowP EN. However, in their case hard chrome had a lesser wear resistance than nickel-boron, with a value of 1.02.

Scratch Test Behavior

Scratch test is a method that allows measuring the adhesion of a coating on a substrate. This method is based on the application of a load by the way of an indenter (usually of the Rockwell C type) on the coating. The load can be either increased step by step or continuously. The scratch test is standardized in EN 1071-3 and compiled by CEN/TC 184 [116]. The adhesion is quantified by the critical load (L_c) which is the load were damage is detected. According to the nature of both substrate and coating, their properties and the adhesion, several kind of damage can be observed. In some cases, various kinds of damage are observed successively. Damage and damage modes are detected either by microscopic observation or measurements of the friction force and of the acoustic emission.

Bull [117] studied the various fracture modes observed in coated systems as a function of the hardness of the substrate and the coating. He classifies the failure modes in 4 main groups: plastic deformation, interfacial fracture, through thickness fracture and bulk fracture.

In the case of hard coatings (electroless nickel are usually considered as part of this group), the usual failure modes are: through thickness cracking (tensile cracking behind the indenter, conformal cracking and Hertzian cracking), spallation (compressive or buckling spallation ahead of the indenter or spallation behind the indenter due to elastic recovery), chipping in the coating.

Recently, Sander [116] discussed the limits of the traditional scratch test, mainly in the case of hard coatings on soft substrates. Bull predicts in this case a trough-thickness fracture, but only large plastic deformation can be observed up to very large loads. Most authors interpret this as good adhesion but Sanders noticed that the system could be damaged by smaller loads in everyday use and theorized that the friction between the indenter and the system had to be taken into consideration. He then developed an alternative scratch test for hard coatings on soft substrate.

Sander's work, while very relevant, has not been put to use yet on electroless nickel coatings because it was published after most of the results obtained on those coatings.

The first recorded use of scratch test to measure the adhesion of electroless nickel coatings was carried out by Yu in 1993 [103]. He measured the adhesion of midP and highP EN (with 9.4 and 14.4 wt.% P) but did not give information about the thickness. He observed critical loads of 15-20N in the as-plated state up to 25-30 after heat treatment at 600°C, with a quite linear evolution of the critical load with heat treatment temperature, which he attributes to interdiffusion between the coating and the substrate. He also noticed that the highP EN coating had a slightly better adhesion that the 9.4 wt.% P coating.

Yu's work did not cause a lot of emulation and the next work about scratch tests of electroless coatings was published by Delaunois [18, 31]: on composite coatings: she used aluminum as a substrate and had to plate it first with EN-P to allow alkaline EN-B plating. Her nickel-boron coatings contained either 6 wt.% B or 4 wt.% B and 7wt.% Tl and had a thickness of 25μm. She observed mainly plastic damage, without cracks and obtained critical loads of 82N in the as-plated condition and 102 N after heat treatment.

The comparison of Yu's and Delaunois' work shows how difficult it is to compare scratch tests results: the nature of the coating and substrate and the surface preparation are far from the only parameters that influence scratch test response. The thickness of the coating has also a great influence on the critical load, and it is not possible to quantify this influence. There is also a great diversity in the presentation of results. This means that it is close to impossible to compare values obtained on coatings with different thickness.

Since Delaunois' work, scratch tests have become relatively popular for the study of electroless nickel-phosphorous, but not really on nickel-boron. The main results, presented hereunder, are summarized in Table 2.

Jin [118] used scratch testing to measure the adhesion of very thin EN-P coatings (1 to 4 μm) on silicon wafers pre-coated by sputtered aluminum. He obtained very low critical loads (0.2 to 2.5 N) and observed an increase of critical load with thickness and with better surface preparation. He also managed to increase the adhesion of the coating by the use of ultrasonic agitation during the zincating process needed to promote electroless nickel plating on aluminum. It is interesting to note that he measured very high values of standard deviation on his critical loads (close to 50% of the average value).

Liu [106, 119] used scratch tests to investigate the adhesion of EN-P on various magnesium alloys. On AZ91, he noticed that adhesion of the coating increased with the roughness of the substrate (that he attributed to mechanical interlocking) but did not observe significant changes in adhesion with heat treatment. All his critical loads values are relatively low (10 to 17N). He did not give information about the thickness of the coatings but they appear to be close to 20μm. He observed severe wedge spallation on the polished substrate (Figure 13). He did also compare the adhesion on pure Mg, AZ31 and AZ91 alloys and observed a better adhesion on AZ31, with an L_c value of 13N for a coating of 60μm.

Table 2. Summary of critical load values obtained from scratch testing

Author	*Coating*	*Thickness*	*Critical load (N)*	
			As plated	Heat treated
Yu [103]	MidP EN (9.4 wt.%)	N.A.	15-20	25-30 (600°C)
Yu [103]	HighP EN (14.4 wt.%)	N.A.	15-20	25-30 (600°C)
Delaunois [18,31]	NiB (on NiP underlayer)	25 μm	82	102
Jin [118]	NiP	1-4 μm	0.2 to 2.5	
Liu [106,119]	NiP	N.A.	10-17	
Trzaska [109]	midP EN (7 wt.%)	15μm	>10	>10
Alexis [121]	MidP EN (5wt.%)	30-40 μm	5N	Various treatments
Vitry [26]	NiB	20 μm	10N	16N
Vitry [26]	NiB	6 μm	21N (complete failure)	13N (complete failure)

Lin [120] observed a better adhesion on heat treated samples due to a reduction of peripheral damage. He increased the load up to 100 N and did not attain critical load.

Novak used [105] scratch test on EN-P coated AlSi10Mg0.3, with heat treatment from 400 to 550°C from 1 to 8 hours. He used a step by step increase of the load, with steps of 8.8

N. He observed a better adhesion of the heat treated samples, except for those treated for 8 h at 550°C.

Trzaska [109] measured adhesion of 15 μm thick 7wt.% P electroless nickel coatings, with a maximal load of 10 N. He reported perfect adhesion, without any observation of chipping, decohesion or acoustic emission, but this is probably due more to the extremely low maximal load than to exceptional properties of his coating.

Alexis [121] carried out a very detailed study on 30 to 40μ m thick coatings with 5 wt.% P and a maximal load of 20 to 30 N (Figure 14).

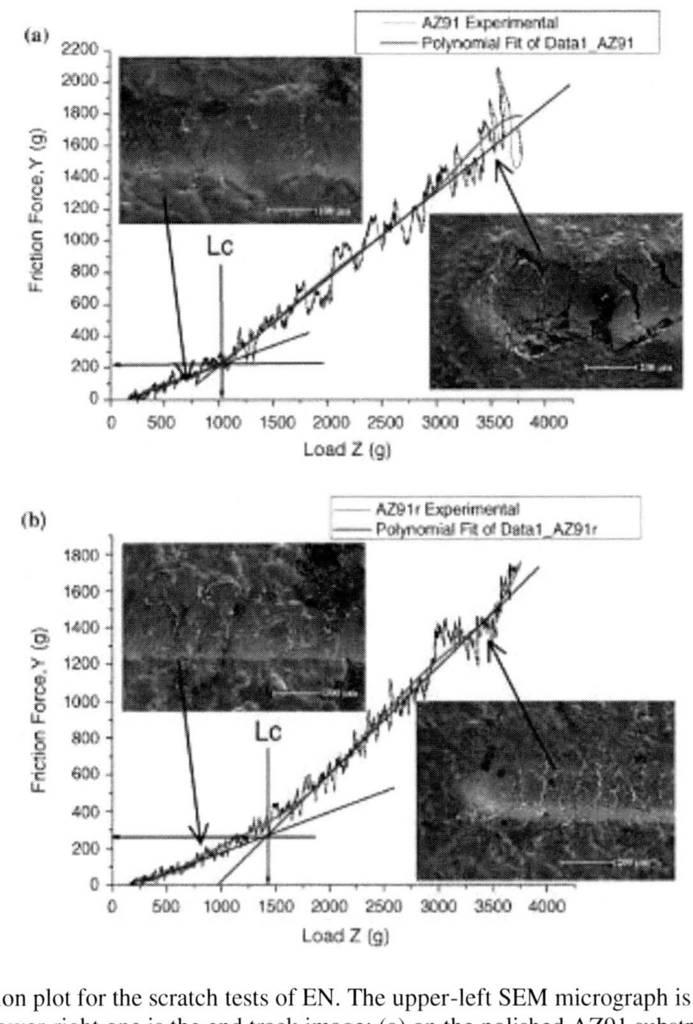

Figure 13. Friction plot for the scratch tests of EN. The upper-left SEM micrograph is the track at the beginning, the lower-right one is the end track image: (a) on the polished AZ91 substrate, (b) on rough substrate [106], with permission.

In the as-plated state, he registered transverse cracks from 5 N and lateral cracks from 12 N. After heat treatment at 420 °C, he observed transverse cracks from 5 N, lateral cracks from 10 N, like in the as-plated state but he also observed spallation from 17 N and adhesive spallation and fragmentation from 24 N. After heat treatment at 600 °C, he observed fragmentation from 13 N and spallation form 20 and 25 N. One of his most interesting findings is that failure occurs by plastic deformation, except for the coating treated at 420 °C where is happens by interfacial failure. This last coating is the only one presenting a fragile behavior: the others (as-plated and treated at 600 °C) present a ductile behavior.

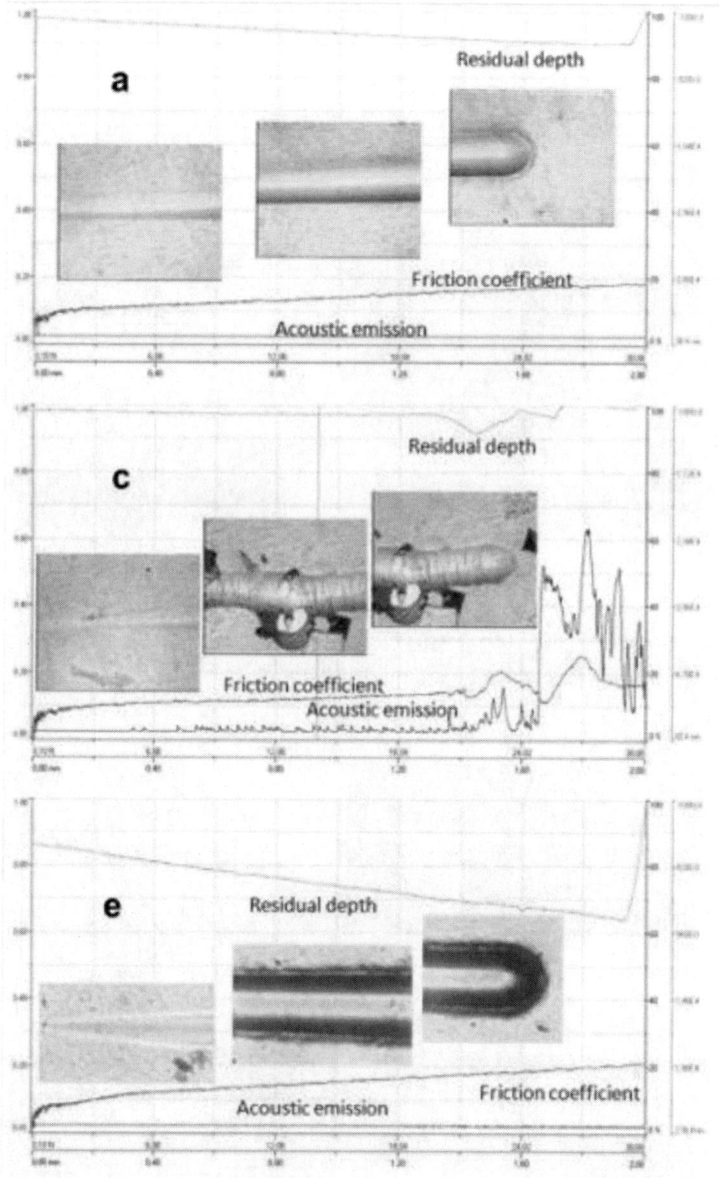

Figure 14. Curves and optical observations of scratches on NiP coatings, respectively, as deposited (a), after treatment at 420 °C (c) and after treatment at 600 °C (e) [121], with permission.

Recently, Vitry [26] studied the adhesion of EN-B on steel, without EN-P pre-treatment (contrary to Delaunois'work [18, 31] and to Vitry's earlier work [25]). She used coatings with a B content of 6wt.% and a thickness close to 20 μm. Her maximal load was 150 N. In the as-plated state, she recorded transversal intercolumnar cracks from 10 N and intercolumnar cracks parallel to the testing direction from 55 N. Those values are relatively higher than that of Alexis [121] and were obtained for thinner coatings. This suggests a better adhesion. After heat treatment, forward intercolumnar chevron cracks at the edge of the scratch, accompanied by cracks at the bottom of the scratch were observed from 16 N and continuous parallel intercolumnar chevron cracks from 30 N, which suggest an improvement of adhesion. Bigger, deeper cracks were observed from 143.5 N. Once again, this suggests better behavior than Alexis' coatings.

In order to reach complete failure of the coating, thinner coatings (6 μm) were also tested. Failure occurred by ductile perforation, from 21 N in the as-plated state and from 13 N after heat treatment (Figure 15).

All the results presented hereinabove confirm that comparing scratch tests obtained by different researchers is very difficult: the number of parameters that influence the results (some of which are not even given in the text) is important and some of them are difficult to control. Nevertheless, those results present common features: the adhesion of electroless coatings appears to be satisfying (but this should be verified using the alternate scratch test developed by Sander [116]) and heat treatment seems to improve adhesion.

Corrosion Resistance

The parameters influencing the properties of electroless nickel coatings are numerous. It is even truer of the corrosion resistance of those coatings. Here follows a list of the influencing parameters, according to Cheong [122]:

- The composition of the coating (P and/ or B content);
- The homogeneity of elemental distribution on the surface and throughout the deposit;
- The surface morphology of the deposit;
- The microstructure of the deposit;
- The test medium (especially the pH);
- The presence of impurities and their distribution in the deposit;
- The nature of the substrate;
- The presence of pores and pits.

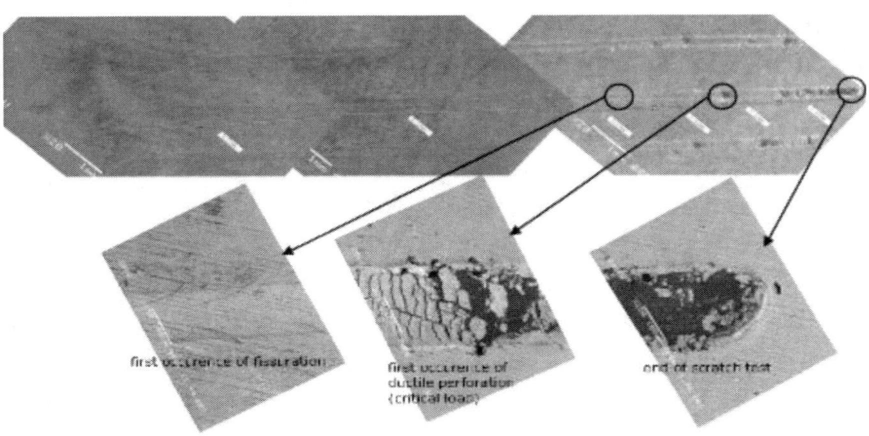

Figure 15. Observation of the scratch tracks up to 25 N on a 6 μm thick as-deposited electroless nickel-boron coating [26], with permission.

This list appears exhaustive but they are other factors that have to be taken into account. Most of them have an indirect influence on the corrosion behavior that occurs because of their effect on one of the pre-cited factors. Those supplementary parameters include:

- The surface state of the substrate. Lee [123] showed, on carbon fiber reinforced polymers plated with electroless nickel that the corrosion resistance of ENP in 3.5 wt.% NaCl solution increased with increasing substrate roughness. The important decrease in corrosion potential observed by Lee (from -363 to -505 mV Vs SCE) with the decrease of roughness (from 0.8 to less than 0.1 μm in Ra respectively) and the lack of general interest for substrate roughness in most publications shows that any comparison between coatings obtained by different groups will be excessively difficult.
- The nature of the stabilizer is also an important parameter. It may not influence the phosphorous or boron content of the coating but it influences the chemistry in a more substle way. Cheong [122] showed, on NiP, that superficial morphology and surface composition aspects (including the presence of impurities linked with the stabilizer) had a great influence on the corrosion resistance of the deposit. Liu [124] also showed that the incorporation of sulfur (due to the use of sulfur-based compounds as stabilizers) decreased the corrosion resistance in 3.5 wt.% NaCl but that the nitric acid corrosion resistance of the coating could be improved by the use of a well-chosen stabilizer, leading to a smooth compact surface.
- The nature of the pre-treatment applied to the coating is also often forgotten. Takacs showed [125] this by using various pre-treatment before midP EN plating of $AlMg_2$ and was able to obtain corrosion potentials from -415 to -514 mv Vs SCE in 0.5 M Na_2SO_4 solution at pH 3. Once again, this suggests that the comparison of results obtained by various authors will not allow to get information about the intrinsic corrosion behavior of electroless nickel coatings.
- Electroless nickel coatings are used to protect other materials from corrosion by the barrier method. Contrary to galvanization, the protecting role of the coating stops when the coating is first breached. It is thus evident that thickness (and in turn plating time) have an influence on corrosion behavior. Manna [126] showed, on a study of rebar surface protection, that the chemistry of ENP without replenishment was modified by plating time, together with the thickness. He obtained, for a single bath chemistry, values of E_{corr} (in a specific electrolyte at pH 7) from -0.35 to -0.40, depending on plating time. Those differences can be easily explained by variations of thickness, surface roughness and chemistry. But those parameters are often overlooked in corrosion studies where they are rarely available. Ashassi-Sorkhabi [73] observed similar results to those of Manna.
- The influence of replenishment is very rarely discussed in studies about electroless nickel plating but, together with thickness, it has a great influence on the corrosion resistance of the coating: the columnar nature of the deposit may lead to intercolumnar porosities and chemical heterogeneities that are privileged sites for corrosion initiation (porosity is in fact the first enemy of electroless nickel in the field of corrosion resistance). Vitry showed that batch replenishment brought a new phase of growth [20, 127], that could be compared to the formation of a second coating on

top of the existing one. This has a barrier effect on the propagation of transverse porosity that are stopped by the replenishment line and thus have lesser chances of reaching the substrate.

The number, and variety, of parameters that play a role in defining the corrosion resistance of electroless nickel coatings suggests caution when describing its evolution. General trends are often given by authors. For example, in its review [1], Hari Krishnan summarizes the subject as follow: alloys with more than 10 wt.% P are more resistant to attack than those with lower P content in neutral and acidic environment and those with low P content (3-4 wt.%) are more resistant to strong alkaline environment. Heat treatment lowers the corrosion resistance due to precipitation of nickel phosphide and the subsequent reduction of the P content of the remaining material. However, in a newer review, Das and Sahoo [35] report another behavior: 'only electroless Ni–high P coating is effective in offering an excellent protection, whereas electroless Ni–low P and Ni–medium P coatings are not recommended for severe environments'. They also say that electroless nickel-boron coatings have a lower corrosion resistance and attribute the reduction of corrosion resistance induced by heat treatment to the formation of grain boundaries in a previously amorphous material.

This shows that describing the corrosion resistance of electroless nickel coatings and comparing nickel-phosphorous and nickel-boron coatings is far from an easy task because there is no absolute for corrosion resistance and this property must always be examined in conditions as close as possible to those encountered by the product during its life-cycle. One interesting work about corrosion resistance of electroless nickel coatings was listed by Wolfgang Riedel in his book 'electroless nickel plating'. He gives corrosion rates (in μm/year) for Kanigen ENP (MidP EN), HighP EN and ENB in various media including foodstuff [17]. He concludes from the data available at the time of his writing that ENP presents a better corrosion resistance than ENB. However, ENB has known lots of developments since that time and the extensive testing work reported by Riedel has not been reproduced since, and the corrosion resistance of ENB coatings may have improved.

Nowadays, electrochemical techniques are the most popular ways of investigating the corrosion behavior of electroless nickel coatings. They present the advantage of being quick, and allow getting quantitative information. When electrochemical impedance spectroscopy is used, it is even sometimes possible to create an equivalent electrical circuit that models the behavior of the coated system. Polarization methods allow measuring corrosion current and potential and it is possible, from the shape of the polarization curve, to identify the mechanisms involved in corrosion.

A lot of researchers studied the corrosion resistance of ENP by electrochemical techniques, on various substrates and in various electrolytes. On Mg alloys, for example, Zhang [128] reported a E_{corr} value of −0.484 V Vs SCE in 3.5% NaCl for a 7.5wt.% P coating while Sun [129] obtained -0.323 V Vs SCE for a 10 wt.% P coating and Zhao [130] -1.316 V Vs SCE (for relatively thin coatings). This is a perfect example that, even on similar substrates, the P content of the coating is not a sufficient parameter to define corrosion resistance of electroless nickel coatings.

On the effect of P (and S) content, Cui observed [131], on steel substrates (but without information about thickness), that the incorporation of P improved the corrosion resistance (measured by EIS, figure 16) while S (from the organic stabilizer) decreased it. This was confirmed by results obtained by Sankara Narayanan [72] who measured the E_{corr} for 24 μm thick coatings on mild steel in 3.5% NaCl and obtained -0.536, -0.434 and -0.411 V Vs SCE for 3.34, 6.7 and 13.3 wt.% P respectively.

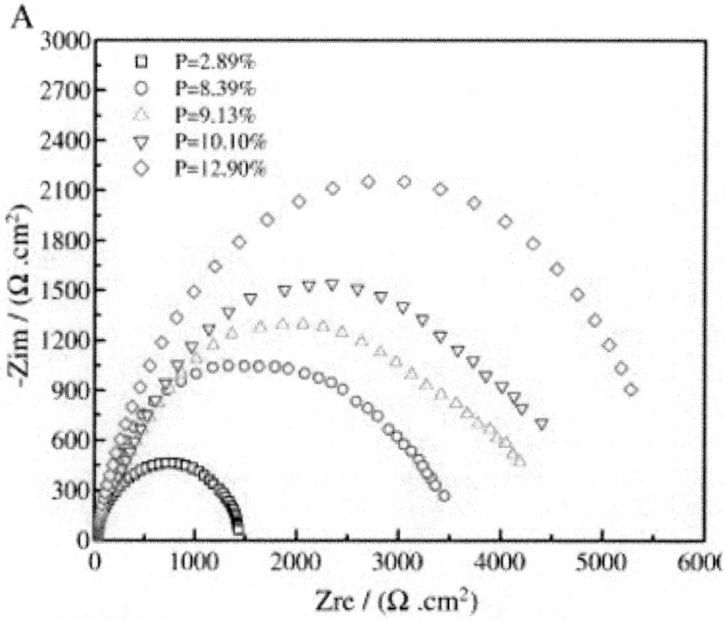

Figure 16. Nyquist plots of electroless nickel coatings in as-plated (A) conditions at various phosphorus contents on the pre-etch coatings in 3.5% sodium chloride solution [131], with permission.

Anik [132], on Mg alloys, showed an important reduction of E_{corr} after heat treatment, in a pH 7 solution of NaH_2PO_4, KOH and K_2SO_4 but did not give information about the P content of his coating. However, Bigdeli [133] reported an improvement of E_{corr} from -0.48 to -0.304 V Vs SCE after heat treatment at 400°C for 1h in the case of a 20 μm thick 9-10 wt.% P coating.

Several authors proposed equivalent circuits for their coatings. They are usually composed of 3 elements: Rs (resistance of the solution), Rc (resistance of the coating) and a constant phase element [72, 131, 133].

There are also studies on corrosion of ENB. Values of E_{corr} in 3.5% NaCl on steel for as-plated coatings are usually close to -.055 V Vs SCE [85, 92, 96] but higher values were obtained recently [27, 134].The values reported for ENB are relatively similar to those obtained for ENP and the best ones are better than Typical E_{corr} of ENP. There is no study that investigates the effect of stabilizer on the corrosion resistance of ENB and the differences observed between Vitry and Das' results ant the other ones may be due to other parameters. However, it is interesting to note that the best results for the corrosion resistance of ENB coatings were both obtained by lead stabilized baths [27, 134]. Contrary to ENP, the corrosion resistance of ENB is not decreased (or not much decreased) by heat treatment [27, 85, 92, 96, 134].

Sankara Narayanan [85] proposed for ENB an equivalent circuit, similar to that used to model ENP, which was also proposed by Vitry [16]. She also proposed a more complex equivalent circuit that can model the behavior of ENB coatings on steel after immersion in NaCl for 12 hours and more [16].

The only study where corrosion resistance was measured on both ENP (9 wt.% P) and ENB (6wt.% B, thallium stabilized) by the same group was carried out by Sankara Narayanan [94]. The obtained, on mild steel, in 3.5 wt.% NaCl, with 20 μm thick coatings, values of -0.35 and -0.51 V Vs SCE for ENP and ENB respectively. After heat treatment, they observed an important decrease of the E_{corr} value for ENP but only a slight one for ENB, leading to -0.49 and -0.52 V Vs SCE respectively.

Conclusion

It is difficult to compare ENP and ENB coatings, firstly because very few researchers did carry out work on both kinds of coatings. Sankara Narayanan and his group [94] provide the best results for a combined study. This is of course not the only reason why a comparison is so difficult: the variety in the composition of coatings does not help for this comparison, nor does the lack of information about chemistry sometimes found in papers. Moreover, a lot of parameters other than the P and B content of the coatings have a great influence on the properties of electroless coatings. There are matters were a definite trend can be established. For example: ENB presents a higher hardness and better resistance to abrasive wear. However, it is not possible to establish such trends for all the properties investigated because, for some of them, ENB coatings have a behavior intermediate to the various types of ENP coatings or just because the present state of research does not allow the identification of trends.

References

[1] Hari Krishnan, K. ; John, S. ; Srinivasan, K.N ; Praveen, J. ; Ganesan, M. ; Kavimani, P.M. *Metal. Mater. Trans.* 2006 ; *37A* ;1917–1926.

[2] Wurtz, A. *Comp. Rendus Acad. Sci.* 1844; *18*; 702.

[3] Paal C.; Frederici, L. *Ber. Deutschen Chem. Gesellschaft* 1931; *64B;* 1766.

[4] Scholder, R.; Heckal, H Z. *UnorganischeAllgemeineChemie*1931; *198;* 329.

[5] Scholder, R.; Haken, H.I.*Ber. Deutschen Chem. Gesellschaft* 1931; *64B;* 2870.

[6] Bretean, P. *Bull. Soc. Chim.* 1911; *9;* 515.

[7] Roux, F.G. *U.S. Patent no. 119612870*, 1916.

[8] Brenner, A.; Riddell, G.E. *Proc. Am. Electropl. Soc.* 1946; *33*, 4-12.

[9] Brenner, A.; Riddell, G.E. *Proc. Am. Electropl. Soc.* 1947; *34;* 156.

[10] Brenner, A.; Riddell, G.E. *U.S. Patent no. 21950532* 1950.

[11] Sherwood, B. *metalfinishing* 2008; *6;* 92-94.

[12] Mayer, L. *Prod. Finish.* 1993; *6;* 48-53.

[13] Knanishu, J. *Armed services technical information agency; Unclassified report n°61-4359;* 1961.

[14] Schlessinger, H.I. ; Brown, H.C. *U. S. patent 2,461,663* 1949.

[15] Warf, J.C. ; Schaltegger, K.H. *US patent 2.726.170* 1955.

[16] Vitry, V. *Electroless Nickel-Boron deposits: Synthesis, formation and characterization; Effect of heat treatments; Analytical modeling of the structural state, PhD Thesis,* University of Mons : Mons BE, 2010.

[17] Riedel, W *Electroless Nickel Plating;*ASM International: Metals Park, OH ; 1991.

[18] Delaunois, F *les dépôts chimiques de nickel-bore sur alliages d'aluminium, PhDthesis,* Faculté polytechnique de Mons : Mons BE, 2000.

[19] Ohno, I. *Mater. Sc. Eng. A* 1991 ; *146 ;* 33-49.

[20] Vitry, V. ; Kanta, A.F. ; Delaunois, F. *Ind. Eng. Chem. Res.* 2012 ; *51 (27) ;* 9227–9234.

[21] Vitry, V. ; Sens, A ; Kanta, A.F. ; Delaunois, F. *Appl. Surf. Sc.* 2012 ; *263 ;* 640-647. (

[22] Kanta, A.F. ; Vitry, V. ; Delaunois F. *JAlloys Compd* 2009 ; *486 ;* L21–L23.

[23] Kanta, A.F. ; Poelman, M. ; Vitry, V. ; Delaunois, F. *J. All. Comp.* 2012 ; *505 (1)* ; 151-156 .

[24] Kanta, A.F. ; Vitry,V. ; Delaunois, F. *Mater. Sc. Forum.* 2012 ; *638-642* ; 846-851.

[25] Vitry, V. ; Delaunois, F. ; Dumortier, C. *Surf. Coat. Technol.* 2008 ; *202 (14) ;* 3316-3324.

[26] Vitry, V. ; Kanta, A.-F. ; Delaunois, F. *Surf. Coat. Technol.,* 2011 ; *206 (7)* ; 1879-1885.

[27] Vitry, V. ; Kanta, A.F. ; Delaunois, F. *Mater. Sc. Eng. B* 2010 ;*175 (3) ;* 266-273.

[28] Kanta, A.F. ; Vitry, V. ; Delaunois, F *Mater. Sc. Forum.* 2010 ; *638-642 ;* 846-851.

[29] Vitry, V. ; Kanta, A.F. ; Sens, A. ; Delaunois, F *Adv. Mater. Res.* 2012 ; *409 ;* 808-813

[30] Vitry, V. ; Kanta, A.F. ; Dille, J. ; Delaunois, F ; *Surf. Coat. Technol.,* 2012 ; *206 (16)* ; 3444-3449.

[31] Delaunois, F ; Lienard, P. *Surf. Coat. Technol.* 2002 ; *160 ;* 139–148.

[32] Delaunois, F. ; Petitjean, J.P. ; Lienard, P. ; Jacob-Duliere, M. *Surf. Coat. Technol.,* 2000, *124,* 201–209.

[33] Anyk, M. ; Korpe, E ; Sen, E *Surf. Coat. Technol.* 2008 ; *202 ;* 1718-1727.

[34] Correa, E. ; Zuleta, A.A. ; Sepúlveda, M. ; Guerra, L. ; Castaño, J.G. ; Echeverría, F. ; Liu, H. ; Skeldon, P. ; Thompson, G.E. *Surf. &Coat. Technol.* 2012 ; *206* ; 3088–3093.

[35] Sahoo, P. ; Das, S.K. *Mater. Design,* 2011 ; *32* ; 1760-1775.

[36] Correa, E. ; Zuleta, A.A. ; Guerra, L. ; Gómez, M.A. ; Castaño, J.G. ; Echeverría, F. ; Liu, H. ; Baron- Wiecheć, A. ; Hasimoto, T ; Skeldon, P. ; Thompson, G.E. *Surf. Coat. Technol.* 2013 ; *232* ; 784-794.

[37] Wang, Z.C. ; Jian, F. ; Yu, L. ; QI, Z.B. ; Tang, Y. ; Song, G.-L. *Surf.Coat.Technol.* 2012 ; *206 ;* 3676-3685.

[38] Bulbul, F. *Met. Mater. Trans.* 2011 ; *17(1) ;* 67-75.

[39] Pal, S.; Verma, N.; Jayaram, V.; Biswas,S.K.; Riddle, Y. *Mater.Sci. Eng. A* 2011 ;*528* ; 8269-8276.

[40] Das, S.K ; Sahoo, P. (2012).Influence of Process Parameters on Microhardness of Electroless Ni-B Coatings. *Advances in Mechanical Engineering* [Article ID 703168], 2012.

[41] Gao, M.S.; Bai, S.L. *Plating Finish.* 1987; *9(1);* 12-14.

[42] Gervasio, D. ; Xu, M. ; Thomas, E. (2005) *Properties of aqueous alkaline sodium borohydride solutions and by-products formed during hydrolysis (presented at the Fuel cell design, fabrication and materials selection workshop, University Illinois Urbana Champaign, july 26-27,*

*2005).*http://fsl.ne.uiuc.edu/Project%20Presentation/fuel%20cell%20project_files/July %20workshop%20presentations/uiuc-talk-25July2005.pdf
[43] Bielinski, J.; Krol,J.; Bielinska, A. *Bull. Electrochem.* 1990; *6;* 828–831.
[44] Werner, C.; Stark, F.J.; Horsthemke, H.; Kraft, H.J. *Electroless Nickel Evolution: From Lead to Metal-free Stabilized Systems, Orale presentation, Eurocorr,* 2013.
[45] Cheong, W.J.; Luan, B.L.; Shoesmith, D.W. *Applied Surface Science* 2004; *229;* 282-300.
[46] Gorker, L.; Shyriak, I. *Metal Finish.* 1996; *6*; 104-105.
[47] Huang, Y.S.; Cui, F.Z. *Surf. Coat. Technol.* 2007; *201;* 5416-5418.
[48] Xiao, Z.; Wang, W.; YE, L.; Sha, Y.; Song, T. *Surf. Coat. Technol.* 2008; *202(20);* 5008-5011.
[49] Baskaran, I.; Sankara Narayanan, T.S.N.; Stephen, A. *Mater. Chem.Phys.*2006; *99;* 117–126
[50] Liu, H.P.; LI, N.; Bi, S.F.; Li, D.Y.; Zou, Z.L. *Chem. Res. Chinese Univ.* 2008; *24(1)*; 101-105.
[51] Zhang, S.; De Baets, J.; Vereeken, M.; Vervaet, A.; Van Calster, A. *J. Electrochem. Soc.* 1999; *146(8);* 2870-2875.
[52] Etcheverry, B. Adhérence, mécanique et tribologie des revêtements composites NiP – Talc multifonctionnels à empreinte écologique réduite, PhDThesis, Institut National Polytechnique De Toulouse ; Toulouse FR ; 2006.
[53] Keong, K.G. ; Sha, W.; Malinov, S. *Surf. Coat. Technol.* 2004; *168;* 263-274.
[54] Riddle, Y.W.; Bailer, T.O. *JOM* 2005; *4;* 40-45.
[55] Rao, Q.L.; Bi, G.; Lu, Q.H.; Wang, H.W.; Fan, X.L. *Appli. Surf. Sc.* 2005 ; *240 ;* 28-33.
[56] Tsai, T.K.; Chao, C.G. *Appl.Surf. Sc.* 2004; *233;* 180-190.
[57] Contreras, A. ; León, C. ; Jimenez, O. ; Sosa, E. ; Pérez, R. *Appl. Surf. Sc.* 2006 ; *253 (2)* ; 592-599.
[58] Das, S.K. ; Sahoo, P. *Proceedings Int. Conf. Mech. Eng.ICME;* 2009 (paper reference: ICME-RT-03).
[59] Yoon, J.W.; Park, J.H.; Shur, C.C.; Jung, S.B. *Microelectronic Eng.*2007; *84,* 2552–2557.
[60] Hou, H.; Li, Y.; Wang, F. *Corrosion Science* 2004; *46;* 1467–1477.
[61] Liu, H.; Bi, S.; Cao, L.; Bai, Q.; Teng, X.; Yu, Y. *Int. J. Electrochem. Sci.* 2012; *7;* 8337 – 8355.
[62] Sha, W.; Nurzairyani, H.J.; Zairin, M.; Wu, X. *Microsc. Anal.* 2010; *6;* 11-14.
[63] Park, S.H.; Lee, D.N. *J. Mater. Sc.* 1988; *23;* 1643-1654.
[64] El Mahallawy, N.; Bakkar, A.; Shoeib, M.; Palkowski, H.; Neubert, V. *Surf. Coat. Technol.* 2008; *202;* 5151-5157.
[65] Taheri, R. *Evaluation of Electroless Nickel-Phosphorus (EN) Coatings, PhD Thesis;* University of Saskatchewan, Saskatoon: Saskatchewan, CA, 2002.
[66] Alirezaei, S.; Monirvaghefi, S.M. ; Saatchi, A. ; Ürgen, M.; Kazmanlı, K. *Tribol. Intl* 2013; *62;* 110–116. .
[67] volume 3 : Alloy Phase Diagrams of ASM handbook online. ASM international, 2005.
[68] Hentschel, T.; Isheim, D.; Kirchheim, R.; Muller, F.; Kreye, H. *Acta Mater.*2000; *48;* 933–941.
[69] Farber, B.; Cadel, E.. Menand, A.; Schmitz, G.; Kirchheim, R., *Acta Mater.*2000; *48;* 789–796.

[70] Kumar, P.S.; Nair, P.K. *Nanostructu. Mater.* 1994 ; *5 ;* 183-198.

[71] Watanabe, T.; Tanabe, Y. *Trans. Jpn. Inst. Metals* 1983; *24;* 396-404.

[72] Sankara Narayanan, .S.N.; Baskaran, I.; Krishnaveni, K.; Parthiban, S.*Surf. Coat. Technol.* 2006; *200(11);* 3438-3445.

[73] Ashassi-Sorkhabi, H.; Rafizadeh, S.H. *Surf. Coat. Technol.* 2004; *176;*318–326.

[74] Jeong, D.H.; Erb, U.; Aust, K.T.; Palumbo, G. *Scripta Mater.* 2003; 48; 1067–1072.

[75] Jiaqiang, G.; Yating, W.; Lei, L.; Bin, S.; Wenbin, H. *Mater. Letters* 2005; *59;* 1665–1669.

[76] Gaevskaya, T.V.; Novotortseva, I.G.; Tsybulskaya L.S. *Metal finishing* 1990; *6;* 100–103.

[77] Guo, Z. ; Keong, K.G.; Sha, W. *J. All. Compds*2003; *358*; 112–119.

[78] Martyak, N.; Drake, K *J. All. Compds* 2000; *312;* 30-40.

[79] Zhang, Y.Z.; WU, Y.Y.; Yao, M. *J. Mater. Sc. Let.* 1998; *17;* 37-40.

[80] Kumar, S.; Nair, P.K. *J. Mater. Proc. Technol.* 1996; *56;* 511-520.

[81] Keong, K.G. ; Sha, W.; Malinov, S. *J. All. Compds* 2002; *334 (1-2);* 192-199.

[82] Keong, K.G. ; Sha, W.; Malinov, S. *J. Mater. Sc.* 2002; *37 (20);* 4445-4450.

[83] Keong, K.G. ; Sha, W. Malinov, S. *Mater. Sc. Eng. A* 2004; *365 (1-2);* 212-218.

[84] Staia, M.H.; Puchi, E.S.; Castro, G.; Ramirez, F.O.; Lewis, D.B. *Thin Solid Films* 1999; *355-356;¸472-479.*

[85] Sankara Narayanan, T.S.N. ; Seshadri, S.K. *J. All. Comp.* 2004 ; *365 ;*197–205.

[86] Li, H.; Li, H.; Dai, W.; Qiao, M. *Appl. Catal.* 2003 ; *A238 ;* 119-130.

[87] Shi, Z. ; Wang, D. ; Ding, Z. ; *Appl.Surf.Sc.* 2004 ; 221 ; 62-68.

[88] Anik, M.; Korpe, E.; Sen, E. *Surf. Coat. Technol.* 2008; *202;* 1718-1727.

[89] Gawne, D.T.; Ma, U. *Wear* 1987; *120;* 125-149.

[90] Agarwala, R.C. ; Agarwala, V. *Sadhana* 2003, *28(3-4),* 475-493.

[91] Yan, M.; Ying, HG.; Ma, TY. *Surf. Coat. Technol.* 2008; *202(4);* 5909-5913.

[92] Dervos, C.T ; Novakovic, J. ; Vassiliou, P. *Mater. Let.* 2004 ; *58* ; 619–623.

[93] Oraon, B. ; Majumdar, G. ; Ghosh, B. *Mater. Design* 2007 ; *28 ;* 2138-2147.

[94] Sankara Narayanan, T.S.N. ; Krishnaveni, K. ; Seshadri, S.K. *Mater. Chem. Phys.,* 2003 ; *82 (3)* ; 771-779.

[95] Abdel Hamid, Z.; Hassan, H.B.; Attya, A.M. *Surf. Coat. Technol.* 2010; *205;* 2348-2354.

[96] Srinivasan, K. N.; Meenakshi, R.; Santhi, A.; Thangavelu, P. R.; John, S. *Surf. Eng.* 2012; *26(3);* 154.

[97] Oraon, B. ; Majumdar, G. ; Ghosh, B. *Mater. Design* 2008 ; *29 (7) ;* 1412-1418.

[98] Evans, W.T. ; Schlesinger, M. *J. Electrochem. Soc.,* 1994 ; *141 ;* 78-83.

[99] Hutchings, I.M. *Tribol. Int.* 1998; *31 (1-3);* 5-15.

[100] Palaniappa, M. ; Seshadri, S.K.*Wear* 2008; *256(5-6);* 735-740.

[101] Staia, M.H.; Castillo, E.J.; Puchi, E.S.; Lewis, B.; Hintermann, H.E. *Surf. Coat. Technol.* 1996; *86-87*; 598-602.

[102] Sharma, A.; Singh, A.K. *Cent. Eur. J. Eng.* 2011; *1(3);* 234-243.

[103] Yun L.; Zhang, X. *Thin Solid Films* 1993; *229;* 76-82.

[104] Taheri, R. ; Oguocha, I.N.A.; Yannacopoulos, S. *Wear* 2001; *249;* 389-396.

[105] Novak, M., Vojtech, D.; Zelinkova, M.; Vitu, T. *Proceedings Metal 2009*, 2009.

[106] Liu, Z. ; Gao, W. *Surf. Coat. Technol.* 2006 ; *200 ;* 3553-3560.

[107] Ramalho, A. ; Miranda, J.C. *Wear* 2007 ; *263(1-6) ;* 592-597.

[108] Krishnaveni, K. ; Sankara Narayanan, T.S.N. ; Seshadri, S.K. *Surf. Coat. Technol.*2005; *190;* 115–121

[109] Trzaska, M. *J. Achiev. Mater. Manufact. Eng.* 2010; *43(1);* 269-275.

[110] Correa, E; Zuleta, A.; Guerra, L.; Gómez, M.A. ; Castaño, J.G.; Echeverría,F.; Liu,H.; Skeldon, P.; Thompson, G.E. *Wear* 2013; *305(1-2);* 115-123.

[111] Kaya, B.; Gulmez, T. Dermikol, M. *Proceedings of the world congress on Engineering and Computer Science* 2008

[112] Brooman, E.W. *Metal. Finish.* 2004; *7*; 74-82.

[113] Apachitei, I.; Duszczyk, J. *Surf. Coat. Technol.* 2000; *132;* 89-98.

[114] Allen,; K.; Torrance, A.A.; *Tribol. Int.* 1986; *19(3);* 123-127.

[115] Riddle, Y.W. ; Bailer, T.O. *JOM* 2005 ; *4 ;* 40–45.

[116] Sander, T.. Tremmel, S.; Wartzack, S. *Surf. Coat. Technol.* 2011; *206(7);* 1873-1878.

[117] Bull, S.J. ; Berasetegui, E.G., *Tribol. Int.* 2006; *39;* 99-114

[118] Jin, J.G.; Lee, S.K.; Kim, Y.H. *Thin Solid Films* 2004; *466;* 272-278.

[119] Liu, Z.. Gao, W. *Int. J. Mod. Phys.* 2006; *20(25-27);* 4637-4642.

[120] Lin, C.J.; Chen, K.C.; He, J.L. *Wear* 2006; *261*; 1390–1396.

[121] Alexis, J.; Gaussens, C.; Etcheverry, B.; Bonino, J.P. *Mat. Chem. Phys*2013; *137;* 723-733.

[122] Cheong, W.J.; Luan, B.; Shoesmith, D. *Corr. Sci.* 2007; *49;* 1777-1798.

[123] Lee, C.K. *Mat. Chem. Phys.* 2009; *144(1);* 125-133.

[124] Liu, H.P.; Li, N.; Bi, S.F.; Li, D.Y.; Zou, Z.L. *Thin Solid Films* 2008; *516(8);*1883-1889.

[125] Takács, D.; Sziráki, L.; Török, T.I.; Sólyom, J.; Gácsi, Z.; Gál-Solymos, K. *Surf. Coat. Technol.* 2007; *201;* 4526-4535.

[126] Manna, M.; Bandyopadhyay, N.; Bhattacharjee, D.*Surf. Coat. Technol.* 2008; *202(14);* 3227-3232.

[127] Vitry, V. ; Sens, A.; Kanta, A.-F.; Delaunois, F.*Surf. Coat; Technol.* 2012; *206(15);* 3421-3427.

[128] Zhang, H.; Wang, S.; Yao, G.; Hoa, Z. *J. All. Compds.* 2009; *474;* 306-310.

[129] Sun, S.; Liu, J.; Yan, C.; Wang, F. *Appl. Surf. Sc.* 2008; *254;* 5016-5022.

[130] Zhao, H.; Huang, Z.; Cui, J. *Surf. Coat. Technol.* 2007; *202;* 133-139.

[131] Cui, G.; Li, N.; Li, D.; Zheng, J.; Wu, Q. *Surf. Coat. Technol.* 2006; *200(24);* 6808-3814.

[132] Anik, M.; Körpe, E. *Surf. Coat. Technol.* 2007; *201;* 4702-4710.

[133] Bigdeli, F.; Allahkaram, S.R *Mater. Design* 2009; *30(10);* 4450-4453.

[134] Das, S.K.; Sahoo, P. *Portugaliae Electrochim. Acta* 2011; *29(4);* 211-231.

In: Comprehensive Guide for Nanocoatings Technology. Vol. 1 ISBN: 978-1-63482-447-7
Editor: Mahmood Aliofkhazraei

Chapter 8

SOL-GEL-BASED POLYMERIC COATINGS ON SURFACES OF MATERIALS

***L. Saeednia*[1] *and R. Asmatulu*[1,•]**

[1]Department of Mechanical Engineering,
Wichita State University, Wichita, KS, US

ABSTRACT

Sol-gel is a wildly used method in many areas of science, engineering and technology, and has been gaining much attention worldwide. One of the most important applications of sol-gel technology is the thin film coating on the surfaces of materials against corrosion, degradation, other environmental effects, and decorations purposes. Sol-gel coating mainly contains four steps, including the gel formation, aging, drying and heat treatment, and can also be in various forms, sizes and shapes. Compared to the conventional coating methods, sol-gel coating process contains considerably less equipment and labor; thus, it is more economic and an easy process for numerous applications. Furthermore, the sol-gel technique is an environmentally friendly method which can be a potential replacement method for the toxic pretreatments of some other conventional methods (such as, hexavalent chromium conversion, surface acid/base treatment, etching, and sand blasting), and they can be in both organic (polymeric) and inorganic (metals, alloys, oxides and composites) forms. Among the sol-gel coating applications, the protective coatings at various thicknesses (from nanoscale to micro scale) have gained significant attention because of the excellent chemical stability at room and elevated temperatures, oxidation control, heat dissipation, and enhanced corrosion and UV resistance. In the present book chapter, we have analyzed all the latest developments in the sol-gel coatings and their applications on various materials (e.g., stainless steel, aluminum, magnesium, copper and their alloys), progressions and challenges, and future trends in the field. We believe that this study will greatly provide several benefits for the readers and companies in the coating and related businesses.

• E-mail: ramazan.asmatulu@wichita.edu.

Keywords: Corrosion, degradation, sol-gel, coating, protection, surface treatments, recent development

1. INTRODUCTION

1.1. Brief History of Sol-Gel

Sol-gel technology has been of great interest worldwide for several decades for various industrial applications. The sol-gel method was first developed by French chemist, Ebelmen, to prepare new oxide material by slow hydrolysis in 1845 [1]. Ceramics and glasses have been prepared by sol-gel technique for a long time and used for different purposes. In 1930, the sol-gel process was used for oxide layer developments on industrial glasses. This achievement was established by Geffcen and Berger from the Shot Company using different metals combined with precursors by dip coating; however, those did not get much attention until D. Roy and R. Roy fabricated more homogeneous glasses using the sol-gel process [2]. Sol-gel method can be used to fabricate various products. Figure 1 shows different types of material can be made from sol-gel method [3].

One of the most attractions of the sol-gel techniques is to provide a thin layer of coating on multiple surfaces of glass, ceramic, metal and/or composite substrates [4].

In 1939, thin silicate layer of a sol-gel coating was fabricated using a dip coating technique by Jenaer Glaswer Schott and Gen. The investigators received the first patent on sol-gel technology from their pioneering research studies [5]. The first commercial application of sol-gel coatings was in the early 1960s.

At that time, several investigations were carried out to produce different ceramics and glasses from the gel precursors at lower temperatures. Among them, the study on synthesis of boron-silicate glass using sol-gel process at low temperature gained most significant attention globally.

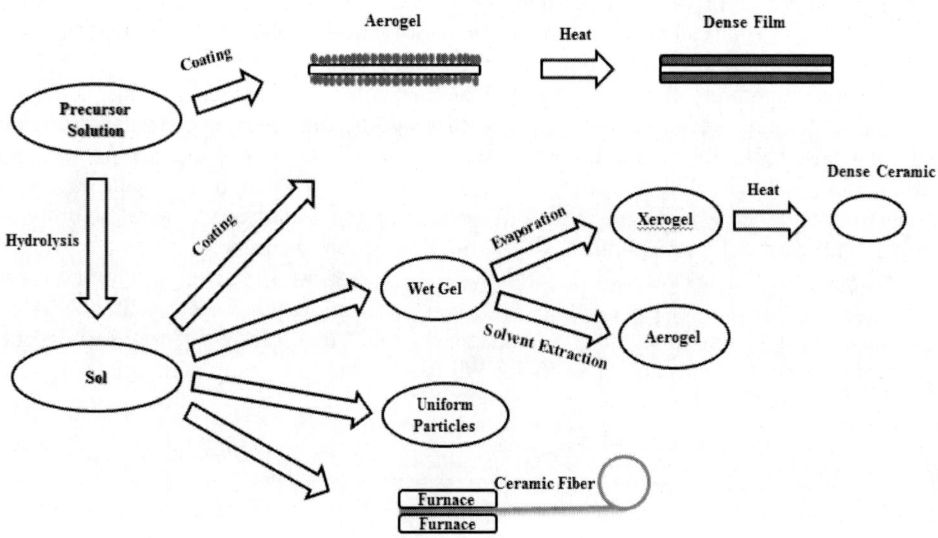

Figure 1. Different Products of Sol-Gel Technology.

The test results opened up new possibilities to further develop new materials and technologies for different industries [6].

According to the various reports, many research studies were carried out on the fabrication of new glasses and ceramics, as well as homogenous thin films and fibers using the sol-gel processes [1, 2, 4, 7]. Therefore, the number of publications in this area grew exponentially at the beginning of the 1980s. At the same time, the first international conference on glasses and glass ceramics obtained from sol-gels were held in Italy. This event brought a significant development in sol-gel technology, and since then the similar conferences with more advanced progresses have been held worldwide. Furthermore, in 1993 the first issue of the *Journal of Sol-Gel Science and Technology* was published which provided wide ranges of pioneering investigations in the sol-gel field. Ever since, sol-gel technology has gained a great deal of attention in materials science, especially in the fields of ceramics, polymer chemistry, organic and inorganic chemistry, physics and many engineering field [8]. The sol-gel coating is still wildly used in numerous applications, such as protective and optical coatings, semi-conductive and superconductive films, magnetic films, thermally conductive layers, high strength thin films, and so on [5].

There have been valuable proceeding published in research on sol-gel technology around the world. Some of the pioneering works in the sol-gel technology are summarized in table 1.

2. Sol-Gel Chemistry

Sol-gel process is defined as the method of producing solid materials from a number of different solutions at lower temperatures compared to the other ceramic and glass processes. Sol-gel technique was initially developed for the production of inorganic materials, such as ceramics and glasses [10].

In the sol-gel process, a system of colloidal particles in a solution (sol) becomes a network material (gel), which is interpenetrated by a liquid/solvent. The colloidal solution is a suspension in which the dispersed particles are so small (1-1000 nm); thus, the gravitational force is mainly insufficient and can be neglected during the process. In this system, short-range forces such as van der Waals attraction, hydrogen bonding and surface charges are major forces/interactions. A gel is a 3D porous solid network at which the pores are filled mainly by the liquid (usually the original solvent of the sol). The sol-gel process basically consists of two chemical reactions: *hydrolysis reaction* followed by the *condensation reaction* [11]. The hydrolysis reaction leads to the formation of the sol, while the condensation reaction leads to the formation of the gel in the sol-gel structure. Figure 2 shows the schematic view of the sol-gel process.

Sol-gel processing is known in two different ways: particulate (or colloidal) and polymeric route. When using organic precursor, a network of colloidal suspension of very small particles (1-100 nm) is developed, and then the gel is simultaneously formed in this process.

While in the case of metal organic method which is more common than the organic method, the sol-gel process starts with a solution of monomeric precursors in a low-molecular weight organic solvent, and develops a polymeric gel [12]. These two different routes of sol-gel process are illustrated in Figure 3.

Table 1. Pioneering developments in sol-gel technology

Year	Author	Innovation
1845	Ebelman	Formation of new transparent material as a result of slow hydrolysis [1]
1930	Geffcen and Berger	Development of sol-gel process for oxide layers [1]
1939	JenaerGlaswer Schott and Gen.	Formation of thin film by a dip-coating process (first patent) [5]
1950	Della and Rustum Roy	fabricated homogeneous silicon tetraethoxide and metal nitrate salt glasses using the sol-gel process [6]
1969	H. Schroeder	Deposition of transparent coating on glass [6]
1971	Dislich	Synthesis of boronsilicate glass by heating oxide powder at low temperature [1]
1984	Avnir et al.	entrapment of the dye molecules in a porous silica cage [8]
1985	Schmidt	Dvelopment of inorganic-organic hybrids [8]
1990	Braun et al.	Immobilization of protein in sol-gel silica matrix [9]
1994	J.w. Fleming et al.	Fabrication of optical fibers using sol-gel [8]
1995	E.J.A. Pope	Development of living ceramics by incorporation of living cells into porous oxide matrices [8]
1997	T. Minami et al.	Production of super-repellent coating [8]

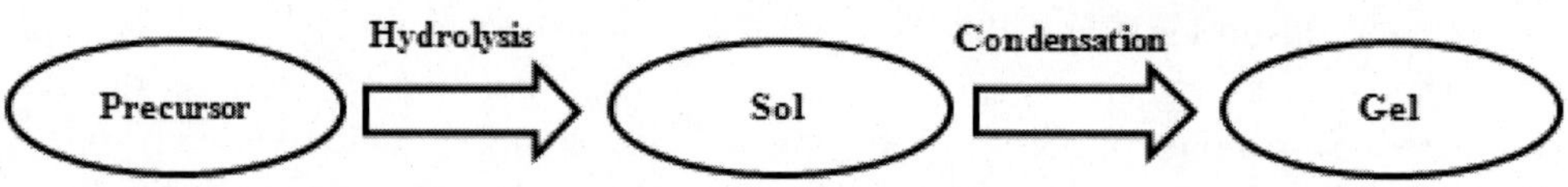

Figure 2. The schematic representation of the sol-gel process reactions.

Metal alkoxides, $M(OR)_n$, are the most commonly used metal-organic precursors. Here, M represents a network-forming element, such as Si, Ti, Zr, Al, Fe, and B, while R is an alkyl group.

During the hydrolysis reaction, water is added to metal alkoxide for a chemical reaction to make a metal hydroxide. Usually, an alcohol is another by-product of the hydrolysis reaction. In the next step, two metal hydroxides react to form a condensation reaction and produce a metal oxide. One molecule of water is also released in this reaction. The general reactions are as follows [13]:

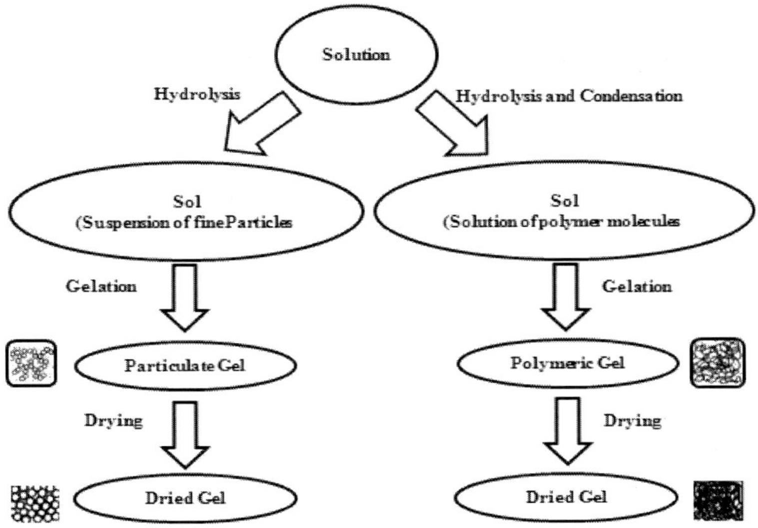

Figure 3. Two routes of sol-gel process.

$$M(OR)_n + (n/2)\ H_2O \longrightarrow M(OH)_n + n\ ROH\ \text{(HYDROLYSIS)}$$
$$M(OH)_n \longrightarrow MO_{n/2} + (n/2)\ H_2O\ \text{(CONDENSATION)}$$
$$M(OR)_n + (n/2)\ H_2O \longrightarrow MO_{n/2} + n\ ROH\ \text{(OVERALL)}$$

There are primarily four stages of sol-gel process, including ***i***) hydrolysis process, ***ii***) condensation for the polymerization of monomers to form chains and particles, ***iii***) particle growing, and ***iv***) formation of polymeric network structure throughout the liquid medium due to the agglomeration which results in thickening of liquid media and formation of gel. Heating and densification processes follow the primary stages of the sol-gel process [13].

Figure 4 shows the sol-gel process with different temperatures along with the microstructures of target products. As is observed, the sol-gel conversion usually takes place at lower temperatures and can result in a formation of different target products, such as coating films, fibers (through drawings or electrospinning) and molded bulky shapes. Coating films are the most important products of the sol-gel method [7]. Once the hydrolysis reaction initiates, the condensation reaction takes place at the same time. Alcohol and water are usually the by-products of both hydrolysis and condensation reactions. These small molecules are driven off during the drying step. Initial reaction conditions, such as pH, temperature, molar ratios of reactants, agitations, catalyst, duration and solvent composition mainly affect the sol-gel process [10].

3. Sol-Gel Derived Coatings

3.1. Advantages of Sol-Gel Coatings

Sol-gel method is one of the simplest methods to apply on the surfaces of various metals, ceramics and glasses.

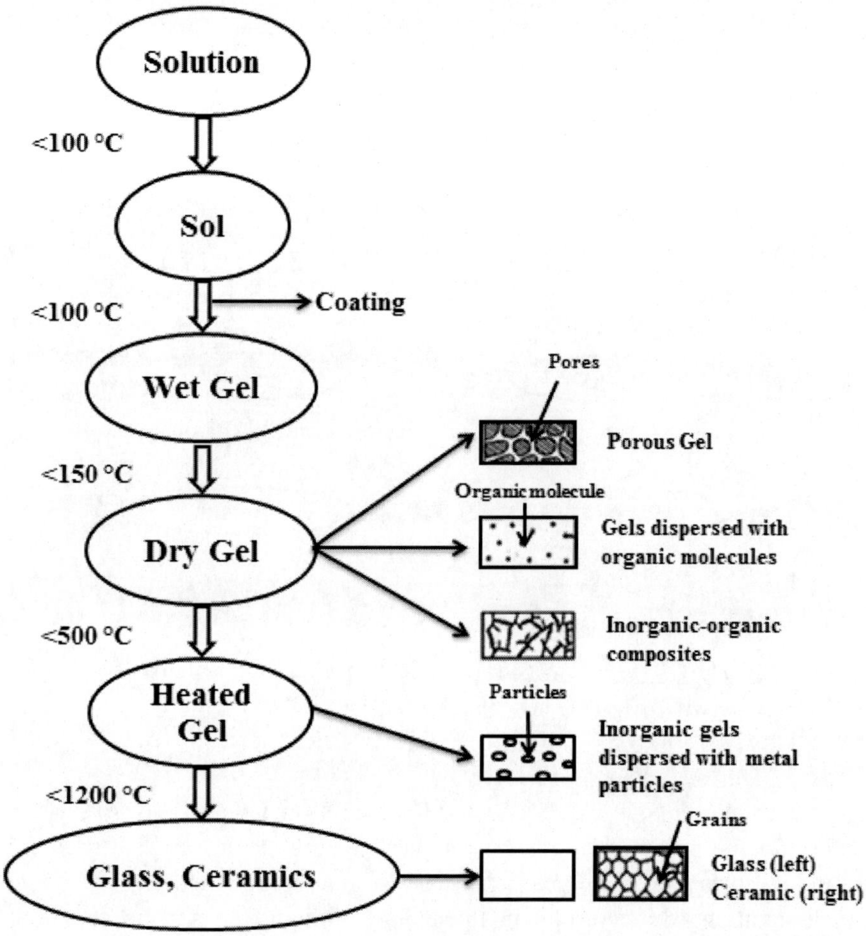

Figure 4. The schematic diagram of the sol-gel process at different temperatures and microstructures.

Formation of a thin film is one of the oldest commercial applications of sol-gel technology. As mentioned previously, the formation of a silica film by sol-gel process brought the first issued patent in 1939 to the Glaswerk Schott and Gen. [10].One of the unique characterizations of this method is to provide any single or multi component oxide coatings. Sol-gel coatings can improve corrosion, optical, electronic, magnetic, and other physical and chemical properties of the materials. For the metals, sol-gel coatings are typically applied to prevent corrosion formations. For instance, wildly used ceramic coating films provide a protection layer against corrosion and oxidation in stainless steel, aluminum, silver, magnesium and copper substrates [14].

Sol-gel coating mainly contains four steps, including the gel formation, aging, drying and heat treatment. Sol-gel process has gained much significant attention compared to other deposition techniques, such as chemical vapor deposition (CVD), physical vapor deposition (PVD), evaporation or sputter coating since its formation requires considerably less equipment, energy and labors, and is potentially an inexpensive process for larger scale applications. The other advantages of sol-gel process are listed below [15]:

- Sol-gel coating shows good adhesion to the solid surface
- Thicker layer of protection can be provided by sol-gel techniques
- Sol-gel coating can be shaped in complex geometric configurations
- High purity products can be produced by sol-gel method since the composition of the precursor and solvent are highly controllable
- Sintering temperature can be as low as 200-600°C depending on the precursors and substrates.

Although the sol-gel coating method provides many advantages, there are still some limitations that restrict its full industrial potential. Low wear resistance, weak bonding and uncontrollable porosity are some of the limitations of the sol-gel process. Moreover, the thermal expansion mismatches of the substrates is one of the most important limitations of the sol-gel coating which may result in a crack formation on thin films, and make the sol-gel method highly substrate dependent. Some of the most important techniques of the sol-gel coatings are briefly explained below [6]:

3.2. Sol-Gel Coating Techniques

3.2.1. Dip Coating Technique

Dip coating process contains the immersion of the substrate in a liquid (or sol-gel liquid) and then the withdrawal of the substrate after a certain period of time (or deposition time). This process occurs under a controlled condition with a known withdrawal speed. The next step is a solvent evaporation which consequently leads to the gelation of the sol. Then, the obtained film is densified by a heat treatment. Here, the substrate concentration, the withdrawal speed and the viscosity of the liquid are important factors affecting the film thickness and uniformity. Using the appropriate viscous liquids, the thickness of the thin film can be varied from 20 nm to 50 nm (or sometimes higher based on the concentration and types of sol-gels, as well). Dip coating technique can be used for many plates and the curved surfaces, as well. One of the major applications of this technique is to develop a scratch resistance coating on plastics substrates used for eyeglass and lenses, for instance [12]. Figure 5 shows the schematic representations of dip coating stages.

3.2.2. Spin Coating Technique

Spin coating process was first considered for paint and pitch coating. In the spin coating process, the substrate is spun at high speed around an axis perpendicular to the coating surface. Spin coating process is mainly used for ceramics and glasses in microelectronics, and other substrates with rotational symmetry, such as optical lenses. This process consists of four stages: deposition of the sol, spin up, spin off and gelation by the solvent evaporation [16]. The coating formed by the spin coating technique is usually homogeneous in thickness. The most important factors affecting the spin coating process are spin speed, temperature and humidity. The porosity, the evaporation of the solvent, and the centrifugal force are vital parameters for the thickness and uniformity of the coatings.

The major applications of spin coating include the film deposition in microelectronics and solar energy industries [17]. Figure 6 shows the schematic view of the spin coating stages.

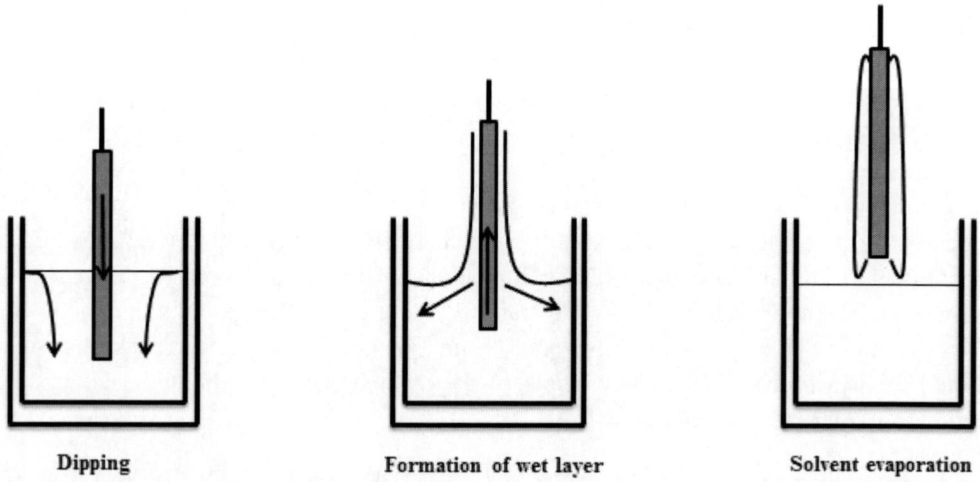

Figure 5. The schematic representations of the dip coating process.

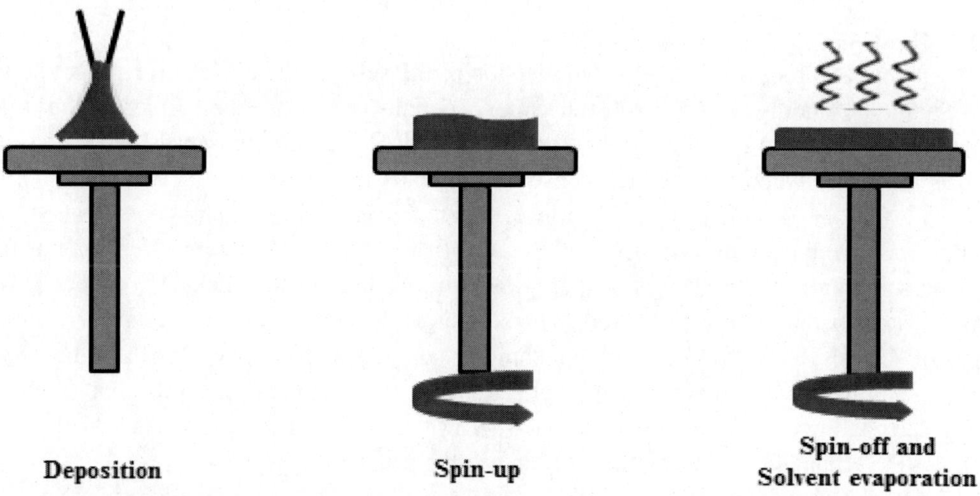

Figure 6. The schematic views of the spin coating stages.

3.2.3. Spray Coating Technique

As it is obvious from its name, spray coating is the technique where the coating solution is sprayed on the substrate surface through the air spray nozzle and other nozzle systems. This technique is wildly used by many industries for the lacquers and other coatings on irregularly shaped surfaces (e.g., lamps, containers, glasses, etc.).

The rate of spray coating is much faster than the dip coating and amount of coating sol waste is also much lesser; therefore, the waste of overall materials is relatively low in this technique. Spray coating can be combined with other coating methods, such as spin coating, for specific industrial applications, as well [18].

Additionally, spray coating is a common method for the deposition of organic layers (e.g., paints) on 3D products and devices, as well. This method can be divided into three different types depending on the ways of formations of the droplets: conventional air spraying, high volume low pressure (HVLP), electrospinning and electrostatic spraying. Figure 7 shows the schematic view of the spray coating process [10].

3.2.4. Flow Coating Technique

In the flow coating technique, the liquid coating material is poured over the clean substrate surface with a high angle.

In this technique, the viscosity of the liquid, the solvent evaporation rate and the angle of the substrate are the major factors that affect the thickness and uniformity of the coating. Large parts that cannot easily be dip coated are used for flow coating. This technique is fast and easy and requires little space, low installation cost and low maintenance cost.

All these techniques include a stage of heat treatment which depends on the different microstructures, quality requirements and practical applications. Controlling the film formation condition is an important situation to reduce the crack formations of the coatings because of the potential internal and external stress accumulations. Figure 8 shows the schematic representation of the flow coating method [19].

4. Purposes of Sol-Gel Coatings

4.1. Corrosion Protections

Metallic materials, such as iron, aluminum, copper, titanium, magnesium and their alloys are used for numerous industrial applications because of their excellent properties, such as electrical and thermal conductivities, stiffness and high strength to weight ratios, etc. However, many of the metals and alloys are usually vulnerable substrates against the corrosion attacks at various environmental conditions (e.g., moisture, oxygen, UV light, high and low pH values, emissions and pollutants).

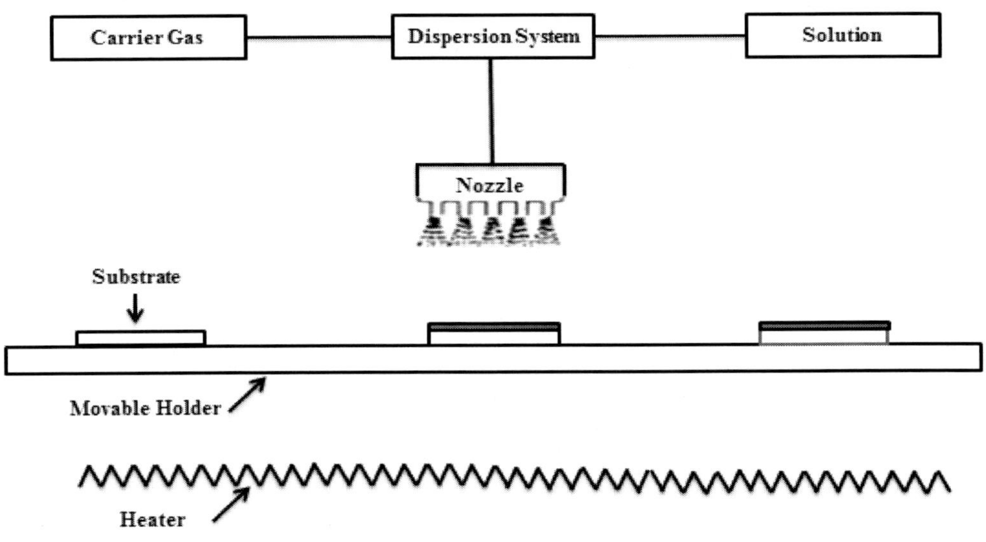

Figure 7. The schematic view of the spray coating process.

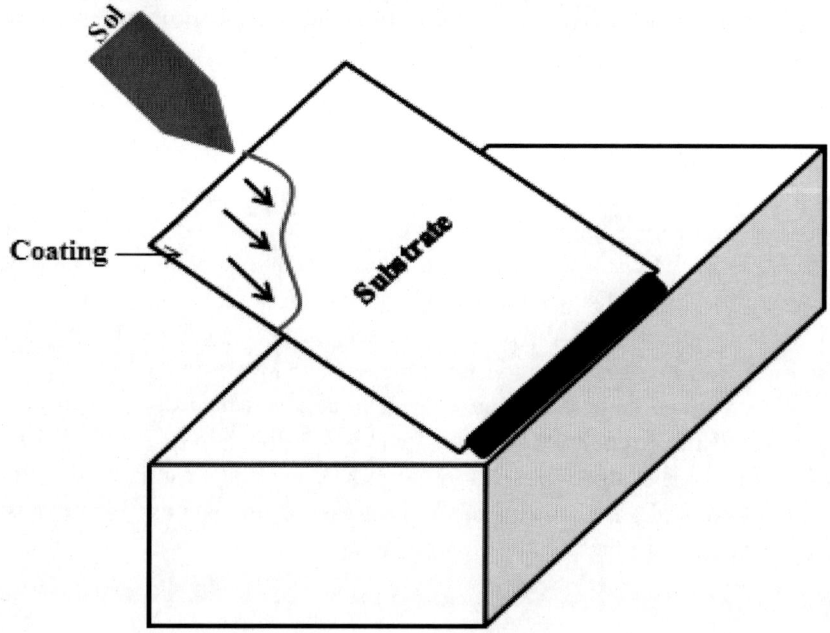

Figure 8. The schematic representation of the flow coating method.

Corrosion is the result of conversion of metal into its original ore states (oxides, sulfides and nitrates) due to the electrochemical and chemical reactions between the metal surface and the environment.

Thermodynamically, many metals and alloys are not sable at normal conditions, so oxidations and corrosions can take place to produce thermodynamically more stable phases. One of the most common corrosion types is oxidation of metals with an oxidant to create physically poor metal oxide structures [20].

Most of the energy and material loss around the world is coming through corrosion of the substrates. It is reported that every year approximately 4% of gross national product (GNP) is spent through corrosion, maintenance and replacements.

In the United States, the effect of corrosion on economy is as large as $300 billion per year, including the other side costs, such as the cost of application of protective coatings, the cost of inspection and repair of corroded materials, as well as the cost of disposal of corroded waste materials [21]. In order to decrease the devastating effects of corrosion, mainly three stages of the corrosion protection techniques have been applied, including ***i***) alloying metals with elements that enhance the corrosion resistance of the surface, ***ii***) adsorption of aqueous inhibitors on metal surface to prevent the oxidation reaction, and ***iii***) deposition of protective thin films [22].

Metallic coating methods include PVD, CVD, electrochemical deposition, hot pressing, galvanizing, plasma spraying and sol-gel processes. The sol-gel coating method has various advantages compared to the other coating techniques. Some of the important advantages of the sol-gel process are given below:

- Processing temperature is usually low, which in turn will reduce the chance of degradation or volatilization of the substrates.

- Coatings can be casted in complex shapes using the liquid precursor. Also, thin films can be produced without requirement of machining or melting.
- Sol-gel coating is considered as "green" coating technology. Since the initial substrate does not make any impurities into the final products, there is no need to wash the coated surfaces in the further processes.
- Other techniques (e.g., laser deposition, annealing) can be associated with the sol-gel processes to create stronger films.

The most commonly used sol-gel methods to apply organic and inorganic coatings on metallic substrates are dip coating and spin coating. Also, recently nozzle spraying, electrospinning and electro-deposition methods have gained significant attentions for the sol-gel applications [10, 23]. For nearly two decades, many applications of sol-gel techniques in the corrosion coating of metals have been introduced by Gugliemi [24]. Metal oxides can provide a chemically inert barrier against the diffusion of corrosive species if they are applied as a submicron thick film on the surfaces. A good example of metal oxide coatings that has been commonly used in different industrial applications is SiO_2. It has high chemical and heat resistances, so preferred to use for the protection of the surfaces in different harsh environmental conditions.

Another hard and chemically stable metal oxide coating for the corrosion protection of metals is ZrO_2 which has a high expansion coefficient, and is very close to the expansion coefficient of many bulk metals. This characteristic of ZrO_2 reduces the chance of crack formations during the curing and heat treatment processes which usually occur at elevated temperatures. Furthermore, one of the most common sol-gel coatings is Al_2O_3 (alumina) because of its relatively cheap precursor, simple preparation and long storage stabilities. $Al_2O_3$is a well-known high performance ceramic with very high wear resistance, great hardness, high thermal conductivity and low electrical conductivity (or insulator) which makes it desirable material for the numerous surface protections [25, 26]. Another excellent anti-corrosion metal oxide film is TiO_2 with excellent chemical stability, heat resistance and low electron conductivity, which is mostly preferred in many catalyst applications. In another approach, these metal oxides can be combined together, and applied on the surfaces in two or multiple-component oxide layers to broaden their application. The broader applications of these ceramic coatings have been also studied by several authors, which will be provided in the following sections [27].

Although these metal oxides can provide a very good coating ability against the corrosion attacks on the surface of metallic substrates, they are still suffering from some problems associated with these inorganic coatings. First of all, these inorganic coating films are not stable and may have porous structure, so in order to stabilize them and reduce the porosity, high temperature treatments are usually required (>600°C). Furthermore, inorganic metal oxide ceramic coatings are brittle, and the thermal expansion coefficient of the ceramic coatings and metallic substrates are generally incompatible; thus, the cracks are easily formed in these coating. Another words, crack-free sol-gel coatings at thicker layers (>1mm) are considerably difficult. To be able to solve these limitations, hybrid organic–inorganic protective coatings were developed. The physical barrier properties of the organic-inorganic hybrid coatings prevent the penetration of the electrolyte into the surface of the metallic substrate. These types of organic-inorganic coatings provide high hardness, scratch resistance, thermal stability, as well as the flexibility, transparency and tunable adhesion of the organic substances. Moreover, it is possible to produce a thicker coating without the formation of cracks [28].

Recently, one of the most important sol-gel materials to improve the corrosion resistance of different metals, such as stainless steel, aluminum and magnesium alloys is hybrid silica sol–gel coatings, which is mainly because of their good adhesions to metallic substrates and good barrier effects against the corrosion. Alkoxysilanes, including tetraalkoxysilane ($Si(OR)_4$) and organoalkoxysilane ($ROnSi(OR)_{4-n}$ or $(RO)_3SiROSi(OR)_3$), are the most commonly used precursors for the hybrid silica sol-gel coating. The preparation of hybrid silica sol-gel coating process includes two main reactions: the hydrolysis of Si–OR and the condensation of Si–OH. The catalyst in these two reactions could be either through OH^- or H^+ ions. If the process is catalyzed by OH^- ions, the rate of condensation and consequently the gelation formation will increase, while in the case of H^+ ions, the rate of hydrolysis will increase with a slow gelation. Hence, acids are generally considered to enhance the stability of hybrid silica sol–gel solution. On the other hand, in acidic solutions some metals, such as Fe, Al and Mg are easily dissolved and form complex metal ions which are undesired in anticorrosion hybrid silica sol–gel coatings since the coating network would be disordered in the solution with high concentration of metallic ions. Therefore, finding a non-acidic catalyst for the preparation of hybrid silica sol–gel coating is necessary in case of applying the coating on active metallic substrates [29, 30].

Most of the early studies investigating the corrosion protection of hybrid sol-gel coating focused on the class I hybrids in which the inorganic and organic components are bonded together by mainly van der Waals and hydrogen bond attraction forces. According to the some research studies, the corrosion resistance of stainless steel was increased by a factor 30 using the class I hybrid coating. Nonetheless, long term failures, such as phase separation and subsequent delamination are possible using the class I hybrids. These drawbacks can be solved using the class II hybrids in which the organic and inorganic components are bonded together covalently for a stable surface formation [22].

Although the hybrid coatings form a uniform, defect-free and relatively dense film barrier on metallic substrates, corrosive ions can still diffuse through micro to nanopores in the hybrid coatings and corrode the metals at the interface between the coating and substrate. This means that hybrid coatings cannot remove porosity of the film completely. In order to solve this problem, hybrid coatings can be doped with some corrosion inhibitors. These inhibitors are divided into inorganics, such as phosphates, vanadates, and rare earth elements, and organics such as triazole and thiazole derivatives, 8-hydroxyquinoline (8HQ). The corrosion inhibitors can be added to the sol–gel system during the synthesis procedure or at the stage of the thin film formation and crosslinking. In order to improve the corrosion protection of hybrid coatings, the inhibitor materials have been largely used. These inhibitors can be organic or inorganic materials, and show relatively low toxicity. Nevertheless, ionizable organic inhibitors that could interact with the surface of the coatings cannot improve the corrosion resistance behavior of the substrates [29].

One of the issues in the sol-gel coating process is that the sol-gel coatings normally require a calcination process at a higher temperature to covalently densify the gel film on the surfaces of substrates. The high temperature may affect the substrate microstructures, and also restrict the application of this technique in some industries (e.g., aircraft, wind turbine, solar energy, microelectronics, and so on). In order to overcome this problem, some researchers used an excimer laser to process coating films developed through the sol-gel method. Using these two technologies together has gained significant attention for the surface modification because of the appropriateness of both techniques for thin film coating. This combined method introduces various advantages, including the accurate controls over laser parameters of energy, speed and wavelength. Moreover, the variability of compositions of sol–gel materials makes the process useful to achieve desired functions. Therefore, treating the surface and interface of the metallic substrates is more likely without affecting the bulk properties of the materials, which are required in many cases of engineering components [25].

4.2. Selected Substrates

Stainless Steel Substrates: Stainless steels (e.g., 316L, 304, austenitic stainless steel)have been known as one of the most common metallic materials and are wildly used in many high load-bearing applications worldwide [30]. Stainless steels exhibit very good mechanical and corrosion resistance properties which increase their potential applications in various fields. However, they may corrode in some extreme conditions in the presence of halide ions and other corrosive species at higher and lower pH values. Recently, many studies have been conducted on the corrosion behavior of sol-gel coatings on stainless steel substrates and the results are highly promising [31].

In the previous studies, different precursors of ZrO_2, TiO_2, SiO_2 and Al_2O_3 were used in the dip coating method to deposit single and multiple-layers of thin films on 316L stainless steel substrate through the sol-gel process. The resultant film had a thickness of 0.4 to 0.6 μm. The scanning electron microscopy (SEM) analysis indicated the anti-corrosive behavior of the coated film and the increased substrate life time up to a factor 8.5[32]. In the 1990s, various studies were conducted on the stainless steels using the different single or multi-component oxide coatings to improve the corrosion resistance of the substrates (e.g., 316L and 304). The dip coating technique was the most commonly used technique for these applications. The resulting films thicknesses were between 0.2 and 2 μm. Since the deposition of a dense sol-gel derived coating was desired, the organic-inorganic hybrid coating without the need of post-deposition annealing gained significant attention by many scientists to protect the surface of the stainless steel substrates against corrosion attacks [33]. In one of the pioneering researches of the organic-inorganic hybrid coatings on stainless steels was in 1995 by Atik et al., which investigated the corrosion behavior of hybrid coating of polymethylmethacrylate (PMMA) and ZrO_2 on 316L stainless steel in 0.5 M of H_2SO_4 solution at room temperature. The hybrid coatings act as a protective layer against the corrosive media and enhance the life time of the stainless steel substrates [34].

Although various types of sol-gel coating techniques including dip coating, spray coating, flow coating, spin coating, and particulate coating are applied to the substrates, the dip coating is one of the most effective coating method for 316L stainless steel substrate. In one of the most recent experiments, excimer laser was used to deposit alumina coating onto stainless steel substrates. In this technique, first sol-gel alumina coating was applied on the substrate surface and then immediately irradiated with the excimer laser to initiate the strong deposition of alumina from the sol onto the substrate surface. The characterization tests on the composition and microstructure of resulting alumina coatings showed that the laser induced deposition of alumina coating has the hardness value of four times higher than the as-dried xero-gel coating [35].

Aluminum Substrates: Aluminum and its alloys are another group of metals with widespread applications in many industries. Aluminum is the third abundant element on the earth. Being abundant, light weight, good electrical and thermal conductivity and low cost are the main characteristics that make aluminum an important construction material for many applications in aerospace, automobiles, household appliances, containers, constructions, and electronic devices. Natural formation of a passive oxide layer on the surface of aluminum substrate is one of the main reasons for some of its applications. Although aluminum alloys are generally passive in aqueous solutions after the surface oxidations, the formations of the passive oxide layers can be deteriorated in some aggressive media, such as chloride and halide, causing a pitting corrosion [18].

Conventional way of protecting aluminum alloys from corrosion was the chromate conversion coating which worked very effectively, but due to the extreme side effects of chromate (specifically *hexavalent chromium*), the U.S. Environmental Protection Agency (EPA) put sever restrictions on the use of chromate in many coating [33-36]. Therefore, the current trend is to find environmentally friendly alternative coatings for the corrosion protection of aluminum alloys, so the sol-gel coating has been evaluated and wildly studied as a promising replacement method of the toxic chromate conversion coatings on aluminum substrates [28].

Inorganic oxide coatings of SiO_2 and ZrO_2 have been studied on a substrate containing%98 aluminum. According to the recent studies, dip coating and spin coating were the most common techniques that have been used to apply the thin film oxide layers on the aluminum substrates. The formation of SiO_2 film by the dipping aluminum substrate into silicic acid solution was studied by Thim et al. in 2000. According to their results, the defects during the heat treatment process caused the decrease in the adhesion of the coated film. They used an anodic polarization pretreatment in order to overcome this limitation and improve the corrosion resistance of the oxide film.

In 2001,Yang et al. studied the effect of a coating film containing SiO_2-ZrO_2 as a corrosion protection on Al 2024-T3 substrate by immersion the substrate into a dilute Harrison's solution (3.5 g/L $(NH_4)_2SO_4$, and 0.5 g/L NaCl). The evolution of the coating was investigated during the immersion. They concluded that the increased impedance could cause the pitting corrosion and accelerate the degradation products on the coating. The authors also assumed that a stable mixed oxide barrier layer might be formed at the interface of the coating layer and the substrate after the initial corrosion, which in turn will prohibit the further pitting corrosion [37].

As mentioned previously, hybrid coatings containing the organic and inorganic part could have the combined properties of organic and inorganic materials, as well. Hybrid coatings enable the formation of thicker, more flexible and crack-free coating film with lower curing temperature.

Furthermore, the addition of some nanoparticles can increase density and mechanical properties of the hybrid films [38]. In one of the studies in 1998, thick, dense and smooth corrosion protective coatings were formed on aluminum substrate by combining Al_2O_3, SiO_2 and ZrO_2 nanoparticles in the hybrid sol-gel matrices. The test results were comparable with the conventional ones.

In 2003, alkoxysilanecompounds were used as precursors in the sol-gel coatings onto an aluminum substrate, and cured by the UV or thermal heat sources. The resultant coatings showed very promising mechanical, chemical and thermal properties and very good adhesion on the substrate surface. According to the recent studies obtained using the hybrid coatings onto aluminum substrate, hybrid films cured with amine showed a great corrosion resistance and strong adhesion on the substrates [29].

Additionally, inorganic inhibitors, such as phosphates, vanadates, rare earth elements, and organic inhibitors, such as triazole and thiazole derivatives, 8-hydroxyquinoline (8HQ), have shown positive influences on the corrosion protection of aluminum alloys. Among these inhibitors, cerium nitrate has gained more attention because of various advantages: cerium ions can be used as cathodic inhibitors since they form insoluble hydroxides, and they have a very low toxicity and are relatively abundant in the nature [26]. In 2009, Paussa et al. studied the development of sol–gel hybrid films doped with cerium element. The authors prepared different thicknesses of cerium doped sol-gel film by controlling the sol concentration, dipping number and withdrawal speed [33].

Copper Substrates: Copper is another metallic substrate which is wildly used in various industrial applications, such as electric, thermal, electronic, construction, architectural and many other daily applications. Copper is used in different shapes including tubes, rods, wires, coils and plates. Low chemical reactivity, easy processing and beautiful appearance of copper makes it suitable for these kinds of applications. Although copper exhibits a very good corrosion resistance, it can be corroded in certain environmental conditions. For instance, in wet environment, the formation of hydroxides and other harmful complexes can increase the corrosion of copper surface [38].

In order to protect copper against corrosion attacks, different techniques have been developed and used on the copper substrates. Among them, sol-gel coating technique exhibit more advantages compared to other techniques. The studies were mainly carried out to investigate the sol–gel coating for the protection of copper surfaces. In one of the earliest studies, SiO_2 sol–gel coatings have been reported as barrier layers on the copper surfaces. However, it was found that when the temperature increased up to 400 °C, the SiO_2 coating layer tended to peel off and eventually accelerate the degradations. This behavior has been attributed to the quite large difference between the thermal expansion coefficients of SiO_2 and copper. Other studies investigated the inorganic oxide coating containing SiO_2-ZrO_2 on copper surface. It was expected to solve the low interface adhesion and stability of the SiO_2 coating by adding ZrO_2 with high thermal expansion coefficient. The results indicated that the produced coating was well adhered and had a uniform, detect-free surface [39].

Inorganic-organic hybrid coatings were also studied for the replacement of the inorganic metal oxide sol-gel coatings. In 2003, Bescher et al. used various compounds (3-glycidoxypropyltrimethoxysilane - GPTMS, methyltrimethoxysilane- MTMS) as precursors. The authors investigated the properties of the hybrid sol-gel coating on the surfaces of copper and bronze substrates, which were prepared using GPTMS+MTMS as precursors. According to the report, the coating could be applied as thick as 5-10 μm and still show a strong adhesion to the substrate surface. Also, the substrate was exposed to high sulfur/humidity conditions for two years and no corrosion formed on the treated surfaces. Furthermore, the sol-gel coatings were top coated with a fluoropolymer layer in the same study to improve the anti-corrosion ability. This enhancement was attributed to the high degradation resistance of the fluoropolymer [2, 39].

Magnesium Substrates: Magnesium is another attractive metal for many industrial applications. Magnesium shows several advantages, which make it appealing material in various applications: being light, (the lightest metal), having high strength to weight ratio, good thermal and electrical conductivity and so on. The density of magnesium is only 65% that of aluminum and 25% that of iron. Therefore, one of the most important applications of magnesium and its alloys are the industries where the weight reduction is of significant interest [40].

Apart from all the advantages, magnesium and its alloys have a disadvantage because they are vulnerable to corrosion since magnesium is one of the most electrochemically active metals. It is also explosive at power forms, and makes many ongoing processes more difficult. Additionally, magnesium and alloys do not show very high wear resistance. Therefore, numerous investigations were carried out to overcome the weaknesses of this important metal. One of the relatively new and promising methods to overcome the disadvantages is to utilize the sol-gel coating technique. Recently, 68 wt.% MAPTS+2% MPTMS+30% SiO_2 combination was used as hybrid coating layer on the surface of magnesium, resulting in a defect-free coating film with the thickness of 23μm acted as a physical barrier against corrosion. In 2006, Khramov et al. discovered that the addition of phosphonate group to silane precursors increased the affinity on the magnesium surface, and improved the corrosion protection of the sol-gel coating film [41, 42].

Metallic Biomaterials Substrates: Stainless steels, titanium alloys and magnesium alloys are also some major examples of metallic biomaterials that are commonly used for many implant applications. The mechanical properties, corrosion resistance and especially the low cost of stainless steel 316L make it an appealing candidate for the orthopedic implantation in the human body. In order to prevent its degradation in the corrosive environment of the body, a thin layer of coating is applied on the stainless steel surface. In contrast, hydroxyapatite is a bioactive material which is chemically very similar to natural bone mineral, but has a low mechanical strength which makes the bulk material inappropriate for the direct implantation use. Therefore, hydroxyapatite is wildly used as a coating layer on the surface of stainless steel or other metallic implants to enhance their life times in the body [43, 44].

5. Challenges and Future Trends

The sol-gel coatings have been investigated widely, and used for various commercial applications in several industries; however, the sol-gel processes and coatings are still facing some challenging problems and difficulties that restrict their large-scale applications in many other industries. Further studies on sol–gel coatings and techniques in order to improve the major drawbacks will cause wider and more practical applications of sol-gel coatings [13].Some of the major issues of the sol-gel coatings are needed to be further investigated:

- Adhesion and delamination of the sol-gel coatings
- Optimization and new synthetic routs for sol-gel coatings
- Integration of sol-gel process with other processes to enhance the overall values of the sol-gel processes
- Comparison studies of the sol-gel with the other coating techniques

- Investigation of new applications of sol-gel processes
- Automation of the sol-gel process in larger industrial applications
- Development of new raw materials and precursors.

5.1. Commercialization

Sol-gel coating procedure is one of the most useful and versatile techniques to be simply applied to a wide range groups of materials. In addition, sol-gel coating is relatively economical method compared to the other film deposition techniques and the capital investment for sol-gel film coating is minimized as well as the improved properties of the coated film [45]. Sol-gel coating can be used to serve different functions, such as: Hardness, scratch resistance, easy clean/anti-sticking, anti-fingerprint, anti-bacteria/anti-microbial, bio-compatibility, higher temperature resistance, anti-reflection/optical and decorative coating. Some of the benefits of sol-gel coating techniques that make it a versatile method for commercial applications are listed below:

- Excessive diversity of accessible precursors
- Wide-ranging of high-purity materials
- Soluble precursors allowing mixing on molecular level
- Homogeneous products
- Allowing to modify material composition and microstructure
- Rather low-priced deposition equipment
- Ability of curing at low temperature or by UV [46]

One of the initial commercial applications of sol-gel technique is Zinc-based paints which have been produced for over 50 years and are applied as corrosion resistant primers and high temperature coatings for steel [47]. Some of most recent application of sol-gel coating are improvement of water repellent and oil resistant properties of water [48], antireflective coatings on large plastic panels in computer screens, car dashboards and TV [49].

Sol-gel coating has also been applied in electrical component coatings such as resistors and molding compounds, as well as dielectrics in microelectronic interlayer dielectric and planarization applications. Another recent versatile application of sol-gel coating is in automotive coatings which provides scratch resistant function [47]. In general, sol-gel technology has been used in widespread applications, including:

- Coatings and thin films used in electronic, optical and electro-optic components
- Antireflection coatings for automotive and architectural applications
- Protective and decorative coatings for glass, metal and other types of substrates
- Dental and biomedical products that serve as carriers for biomolecules
- Fiber optic sensors applications
- Thermal insulation applications

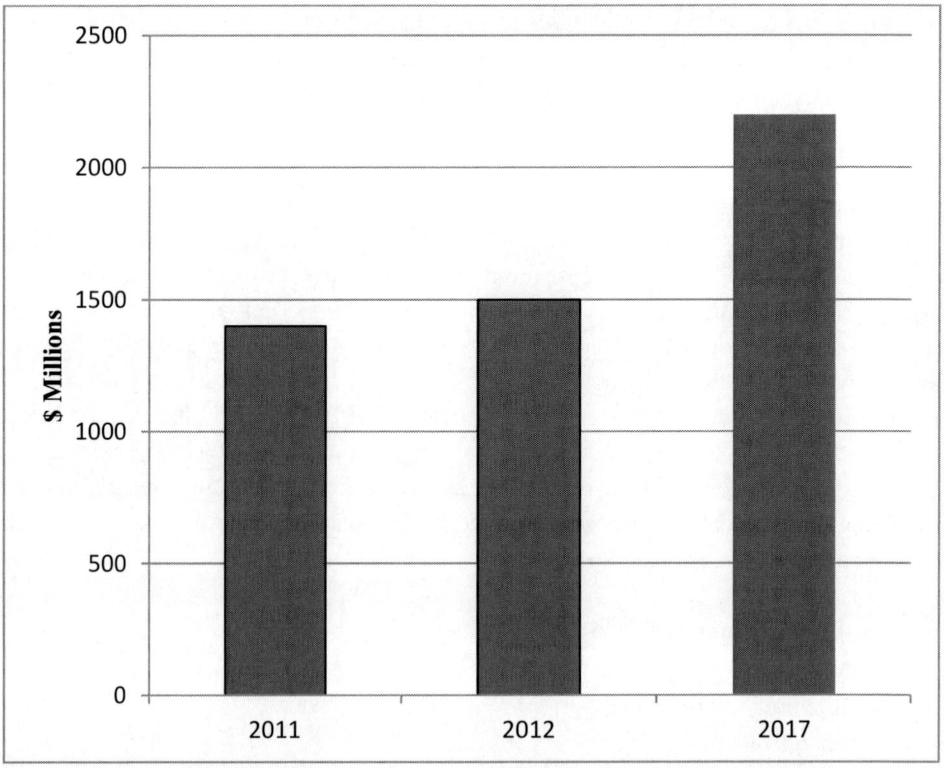

Figure 9. global Market of Sol-Gel Products.

- Abrasives applications
- Applications as passive and nonlinear active waveguides and lasers
- Applications as membranes for separation and filtration

However, US has been a leader in sol-gel based research, Japan is leading in commercialization the sol-gel technology and has been marketing the sol-gel products faster than US.

It is also noticeable that among European countries, Germany is the most competitive one in sol-gel technology. The commercial application of sol-gel as well as the market for sol-gel based products is drastically increasing. According to a BCC research, the market for the products made using sol-gel technologies is predicted to have a major growth in the next few years. Figure 9 shows the global market for sol-gel products in 2011, 2012 and 2017 [50].

CONCLUSION

Sol-gel technology is a promising technology to produce various materials and devices. One of the most important applications of sol-gel technology is to produce thin oxide films on the surfaces of many substrates to enhance their properties. Sol-gel coating has been gained a magnificent attention in the recent years to employ as protective layer against corrosion on the surfaces of metallic and other oxide substrates. The sol-gel protective coating layers can enhance the corrosion resistance of the metals in an extreme corrosive environment.

Conventional sol-gel coating materials consist of inorganic metal oxides, but due to the drawbacks of metal oxide inorganic coatings, some organic materials can be added to the inorganic metal oxide parts to improve the overall corrosion resistances of the surfaces. According to the recent research studies, hybrid inorganic-organic sol-gel coatings showed the excellent characteristics of corrosion protection on metals, and are very promising in this field. Another important use of the sol-gel coatings is to eliminate the *hexavalent chromium* coating because of the major toxic side effects of this conversion coating. In this book chapter, several sol-gel techniques are explained in detail, and the most important work in these areas have been reviewed, which will be very useful to the readers and other participants in the field.

REFERENCES

[1] Y. Dimitriev, Y. Ivanova, and R. Iordanova, "<HISTORY OF SOL-GEL SCIENCE AND TECHNOLOGY.pdf>," *Journal of the University of Chemical Technology and Mettalurgy,* vol. 43, p. 12, 2008.

[2] D. Wang and G. P. Bierwagen, "Sol–gel coatings on metals for corrosion protection," *Progress in Organic Coatings,* vol. 64, p. 12, 2009.

[3] N. K. Chaudhury, R. Gupta, and S. Gulia, "Sol-gel Technology for Sensor Applications," *Defence Science Journal,* vol. 57, p. 13, 2007.

[4] G. C.-d. l. Torre, M. A. Espinosa-Medina, A. Martinez-Villafañe, J. G. Gonzalez-Rodriguez, and V. M. Castaño, "Study of Ceramic and Hybrid Coatings Produced by the Sol-Gel Method," *The Open Corrosion Journal,* vol. 2, p. 7, 2009.

[5] C. J. Brinker and A. J. Hurd, "Fundamentals of sol-gel dip-coating," *Journal de Physique III,* vol. 4, p. 12, 1994.

[6] Sol-Gel Technology for Thin Films, Fibers, Preforms, Electronics, and Specialty Shapes. Mill Road, Park Ridge, New Jersey, United States of America: Noyes Publications, 1988.

[7] S. Sakka, "Preparation and Properties of Sol-Gel Coating Films," *Journal of Sol-Gel Science and Technology,* vol. 2, p. 5, 1994.

[8] J. D. Mackenzie, "Sol-Gel Research—Achievements Since 1981 and Prospects for the Future," *Journal of Sol-Gel Science and Technology,* vol. 26, p. 5, 2003.

[9] V. Glezer and O. Lev, "Sol-Gel Vanadium Pentaoxide Glucose Biosensor," *Journal of the American Chemical Society,* vol. 115, p. 2, 1993.

[10] S. M. Attia, J. Wang, G. Wu, J. Shen, and J. Ma, "Review on sol-gel derived coatings process, techniques and optical application.," *Journal of Materials Science and Technology,* vol. 18, p. 8, 2002.

[11] J. E. Mellor, "Investigation of a Sol-Gel Coating Technique for Polarized He Target Cells," Bachelor of Science, Physics, College of William and Mary in Virginia, 2001.

[12] C. J. Brinker, G. C. Frye, A. J. Hurd, and C. S. Ashley, "Fundamentals of Sol-Gel Dip Coating," *Thin Solid Films,* vol. 201, p. 12, 1991.

[13] M. Niederberger and N. Pinna, Metal Oxide Nanoparticles in Organic Solvents: Synthesis, Formation, Assembley and Application: Springer 2009.

[14] C. J. Brinker and G. W. Scherer, Sol-Gel Science: The physics and chemistry of sol-gel processing: Academic Press INC., 1990.

[15] J. A. Schwartz, "Methods for Preparation of Catalytic Materials," *Chemical Reviews,* vol. 95, p. 34, 1995.

[16] S. N. B. Parija, and S. Panigrahi, "Fundamental understanding and modeling of spin coating Process: A Review " *Indian Journal of Physics,* vol. 83, p. 10, 2009.

[17] K. Vorotilov, V. Petrovsky, and V. Vasiljev, "Spin coating process of sol-gel silicate films deposition: Effect of Spin speed and processing temperature," *Journal of Sol-Gel Science and Technology,* vol. 5, p. 11, 1995.

[18] A. Younis, "Protection of Aluminum Alloy (AA7075) from Corrosion by Sol-Gel Technique," Chemnitz University of Technology, 2011.

[19] U. C. Nwaogu and N. S. Tiedje, "Foundry Coating Technology: A Review," *Materials Sciences and Applications,* vol. 02, p. 18, 2011.

[20] A. J. Shah and G. B. Bhagchandani, "Anti Corrosive Rubber Coating," *International Journal of Scientific Engineering and Technology,* vol. 2, p. 5, 2013.

[21] H. Rahimi, R. Mozaffarinia, and A. Hojjati Najafabadi, "Corrosion and Wear Resistance Characterization of Environmentally Friendly Sol–gel Hybrid Nanocomposite Coating on AA5083," *Journal of Materials Science and Technology,* vol. 29, p. 6, 2013.

[22] V. H. V. Sarmento, M. G. Schiavetto, P. Hammer, A. V. Benedetti, C. S. Fugivara, P. H. Suegama, et al., "Corrosion protection of stainless steel by polysiloxane hybrid coatings prepared using the sol–gel process," *Surface and Coatings Technology,* vol. 204, p. 13, 2010.

[23] R. Asmatulu, H. Haynes, M. Shinde, Y. H. Lin, Y. Y. Chen, and J. C. Ho, "Magnetic Characterizations of Sol-Gel-Produced Mn-Doped ZnO," *Journal of Nanomaterials,* vol. 2010, p. 4, 2010.

[24] J.-B. Cambon, F. Ansart, J.-P. Bonino, and V. Turq, "Effect of cerium concentration on corrosion resistance and polymerization of hybrid sol–gel coating on martensitic stainless steel," *Progress in Organic Coatings,* vol. 75, p. 9, 2012.

[25] Y. Adraider, Y. X. Pang, F. Nabhani, S. N. Hodgson, and Z. Y. Zhang, "Laser-induced deposition of sol–gel alumina coating on stainless steel under wet condition," *Surface and Coatings Technology,* vol. 205, p. 5, 2011.

[26] N. Avci, P. F. Smet, J. Lauwaert, H. Vrielinck, and D. Poelman, "Optical and structural properties of aluminium oxide thin films prepared by a non-aqueous sol–gel technique," *Journal of Sol-Gel Science and Technology,* vol. 59, p. 20, 2011.

[27] G. Schottner, "Hybrid Sol-Gel-Derived Polymers: Applications of Multifunctional Materials," *Chemistry of Materials,* vol. 13, p. 14, 2011.

[28] B. J. Basu, A. Srinivasan, J. Manasa, and V. K. W. Grips, "Improved corrosion protection of aluminium alloy AA 2024 by sol–gel hybrid coatings after surface pretreatment by silanisation," *Surface Engineering,* vol. 28, p. 7, 2012.

[29] S. Zheng and J. Li, "Inorganic–organic sol gel hybrid coatings for corrosion protection of metals," *Journal of Sol-Gel Science and Technology,* vol. 54, p. 14, 2010.

[30] S. Peng, W. Zhao, Z. Zeng, H. Li, Q. Xue, and X. Wu, "Preparation of anticorrosion hybrid silica sol–gel coating using Ce(NO3)3 as catalyst," *Journal of Sol-Gel Science and Technology,* vol. 66, p. 6, 2013.

[31] P. D. L. Neto, M. Atik, L. A. Avaca, and M. A. Aegerter, "Sol-Gel Coatings for Chemical Protection of Stainless Steel," *Journal of Sol-Gel Science and Technology,* vol. 2, p. 6, 1994.

[32] C. Y. Zheng, F. L. Nie, Y. F. Zheng, Y. Cheng, S. C. Wei, L. Ruan, et al., "Enhanced corrosion resistance and cellular behavior of ultrafine-grained biomedical NiTi alloy with a novel SrO–SiO2–TiO2 sol–gel coating," *Applied Surface Science,* vol. 257, p. 7, 2011.

[33] T. P. Chou, C. Chandrasekaran, S. J. Limmer, S. Seraji, Y. Wu, M. J. Forbess, et al., "Organic-Inorganic hybrid coatings for corrosion protection," *Journal of Non-Crystalline Solids,* vol. 290, p. 10, 2001.

[34] G. Ruhi, "Hot Corrosion Behavior of Sol-Gel Nano Structured Zirconia Coated 9Cr1Mo Ferritic Steel in Alkali Metal Chlorides and Sulphates Deposit Systems at High Temperatures," *Journal of Surface Engineered Materials and Advanced Technology,* vol. 03, p. 6, 2013.

[35] L. Paussa, N. C. Rosero-Navarro, F. Andreatta, Y. Castro, A. Duran, M. Aparicio, et al., "Inhibition effect of cerium in hybrid sol-gel films on aluminium alloy AA2024," *Surface and Interface Analysis,* vol. 42, p. 7, 2010.

[36] R. Asmatulu, *Nanotechnology Safety*, 2013.

[37] A. J. Vreugdenhil, V. N. Balbyshev, and M. S. Donley, "Nanostructured Silicon Sol-Gel Surface Treatments for Al 2024-T3 Protection," presented at the 78th Annual Meeting of the Federation of Societies for Coatings Technology, Chicago, IL, 2000.

[38] J. Nikkola, J. Mannila, R. Mahlberg, J. Siivinen, M. Kolari, and A. Mahiout, "Sol–gel based protective coatings for copper products," *Journal of Coatings Technology and Research,* vol. 5, p. 10, 2008.

[39] A. J. Atanacio, B. A. Latella, C. J. Barbé, and M. V. Swain, "Mechanical properties and adhesion characteristics of hybrid sol–gel thin films," *Surface and Coatings Technology,* vol. 192, p. 11, 2005.

[40] H. Hornberger, S. Virtanen, and A. R. Boccaccini, "Biomedical coatings on magnesium alloys - a review," *Acta Biomaterialia,* vol. 8, p. 14, Jul. 2012.

[41] A. L. K. Tan, A. M. Soutar, I. F. Annergren, and Y. N. Liu, "Multilayer sol–gel coatings for corrosion protection of magnesium," *Surface and Coatings Technology,* vol. 198, p. 5, 2005.

[42] S. V. Lamaka, M. F. Montemor, A. F. Galio, M. L. Zheludkevich, C. Trindade, L. F. Dick, et al., "Novel hybrid sol–gel coatings for corrosion protection of AZ31B magnesium alloy," *Electrochimica Acta,* vol. 53, pp. 4773-4783, 2008.

[43] S. Jafari, M. M. Taheri, and J. Idris, "Bioactive Coating on Stainless Steel 316 L through Sol-Gel Method," *Advanced Materials Research,* vol. 383-390, p. 6, 2011.

[44] M. M. Atabaki, R. a. Jafar, and J. Idris, "Sol–Gel Bioactive Glass Coating for Improvement of Biocompatible Human Body Implant," *Metalurgija – Journal of Metallurgy,* vol. 16, p. 15, 2010.

[45] A. H. Choi and B. Ben-Nissan, "Advancement of Sol-Gel Technology and Nanocoatings in Australia," *Journal of the Australian Ceramics Society,* vol. 50, p. 16, 2014.

[46] L. Wu. Industry seminar and roundtable discussion on coatings for polymers [Online].

[47] B. Arkles, "commercial applications of sol-gel derived hybrid materials," *Materials Research Society Bulletin,* p. 6, 2001.

[48] M. Yamane, "Unique commercial applications of the sol-gel process in Japan," *Journal of Sol-Gel Science and Technology,* vol. 40, p. 7, 2006.

[49] A. Beganskiene, S. Sakirzanovas, I. Kazadojev, A. Melninkaitis, V. Sirutkaitis, And A. Kareiva, "Sol-gel derived antireflective coating with controlled thickness and reflective index," *Materials Science-Poland,* vol. 25, p. 8, 2007.

[50] P. Wray. Continued growth of sol-gel market predicted [Online].

In: Comprehensive Guide for Nanocoatings Technology. Vol. 1 ISBN: 978-1-63482-447-7
Editor: Mahmood Aliofkhazraei

Chapter 9

MULTIPLE LAYER COATINGS

Preeta Bhore
Department of Electronics Engineering,
Watumull Institute of Electronics Engineering
and Computer Technology, Mumbai University,
Worli, Mumbai, India

ABSTRACT

The use of antireflection coating and high reflection coating on laser diode facets provide the necessary passivation and change the reflectivity so that power emission is predominantly from one end. Design of antireflection coating is optimized for single layers of Al2O3 and high reflection coating is optimized for stacks of Al2O3andSi Bragg reflector. Electron beam evaporation has been used to deposit uniform Al2O3/ Si film under high vacuum with pressure 10-6 mbar and temperature 150^0C-1600°C. Experimental parameters like refractive index of the material and thickness of deposition are necessary to match with simulated AR/HR curves. LI measurements of a coated laser diode are compared with an uncoated one.

CHAPTER CONTENTS

I. Laser Diode Facet Coatings and Optimization

Introduction

Laser diodes operating at 980 nm are the critical components of the fiber communication system. For most of the applications we use light coupled to optical amplifiers, pumped by these laser diodes.

As such, the ends of laser cavity are cleaved faces, which act like mirrors because of the refractive index difference between GaAs and air. The reflectivity of pure cleaved GaAs facets is nearly 30%. Cleaved lasers emit light from both directions but for all practical applications, we want maximum light from one end. This is discussed in the first module.

Changing facet reflectivity, which is obtained with optical coatings, can accomplish this. In the Second Module, laser fundamentals, conditions for lasing, and the structural details are explained. The Third Module deals with the Principle and theory behind laser reflectance – followed by design criteria for the coatings in the Fourth Module.

The next section deals with theoretical simulations done with analysis to suit ranges of wavelengths. Practical experiments with related graphs are explained in the next Module, (Experimental Measurements). Comparison between theoretical and practical results with conclusions is also included.

References used during the project and the data needed during the experimental work is at the end of chapter in which the flow chart and the program of simulations are done with simulated results.

Problem Formulation

For the analysis, we consider the laser cavity for length l, and with mirrors of reflectivity, Rl and R2, at two ends. As such, the ends of the laser cavity are cleaved faces, which act like mirrors because of the refractive index difference between GaAs and air.

It is necessary to cover the mirror faces of the laser cavity for a) passivation b) to arrange light emission from predominantly one face.

To accomplish this face, it is covered with a dielectric coating so that it has low reflectivity (Anti Reflection AR).The other end is covered with multiple dielectric layers so that the reflectivity is high (HR). In this structure, one end has reflectivity R1, and the other R2.

The condition of laser threshold is expressed in terms of gain, as follows.

$$g_{th} = \gamma + \frac{1}{21} * \ln \frac{1}{R_1 R_2}$$

where 1st term represents the combined optical loss of the cavity, 2nd term, the useful laser output.

The main part of our problem is to design these AR and HR coatings and check the performance of the laser before and after coating.

II. Laser Fundamentals – A Review

A laser is a device that extracts stimulated emissions from a semiconductor system with an inverted population. Basic ingredients:

i). Light amplification in a material with optical gain.
ii). Optical feedback with a resonant cavity and
iii). Pumping mechanism

Light Emitting Diodes and Laser Diodes

LED's and Laser Diodes are used as sources of electromagnetic radiations.

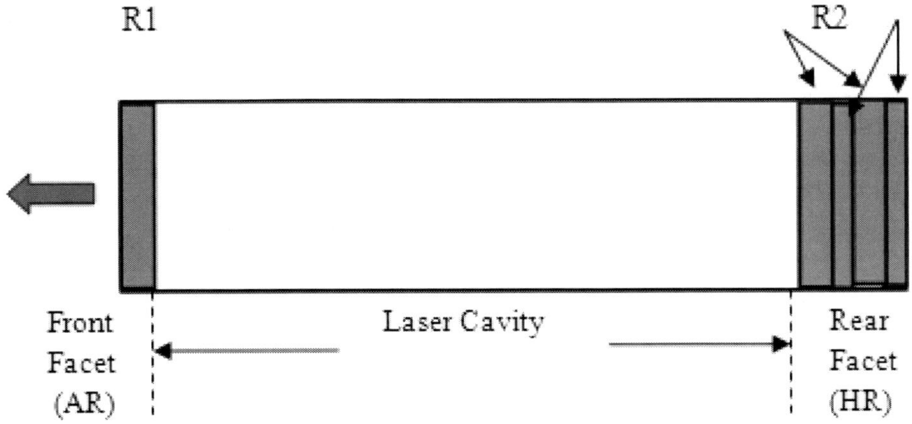

Figure 1. Laser Diode with front and back facet.

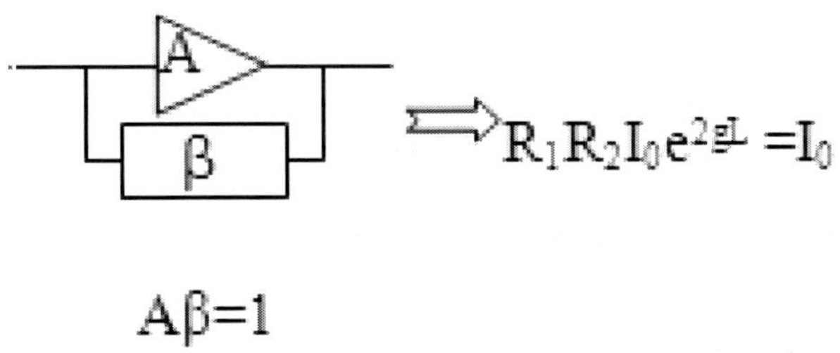

Figure 2. Light Amplification.

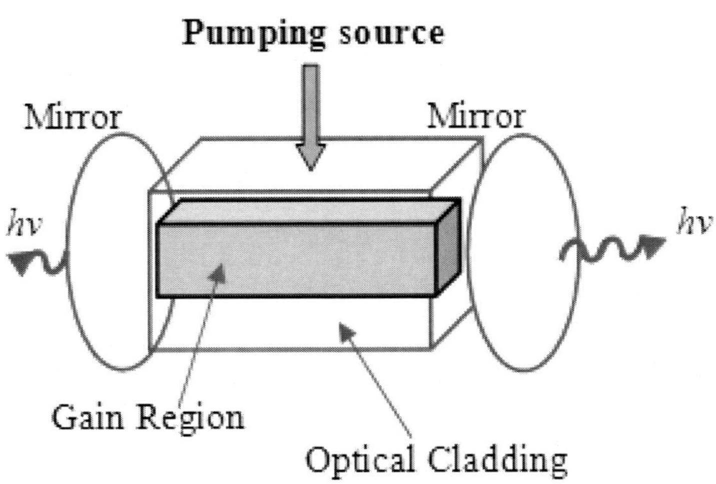

Figure 3.Basic ingredients of Laser.

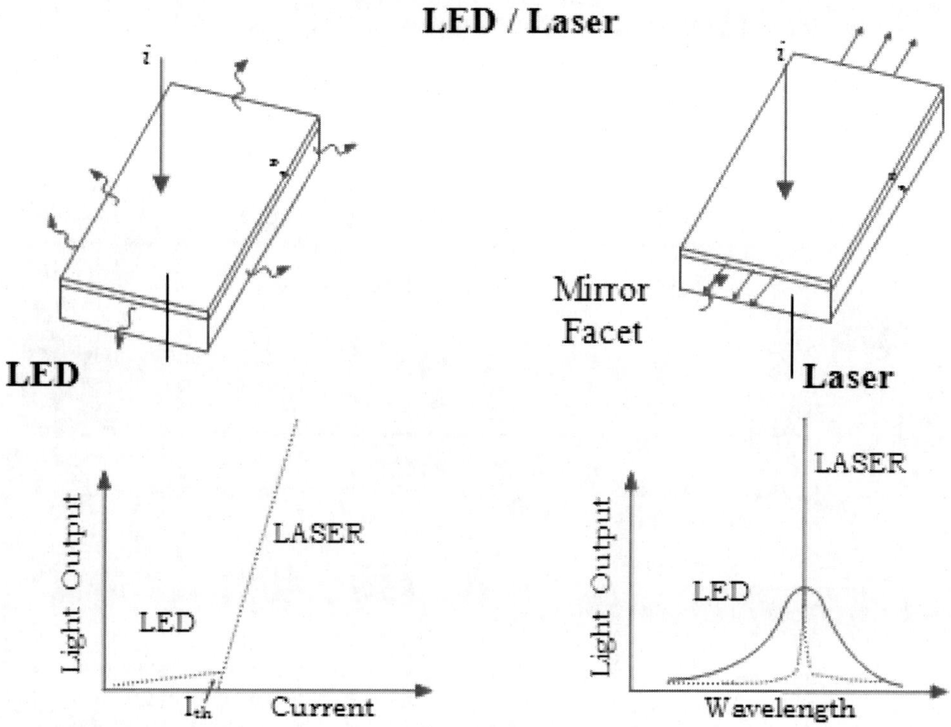

Figure 4. LED VS. LASER.

When used in conjunction with photo diodes, detectors, and fiber optics transmission cables, this source provides means to create an optical communication system. The light from the LED is incoherent. That means the photons are randomly emitted due to the recombination process, which is uncorrelated. A different type of source with coherent source results from laser diodes. In a laser diode the current flowing through the diode is increased until the number of electrons in the conduction band becomes larger than the number left in the balance band. This produces a population inversion, with more electrons at the higher energy level and fewer at lower levels.

An optical cavity is created by cleaving the ends of the semiconductor crystal so that the excited electrons are coupled, and their transition back to the lower energy state, in a coordinated process, results in photons being emitted in phase. The laser diodes thus function as a coherent light source.

Because of the directional nature of the cavity and the coherent nature of the radiation, output from laser diodes diverge much less than the light from an LED. Thus coupling is made quite efficiently.

The laser diodes are temperature sensitive and often must be cooled with built-in thermo-electric coolers.

A fiber optic communication system can consist of LED and Laser diodes source, the fiber cable, and a photo diode detector.

One of the most common methods to generate more output power from a semiconductor diode laser is to increase the width of emitting area. However, if the width of a single emitter is increased, filamentation and instabilities increase severely.

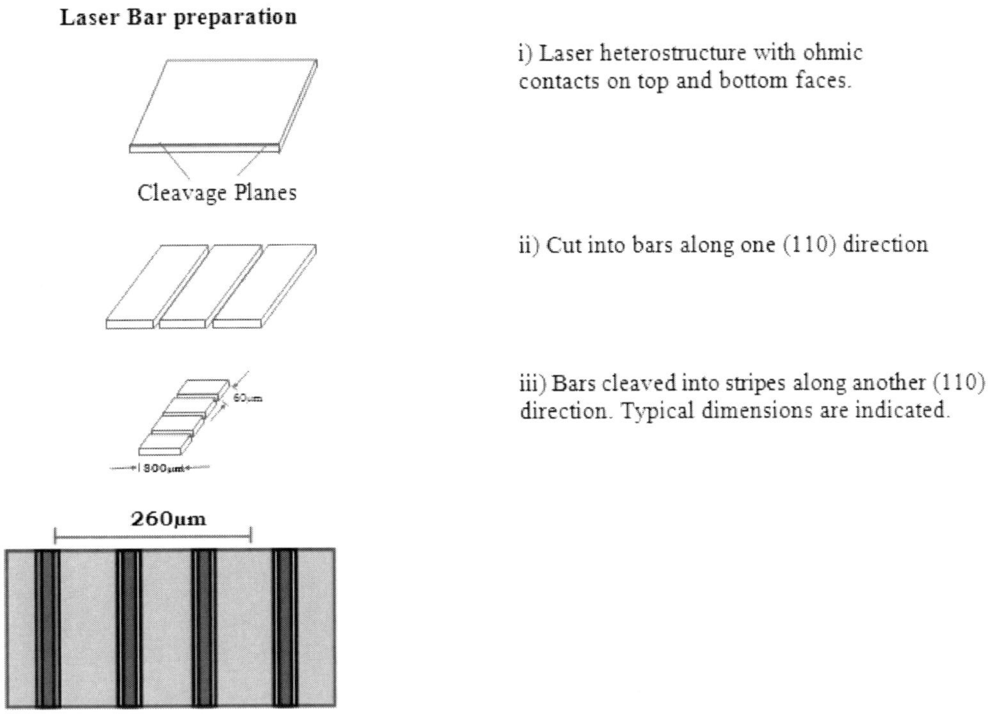

Figure 5. Formation of broad-area lasers by cleaving.

This means that optical power is no longer distributed homogeneously along the facet. So called hot spots are likely to occur, which leads to a fast degradation of the diode laser. As a matter of fact the output power of single stripe emitters cannot be enhanced significantly for stripe widths exceeding about 200 μm. A practical way to increase the output power is monolithic integration of many emitters into one diode laser bar. These emitters are isolated optically and electrically from each other. Each emitter can be driven at a moderate output power, which does not lead to an accelerated degradation of the output facet. By summing the power of all single emitters of the laser bar, a total output power of 100 watts can be achieved. In spite of high power loss (about 50% of electrical power), in such integrated array structures, the thermal management – the soldering procedure and the heat sink architecture play an important role in achieving a stable high power operation with long life times of the order of 10,000 h. In recent years, a standard length of laser bars of 10 mm has been established. The resonator length is typically 600 μm and 1000 μm and the distance between single emitters is about 50 μm and 200 μm. The ratio of active to the whole area of laser bar is about 30-80%.

Double heterostructure systems have emerged as contenders for use in optical fiber communication. These are $In_{(1-x)}Ga_xAs_yP_{(1-y)}/InP$ and $In_xGa_{(1-x-y)}$ $Al_{0.48}As/In_{0.52}Ga_yAl_{0.48}As$ systems, which can be easily grown by LPE, OMVPE.

Band Gap of Some Semiconductors

Figure 10 shows the band gap energies of some common semiconductors relative to visible, infrared and ultraviolet portion of the spectrum. We observed that GaAs, Si, Ge and InSb lie outside the visible region of the infrared.

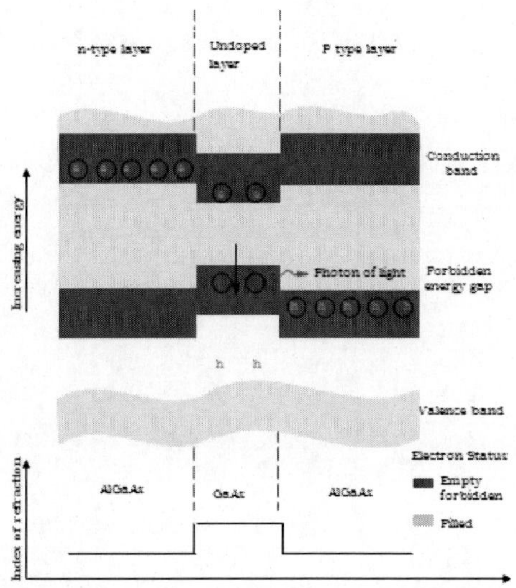

- Sandwich layer of low band gap material e.g. (GaAs) in between layers of higher band gap material (e.g. AlGaAs)
- Active layer is low-band gap material

Figure 6. Schematic cross section of laser structure.

	Layer	Thickness
	p+ : GaAs Contact	0.25 μm
	p : $Ga_{0.5}In_{0.5}P$ cladding	1 μm
InGaAs DQW	$Ga_{0.71}In_{0.29}As_{0.4}P_{0.6}$ waveguide	0.25 μm
7 nm		
	$Ga_{0.71}In_{0.29}As_{0.4}P_{0.6}$ waveguide	0.25 μm
	n : $Ga_{0.5}In_{0.5}$ cladding	1 μm
	n : GaAs buffer	
	n : GaAs substrate	

Figure 7. Laser Structure.

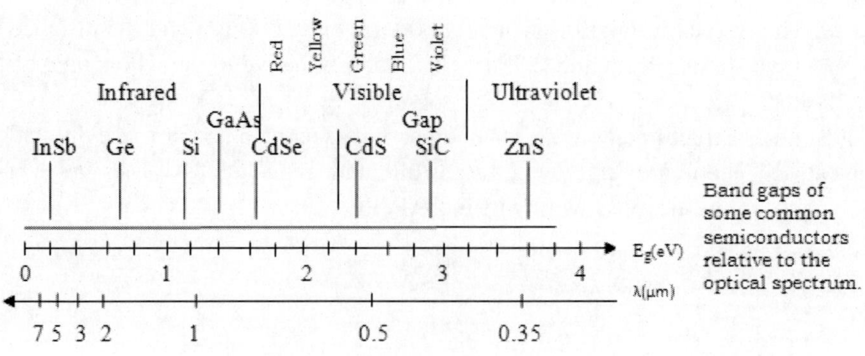

Figure 8. Band gaps of some common semiconductors relative to the optical spectrum.

Others such as GaP and CdS have bandgaps wide enough to pass photons in the visible range. It is important to note that a semiconductor absorbs photons with energies equal to the band gap or even larger. Thus, Si absorbs not only band gap light (1μm) but also shorter wavelengths, including those in the visible part of the spectrum. When carriers are injected across the forward bias, junction current is usually accounted for by recombination in the transition region and near the junction. In a semiconductor with an indirect band gap, considerable light may be given off from the junction under forward bias. The effect of injection electroluminescence provides an important application of diodes as generators of lights.

III. Reflectance of Laser Diode at the Interface

Whenever light passes from one medium into a medium of different optical properties, most notably the refractive index, part of the light is reflected (between zero and 100%), and part of the light is transmitted (between 100% and zero).The intensity ratio of the reflected and transmitted components is mainly a function of the difference in the refractive index between the two materials and the angle of incidence. In principle, the surface of any optical element can be coated with thin layers of various materials (termed thin films) in order to ensure the desired reflection. This depends on the nature of the material from which the optic is fabricated as well as the wavelength of incident light and the angle of the incidence. As stated earlier, external mirrors are not needed in a diode laser and the cleaved planes at the two laser ends essentially provide the required reflection to form the cavity. The reflectance at the semiconductor/air interface is given by

$$R = [(nr_2-nr_1)/(nr_2+nr_1)]^2$$

Where nr_2 and nr_1 are the refractive indices of the semiconductor and air, respectively, typically with $nr_2 = 3.6$ and $nr_1 = 1$, R=0.32.

e.g.,

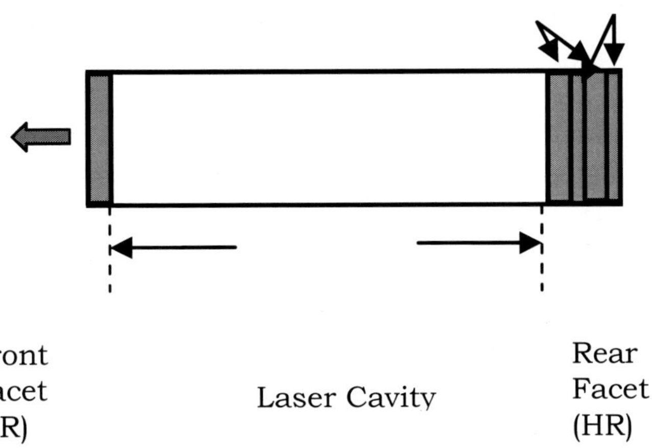

Figure 9. Facet Coating.

Table 1. Facet Coating

Sr. No.	Coating	R_1	R_2	g_{th}
1.	No Coating	0.32	0.32	21.4
2.	One Side AR	0.32	0.1	27.2
3.	One Side HR	0.05	0.9	16.2
4.	One Side AR; Other side HR	0.05	0.9	25.5

Inference: Mirror loss = f (Reflectivity R)
Changing R i) changes Loss ii) changes threshold.

Facet Coatings

One of the most crucial aspects of high power semiconductor laser technology is that of generating long-term stable laser mirrors at the facets of the cleaved wafer. The heat energy in the facet region is concentrated in a small volume and the power density reaches a very high value at the output mirror, which could be nearing the damage threshold of a laser facet. This can be avoided and the mirrors are realized by a three-step process.

- cleaving the processed wafers into laser bars,
- passivation of the cleaved surfaces,
- coating for the desired reflectivity.

This process has to fulfill three main requirements. First, the desired reflectivity, typically greater than 90% at the rear facet, and values between 3% and 20% at the front facet, should be realized. Second, the process has to yield a high stability of the facet to ensure a device lifetime of several thousand hours. Third, it must be reproducible and cheap technology, suitable for mass production.

IV. Design of Anti-Reflection and High Reflection Coating

The principle of single layer AR coating is as follows. The substrate is coated with a thin layer of material so that reflections from the outer surface of the film and the outer surface of the substrate cancel each other (destructive interference). The requirement is exact cancellation of reflected beams with a single layer coating so that the reflections are exactly 180^0 out of phase and of the same intensity.

The thickness of a single AR film must be an odd number of λ /4 in order to achieve the correct phase or cancellation. Film thickness d.

This is shown by the accompanying Figure 10, which explains the mechanism of a perfect single layer AR coating. There is a $\pi/2$ phase shift for reflections at both interfaces because they are reflections from a rare medium interface to a dense medium interface.

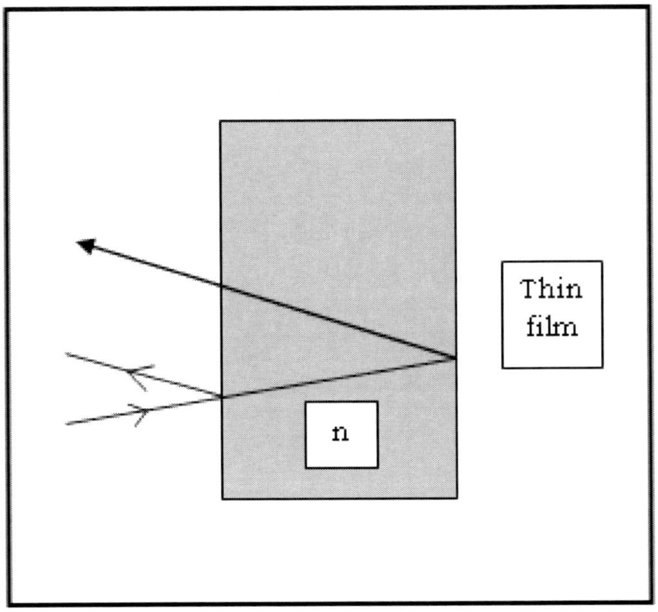

Figure 10. Reflection of thin film.

These identical phase shifts cancel out. The net phase shift between the two reflections is therefore determined solely by optical path difference 2d x n_s where 'd' is the physical thickness of the coating layer and 'n_s' is the refractive index of the coating material.

The phase shift is therefore 2dn/λ.

A single layer of AR coating is generally deposited with a thickness of λ/4 where λ is the desired wavelength for peak performance. The phase shift is 180° and the reflections are in a condition of exact destructive interference.

Refractive Index

The intensity of a reflected beam of a single surface at a normal incidence is given by

$$[(1\text{-}n)/(1+n)]^2 * \text{incident intensity}$$

where n is the ratio of refractive indices of two materials at interface.

For the two reflected beams to be equal in intensity, it is necessary that the refractive index ratio be the same at both interfaces.

$$n_{air}/n_{film} = n_{film}/n_{substrate}$$

i.e., the three refractive indices must form a geometric progression.

Since the refractive index of air is 1.0, AR film ideally should have a refractive index of sqrt ($n_{substrate}$)

Al_2O_3 is a widely used thin film material for optical coating satisfying the above criteria.

Considering the incident medium as air, the reflection coefficient at each interface is negative.

Phase change of 180°

$\Phi = \omega t$, $\omega = 2\pi f$

$\omega = 2\pi * v/\lambda * 2D/v$, where $f = v/\lambda$ and $t = 2D/v$

$\lambda = 4D$

$D = \lambda/4$ condition for complete cancellation of reflectance at a wavelength, i.e., to get zero reflectance.

Now,

$(n_0 - n_l)/(n_0 + n_l) = (n_l - n_m)/(n_l + n_m)$

$n_l = (n_0 n_m)^{1/2}$

where n_0 = Refractive index of air
n_l = R.I of thin film
n_m = R.I. of substrate

Condition for a perfect single layer antireflection coating is therefore the quarter wave optical thickness of material with

Optical admittance = $(n_{sub})^{1/2} = (3.6)^{1/2}$

Available materials are SiO_2, Al_2O_3, TiO_2.

Al_2O_3 is a widely used thin film material for optical coating, satisfying the above criteria as discussed before.

Design of High Reflection Coating

High reflectance dielectric layers work on the same principles as dielectric antireflection coating.

Quarter wave thickness of various materials is applied to the substrate to form a dielectric multilayer, the basic element of which is the QW stack. The QW stack is a stack of alternate high and low refractive index films, each one having an optical thickness of QW.

Reflections at low-high index boundaries undergo a halfwave (180°) phase shift, whereas light reflected at high/low index boundaries experience no phase shift. So, reflected waveforms in phase undergo constructive interference.

These stacks achieve a very high reflectance surrounding the wavelength for which the optical thickness of both layers is an exact quarterwave.

The reflectance is very sensitive to –

1) The ratio of the refractive indices of the two materials.
2) The number of pairs of high and low index layers.

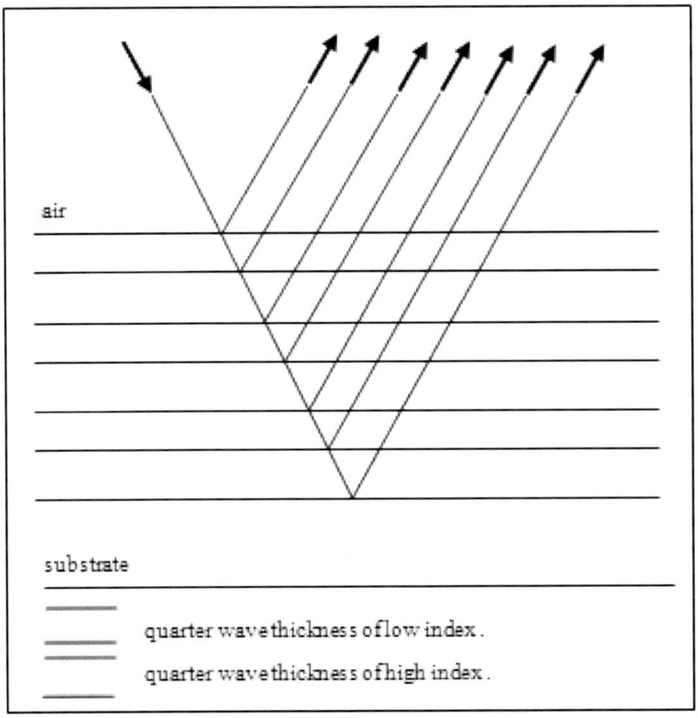

Figure11. Single Quarterwave stack.

Increasing either increases the reflectance.

V. Reflectance of Thin Film

Macleod Method

The Macleod method provides a variety of tools to support the design process. Editing tools make design manipulation easy. These tools include: increasing the number of layers in the design, changing the material in a design, by formula, and scaling thickness.

The substrate for the coatings may absorb light as a function of wavelength. The Macleod method is a tool used in calculation of the performance of many coatings. It has the tolerance capacity to investigate the sensitivity of a design to manufacturing errors. Alternative designs may be compared and the best design is selected. Although designs may be similar, their sensitivity to errors may not be, and are shown by the tolerance feature.

The Reflectance of a Thin Film

Whenever a thin film of material covers the surface of a substrate, the presence of two (or more) interfaces means that a number of beams will be produced by successive reflections and the properties of the film will be determined by the summation of these beams. A single extension of the above analysis occurs in the case of thin, plane parallel films of material covering the surface of a substrate. The presence of two (or more) interfaces means that a number of beams will be produced by successive reflections and the properties of the film will be determined by the summation of these beams. We say that the film is thin when interference effects can be detected in the reflected or transmitted light, which is when the path difference between the beams is less than the coherence length of the light and thickness, when the path difference is greater than the coherence length. Normally, we will find that the films on the substrate can be treated as thin films, while the substrate supporting the films can be treated as thick films.

Symbol + (that is for positive-going wave) and waves in the opposite direction by – (that is for negative-going wave).

The interface between the film and the substrate, denoted by the symbol **b**, can be treated in exactly the same way as the simple boundary already discussed. We consider the tangential components of the field. There is no negative-going wave in the substrate and the waves in the film can be summed into one resultant positive going wave and one resultant negative-going wave.

At this interface, then, the tangential components of E and H are

$$\begin{aligned} E_b &= E_{1b}^{+} + E_{1b}^{-} \\ H_b &= \eta_1 E_{1b}^{+} - \eta_1 E_{1b}^{-}, \end{aligned} \tag{A}$$

where we neglect the common phase factors, and where E_b and H_b represent the resultants. Hence,

$$\begin{aligned} E_{1b}^{+} &= \frac{1}{2}(H_b/\eta_1 + E_b) \\ E_{1b}^{-} &= \frac{1}{2}(-H_b/\eta_1 + E_b) \\ H_{1b}^{+} &= \eta_1 E_{1b}^{+} = \frac{1}{2}(H_b + \eta_1 E_b) \\ H_{1b}^{-} &= -\eta_1 E_{1b}^{-} = \frac{1}{2}(H_b - \eta_1 E_b) \end{aligned} \tag{B}$$

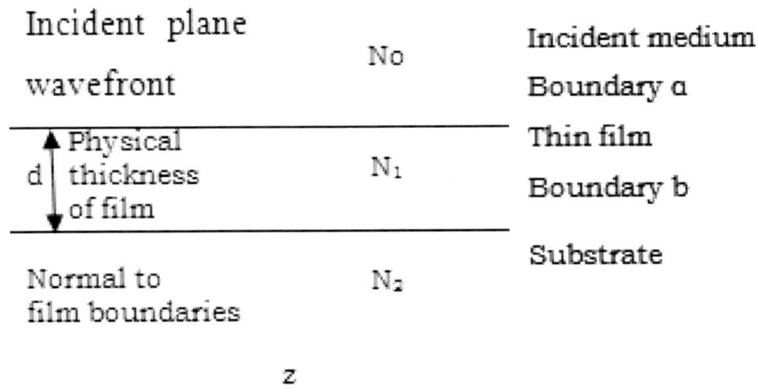

Figure 12.Plane wave incident on a thin film.

The fields at the other interface a at the same instant and at a point with identical x and y coordinates can be determined by altering the phase factors of the waves to allow for a shift in the z coordinate from O to d. The phase factor of the positive-going wave will be multiplied by exp(iδ) where

$$\delta = 2\pi N_1 d \cos \vartheta_1 / \lambda$$

and ϑ_1 may be complex, while the negative-going phase factor will be multiplied by exp (-iδ).

The values of E and H at the interface are now,

$$E_{1a}^{+} = E_{1b}^{+} e^{i\delta} = \frac{1}{2}(H_b/\eta 1 + E_b)e^{i\delta}$$

$$E_{1a}^{-} = E_{1b}^{-} e^{-i\delta} = \frac{1}{2}(-H_b/\eta 1 + E_b)e^{-i\delta}$$

$$H_{1a}^{+} = H_{1b}^{+} e^{i\delta} = \frac{1}{2}(H_b + \eta 1 E_b)e^{i\delta}$$

$$H_{1a}^{-} = H_{1b}^{-} e^{-i\delta} = \frac{1}{2}(H_b - \eta 1 E_b)e^{-i\delta}$$

So that

$$\begin{aligned} E_a &= E_{1a}^{+} + E_{1a}^{-} \\ &= E_b\left(\frac{e^{i\delta} + e^{-i\delta}}{2}\right) + H_b\left(\frac{e^{i\delta} - e^{-i\delta}}{2\eta 1}\right) \\ &= E_b \cos\delta + H_b \frac{i \sin\delta}{\eta 1} \\ H_a &= H_{1a}^{+} + H_{1a}^{-} \\ &= E_b \eta 1\left(\frac{e^{i\delta} - e^{-i\delta}}{2}\right) + H_b\left(\frac{e^{i\delta} + e^{-i\delta}}{2}\right) \\ &= E_b i \eta 1 \sin\delta + H_b \cos\delta. \end{aligned}$$

This can be written in matrix notation as:

$$\begin{bmatrix} E_a \\ H_a \end{bmatrix} = \begin{bmatrix} \cos\delta & (i\sin\delta)/\eta_1 \\ i\eta_1\sin\delta & \cos\delta \end{bmatrix} \begin{bmatrix} E_b \\ H_b \end{bmatrix} \tag{E}$$

Since the tangential components of **E** and **H** are continuous across a boundary and since there is only a positive-going wave in the substrate, this relationship connects the tangential components of **E** and **H** at the incident interface with the tangential components of **E** and **H,** which are transmitted through the final interface. The 2 x 2 matrix on the right-hand side of the equation is known as the characteristics matrix of the thin film.

We define the input optical admittance of the assembly by analogy with equation as:

$$Y = H_a/E_a$$

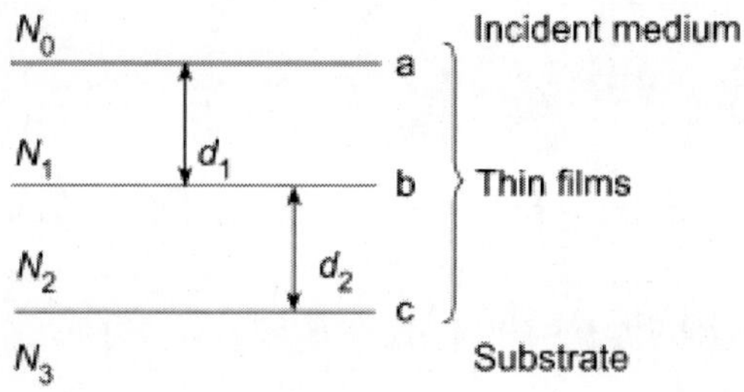

Figure 13.Notation for two films on a surface.

when the problem becomes merely that of finding the reflectance of a simple interface between an incident medium of admittance η_0 and a medium of admittance Y, i.e.,

$$\rho = \frac{\eta_0 - Y}{\eta_0 + Y}$$

$$R = \left(\frac{\eta_0 - Y}{\eta_0 + Y}\right)\left(\frac{\eta_0 - Y}{\eta_0 + Y}\right)^*. \tag{F}$$

we can normalize equation E by dividing by E_b to give

$$\begin{bmatrix} E_a/E_b \\ H_a/E_b \end{bmatrix} = \begin{bmatrix} B \\ C \end{bmatrix} = \begin{bmatrix} \cos\delta & (i\sin\delta)/\eta_1 \\ i\eta_1\sin\delta & \cos\delta \end{bmatrix} \begin{bmatrix} 1 \\ \eta_2 \end{bmatrix} \tag{G}$$

and B and C, the normalized electric and magnetic fields at the front interface, are the quantities from which we will be extracting the properties of the thin-film system. Clearly, as an equation of optical admittance we can write

$$Y = \frac{H_a}{E_a} = \frac{C}{B} = \frac{\eta_2 \cos\delta + i\eta_1 \sin\delta}{\cos\delta + i(\eta_2/\eta_1)\sin\delta} \tag{H}$$

and from (equation H and F) we can calculate the reflectance.

$$\begin{bmatrix} B \\ C \end{bmatrix}$$

is known as the characteristic matrix of the assembly.

The Reflectance of an Assembly of Thin Films

Let another film be added to the single film of the previous section so that the final interface is now denoted by c, as shown in Figure 8. The characteristic matrix of the film nearest the substrate is

$$\begin{bmatrix} \cos\delta_2 & (i\sin\delta_2)/\eta_2 \\ i\eta_2 \sin\delta_2 & \cos\delta_2 \end{bmatrix} \tag{I}$$

and from equation H.

$$\begin{bmatrix} E_b \\ H_b \end{bmatrix} = \begin{bmatrix} \cos\delta_2 & (i\sin\delta_2)/\eta_2 \\ i\eta_2 \sin\delta_2 & \cos\delta_2 \end{bmatrix} \begin{bmatrix} E_c \\ H_c \end{bmatrix}.$$

We can apply equation E again to give the parameters at interface a, i.e.

$$\begin{bmatrix} E_a \\ H_a \end{bmatrix} = \begin{bmatrix} \cos\delta_1 & (i\sin\delta_1)/\eta_1 \\ i\eta_1 \sin\delta_1 & \cos\delta_1 \end{bmatrix} \begin{bmatrix} \cos\delta_2 & (i\sin\delta_2)/\eta_2 \\ i\eta_2 \sin\delta_2 & \cos\delta_2 \end{bmatrix} \begin{bmatrix} E_c \\ H_c \end{bmatrix}$$

and the characteristics matrix of the assembly, by analogy with equation G is,

$$\begin{bmatrix} B \\ C \end{bmatrix} = \begin{bmatrix} \cos\delta_1 & (i\sin\delta_1)/\eta_1 \\ i\eta_1 \sin\delta_1 & \cos\delta_1 \end{bmatrix} \begin{bmatrix} \cos\delta_2 & (i\sin\delta_2)/\eta_2 \\ i\eta_2 \sin\delta_2 & \cos\delta_2 \end{bmatrix} \begin{bmatrix} 1 \\ \eta_3 \end{bmatrix}.$$

Y is as before, C/B, and the amplitude reflection coefficient and the reflectance are from equation F,

$$\rho = \frac{\eta_0 - Y}{\eta_0 + Y}$$

$$R = \left(\frac{\eta_0 - Y}{\eta_0 + Y}\right)\left(\frac{\eta_0 - Y}{\eta_0 + Y}\right)^*.$$

This result can be immediately extended to the general case of an assembly of q layers, when the characteristics matrix is simply the product of the individual matrices taken in the correct order, i.e.,

$$\begin{bmatrix} B \\ C \end{bmatrix} = \left\{\prod_{r=1}^{q}\begin{bmatrix} \cos\delta_r & (i\sin\delta_r)/\eta_r \\ i\eta_r\sin\delta_r & \cos\delta_r \end{bmatrix}\right\}\begin{bmatrix} 1 \\ \eta_m \end{bmatrix}$$

where

$$\delta_r = \frac{2\pi N_r d_r \cos\vartheta_r}{\lambda}$$

VI. Experimental Breakups

The entire range of experiments is divided as follows:

a. Design proper coating

1. Choose the material.
2. Choose the thickness.
3. # Layers?
4. Which one first?
5. How is the process?

b. Deposition

1. e^- beam evaporation
2. Optimization of system parameter.
3. Thickness monitoring.
4. Conditions (Different)
5. Table of various experiments

c. Characteristics

1. Measure Reflectivity.
2. Reflectivity Setup.
3. Calculated and experimental spectra.
4. Matching the spectra.

d. Experiment on

Laser Diode

e. Comparative Performance

Uncoated laser diodes vs. coated laser diode
Consider optoelectrical properties for an uncoated laser diode with the coated one.

Experimental Work

Coating design should be optimized for minimal overall thickness. The method of electron beam evaporation is to produce thin, homogeneous, uniform, and pure films of dielectric materials. Such films cannot normally be produced in the atmosphere. Instead, operation is performed in a vacuum because of the atmospheric gases of the film. Vacuum coating thus has functional properties in the field of semiconductor optoelectronics. As shown, a high-speed vacuum pumping system integrated with necessary pipes and valves, and operated manually is rated for high throughput and produces clear vacuum10^{-6} mbar in the chamber. The experiments are done for anti-reflection and high reflection coatings on GaAs using A12O3 and Si. The reflectance has been measured as a function of wavelength by the set-up explained in the next section. Results of single layer A12O3, deposited on GaAs with different thicknesses on each sample are also compared with calculations done for AR structures. The same module parameters are repeated for HR coatings. The AR coating design is optimized for minimal reflectivity for the front mirror with single A12O3 layers having a thickness of $\lambda 0/4n$. Various thicknesses of substrate and their reflectivities were measured with set-up to find the optimal thickness for minimum reflectivity.

Aluminum Oxide, $AL_2O_3$for Optical Coating

Introduction

Aluminum oxide, Al_2O_3, is a medium index, low absorption material usable for coatings in the near-UV (<300 nm) to IR (~5μm) regions. Typical applications include near-UV laser AR and dielectric mirror designs. Alumina can be used in combination with silicon dioxide layers to form multilayer structures with high damage thresholds for UV laser applications.

Reflectance Measurement

The light signal is chopped at a particular frequency using a chopper. The same chopper frequency is also applied to the reference input of the locking amplifier.

A monochromator is placed before the chopper.

The monochromator consists of diffraction grating arrangements, for selecting the wavelength. Also, a filter is required to pass the light beyond a particular wavelength to avoid light from a different order of diffraction. At the output, a photo detector is used. It converts the optical signal into an electrical signal. An electrical output from the photo detector is given to the lock-in-amplifier and the display unit of the lock-in-amplifier gives the required rms value of the signal. Coating on back mirrors having high reflectivity consist of Bragg's stacks (layers of high and low refractive indices) having thickness of each layer $\lambda_0/4n$.AR and HR coatings were then deposited on facets of edge emitting lasers, and their electro-optical characteristics were compared before and after facet coatings.

Evaporation Parameters

Evaporation Temperature	~2100°C
Source Container	No liner for E-beam. Electron beam should be swept.
Rate	2-5 Å/sec.
Partial Pressure of Oxygen	1 x 10^{-5}Torr.
Substrate Temperature	200°C to 300°C

Physical Properties of Solid Material

Molecular Weight	101.96
Melting Point	2020°C
Color	Clear to white
Crystal Density	3.97 g/cc

VII. Comparison of Simulated and Experimental Results

Performance Curve

The reflection Vs wavelength curve of a single dielectric layer has a U shape, specifying the minimum wavelength of the layer, 1 QW thick. The curve for a stack of multilayers has a shape inverted to that of the antireflection coating. Both the high and low index layers of the multilayer are exactly one QW thick. Outside the region of high reflectance, the reflectance slowly reduces towards zero in an air oscillatory manner. The width and height of the high reflectance region of the curve are the functions of the refractive index of the two materials used in the layers, together with the number of layers actually included in the stack.

The peak reflectance can be increased either by including more layers in the stack or by using materials with a higher refractive index ratio.

Experimental Results

Electron beam evaporation has been used to deposit uniform Al_2O_3 and Si films.

We discuss the optimization of experimental parameters.

n+GaAs and SIGaAS of (Wafer Tech and Russian) lapped substrates of dimensions 11 mm X 7 mm and thickness about 400 micrometers, were first cleaned by TCE, acetone and methanol rinse and heated in a high vacuum coated unit at a pressure 4-6 X 10^{-6} m bar, or better achieved by diffusion pump and rotary pumps assembled during evaporation.

Deposited thickness varied in the range of 575-1550 Å for a set of experiments needed for optimization of wavelengths 980 nm and 900 nm. Deposition rate for Al_2O_3 was 2-5 Å/sec and that of Si 2-4 Å/Sec.

Simulation for all the different thicknesses were carried out and compared with experimental values.

Measurement of reflectance involved use of (constant current 5 Amp), halogen tungsten light source, monochromator, chopper (at 89.5Hz), and lock-in amplifier and detector uniconnected in places as shown, light focused by lenses and filtered by 690 nm filter. Substrate was near to normal incidence (2-3 degrees).

Results of AR Coating

Simulated and experimental results

Thickness in Å Al_2O_3	Wavelength(nm)	Reflectance (Min.)	
		Simulated	Experimental
1200	880	0.0240	0.0238
1210	880	0.0345	0.0340
1265	890	0.0315	0.0300
1400	1000	0.0230	0.0275
1440	1000	0.0211	0.0250
1550	1120	0.0280	0.0275

A Set of Simulated and Experimental Graphs along with Some Typical Data Antireflection Coating

Antireflection Coating 1200 Å

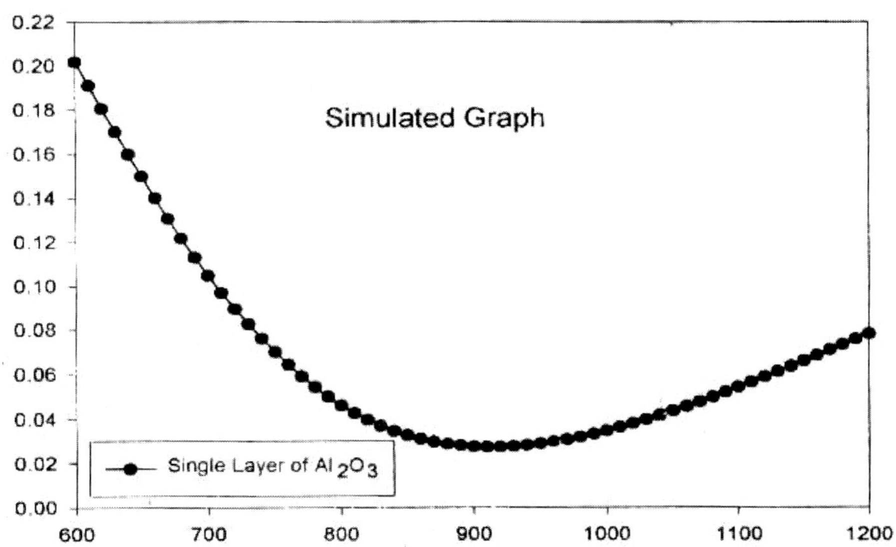

Antireflection Coating 1200 Á

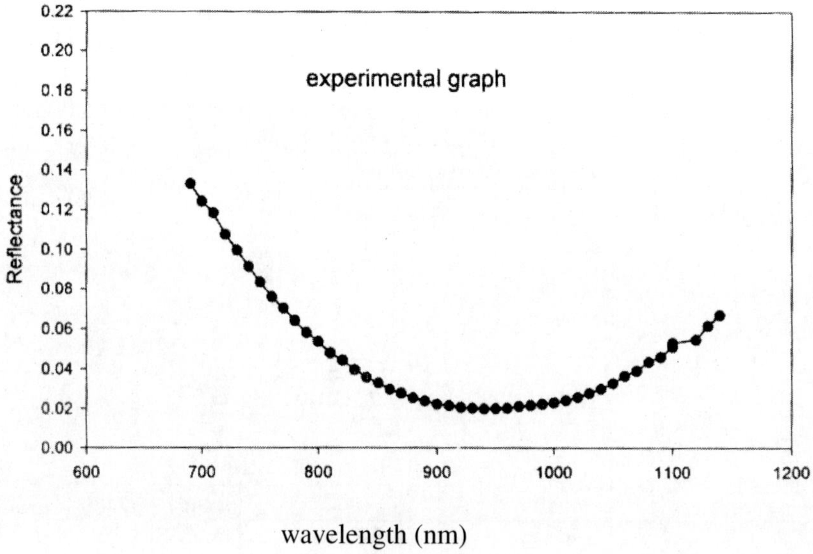

AR GaAs (Russian polished) 1200 Å, Experimental

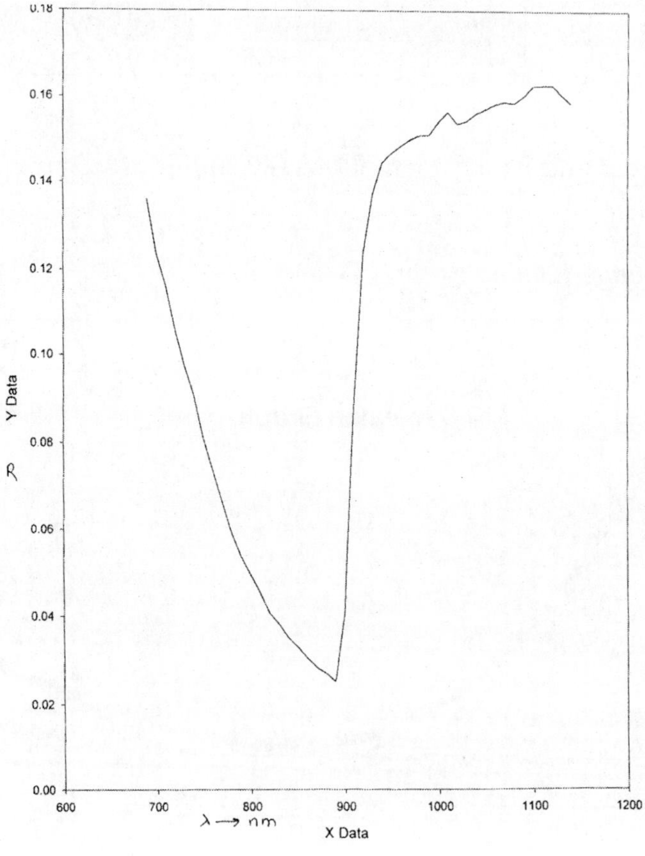

Experimental Antireflection coating d=1400 Å after lapping

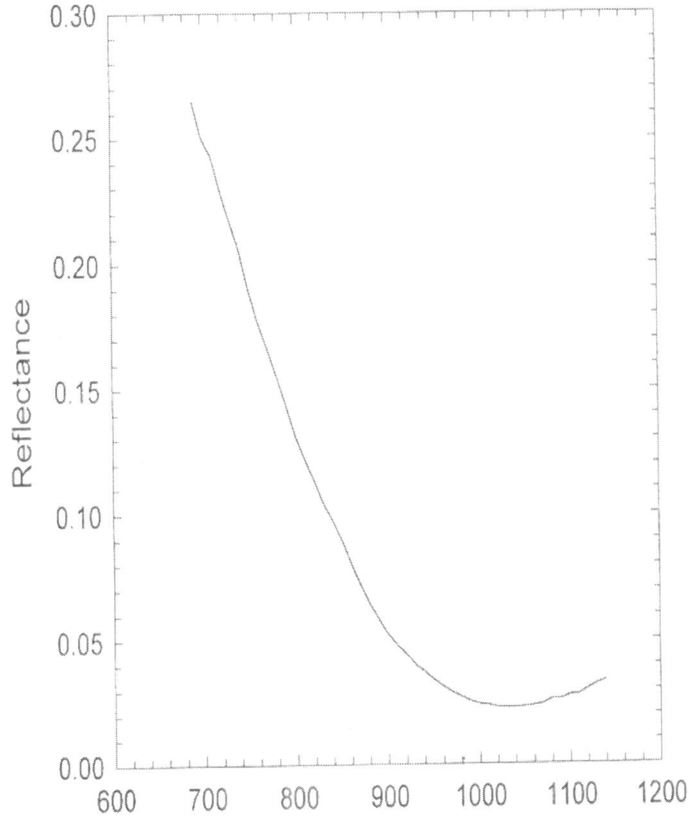

Results of HR coating (Simulated and Experimental)

No. of pairs	Thickness (Å)		Reflectance (Min.)	
	Al_2O_3	Si	Simulated	Experimental
1 pair	1400	600	0.7755	0.6475
1 pair	1450	700	0.7920	0.6736
1 pair	1200	575	0.7845	0.7150
2 pair1.	1400	700	0.9487	0.8331
	1400	700		
2.	1400	650	0.8605	0.7181
	1400	650		
3.	1200	575	0.9555	0.9510
	1200	575		
3 pair1.	1200	575	0.9885	0.9520
	1200	575		
	1200	575		

A Set of Simulated and Experimental Graphs along with Some Typical Data on High Reflection Coating

Simulated High Reflectance

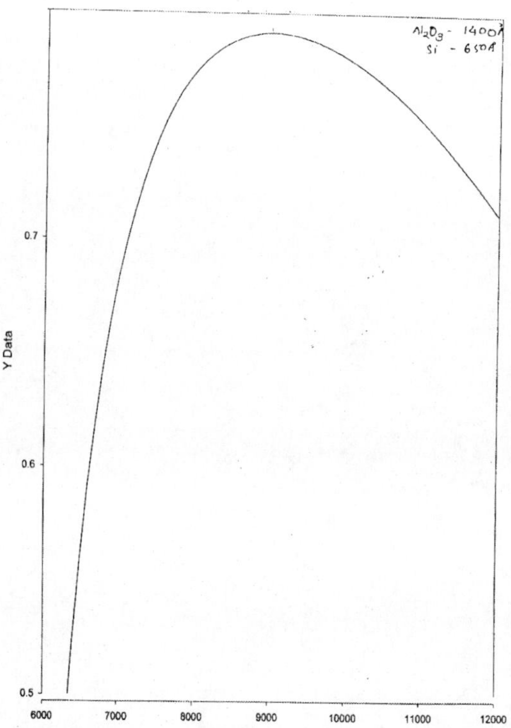

Experimental HR n^+GaAs (Pink) [1200 Å (Al_2O_3), 575 Å (Si)]

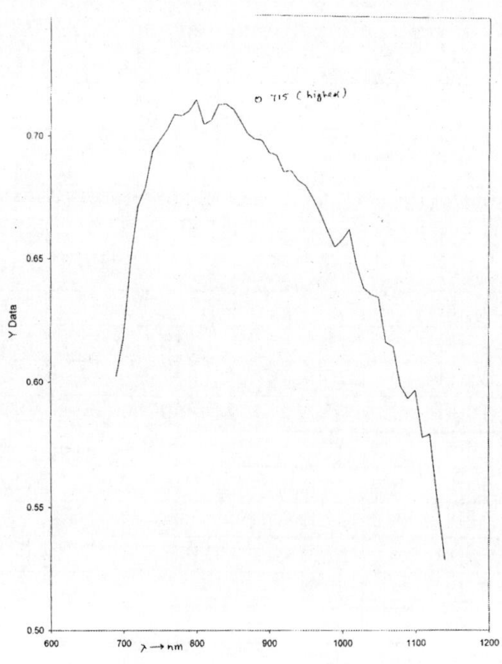

Experimental HR 2 pairs [1200 Å (Al$_2$O$_3$), 575 Å (Si)]

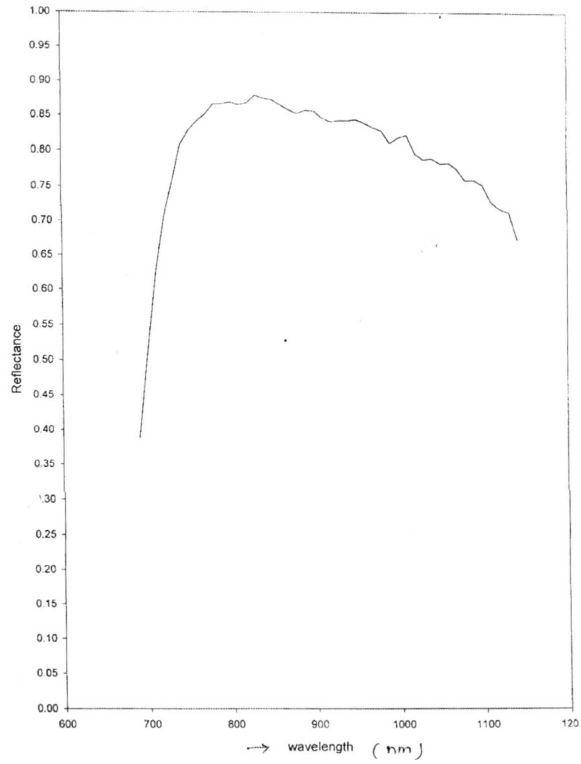

Experimental HR [1020 Å (Al$_2$O$_3$), 583 Å (Si)]

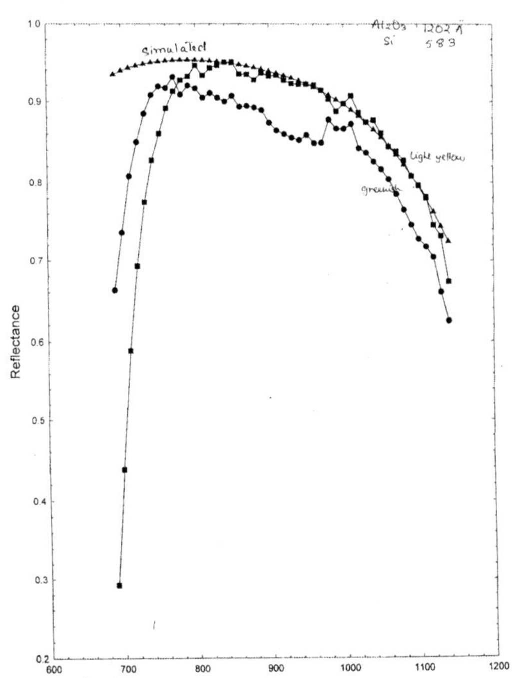

Simulated HR [1220 Á (Al_2O_3), 590 Á (Si)]

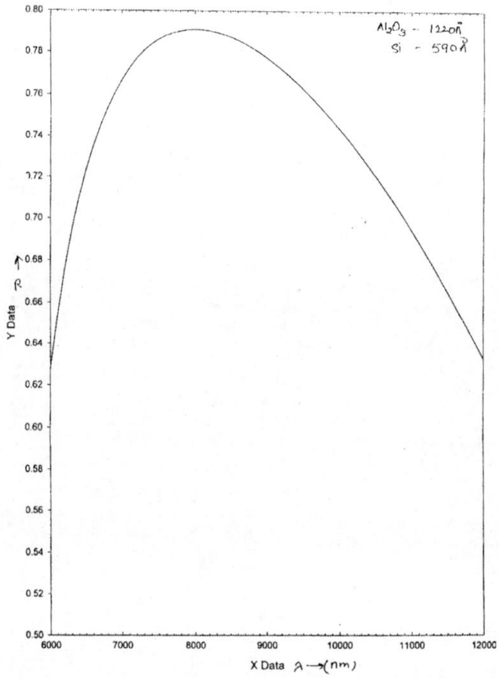

Simulated HR-1 pair [1400 (Al_2O_3), 650 (Si) Á]

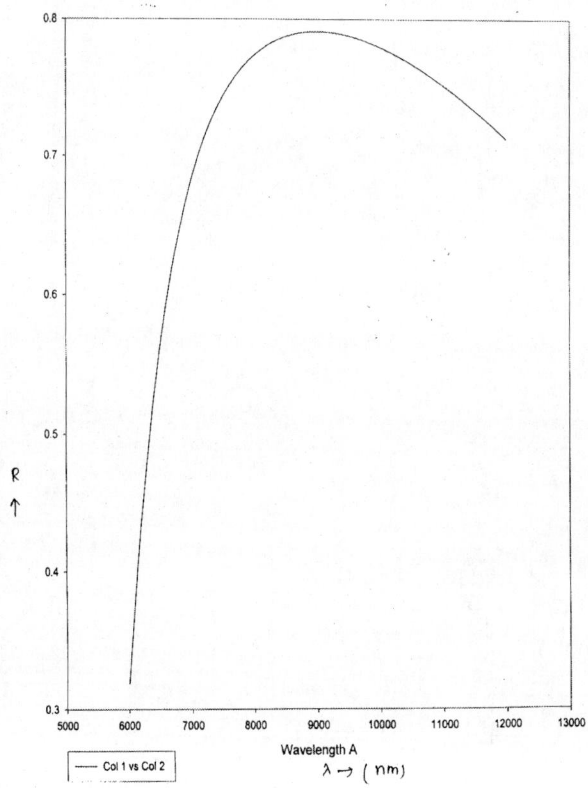

Experimental HR [1400 Á (Al_2O_3), 650 Á (Si)]

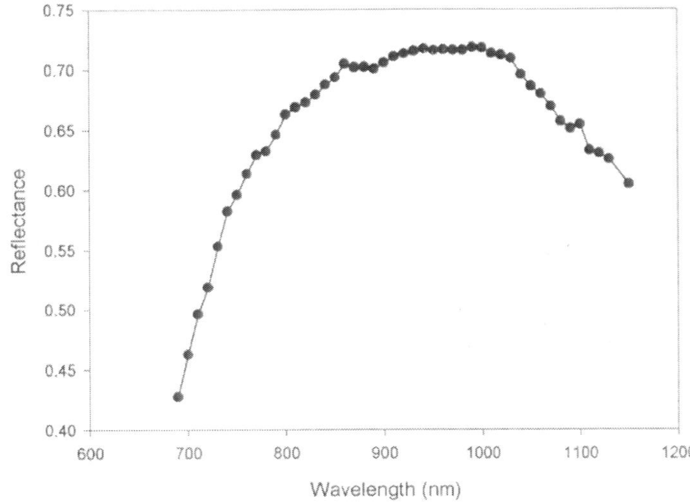

Simulated HR [1200 Á (Al_2O_3), 650 Á (Si)]

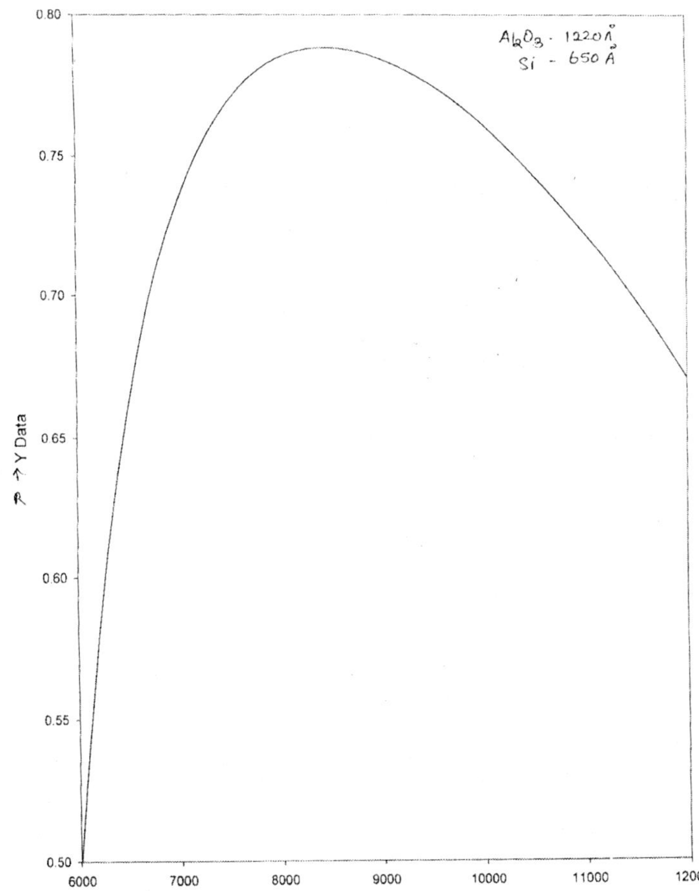

Experimental HR n^{+}GaAs (Pink) [1400 Å (Al_2O_3), 711 Å (Si)]

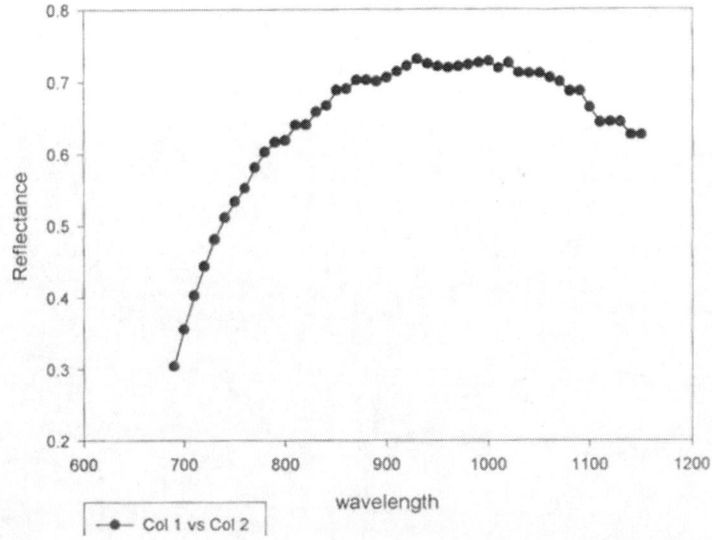

Simulated HR 3 Pairs

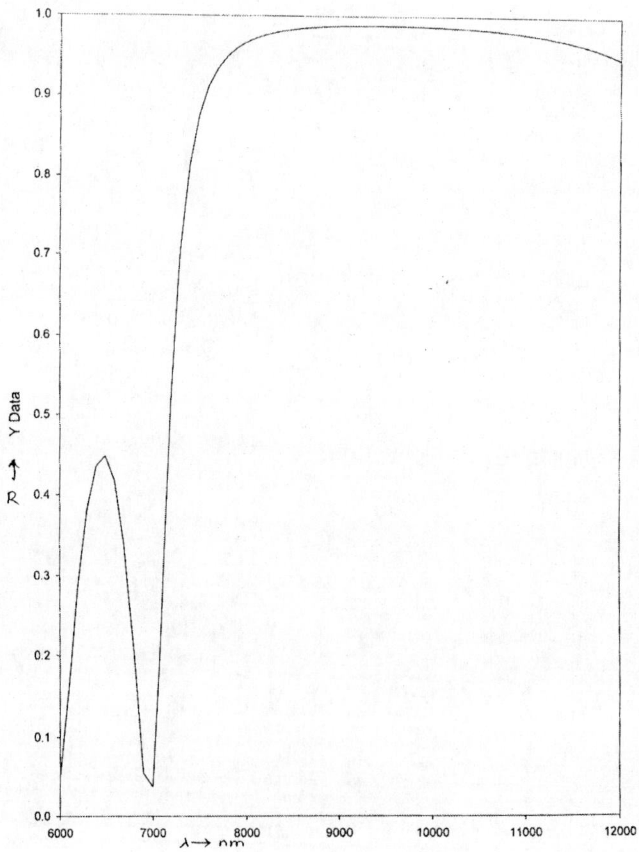

HR Simulated

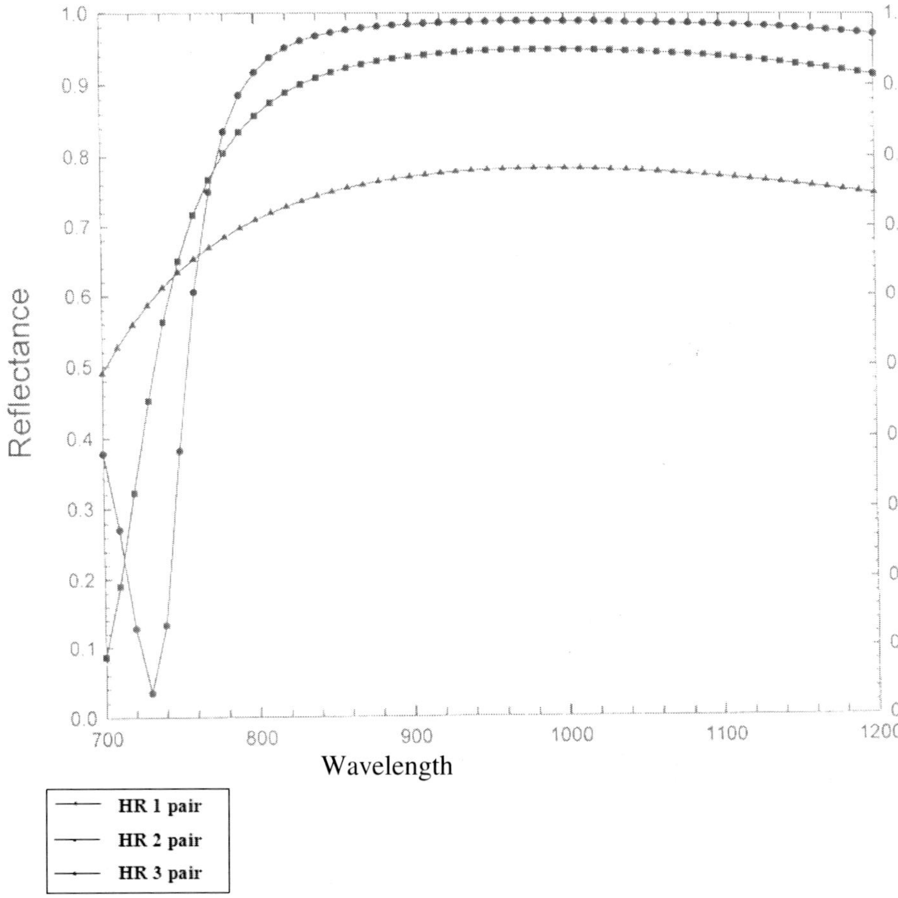

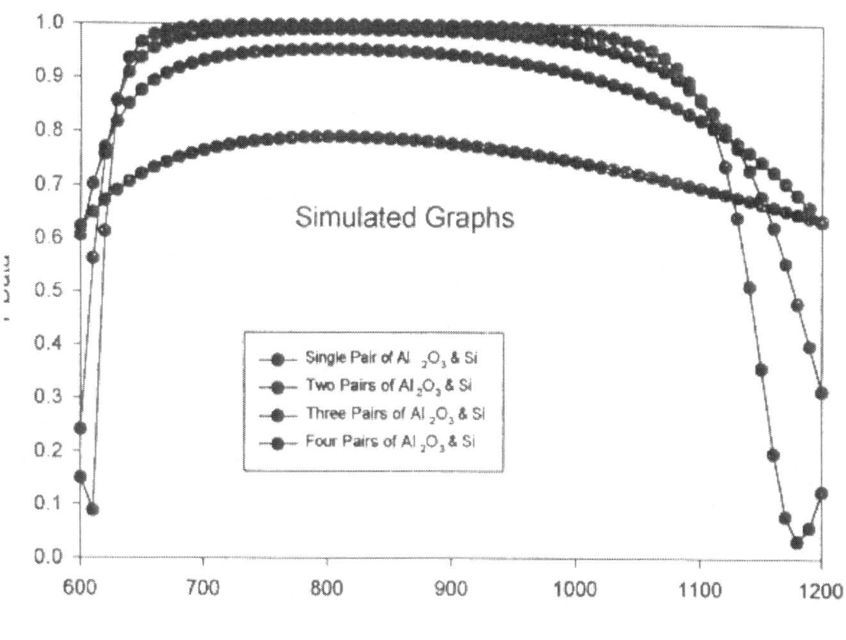

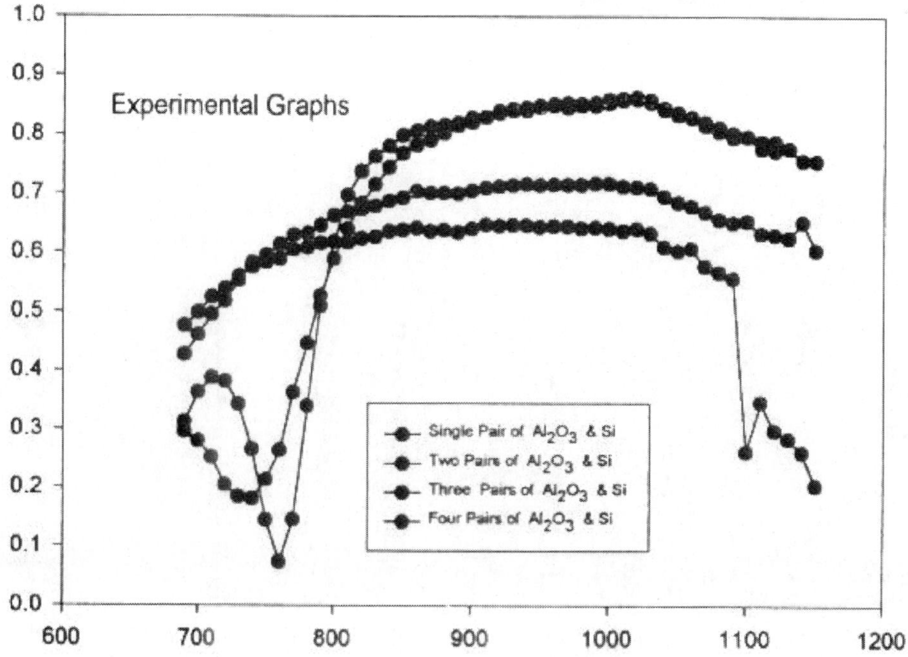

Experimental Measurements Facet Coating

Coating on		Threshold Current density (mA/cm^{2})	Output Power per cm	
A	B		(A)	(B)
-	-	690.0000	0-325	0-347
AR	-	910.0000	0-148	0-68
-	HR	550.0000	0-158	0-265
AR	HR	780.0000	0-490	0-42

550 μm and 750 μm broad area laser devices of 100 μm wide stripe lasing @ ~900nm were used for the coatings measurements. The devices were checked for its I-V characteristics before and after coatings. One such is shown.

Measurement was done using 5 μs current pulse, 1% duty cycle high pulse power resulting in gross damage in the form of spherical mounds visible at the mirror surface where the optical flux density was the greatest. The damage was markedly reduced as the pulse length was reduced from 10 μs down to 1 μs and the duty cycle changed from 10% to 1%.

Removal of affected devices from the bar continued to show the same I-V characteristics as before otherwise the damage proved all other devices leaky. Typical resistance of 2 to 5 ohms was observed from the characteristics. Measurement of L-I used a conventional pulse driver on the laser fixed on a metal plate supported by a thermoelectric cooler.

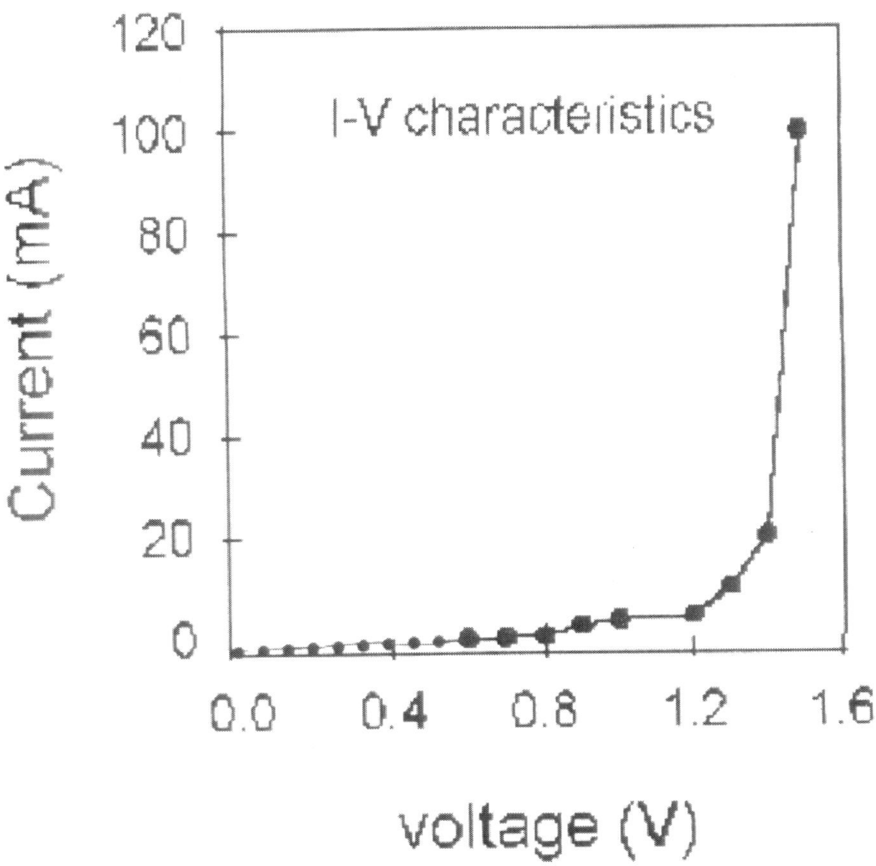

Figure 20. IV Characteristics of Laser Diode.

Results of Facet Coating

Antireflection coating of 1200 Å thickness was applied from the facet. Laser threshold, as obtained from the knee in the L-I curve or extrapolation of the L-I curve to zero light was observed before and after coatings were applied, and output power as a function of driven current (mA) was plotted for front and back facets separately.

As indicated from L-I characteristics, there was an increase of threshold and increased power from the front facet, which was AR coated, compared to lower power from the other end (uncoated), the threshold being higher in both ends.

Only the HR coating on the back mirror reduced the threshold to a noticeable extent and output power was reduced for the coated facet but was increased for the uncoated facet. Later AR coatings on the front facet (A) and the (HR) coating (B) resulted in increased threshold and increased output from the AR end and increased threshold and very low output from the HR end. A 500 μm laser was lased at 450-500 mA/cm^2 and 750 μm at 675-700 mA/cm^2 thus verifying increases of threshold as the length increased, and was easier to handle and tie in the holder in which the laser bars were stacked with two dummy bars on either side. (An observation with accd camera showed the facet before and after the coats along with dummy bars.)

Result of facet coating:

—•— Before coating

—•— Front facet (AR)

—•— Rear facet (HR)

CONCLUSION

We have grown single layer AR coatings with Al_2O_3 and multiplayer HR coating with Al_2O_3/Si on GaAs substrate using the e-beam evaporation method for optimization of different parameters such as thickness to be deposited, evaporation rate, etc.

A low reflectance of 0.02 using a single layer coating of Al_2O_3 and high reflectance with max value 0.95, with 3 layers using Al_2O_3/Si, was observed experimentally.

Coatings were deposited on facets of laser diodes.

Facet reflectivity has thus shown its influence on the amount of spontaneous emission on lasing action.

Decreasing facet reflectivity with AR coating increased the pumping required to reach the lasing threshold, and increasing the facet reflectivity with HR coating decreased it.

FURTHER WORK

In the present work, there was enough experimental data for calculating reflectance for a GaAs wafer, polished on both sides. We said earlier that the backside of a GaAs wafer had to be rough, to match the experiment with simulations. Reflectance behavior in the vicinity of the band gap is available in the data for samples coated with different layers, specified above.

These should be additional features, due to absorption in GaAs substrate. Due to time limitations theoretical values could not be calculated. They are expected in the near future.

APPENDIX

Materials for Semiconductor Lasers

Compound Semiconductors

- III-V materials : One element from Group III other from Group V
- Similarly II-VI etc.

Semiconductor laser history Jth>⇩3 orders of magnitude since 1962.
Various breakthroughs made possible by advancements in crystal growth techniques.

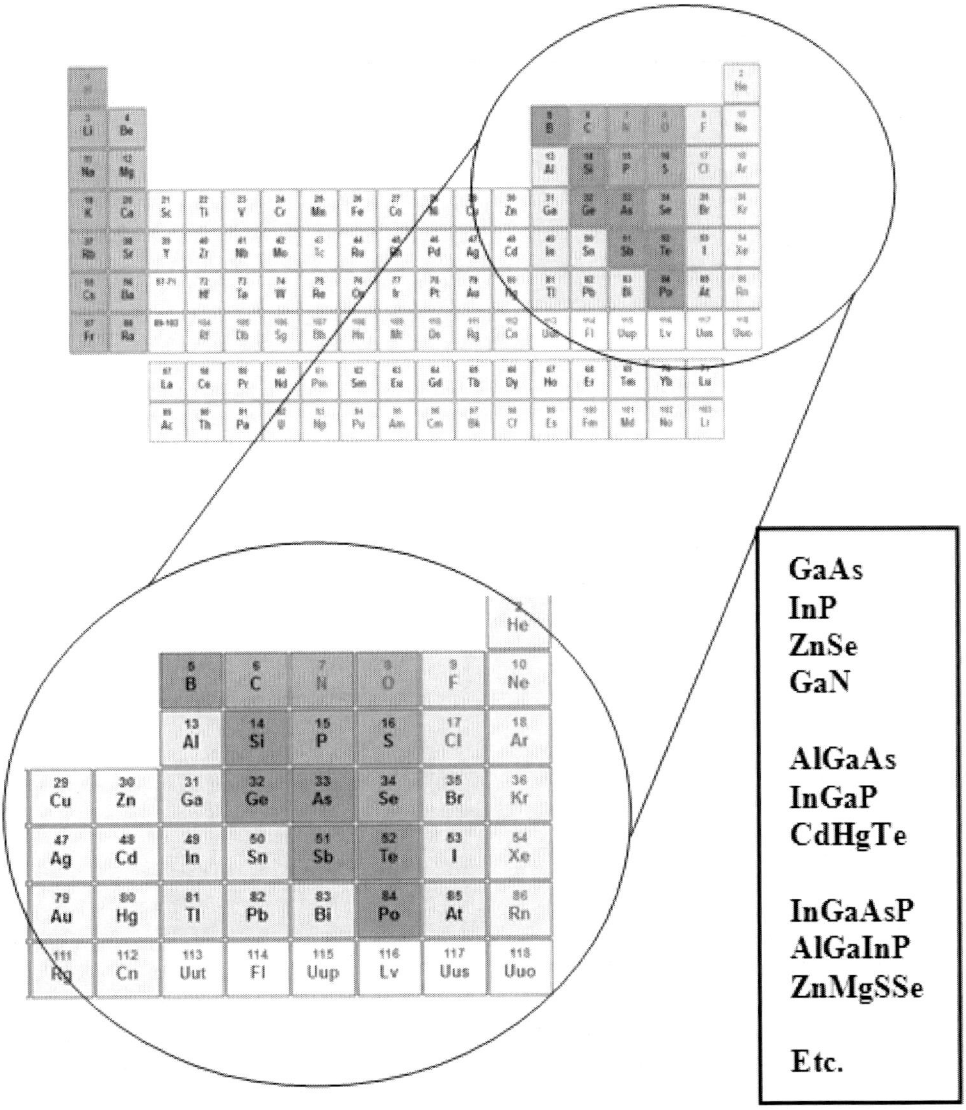


```
C   PROGRAM NAME : AR. for : 30-01-2002; PREETA TSEC
C   ABSTRACT : REFLECTANCE CALCULATION AR
C   ------------------------------------------------------------------
    COMPLEX B,C,Y,rho,i,sum
    COMPLEX M(2,2,25), PROD (2,2,25), L(2,2), sb(2,1)

    INTEGER I1,J1,K1,k2,n1
    REAL d1, la, tht, eta (25), pi, delta (25), R,N (25)
    OPEN (UNIT=1, FILE='STAR')
    i = (0,1)
    pi = 3.141592654

c   *************** DRAW STRUCTURE HERE ************
```

```
c    AIR
c    -------------
c    layer     ****A            #1
c    -------------
c    sub : GaAs      400 μm          #2
c    ~~~~~~~~~~~~~~~~~

c    ************** ENTER DATA BELOW **************

     n1      = 1

     N(1)    = 1.62
     N(2)    = 3.51

     la      = 5000
     c       d (2)   = substrate

     sb (1,1) = 1
     sb (2,1) = 3.51

c    ******************************************
     DO 60 dl = 500,1000,50
     DO 10 k= 1,n1

     eta (k) = N (k)
     delta (k)= 2*pi*N(k)*d1/la

     M(1,1,k)        =        cos (delta (k))
     M(1,2,k)=       i*sin(delta (k))/eta (k)
     M(2,2,k)=       cos (delta (k))
     M(2,1,k)=       i*eta (k) * sin (delta (k))

10   CONTINUE
     PROD (1,1,1)    =        M (1,1,1)
     PROD (1,2,1)    =        M (1,2,1)
     PROD (2,1,1)    =        M (2,1,1)
     PROD (2,2,1)    =        M (2,2,1)

     DO 50 K3 = 2, nl
     DO 40 I1= 1,2
      DO 30 JI = 1,2
     SUM = 0
     DO 20 K1 = 1,2
     SUM = SUM + PROD (I1,K1,K3-1) *M(K1,J1,K3)

20           CONTINUE
```

```
          PROD (I1, J1, K3) = SUM

30 CONTINUE
40  CONTINUE
50 CONTINUE

   L(1,1)   = PROD (1,1,nl)
   L(1,2)   = PROD (1,2,nl)
   L(2,1)   = PROD (2,1,nl)
   L(2,2)   = PROD (2,2,nl)

   B = L (1,1) *sb(1,1) + L(1,2) +sb (2,1)
   C = L (2,1) *sb (1,1)+ L(2,2) +sb (2,1)
   Y = C/B
   rho = (1 - Y)/(1 + Y)
   R = (cabs (rho) ) **2
   WRITE (1, *) d1, R
60 CONTINUE
   END
C  PROGRAM NAME : HR3. for : 30-01-2002; PREETA TSEC
C  ABSTRACT : REFLECTANCE CALCULATION HR
C  ---------------------------------------------------------------
   COMPLEX B,C,Y,rho,i,sum
   COMPLEX M(2,2,25), PROD (2,2,25), L(2,2), sb(2,1)

   INTEGER I1,J1,K1,k2,n1
   REAL d(25)1, la, tht, eta (25), pi, delta (25), R,N (25)
   OPEN (UNIT=1, FILE='STAR')
   i = (0,1)
   pi = 3.141592654

c  ************ DRAW STRUCTURE HERE ************
c  AIR              #0
c  -------------
c  Si       700 A   #1
c  -------------
c           A1203   1400 A  #2
c  -------------
c  Si       700 A   #3
c  -------------
c  A1203  1400 A  #4
c  -------------
c   Si      700 A   #5
c  -------------
c  A1203  1400 A  #6
c  -------------
c  sub :GaAs        400 µm #7
```

```
c     ~~~~~~~~~~~~~~~~~

c     ************** ENTER DATA BELOW ****************
      n1      = 6

      N(1)    = 3.5
      N(2)    = 1.62
      N(3)    = 3.5
      N(4)    = 1.62
      N(5)    = 3.5
      N(6)    = 1.62
      N(7)    = 3.51

      d(1)    = 700
      d(2)    = 1400
      d(3)    = 700
      d(4)    = 1400
      d(5)    = 700
      d(6)    = 1400
c     d(7)    = substrate

      sb (1,1) = 1
      sb (2,1) = 3.51
c     ******************************************

      D0 60 la = 2000, 20000, 100
      DO 10 k = 1, nl
      eta (k)= N (k)
      delta (k) = 2*pi*N (k)*d (k)/la

      M(1,1,k)=        cos (delta (k))
      M(1,2,k)=        i*sin(delta (k))/eta (k)
      M(2,2,k)=        cos (delta (k))
      M(2,1,k)=        i*eta (k) * sin (delta (k))

10    CONTINUE
      PROD (1,1,1)    =       M (1,1,1)
      PROD (1,2,1)    =       M (1,2,1)
      PROD (2,1,1)    =       M (2,1,1)
      PROD (2,2,1)    =       M (2,2,1)

      DO 50 K3 = 2, nl
      DO 40 I1= 1,2
       DO 30 JI = 1,2
      SUM = 0
      DO 20 K1 = 1,2
      SUM = SUM + PROD (I1,K1,K3-1) *M(K1,J1,K3)
```

```
20  CONTINUE
    PROD (I1, J1, K3) = SUM
30  CONTINUE
40  CONTINUE
50  CONTINUE

    L(1,1)  = PROD (1,1,nl)
    L(1,2)  = PROD (1,2,nl)
    L(2,1)  = PROD (2,1,nl)
    L(2,2)  = PROD (2,2,nl)
    B = L (1,1) *sb(1,1) + L(1,2) +sb (2,1)
    C = L (2,1) *sb (1,1)+ L(2,2) +sb (2,1)
    Y = C/B
    Rho = (1 - Y)/(1 + Y)
    R = (cabs (rho) ) **2
    WRITE (1, *) la, R
60  CONTINUE
    END
```

Flow Chart for Simulation of Reflectance

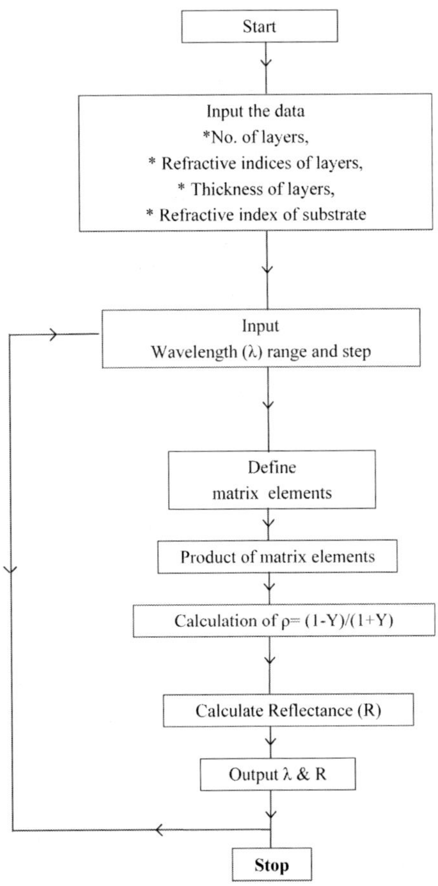

High Reflectance Coatings

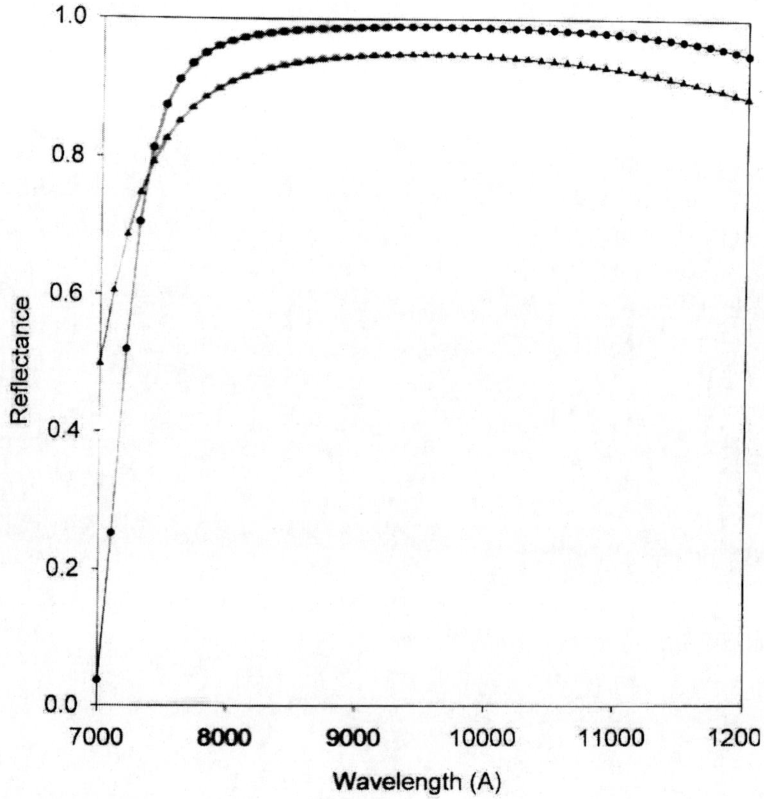

Simulated Antireflection

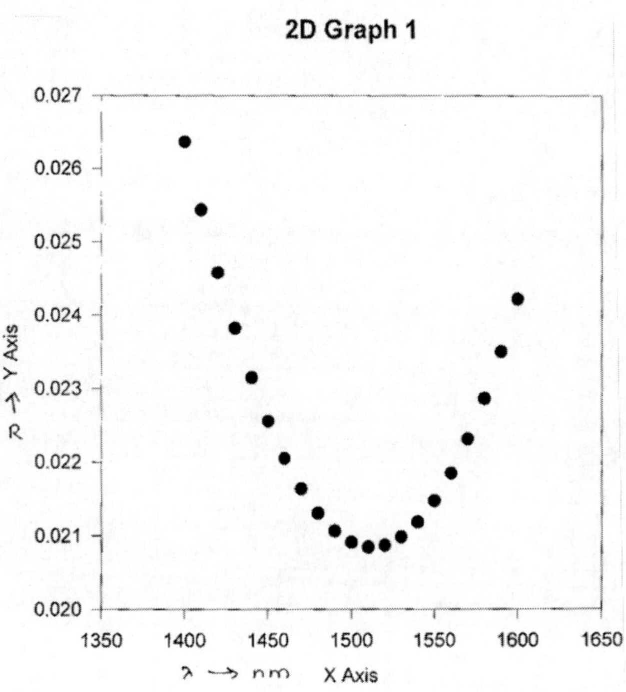

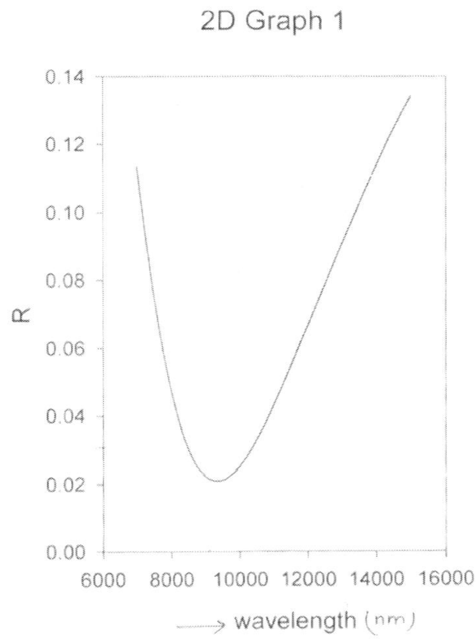

Box D
2D Graph
Device – I – V Laser Diode

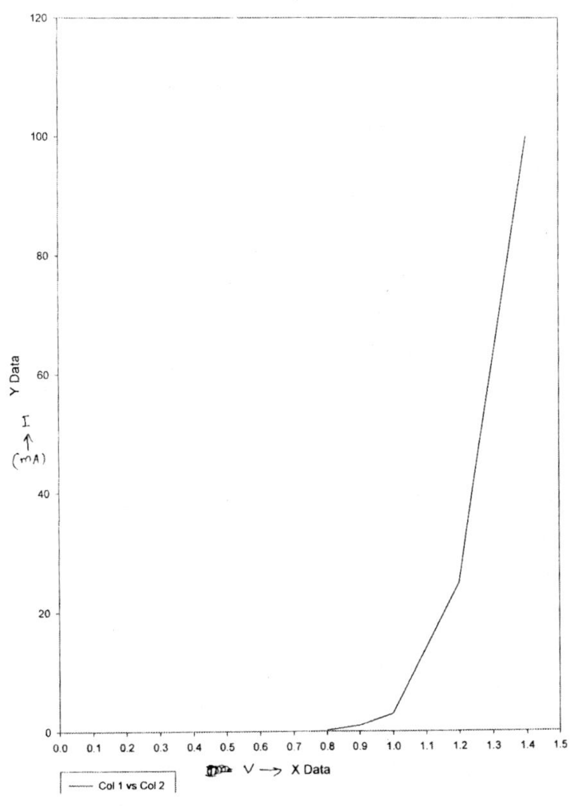

Data 1

mA	μw	μw	μw	μw	μw	μw	μw
0.000	0.000	0.000	0.000	0.000	0.000	0.000	0.000
100.000	1.000	2.000	1.000	2.000	2.000	2.000	2.000
200.000	1.000	4.000	1.000	3.000	3.000	3.000	2.000
300.000	2.000	4.000	3.000	4.000	3.000	4.000	4.000
400.000	2.000	5.000	5.000	5.000	5.000	4.000	5.000
500.000	3.000	9.000	6.000	6.000	5.000	6.000	6.000
600.000	4.000	24.000	10.000	6.000	10.000	16.000	8.000
700.000	5.000	94.000	30.000		25.000	112.000	33.000
725.000	8.000	180.000	41.000		49.000	216.000	76.000
750.000	16.000	250.000	55.000		75.000	250.000	85.000
775.000	25.000		75.000		106.000		165.000
800.000	35.000		104.000		140.000		192.000
825.000	61.000		135.000		198.000		
850.000	96.000		164.000		258.000		
875.000	138.000						
900.000	187.000						
	4th		8th		10th		11th

2D Graph
D 22 750 μm device (AR & HR)

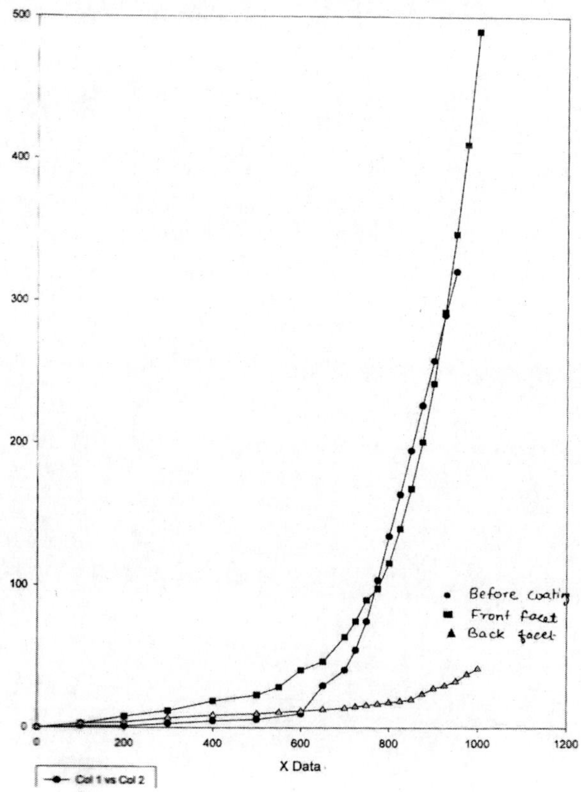

REFERENCES

[1] Introduction to Power diode lasers (Peter Unger, Dept. Of Optoelectronics, University of Optoelectronics, University of Ulm, 89069, Ulm, Germany).

[2] R. Diehl (Ed). High Power lasers Pages 1-58 (2000) Springer – Verlag Berlin Heidelerg, 2000.

[3] P. Bhattacharya: Semiconductor optoelectronics devices 2nd edition (Prentice hall, London, 1997).

[4] Ben Streetman – Solid State Electronics Devices (PHI 4th Edition).

[5] H.C. Casey, M.B. Panish: Heterostructures lasers-part A (fundamental Principles), part B (Materials and operating characteristics (Academic Press, Orlando, Florida 1978).

[6] H.A. Macleod: Thin Film Optical Filters, 2nd edition (Adam Hilger, Bristol, 1980).

[7] T. Ikegami: Reflectivity of mode at facet and oscillation mode in double heterostructures injection lasers, *IEEE J. Quantum Electronics,* 8, 470-476 (1972).

[8] Zahid Hasan Mahmood (Dept. of Electronics University of Dhaka. 1000, Bangladesh Study of Reflectivity of Mirror Facets of Oxide isolated lasers Fabricated on GaAs-AlxGa (1-X) As wafer, photonics.

[9] C.J. Panchanl, C.H. Jariwalaand B.M. Arora Optimization of facet reflectivities of semiconductor diode laser for high power high reliable Operation photonics, 2001.

[10] D.F. Welch A Brief Theory of High Power semiconductor lasers, Page II 1470-78, *IEEE journal of Quantum Electronics* – NOV/DEC 2000, Vol 6.

[11] A.P. Shah, M. R.,Gokhale, A. Bhattacharya, B.M. Arora CW Operation from MOVPE grown Al-free 0.98 μm emitting laser diodes fabricated by a shadow mask process. (photonics-2000 vol 1).

[12] MellesGriot Optical coating, pages 4.2 – 4.34 Optics Guide.

[13] http://www.thinfilmcenter.com/Emacleod.asp The essential Macleod Optical Thin film design and Analysis Program.

[14] http://www.cerac.com/pubs/proddata/a2013. htm Al_2O_3 for optical coating.

[15] http://www.hytekopto.com/optoelectronics. htm TEC controllers, Laser Diode Drivers, DC Heaters.

In: Comprehensive Guide for Nanocoatings Technology. Vol. 1 ISBN: 978-1-63482-447-7
Editor: Mahmood Aliofkhazraei

Chapter 10

ULTRATHIN FILMS

Raj Kumar Gupta** and V. Manjuladevi
Department of Physics,
Birla Institute of Technology and Science,
Pilani, Rajasthan, India

Abstract

From the past few decades, the field of ultrathin films has grown remarkably. A material in the form of ultrathin film gains enormous surface-to-volume ratio and thereby enhances the properties like catalysis, reactivity and adhesion remarkably. The efficiency and sensitivity of the devices employing ultrathin films increases manifold. In this chapter, we will discuss three different types of ultrathin films viz. Langmuir monolayer, Langmuir-Blodgett films and self-assembled monolayer. Since the thickness of these films are mono-molecular, it can be considered as two-dimensional system. Here we will discuss the fabrication of the ultrathin films, its properties, effect of external parameters on its properties, its application, and challenges and opportunities in the field.

PACS: 73.50.Td, 68.18.-g, 68.47.Pe, 68.35.Rh, 64.75.Yz, 82.70.Uv

Keywords: Ultrathin films, Langmuir monolayer, Langmuir Blodgett films, Self assembled monolayers, air-water interface, air-solid interface, surface phases, phase transitions

Introduction

Thin films science and technology is one of the important fields of research and development. The reduction in the dimension of bulk material and approaching to a limit of two-dimensional (2D) system by fabricating ultrathin films of the material provides a remarkable increase in surface-to-volume (SV) ratio as compared to the bulk material. Such huge increase in SV ratio increases the activities of the material enormously and thereby the material properties like catalysis, reactivity, adhesion, etc enhances remarkably. Such changes in the material properties in the state of ultrathin films promises a wide range

*E-mail address: raj@pilani.bits-pilani.ac.in

of applications starting from the basic science research to various device fabrication. It has been realized that the field of ultrathin films has a close association with nanoscience and nanotechnology, and therefore due to such multidimensional application, the field is strengthening everyday. The ultrathin films can be fabricated either top-down or bottom-up mechanism. In this chapter, we will discuss bottom-up mechanism for the fabrication of the ultrathin films. In the bottom-up mechanism, the films are fabricated by arranging the molecules in the defined molecular states in a single layer onto some substrates. Here, we will discuss the ultrathin films fabricated through Langmuir monolayer, Langmuir-Blodgett and self assembly techniques. We find that the molecular states in the thin films are essentially governed by the intermolecular and molecule-substrate interactions. We will discuss the role of external parameters like temperature and ion-contents on the properties of ultrathin films.

The work on ultrathin films has started long back. In 1774 Franklin has observed the phenomenon of calming of ocean wave by spreading oil on water surface. This was due to reduction in surface tension of water by spreading oil on its surface [1]. Pockel was successful in forming monolayer at the air-water interface [2, 3]. Thereafter, Rayleigh, Hardy and Devaux have contributed to the progress in the field of monolayer at the air-water interface. Irving Langmuir has developed the field by making a systematic measurement and recording the isotherms of amphiphilic molecules at the air-water interface [4, 5]. He has been awarded with Nobel Prize in 1932 in chemistry for his contribution in the field of monolayers at the air-water interface. The stable mono-molecular ultrathin film at the air-water interface is named after him, and it is called as Langmuir monolayer. Langmuir's work has initiated a new era of surface science which is vital not only for basic science but also its potential application. In 1943 Katherine Blodgett has developed Langmuir-Blodgett (LB) technique where a Langmuir monolayer from the air-water interface was transfered layer by layer onto solid support by a vertical dipping mechanism [6, 7]. In 1946, Zisman [8] has formed a self assembled monolayer (SAM) on metal surface by the adsorption of surfactant molecules from a solution. In 1980, Sagiv [9] formed the SAM by the chemical adsorption of alkylsilanes on SiO_2 substrate. Allara and Nuzzo have studied systematically the SAM of organothiols and proposed its numerous industrial application. Thereafter, the field is growing remarkably. Numerous industrial applications are being proposed based on the research and development in the field of ultrathin films.

1. Langmuir Monolayer

There are numerous organic molecules available in nature viz., fatty acids, cholesterol, lipids and their derivatives which possess both polar hydrophilic headgroup and non-polar hydrophobic tailgroup. Such molecules are known as amphiphilic molecules (Figure 1). The polar headgroup can form hydrogen bonds easily with water molecules whereas the water-insoluble hydrophobic part can interact with each other through Van der Waals attraction. Hence, at the air-water interface, the hydrophilic headgroup adsorbed onto the water whereas the hyrdophobic part stays away from the water surface. The amphiphilic molecules with a proper balance between the hydrophilicity and hydrophobicity, when dispersed onto water surface form a single layer of stable mono-molecular thin film at the air-water interface. Table 2 shows a list of hydrophilic headgroups of the molecules with its

Table 1. Historical Development: Ultrathin films

Milestones	References
(a) Calming of ocean waves by spreading of oil on water surface	[1]
(b) Formation of monolayer at the air-water interface	[2,3]
(c) Measurement and recording of isotherms of amphiphilic molecules at air-water interface	[4,5]
(d) Transfer of Langmuir monolayer on solid substrates	[6,7]
(e) Self-assembled monolayer on solid substrates	[8,9]

corresponding dipole moment. Amphiphilic molecules consisting of such hydrophilic parts may form a stable monolayer at the air-water interface. The stability of the film at the inter-

Table 2. List of hydrophilic headgroups with its corresponding dipole moments. [10]

Hydrophilic headgroup	Dipole moment (Debye)
(a) -COOH	1.75
(b) -OH	1.68
(c) $-NO_2$	3.66
(d) $-NH_2$	1.17
(e) $-CN_2$	4.02

face is essentially achieved due to reduction in surface free energy at the air-water interface by adsorption of hydrophilic headgroup onto the water surface and repulsion of tailgroup away from the water surface. The hydrophilic headgroup can form hydrogen bonding easily with the water molecules and therefore gets adsorbed to the water surface without the expense of energy. The hydrophobic tailgroup can interact with each other via attractive Van der Waals force. The presence of hydrophobic tailgroup at the interface, perturbs the hydrogen bonding network of the water molecule. Such perturbation sometime increases overall entropy leading to a stabilization of the monolayer. The stable mono-molecular thick layer at the air-water interface is known as Langmuir monolayer [10]. A schematic of the Langmuir monolayer formed by amphiphilic molecules at the air-water interface is shown in Figure 2.

A Langmuir monolayer can be considered as a two-dimensional (2D) system where the amphiphilic molecules are constrained to move on the smooth water surface. The complete thermodynamics of such 2D system can be studied by measuring the surface pressure (π) - area per molecule (A_m) isotherms at different temperatures. The Langmuir monolayer exhibits a variety of surface phases which are dependent on surface density, temperature, ion-content of the subphase etc. These phases are differentiated with respect to their ordering and tilt of the hydrophobic tail on the water surface. The surfaces phases can be tweaked

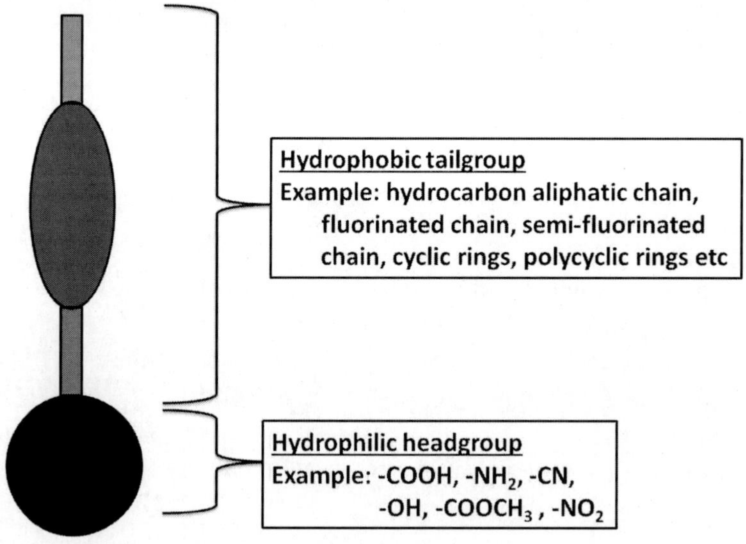

Figure 1. Schematic of an amphiphilic molecules showing the possible hydrophilic headgroup and hydrophobic tailgroup.

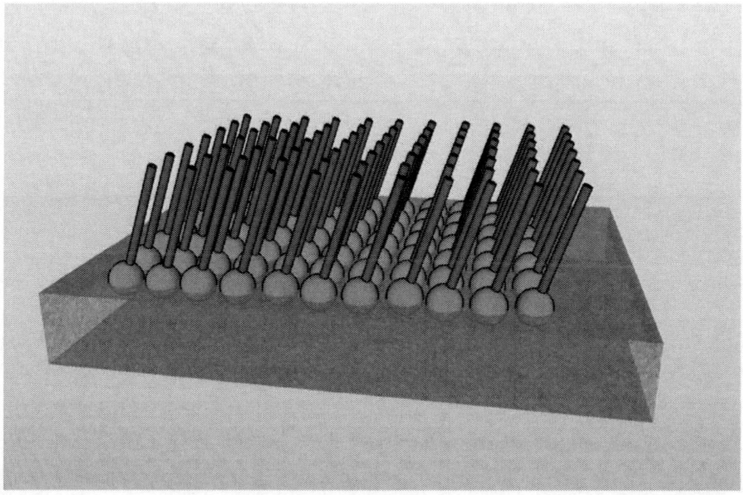

Figure 2. Schematic of Langmuir monolayer at air-water interface formed by amphiphilic molecules.

by altering the intermolecular interactions. Therefore, the Langmuir monolayer provides a system for studying molecular interaction at the air-water interface. The intermolecular interactions can be altered by changing the nature of amphiphilic molecules, altering the pH and ion-contents in the subphase.

Conventionally, when spread at the air-water interface, the hydrophilic part of the amphiphilic molecules gets anchored to the water surface whereas the hydrophobic part stay away from the water surface. In contrary, the report on the Langmuir monolayer of disubstitued urea lipid molecule showed non-traditional alignment of the molecules at the air-water interface. The hydrophilic part of the molecule is found to be suspended in air whereas the hydrophobic chain part is in contact with water [11]. Such non-traditional Langmuir monolayer stabilizes due to strong hydrogen bonding network between the urea moiety suspended in air.

The surface tension of water reduces due to incorporation of additives. The amount of reduction in surface tension of water due to such addition is known as surface pressure (π). As the surface density of the amphiphilic molecules at the air-water interface changes, π also changes. The variation of π as a function of area per molecule (A_m) (inverse of surface density) at a constant temperature is known as $\pi - A_m$ isotherm. Figure 3 shows a schematic of the isotherm and the possible molecular conformation in the different phases of the Langmuir monolayer. The isotherm indicates the rise in π with the decrease in A_m.

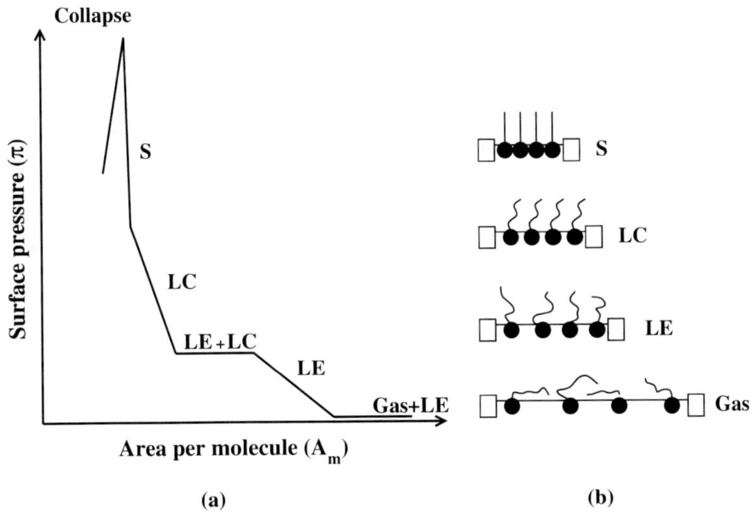

Figure 3. Schematic of (a) isotherm of Langmuir monolayer, (b) possible arrangements of molecules at the interface in different surface phases. Reprinted from [12].

The kinks in the isotherm indicate phase transitions. The plateau in the isotherm represents a co-existence of two phases. The sudden drop in π is due to collapse of the monolayer. When the surface density is very low, the molecules do no interact with each other. This results in zero surface pressure which is analogous to a 2D gas phase. On increasing the surface density, the molecules interact with each other and condense to form a 2D liquid-like phase. In the liquid-like phase, there are no positional and orientational order among the molecules. Such phase is known as liquid-expanded (LE) phase. On decrease in A_m, the molecules may exhibit short range orientational or positional order. Such phase is known as liquid-condensed (LC) phase. On further decrease in A_m, the molecules show long range position and orientation orders. This is 2D solid (S) phase. On further decrease in A_m, the monolayer becomes unstable and it collapses. This is indicated by the sudden drop in π in the isotherm.

The average area occupied by the molecules in a phase is determined by extrapolating the corresponding region of the isotherm to the zero surface pressure on the A_m axis. The extrapolation of the steep region (*e.g.* S phase in Figure 3) of the isotherm to zero surface pressure is called limiting area per molecule (A_o). This is the minimum area to which the molecules can be compressed on the water surface without collapsing the monolayer. The orientational state (tilt or untilt) of the molecules in a phase can be estimated qualitatively by comparing the extrapolated area per molecule with that of molecular cross-sectional area in the bulk single crystal. Table 3 shows the list of commonly studied amphiphilic molecules at the air-water interface.

Table 3. List of some materials which can form stable Langmuir monolayer at the air-water interface

Materials	References
(a) Fatty acids	[13–17]
(b) Cholesterol and derivatives	[18–21]
(c) Phopholipids	[22–25]
(d) Proteins	[26–29]
(e) Polymers	[30–32]
(f) Liquid Crystals	[33–37]
(g) Nanomaterials	[38–42]

1.1. Fabrication of Langmuir Monolayer

The schematic of the experimental setup for the formation of Langmuir monolayer and recording of $\pi - A_m$ isotherms is shown in Figure 4. The trough is made up of teflon material (1) which is both oleophobic and hydrophobic. The trough in general is equipped with two barriers (2) whose lateral motion changes the area available for the molecules and thereby changes the surface density. The water used in the formation of Langmuir monolayer is ultrapure ion-free having a resistivity $\geq$ 18 MΩ-cm. The surface pressure is being measured using a Wilhelmy plate (filter paper) and a pressure sensor (5). The pressure sensor and the barriers are synchronized, and the data like π and A_m is recorded directly into the computers. The temperature of the aqueous subphase is maintained at a constant value by circulating heated (or cooled) water through the internal chamber of the LB trough. The temperature of the subphase is monitored using a resistance temperature device (RTD) and a digital multimeter. A solution of known concentration of the amphiphilic molecules is made in some organic solvent. The choice of the appropriate solvent for dispersing the molecules is such that it should be (i) volatile and evaporate faster from the surface of water, (ii) disperses the amphiphilic molecules uniformly and completely (iii) immiscible with water. The commonly used solvents for dispersing the molecules are chloroform, benzene, toluene, hexane+ethanol, toluene+methanol, etc [10]. A list of solvents with its respective boiling point is shown in Table 4. The solution of the amphiphilic molecules is spread at the air-water interface using a calibrated microsyringe. The solvent evaporates from the surface of water leaving behind the dispersed molecules. The A_m can be varied either by varying the area available for the molecules in the trough or by varying the number of molecules within a defined area. Mostly, A_m is varied by varying the area in the trough by the lateral motion of barrier(s).

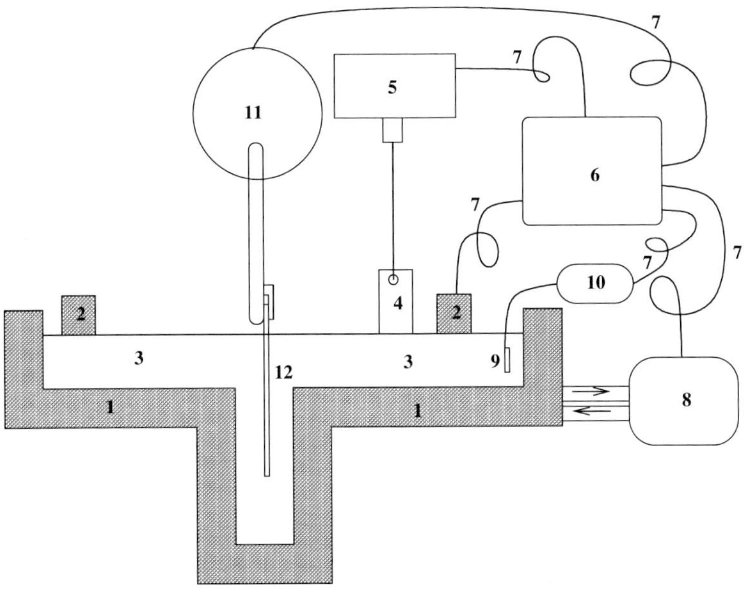

Figure 4. Schematic of the Langmuir-Blodgett trough for the measurement of $\pi - A_m$ isotherm and deposition of LB films. The different parts are: (1)Teflon trough, (2) barriers, (3) aqueous subphase, (4) Wilhelmy plate, (5) surface pressure sensor, (6) computer, (7) connecting wires, (8) water circulator for controlling temperature of water subphase, (9) resistance temperature device, (10) digital multimeter, (11) dipper, (12) substrate. Reprinted from [12].

Langmuir monolayer provides an ideal system for mimicking the biological membrane [25]. The molecular interaction at the fundamental level can be studied. Langmuir monolayer provides a mechanism to control the surface phases. Such an engineering at the nanometer length scale is vital for the creation of materials with unique properties.

The Langmuir monolayer exhibits numerous surface phases. The phases are dependent on various factors including the temperature and ion-contents of the subphase. The Langmuir monolayer of fatty acid exhibits 17 different phases [43]. Such phases are demarcated with respect to the tilt, azimuth variation of the molecules on the water surface. The external perturbation can change the nature of intermolecular interaction at the interface. The intermolecular and molecule-water interaction are the dominant factor for governing the surface phases. Such interactions can be altered by altering the nature of headgroup and the tailgroup of the amphiphilic molecules. We will discuss the effect of headgroups and tailgroups on the Langmuir monolayer.

1.2. Effect of Headgroups

Figure 5 shows the isotherms of octadecanoic acid ($C_{17}H_{35} - COOH$), octadecanol ($C_{18}H_{37} - OH$), octadecanethiol ($C_{18}H_{37} - SH$) and octadecylamine ($C_{18}H_{37} - NH_2$). These molecules are structurally similar but with different headgroups. The polarity of the headgroups of these molecules can vary in decreasing order as: $-OH \rightarrow -COOH \rightarrow -NH_2 \rightarrow -SH$. There is a significant effect of the polarity of the headgroup on the

Table 4. List of solvents with its boiling point for preparing the solution and dispersing the amphiphilic molecules at the air-water interface

Solvents	Boiling point (°C)
(a) Ethyl ether	35
(b) Acetone	56
(c) Chloroform	61
(d) n-hexane	69
(e) Alcohol	78
(f) Benzene	80
(g) Toluene	110
(h) Dimethyl Formamide	153

isotherms. It can be observed from the isotherms that monolayer with the stronger polar headgroups are more compressible yielding a higher value of collapse surface pressure. This is due to the fact that higher polar group may possesses higher binding energy with the water molecules as compared to less polar molecules. In an interesting work, the role of headgroup in the formation and properties of Langmuir monolayer of cholesterol (Ch) and thiocholesterol (TCh) is systematically studied [44]. The molecules are structurally similar except the headgroup of Ch and TCh are -OH and -SH, respectively. The -SH group is weakly acidic, however the TCh can be considered as predominantly a hydrophobic molecule. As expected, since the -OH is highly polar compared to -SH, the monolayer of Ch is very stable. However, the TCh do not spread onto the water surface to yield a stable Langmuir monolayer. The monolayer of TCh was stabilized by incorporating them in the matrix of Ch monolayer. It was found the mixed monolayer of TCh-Ch system was stable till 0.75 mole fraction of TCh in the Ch monolayer. Such study made it viable for the formation of ultrathin films of Ch and TCh on variety of substrates using the LB technique.

1.3. Effect of Tailgroups

The hydrophobic tail group can interact with each other through Van der Waals interaction. Such attractive interaction sometime dominates and plays pivotal role in stabilizing a Langmuir monolayer. There are numerous report where pure hydrophobic materials form a stable Langmuir monolayer [45–48]. Such stabilization is achieved by the attractive Van der Waals interaction between the tailgroups and the overall gain in entropy of the system due to hydrophobic effect. In 2002 Tabe $et.al.$ have demonstrated hydrophobic molecules to form a stable Langmuir monolayer where the stability was attained due to the gain in entropy rather than energetics [45]. The Van der Waals interaction between the hydrophobic moiety can be altered. There are reports on the study on the Langmuir monolayer of fluorinated (F-chain) and semi-fluorinated chains [49]. The hydrogenated chain (H-chain) are found to be less hydrophobic as compared to F-chain. Therefore, F-chain can perturb the conformation of the water molecules at the interface more significantly than H-chain leading to a gain in overall entropy. Such enhanced perturbation can increase the possibility to attain a stable

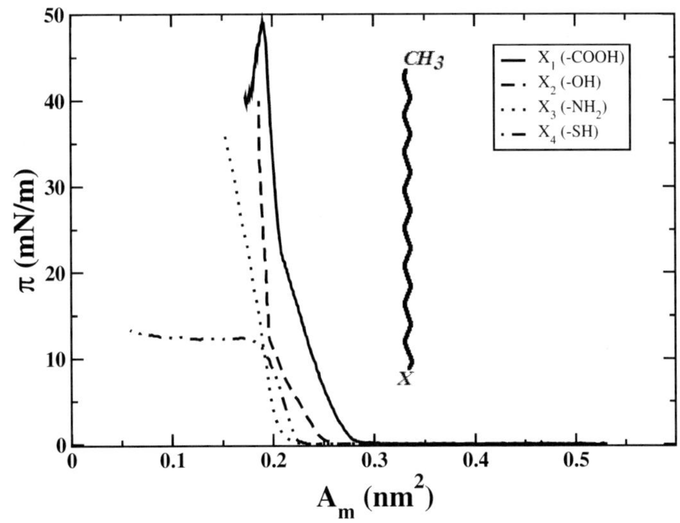

Figure 5. Surface pressure - area per molecule isotherm of octadecanoic acid ($C_{17}H_{35} - COOH$), octadecanol ($C_{18}H_{37} - OH$), octadecanethiol ($C_{18}H_{37} - SH$) and octadecylamine ($C_{18}H_{37} - NH_2$). Reprinted from [12].

Langmuir monolayer of F-chain organic molecule [50]. The non amphiphilic fluoroalkane ($C_{20}F_{42}$) forms a stable Langmuir monolayer at the air-water interface whereas the stable Langmuir monolayer of the H-alkane obtained only for a length one and half times greater than $C_{20}F_{42}$ [48, 51]. The F-chains are more bulkier and stiffer than that of H-chain. The low polarizability of fluorine results in very weak Van der Waals interaction. However, the F-chains are more hydrophobic as compared to H-chains [52].

The F-chain amphiphiles are interesting to investigate as it possess relatively smaller headgroup. The stability of Langmuir monolayer of such molecules can be addressed by considering chain-chain interactions. The Langmuir monolayer of F-chain acids ($C_8F_{17}COOH$, $C_{10}F_{21}COOH$) have been investigated [53, 54]. This is in contrast with that of H-chain fatty acids where it is essential to have a stable Langmuir monolayer only if CH_2 unit is greater than 13 [10]. This indicates that CF_2 unit is more hydrophobic as compared to CH_2. The stiffness of the F-chain has greater impact on the phases of their Langmuir monolayer. The monolayer of F-chain acids or alcohol undergo a direct phase transition from gas to untilted condensed (L_2) phase. The liquid expanded (LE) phase does not appear in the phase diagram. Non appearance of any tilted phase in the Langmuir monolayer of F-chain acids and alcohol is due to stiffness of F-chain [53, 55, 56]. In order to reduce the stiffness in the chain, amphiphiles with semifluorinated chain are also studied [55]. With the introduction of CH_2 spacer in the F-chain acids, the LE phase reappears.

1.4. Effect of Ion-Contents in the Subphase

The presence of ions in the subphase has the significant effect on the Langmuir monolayer. Ions are incorporated into the subphase in order to increase the stability of the Langmuir monolayer. It also enhances the interaction between the substrate and the molecules and thereby supports multilayer formation during LB film deposition. The fatty acid at the air-water interface tends to dissociate into COO^- and H^+ in the basic medium [10,57]. The effect of presence of mono-, di- and tri-valent cations in the subphase on the Langmuir monolayer of fatty acid has been studied [10]. These cations form complexes with the fatty acid only at the air-water interface, and therefore increases the stability of the Langmuir monolayer. It is found that due to incorporation of cations, the monolayer of fatty acid condenses, and the collapse surface pressure increases significantly.

Octadecanthiol (ODT) is the most extensively employed molecule for the formation of self-assembled monolayer on the surface of noble metals. Although, the -SH group of ODT is slightly acidic, the hydrophobic and hydrophilic balance is sufficient enough to form a stable monolayer at the air-water interface [58]. The monolayer exhibited gas and L_2 phases. The monolayer of ODT is found to be extremely sensitive to the presence of ions in the subphase.

Figures 6 shows the $\pi - A_m$ isotherms of the Langmuir monolayer of the ODT for the different concentration of NaOH in the subphase. The presence of even very small quantity of NaOH has a marked effect on the monolayer of the ODT molecules. The isotherms shift systematically towards higher A_m with increase in concentration ($\leq 10^{-3}$ M) of NaOH in the subphase (Figure 6(a)). Above the concentration of 10^{-3} M, the ODT monolayer gets destabilized completely and the isotherms show a continuous and a gradual rise in the surface pressure (Figure 6(b)). The acidic head group (-SH) of the ODT molecule may tend to dissociate on the alkaline subphase. The shift in isotherms towards higher A_m is may be due to the electrostatic charge repulsion between such dissociated components [10,59].

The effect of incorporation of divalent Cd^{2+} ions in the subphase on the monolayer of ODT has been studied [58]. The isotherms of the ODT molecules on the aqueous subphases having different molar concentrations of $CdCl_2$ are shown in Figure 7. We find a drastic change in the π-A_m isotherm of the ODT due to the presence of small amount of $CdCl_2$ in the aqueous subphase.

It is known that a Cd^{2+} ion makes complex with two ODT molecules in order to achieve electroneutrality. In the complex unit, it has been proposed that the orientation of the aliphatic chain of the two ODT molecules are in opposite direction with the Cd^{2+} ion at the center [60]. Therefore, the reduction in A_o due to incorporation of Cd^{2+} ion is due to the formation of complex at the air-water interface which may reduce the area per molecule.

For a very low concentration (8.7×10^{-8} M) of $CdCl_2$ in the aqueous subphase, the trend of the isotherm was similar to the one obtained for the ion-free water. However, there is a reduction in the value of A_o (=16.5 Å^2). The isotherm for the concentration of 10^{-6} M shows similar trend but the surface pressure continues to increase after an initial collapse at around 13 mN/m. The value of A_o is found to be 10.5 Å^2. The isotherms for higher concentrations show a different behavior. They show a gradual rise in the surface pressure to very large values ($\geq$ 60 mN/m). The trend of the isotherms does not change on further increase in the concentration of $CdCl_2$. Here, the values of A_o lie in the range of 10.5 to 11.5 Å^2. These values are about half of those corresponding to the normally oriented molecules in the condensed phase of ODT at the air-water interface. This indicates the formation of bilayer of the complex of ODT and $CdCl_2$ [60]. The A_o value (= 16.5 Å^2) for the lower concentration (8.7×10^{-8} M) is more than for a bilayer. This can be due to the formation of the complex of the ODT molecules in a fraction of the monolayer with $CdCl_2$.

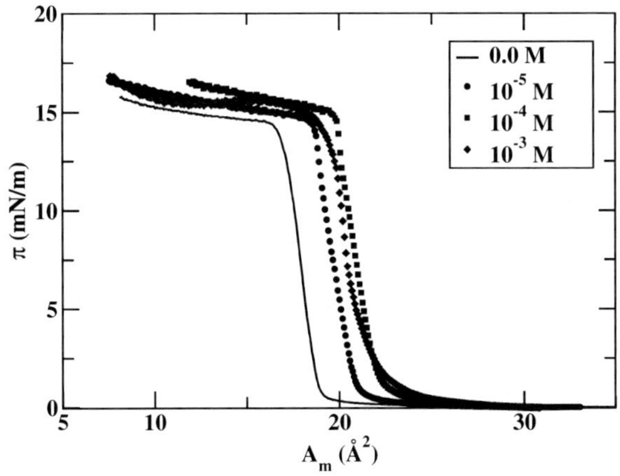

(a)

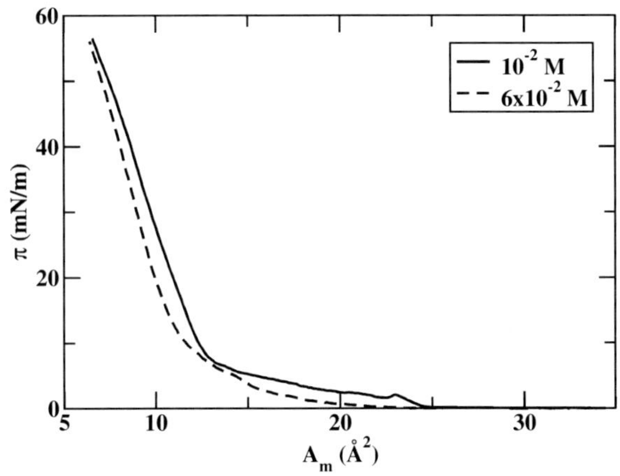

(b)

Figure 6. Surface pressure (π) - area per molecule (A_m) isotherms of the Langmuir monolayer of ODT molecules with different molar concentrations of NaOH in the subphase. (a) shows the isotherms for lower NaOH concentrations ($\leq 10^{-3}$ M) in the subphase. (b) shows the isotherms for higher NaOH concentrations ($>10^{-3}$ M) in the subphase. Reprinted with permission [58].

1.5. Characterization Techniques of Langmuir Monolayer

The Langmuir monolayer at the air-water interface can be characterized using numerous advanced techniques. Some of the important techniques are listed in Table 5. The detail discussion of such techniques are discussed in the cited references.

1.6. Challenges and Opportunities

The Langmuir monolayer of numerous molecules at the air-water interface have been extensively investigated. The different possible phases at the air-water interface have been identified and observed experimentally and theoretically [43]. It will be interesting to study the change in phase behavior by changing thermodynamical conditions for a Langmuir mono-

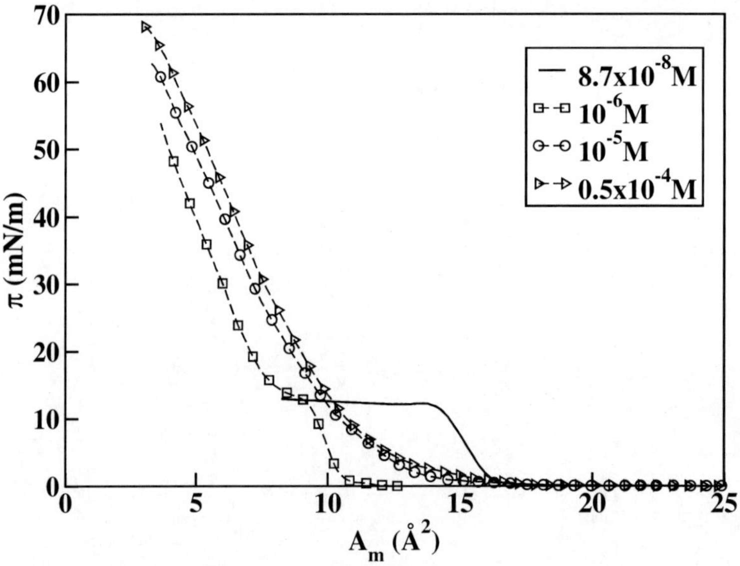

Figure 7. The surface pressure (π) - area per molecule (A_m) isotherms of the ODT monolayer for different molar concentrations of $CdCl_2$ in the aqueous subphase. Reprinted with permission [58].

layer. There are few attempts where the thermodynamical conditions have been changed by changing the nature of the interface. The Langmuir monolayers at the oil-water interface have been formed and studied [98–101]. Kraack *et.al.* have studied the Langmuir monolayer of the stearic acid at air-mercury interface. They found that the molecules undergo a transition from surface planar to surface normal on increasing the surface density. They identified 2D gas phase, and condensed single- and double-layered phases of flat lying molecular dimers [102, 103]. The studies on Langmuir monolayer enable us to understand the molecular interactions at the fundamental level. Numerous reports indicate that the stability of a Langmuir monolayer can be achieved by the interplay of molecular interactions and the overall entropy gain at the given interface. Therefore, in order to understand more about the system, the field of Langmuir monolayer needs a systematic study on the role of different interface on the phase behavior and its stability.

2. Langmuir-Blodgett Film

The material properties are dependent on the states of the constituent molecules and its dimensionality. The Langmuir monolayer exhibits numerous surface phases. Therefore, it highlights the possibility of material engineering if the precise number of layers of such phases are transfered onto the relevant solid substrates. In 1934 Katherine Blodgett have demonstrated that the Langmuir monolayer at an interface can be transferred layer-by-layer onto solid substrates by a vertical deposition mechanism known as Langmuir-Blodgett (LB) technique [104, 105]. The schematic of the experimental setup is shown in Figure 4. The LB films can be transfered by attaching a solid substrate to the dipper which can be made to dip in (downstroke) and out (upstroke) of the surface of water using a computer controlled

Table 5. Experimental techniques for characterizing ultrathin Langmuir films

Technique	Information obtained	References
(a) Surface pressure-area isotherm	Information on monolayer formation, molecular area, monolayer phases, compressibility, monolayer stability, interaction of dopant in subphase with monolayer, level of mixing in mixed monolayers.	[10,61–63]
(b) Surface potential-area isotherm	Orientation of molecular dipoles, molecular aggregation, course of chemical/photochemical reaction, interaction of species in the subphase with the monolayer.	[64–67]
(c) Second harmonic generation	Orientation of molecules within the monolayer, surface density of the monolayer.	[68–70]
(d) Sum-frequency generation	Orientational and conformational changes, surface density of the monolayer, structure of the interface, phase transitions.	[71–73]
(e) Conductance measurement	H-bonding, proton hopping, interaction between head groups.	[74,75]
(f) Brewster angle microscopy	Monolayer phases, phase co-existence, domain formation, aggregation, morphological features, tilt-azimuthal variation	[76–80]
(g) Fluorescence microscopy	Monolayer phases, phase co-existence, domain formation.	[81–84]
(h) UV-visible spectroscopy	Intermolecular interaction, aggregation, presence of chromophores.	[85–87]
(i) Infrared spectroscopy	H-bonding, orientation of head/tail groups, presence of functional groups.	[88–91]
(j) X-ray diffraction	Monolayer thickness, nanostructure of the surface and interface, monolayer and counterion interactions, positional order, in-plane lattice structure, monolayer phases.	[92–96]
(k) Rheological measurement	Viscoelastic behaviour of the monolayer, shear modulus, microstructural phases.	[97]

programme. Every stroke of the dipper may deposit a layer of film on the substrate (Figure 8(a)). In order to deposit the LB film in a given phase of Langmuir monolayer, the corresponding surface pressure is controlled using a feedback mechanism during the transfer of the monolayer on the solid substrate. As the monolayer transfered to the solid substrate during LB deposition, the surface pressure changes. Here the feedback mechanism adjust the barriers to regain the defined surface pressure (also known as target surface pressure). The number of layers supported onto the solid substrate depends on the molecule-substrate interaction. Depending on the arrangement of hydrophilic and hydrophobic groups, there are three different types of LB depositions viz. X-,Y- and Z-type. They are shown in Figure 8(b).

The transfer efficiency of the LB film can be estimated by calculating transfer ratio (τ).

$$\tau = \frac{\text{area of monolayer transfered from the water surface}}{\text{area of the substrate to be deposited}} \tag{1}$$

The transfer ratio equal to one indicates 100% efficient transfer of the monolayer. The

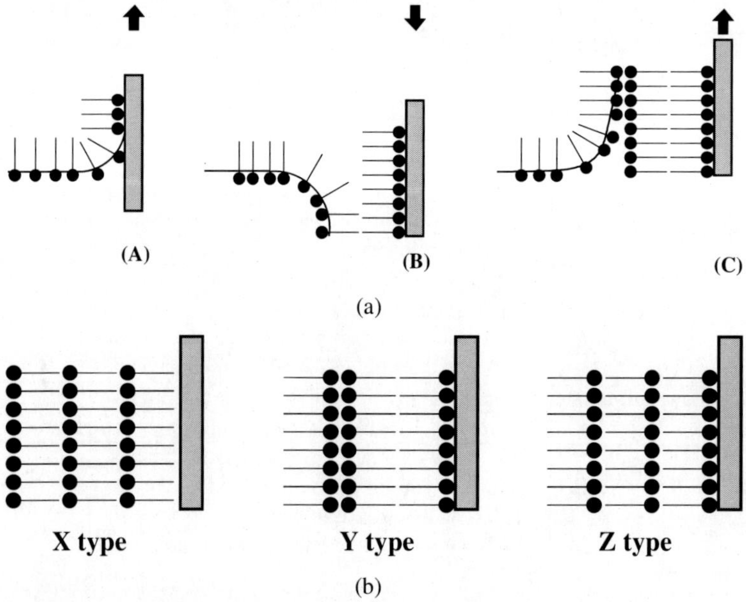

Figure 8. (a) Mechanism of formation of multilayer during Langmuir-Blodgett film deposition: (A)First upstroke, (B) first downstroke and (C) second upstroke (b) Types of LB film. Reprinted from [12].

positive transfer ratio indicate adsorption whereas the negative value of τ indicates desorption. The parameters affecting the nature of aggregation of molecules in the LB films are target surface pressure, rate of deposition, temperature, humidity and ion-content in the subphase [106]. The stability of the Langmuir monolayer at the air-water interface is vital for the efficient and defectless deposition of LB film [44].

The monolayer from the air-water interface can be transferred onto substrates by two other different techniques viz., horizontal transfer and Schaefer's methods. In the horizontal transfer method, the substrate is kept immersed horizontally in the subphase and then the monolayer is formed at the interface. Once the target surface pressure is attained, the aqueous subphase is siphoned out very slowly from the other side of the barriers. As the water layer recedes, the monolayer gets adsorbed onto the substrate. In the Schaefer's method, a horizontally aligned hydrophobic substrate is allowed to touch the monolayer on the water surface. The hydrophobic part of the molecules get adsorbed to the substrate. To facilitate the drainage of water, the substrates in either case can be tilted by a small angle prior to the adsorption.

Any solid non-hygroscopic substrate can be employed for the deposition of LB film. The substrates most commonly used for the LB depositions are silicon, glass, mica, and highly oriented pyrolytic graphite. The wettability of the substrates ($e.g.$ silicon or glass) can be changed by chemically treating the substrates. A hydrophilic silicon or glass substrate can be converted to hydrophobic by silanization [20]. The adhesion and hence the aggregation of molecules on such substrates are different as compared to its hydrophilic counterpart. In 2004, it has been demonstrated that the adhesion and aggregation of cholesterol molecules in the LB films are remarkably different on the hydrophobic substrate than compared to hydrophilic surface. The LB film of cholesterol deposited in condensed phase

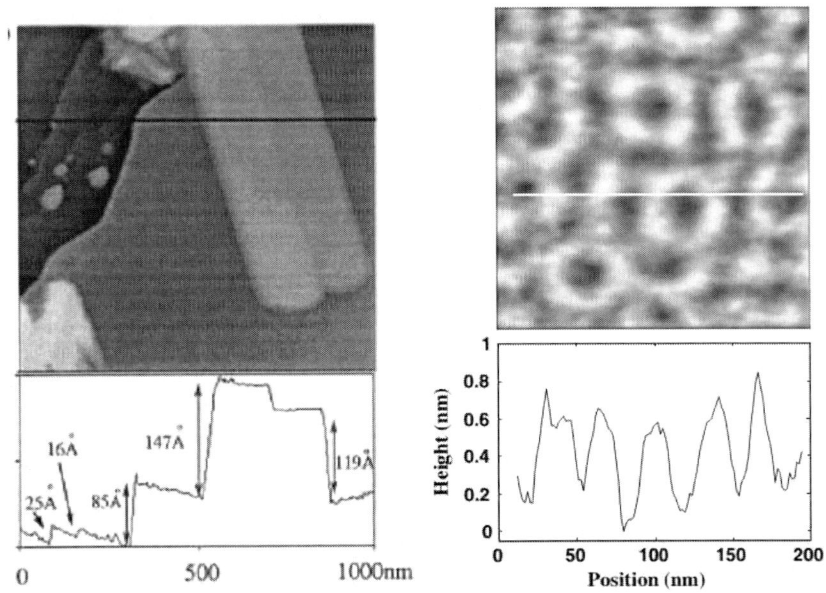

(a) (b)

Figure 9. Atomic force microscope images of LB film of cholesterol deposited at (a) 20 mN/m on hydrophilic mica substrate. Reprinted with permission [21] (b) 30 mN/m on hydrophobic glass substrate. Reprinted with kind permission of The European Physical Journal [20].

on hydrophilic mica reveal flat and more than one layer thick domains (Figure 9(a)). The attempt to form multilayer of cholesterol on hydrophobic substrate by LB technique reveals both adsorption and desorption. The cholesterol molecules aggregate on the hydrophobic surface yielding uniformly distributed donut shapes (Figure 9(b)) [20]. Table 6 shows a list of substrates with pre-deposition treatment which can be used for LB film deposition. Such treatment can be given to the substrates for improving the adhesion and thereby reducing the defects in the resulting LB films.

The structure of the LB films are dependent on the fabrication temperature. Watanabe *et.al.* [107] have demonstrated a control over the shape and size of the microdomains formed in the LB film of mixed monolayer system of hydrogenated and semi-fluorinated fatty acids. The shape and sizes of the domains are found to be dependent on the subphase temperature during the deposition process. Figure 10 shows the AFM images of the LB films of the mixed monolayer of the fatty acids deposited at 10, 20 and 30 $^{\circ}C$. Such control over the structure is important for the creation of template for a large scale patterning different materials. The methodology in their study will be important in the fabrication of devices such as high density memories, sensors and switches. There are numerous study on the effect of temperature on the LB films [108–111].

The presence of ions in the subphase during the deposition affect the LB films significantly. The effect on the surface morphology and structure of the LB films of ODT due to the presence of divalent cations in the subphase during the LB deposition is shown in Fig-

Table 6. List of substrates with pre-deposition treatment in order to obtain the desired wettability on the surface. *Piranha solution is a mixture of 3:1 concentrated sulphuric acid and 30% hydrogen peroxide. Warning: Piranha is very dangerous and it has to be handled carefully

Substrates	Pre-deposition treatment	Nature of the surface
(a) Glass slide	Boil in piranha solution* for 30 mins followed by thorough rinsing successively with de-ionized water, absolute alcohol and acetone. Dry by blowing hot air.	Hydrophilic
(b) Gold film on glass or Si	Treat with cold piranha solution for 5 mins followed by thorough rinsing successively with de-ionized water, absolute alcohol and acetone. Dry by blowing hot air.	Hydrophilic
(c) Mica	Remove few top layers by employing a very sharp knife	Hydrophilic
(d) Highly oriented pyrolytic graphite	Remove few top layers of graphene using adhesive tape	Hydrophobic
(e) Si wafer	same as in (a)	Hydrophilic
(f) Hydrophilic glass slide or Si wafer	Immersed in 10 % v/v chloroform solution of hexamethyldisilazane for 12 hours and then rinse thoroughly with chloroform	Hydrophobic
(g) GaAs epitaxial layer	1 min in Br_2/methanol (1:2000 by volume), 5 mins in concentrated HCl followed by methanol rinse	Hydrophilic
(h) GaP single crystal	3 mins in cold piranha followed by 1 min in alkaline solution	Hydrophilic
(i) InSb, CdTe, CdS, ZnSe	Bromine in methanol (0.5-1.0% by vol)	Hydrophilic

ure 11. The aggregation and nucleation of ODT molecules in the LB films on Si substrates deposited from ultrapure water shows streak-like domains (Figure 11(a)). The thickness of such streaks correspond to bilayer of the ODT molecule. The surface structure obtained for the LB film deposited from the subphase containing divalent Cd^{2+} showed dendrite-like structure (Figure 11(b)). The streak-like domains have grown to wider dendrites due to incorporation of Cd^{2+} ion in the subphase. The thickness of the dendrites also corresponds to be bilayer. Therefore, the presence of Cd^{2+} ions facilitates the condensation of molecules and formation of the bilayer structures.

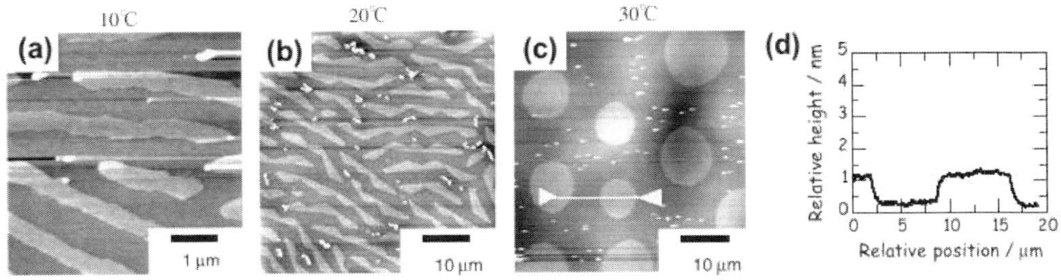

Figure 10. Atomic force microscope images of LB film of mixed monolayer of hydrogenated and semi-fluorinated fatty acids deposited from the subphase having temperature (a) 10 $°C$ (b) 20 $°C$ (c)30 $°C$. (d) shows the height profile along the line shown in (c). Reprinted with permission [107].

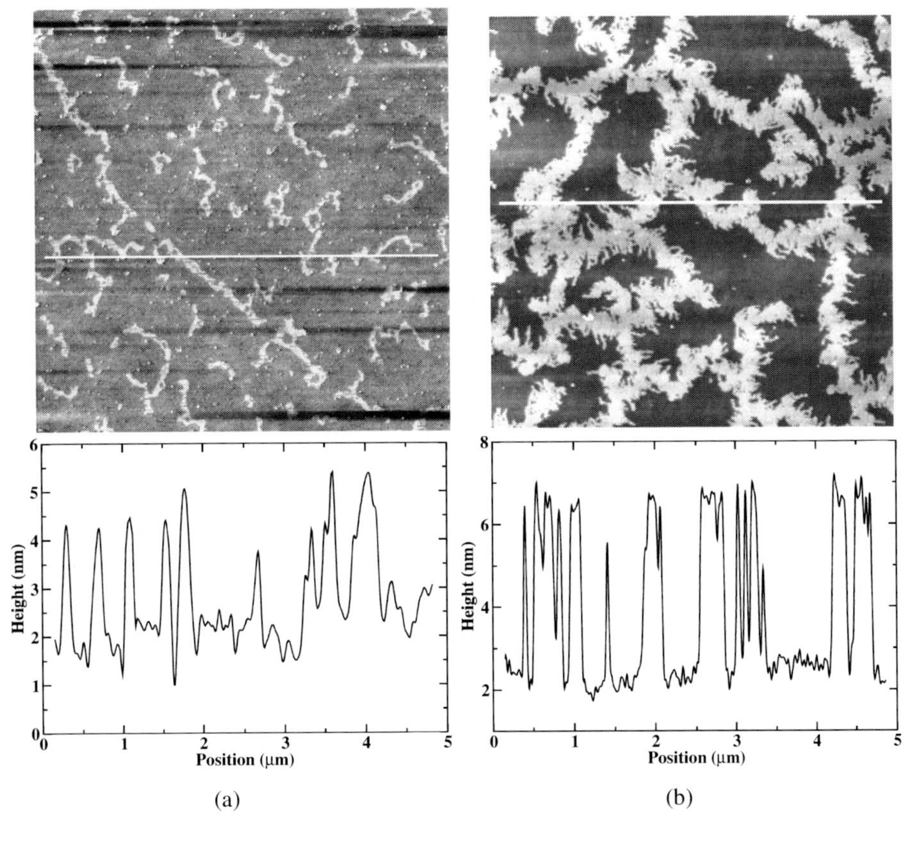

Figure 11. Atomic force microscope images of LB film of octadecanethiol deposited on Si substrate at a target surface pressure of 8 mN/m from subphase containing (a) ultrapure water (b) cadmium chloride (divalent cation, Cd^{2+}). Reprinted with permission [58].

2.1. Application

The possible applications of LB films rely on the unique blend of properties of ultrathin films and a high degree of molecular ordering. It provide flexibility to tailor the surface and hence the properties of the materials. The LB film can be used as a passive layer in metal-insulator-semiconductor devices, for micro/nanolithography, in surface acoustic wave devices, and as lubricants. Since the LB films do not possess center of symmetry, it can be potentially be used for non-linear optical devices. It can also be used for photovoltaics, ultrafiltration membrane and molecular electronic devices. The efficient functional layer of chemical and biological sensors employing the molecular specific interactions can be fabricated using LB technique [61].

2.2. Challenges and Opportunities

The LB films on variety of substrates have been studied. Various interesting nanostructure and their functionality have been investigated. The surface functionalization can be altered by depositing the LB film of appropriate molecule. The alignment of nanomaterials in a defined geometry on the substrate is important for many device fabrication. The materials can be aligned onto any solid substrate by the LB technique. Efforts are being invested in this direction and numerous devices are being proposed on successful deposition and alignment of nanomaterials on substrates. Alternate layers during LB deposition can be potentially employed for the development of materials with unique properties. There are some attempt to deposit the alternate layers of different materials by LB technique. However, there are only a few report on such systems.

3. Self-Assembled Monolayers

The self-assembled monolayer (SAM) is one of the low cost and highly reproducible techniques to obtain a single layer of highly organized organic films at the air-solid interface. The properties of the surfaces can be tailored easily by self-assembly. The concept was introduced by Zisman in 1946 and thereafter the field has grown remarkably. It is known from literature that there are two different classes of organic molecules which are essentially employed to obtain the SAM viz. organothiols and organosilanes. Self assembly by surfactant molecules leading to various supramolecular assemblies is another field of great importance [112, 113] where the supramolecular structures are controlled by tuning the molecular interactions. In this chapter we will be discussing the SAM of organothiol molecules on solid substrate. The ultrathin films obtained through self-assembly possess numerous advantages as compared to the one obtained through molecular beam epitaxy and chemical vapour deposition, as self assembly offers highly ordered and customizable surfaces in a very simple and low cost deposition process. SAM provides a greater flexibility to alter the surface functionality by altering the functional end of the film forming molecules. Therefore a variety of surfaces with different functionality can be created. Such flexibility made it viable that SAM can be potentially used for numerous industrial applications. The SAM of organothiols on the surface of noble metals and organosilanes on hydroxylated surfaces can be fabricated through the solvent route wherein very clean substrate is dipped into a solution of SAM forming molecules. Depending on the incubation time, concentration of the solution, and temperature a highly organized SAM can be obtained. The ordering and the tilt of the molecules can take place due to the interplay of the intermolecular and molecule-substrate interactions. The schematic of SAM is shown in Figure 12.

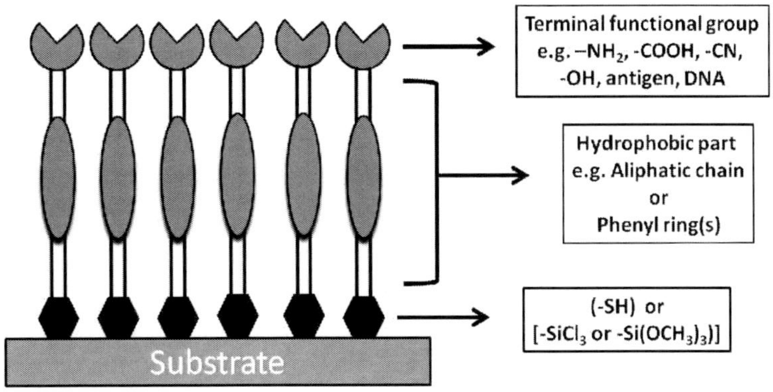

Figure 12. Schematic of self-assembled monolayer deposited on solid substrate.

The study on the kinetics of adsorption of alkanethiol on Au(111) during the fabrication of SAM indicated two step process. Firstly a fast kinetics (few minutes) which is due to the chemisorption of thiol (-SH) group on the gold surface and secondly the slow kinetics which is due to the surface crystallization process governed by the intermolecular interaction of alkyl chains leading to a two-dimensional (2-D) crystal [114]. It has been found that the

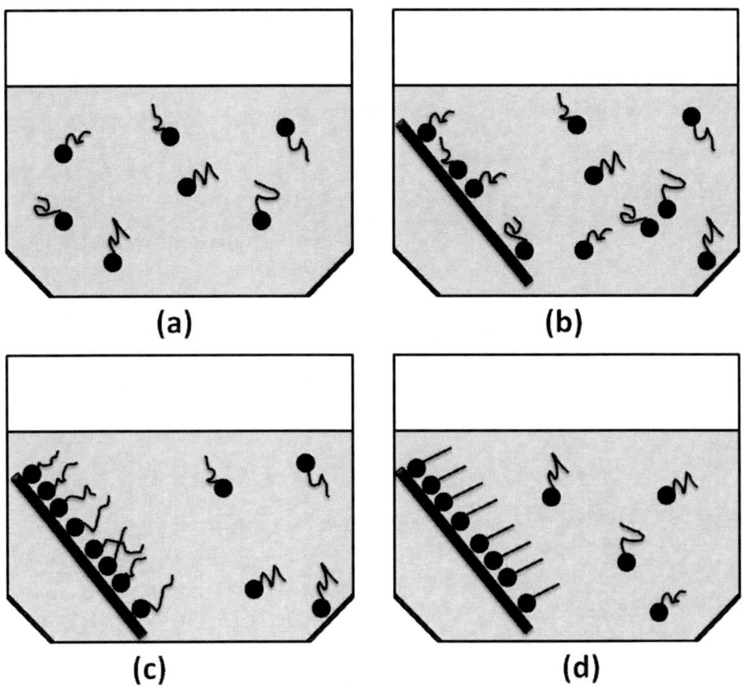

Figure 13. Schematic of formation of SAM of organothiols on gold substrate. (a) Solution of organothiol molecules, (b) initial kinetics indicating the chemisorption of thiol to gold surface (c) increasing the surface coverage and (d) 2D crystallization due to incubation (slower kinetics).

first process depends on the concentration of the organothiol molecules in the solvent and it takes longer time for such adsorption for the lower concentration of the molecules in the solvent [114]. The second process dictating the ordering of the molecules in the SAM is found to be faster if the chain length of the alkylthiol increases [114, 115]. Figure 13 shows the formation mechanism of SAM of thiol molecules on gold substrate.

The quality of the ultrathin SAM also depends on temperature, substrate, and solvent. Due to chemisorption (i.e. Au-S binding), the SAM is found to be quite stable as compared to other techniques like LB where the molecules get attached to the substrate via weaker physical adsorption. The bonding of the thiols onto the gold surface is very strong (∼ 40 kcal/mol) [116]. The thermal stability of the SAM of alkanethiols has been studied. It has been found that the SAM destabilizes due to desorption of sulfur group for a temperature range of 170 - 230 °C [117–119].

The schematic of phase diagram of the decanethiol on Au(111) is shown in Figure 14. The phase diagram is obtained for surface coverage at different temperature. The lowest surface coverage phase is the striped phase (S) where the molecules are lying down whereas the highest coverage phase is standing-up structure (C). The intermediate structure (IS) phase is not clear and it is still debatable. It is important to note that the C phase does not form immediately after the S phase. The S phase undergoes some transformation via IS and finally yields the C phase. The melting transition is found between C and L phases [120].

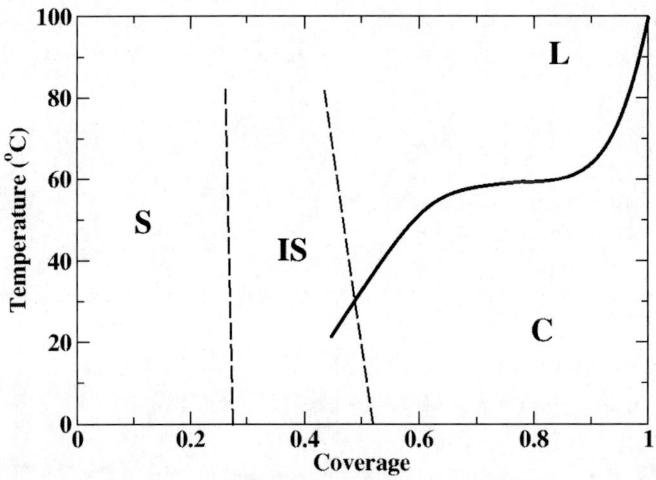

Figure 14. Schematic phase diagram of decanthiol on Au(111). The phases appear are: stripe (S), intermediate structure (IS), standing-up C(4×2) (C) and liquid (L). The broken lines indicate the partial phase boundaries of IS. The solid line between the C and L indicates the melting transition. Reprinted with permission [120].

The structure of SAM has been studied using STM [121–123], xray diffraction [124], AFM [125], electron diffraction [126–128]. The electron diffraction studies on SAM of alkanethiol on Au(111) surface indicate a hexagonal symmetry of the adsorbed sulfur atoms with S-S spacing to be 4.97 Å and the calculated area per molecules to be 21.4 Å^2. The structure of SAM of alkanethiol on Au(111) was studied using AFM. It was found that the film is commensurate with the underlying gold lattice. The SAM is found to exhibit $\sqrt{3} \times \sqrt{3}R30°$ structure [125] as shown in Figure 15.

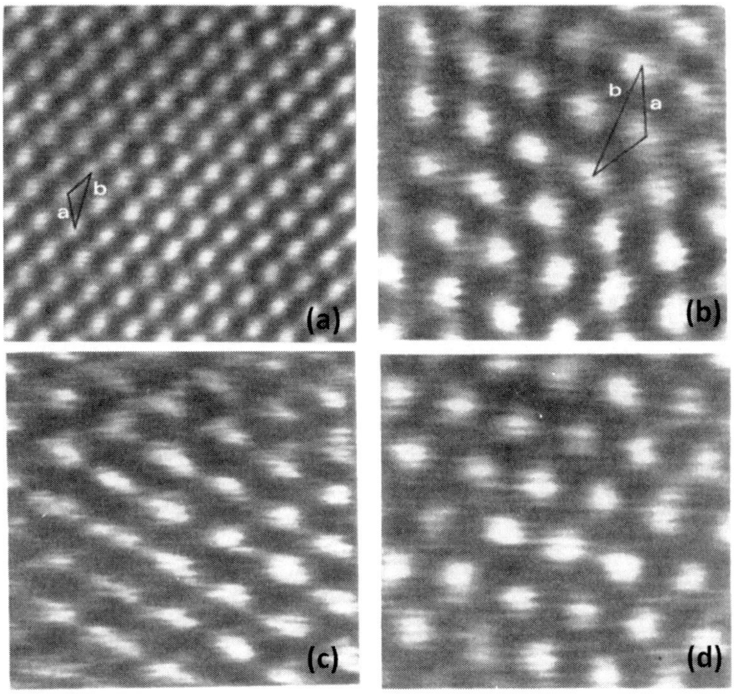

Figure 15. Atomic force microscope images of (a) Au(111) grown on mica, nearest neighbor and next nearest neighbor spacings, a and b are 0.29 and 0.5 nm, respectively (b) SAM of octadecanethiol (c) decanethiol and (d) hexanethiol. The scan area of all the images are 3.02×3.02. The nearest neighbor and next nearest neighbor spacings, a and b are (b)0.52 and 0.9 nm, (c) 0.5 and 0.91 nm and (d) 0.51 and 0.92 nm, respectively. Such spacing of the molecules in SAM on Au(111) corresponds to $\sqrt{3} \times \sqrt{3}R30°$. Reprinted with permission [125].

A 2-D liquid like structure was also obtained for the SAM of very short chain alkanethiols (e.g. butane and hexane) [129]. Using reflection absorption infrared spectroscopy, the tilt of alkanethiols on Au(111) is reported to be 28° [130]. The structure of the SAM of other organosulfur molecules is discussed in detail in the review articles [122, 131].

The surface properties can be tailored precisely by appropriately choosing the terminal functional group of the organothiol molecules (Figure 12). Numerous organothiol molecules with different functional groups have been studied. Some of them are - CF_3, CH=CH, C≡CH, Cl, Br, CN, OH, OCH_3, NH_2, $N(CH_3)_2$, SO_3H, COOH, $COOCH_3$, $COONH_2$ [116, 118, 132–139], antibody [140], DNA, ferrocenyl [141–146] and biotinyl [147–151].

3.1. Application

A control over the surface structures and establishing the relation between those structure and its functionality in the field of ultrathin film is not only interesting for basic science research but also extremely important for technological application point of view. For instance the surface can be engineered with a different degree of wettability starting from teflon like hydrophobic surface to a complete hydrophilic surface. This can be achieved by altering the composition of a SAM of mixed components viz alkanethiol and OH-terminated thiol. With increasing the proportion of OH-terminated thiol, the wettability can be increased. In an interesting work a gradient in hydrophilicity on the surface of has been created by controlling the rate of immersion of gold substrate into the solution of ogranothiols [152]. Figure 16 shows the optical photograph indicating a gradient in hydrophilicity along the length of the treated substrate. A wetting transition in SAM of the mixed component has been reported [153, 154].

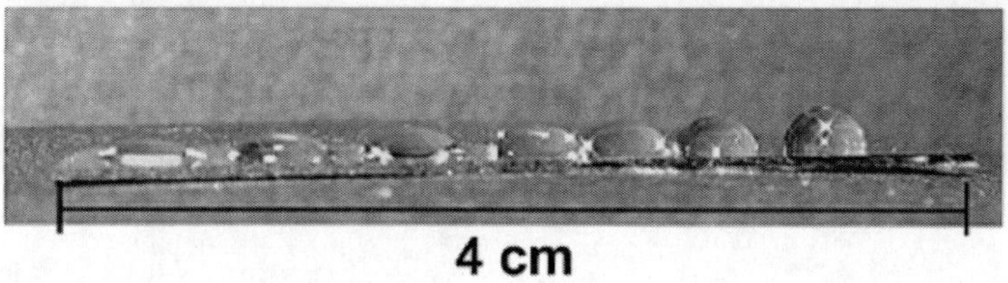

Figure 16. Optical photograph displaying gradient in wetting along the length of the substrate deposited with SAM at different rate of immersion. Reprinted with permission [152].

Hydrophobic SAM of long chain alkanethiols can be used to protect metals from corrosion by aqueous wet chemical etchants [155].

SAM can be used for numerous industrial applications. Some potential application of SAM are in the field of wetting, lubrication, corrosion inhibition, adhesion, lithography, functional films for chemical and biosensors, micro and nanostructured surfaces [156–160]. The defects in SAM can potentially be employed for the fabrication of selective permeation device and array of ultramicroelectrodes [161]. SAM can be used as a control layer for charge transport in many devices including photovoltaics, transistors and catalysts [162, 163]. The SAM can be used for fabrication of FET and molecular electronics [164, 165].

3.2. Challenges and Opportunities

The field of SAM is growing rapidly owing to its potential application in a wider area. Some work has been done for the development of SAM which is switchable due to an external stimulus like electromagnetic field or electrochemical potential [166, 167]. Efforts are required to build a SAM which can switch with some other stimuli (besides light and electrochemical potential) felt due to the presence of some chemical species. Such SAM can be used for the development of sensors for detecting multicomponent species in the solution.

Table 7. Experimental techniques for characterizing LB films and SAM

Technique	Information obtained	References
(a) Scanning probe microscopy	Structures with subnanometer resolution in the spatial domain, depending on probe the local information like conductance, friction, magnetic properties, contact potential, electrochemical properties	[168–170]
(b) Electron microscopes (scanning and transmission)	Structures with subnanometer resolution in the spatial domain	[171, 172]
(c) Grazing angle x-ray diffraction	Film thickness, nanostructure of the surface and interface, positional order, in-plane lattice structure, phases.	[173–175]
(d) Reflection absorption infrared spectroscopy	Orientation of molecules in the film, presence of functional group,H-bonding	[176–179]
(e) Sum-frequency generation	Orientation and conformation of molecules, surface density, surface melting, diffusion, electronic states	[180–182]
(f) Current-Voltage measurement	Electronic band structure of the film	[183]
(g) Surface plasmon resonance	Dielectric and optical properties of film	[184, 185]
(h) Ellipsometry	Optical properties, thickness of the film	[186, 187]
(i) Second harmonic generation	Orientation of the molecules, ordering, defects	[188, 189]
(j) X-ray photoelectron spectroscopy	Composition, chemical species	[190, 191]
(k) Surface Raman Spectroscopy	Structural information, identification of species, chemical composition	[192]

3.3. Characterization Techniques for LB Films and SAM on Solid Substrates

The LB films and SAM on solid substrates can be characterized using the techniques listed in the Table 7.

Acknowledgments

We are thankful to University Grants Commission, India (DRS SAP) and Department of Science and Technology, India for their support.

References

[1] B. Franklin, *Philos. Trans. R. Soc.* London 64, 445 (1774).

[2] A. Pockels, *Nature* 43, 437 (1891).

[3] A. Pockels, *Nature* 50, 223 (1894).

[4] I. Langmuir, *J. Am. Chem. Soc.* 39, 1848 (1917).

[5] I. Langmuir, *Trans. Faraday Soc.* 15, 62 (1920).

[6] K. Blodgett, *J. Am. Chem. Soc.* 57, 1007 (1935).

[7] K. Blodgett, *Phys. Rev.* 51, 964 (1937).

[8] W. C. Bigelow, D. L. Pickett and W. A. Zisman, *J. Colloid Interface Sci.* 1, 513 (1946).

[9] J. Sagiv, *J. Am. Chem. Soc.* 102, 92 (1980).

[10] G.L. Gaines Jr., *Insoluble Monolayers at Liquid-Gas Interfaces*, (Interscience, New York, 1966).

[11] Q. Huo, S. Russev, T. Hasegawa, J. Nishijo, J. Umemura, G. Puccetti, K. C. Russell and R. M. Leblanc, *J. Am. Chem. Soc.* 122, 7890 (2000).

[12] R. K. Gupta and V. Manjuladevi, *Molecular Interactions at Interfaces*, in A. Meghea (ed), *Molecular Interactions* (InTech, Croatia, 2012).

[13] E. G. Finer and M. C. Philips, *Chem. Phys. Lipids* 10, 237 (1973).

[14] G. A. Overbeck and D. Mobius, *J. Phys. Chem.* 97, 7999 (1993).

[15] D. K. Schwartz and C. M. Knobler, *J. Phys. Chem.* 97, 8849 (1993).

[16] M. K. Durbin, A. Malik, R. Ghaskadvi, M. C. Shih, P. Zschack and P. Dutta, *J. Phys. Chem.* 98, 1753 (1994).

[17] A. M. Bibo and I. R. Peterson, *Adv. Mater.* 2, 309 (1990).

[18] E.E. Berring, K. Borrenpohl, S.J. Fliesler and A.B. Serfis, *Chem. Phys. Lipids* 136, 1 (2005).

[19] E. Takano, Y. Ishida, M. Iwahashi, T. Araki and K. Iriyama, *Langmuir* 13, 5782 (1997).

[20] R. K. Gupta and K. A. Suresh, *Euro. Phys. J. E,* 14, 35 (2004).

[21] S. Lafont, H. Rapaport, G.J. Sömjen, A. Renault, P.B. Howes, K. Kjaer, J. Als-Nielsen, L. Leiserowitz and M. Lahav, *J. Phys. Chem. B* 102, 761 (1998).

[22] Y. F. Dufrene, W. R. Barger, J. B. D. Green and G.U. Lee, *Langmuir* 12, 4779 (1997).

[23] A. Gopal and K. Y. C. Lee, *J. Phys. Chem. B* 105 (42), 10348 (2001).

[24] P. Moraille and A. Badia, *Langmuir* 18, 4414 (2002).

[25] H. Möhwald, *Phospholipid monolayers* in R. Lipowski and E. Sackmann (eds), *Handbook of Biological Physics*, (Elsevier, Amsterdam, 1995).

[26] R. Glomm, S. Volden, O. Halskau, Jr. and M. H. G. Ese, *Anal. Chem.* 81, 3042 (2009).

[27] E. Kiss, K. Dravetzky, K. Hill, E. Kutnyanszky and A. Varga, *J. Colloids Interf. Sci.* 325, 337 (2008).

[28] S. Krol, M. Ross, M. Sieber, S. Kunneke, H. J. Galla and A. Janshoff, *Biophysical Journal* 79, 904 (2000).

[29] Z. Khattari, Y. Ruschel, H. Z. Wen, A. Fischer and T. M. Fischer, *J. Phys. Chem. B* 109, 3402 (2005).

[30] J. Kim and T. M. Swager, *Nature* 411, 1030 (2001).

[31] T. Bjornholm, D. R. Greve, N. Reitzel, T. Hassenkam, K, Kjaer, P. B. Howes, N. B. Larsen, J. Nogelund, M. Jayaraman, P. C. Ewbank and R. D. McCullough, *J. Am. Chem. Soc* 120, 7643 (1998).

[32] G. L. Gaines, *Langmuir* 7, 834 (1991).

[33] R. K. Gupta, V. Manjuladevi, C. Karthik, S. Kumar and K. A. Suresh, *Col. Surf. A* 410, 91 (2012).

[34] C. Karthik, V. Manjuladevi, R. K. Gupta and S. Kumar, *J. Mol. Struct.*, DOI:10.1016/j.molstruc.2014.04.036, (2014).

[35] R. K. Gupta and V. Manjuladevi, *Israel J. Chem.* 52, 809 (2012)

[36] I. Giner, I. Gascon, J. Vergara, M. C. Lopez, M. B. Ros and F. M. Royo, *Langmuir* 25, 12332 (2009).

[37] M. N. G. de Mul and J. A. Mann Jr., *Langmuir* 10, 2311 (1994).

[38] R. K. Gupta, K. A. Suresh and S. Kumar, *Phys. Rev. E* 78, 032601 (2008)

[39] L. Feng, H. Li, F. Li, Z. Shi and Z. Gu, *Carbon* 41 , 2385 (2003).

[40] C. P. Collier, R. J. Saykally, J. J. Shiang, S. E. Henrichs and J. R. Heath, *Science* 277, 1978 (1997)

[41] C. Jehoulet, Y. S. Obeng, Y. T. Kim, F. Zhou and A. J. Bard, *J. Am. Chem. Soc.* 114, 4237 (1992).

[42] V. Santhanam, J. Liu, R. Agarwal and R. P. Andres, *Langmuir* 19, 7881 (2003).

[43] V. M. Kaganer, H. Möhwald and P. Dutta, *Rev. Mod. Phys.* 71, 779 (1999).

[44] R. K. Gupta, K. A. Suresh, *Col. Surf. A* 320, 233 (2008).

[45] Y. Tabe, T. Yamamoto, I. Nishiyama, K. M. Aoki, M. Yoneya and H. Yokoyama, *J. Phys. Chem. B* 106, 12089 (2002).

[46] G. L. Gaines, *Langmuir* 7, 3054 (1991).

[47] A. El. Abed, M. -C. Faurè, E. Pouzet and O. Abillon, *Phys. Rev. E* 65, 051603 (2002).

[48] M. Y. Li, A. A. Acero, Z. Huang and S. A. Rice, *Nature* 367, 151 (1994).

[49] M. P. Krafft and M. Goldmann, *Curr. Opin. Col. Interf. Sci.* 8, 243 (2003).

[50] B. Binks, P. Fletcher, S. Kotsev and R. Thompson, *Langmuir* 13, 6669 (1997).

[51] H. Rapaport, K. Kjaer, J. Als-Nielsen, I. Weissbuch, M. Lahav and L. Leiserowitz, *Chem. Rev.* 101, 1659 (2001).

[52] J. G. Riess, *Tetrahedron* 58, 4113 (2002).

[53] B. -Y. Zhu, P. Zhang, R. -X. Wang, Z. -F. Liu and L. Lai, *Col. Surf. A* 157, 63 (1999).

[54] T. Kato, M. Kameyama, M. Ehara and K. -I. Iimura, *Langmuir* 14, 1786 (1998).

[55] S. W. Barton, A. Goudot, O. Bouloussa, F. Rondelez, B. Lin, F. Novak, A. Acero and S. A. Rice, *J. Chem. Phys.* 96, 1343 (1992).

[56] T. Takiue, D. Vollhardt, *Col. Surf. A* vol 198-200, 797 (2002).

[57] M.C. Shih, T. M. Bohanon, J. M. Mikrut, P. Zschack and P. Dutta, *J. Chem. Phys.* 96, 1556 (1992).

[58] R. K. Gupta, K. A. Suresh, R. Guo and S. Kumar, *Anal. Chemica Acta* 568, 109 (2006).

[59] M. Yazdanian, H. Yu and G. Zografi, *Langmuir* 6, 1093 (1990).

[60] A. Itaya, M. Van der Auweraer and F. C. De Schryver, *Langmuir* 5, 1123 (1989).

[61] G. Roberts, *Langmuir-Blodgett Films*, Plenum Press, New York, 1990.

[62] A. Ulman, *An Introduction to Ultrathin Organic Films* (Academic Press, Boston, 1991).

[63] M. C. Petty, *Langmuir-Blodgett films: An introduction* (Cambridge University Press, New York, 1996).

[64] N. K. Adam, J. F. Danielli and J. B. Harding, *Proc. R. Soc. London Ser. A* 147, 491 (1934).

[65] H. C. Parreira, *J. Coll. Inter. Sci.* 20, 742 (1965).

[66] J. Caspers, E. Goormaghtigh, J. Ferreira, R. Brasseur, M. Vanderbranden and J. Ruysschaert, *J. Coll. Inter. Sci.* 91, 546 (1983).

[67] A. Dhanabalan, L. Gaffo, A. M. Barros, W. C. Moreira and O. N. Oliveira Jr., *Langmuir* 15, 3944 (1999).

[68] T. F. Heinz, H. W. K. Tom and Y. R. Shen, *Phys. Rev. A* 28, 1883 (1983).

[69] Th. Rasing, Y. R. Shen, M. W. Kim and S. Grubb, *Phys. Rev. Lett.* 55, 2903 (1985).

[70] K. B. Eisenthal, *Chem. Rev.* 96, 1343 (1996).

[71] P. Guyot-Sionnest, J. H. Hunt and Y. R. Shen, *Phys. Rev. Lett.* 59, 1597 (1987).

[72] G. R. Bell, Z. X. Li, C. D. Bain, P. Fischer and D. C. Duffy, *J. Phys. Chem. B* 102, 9461 (1998).

[73] S. R. Goates, D. A. Schofield and C. D. Bain, *Langmuir* 15, 1400 (1999).

[74] M. Prats, J. Teissie and J. F. Tocanne, *Nature* 322, 756 (1986).

[75] H. Morgan, D. M. Taylor and O. N. Oliviera, *Chem. Phys. Lett.* 150, 311 (1988).

[76] S. Henon and J. Meunier, *Rev. Sci. Instrum.* 62, 936 (1991).

[77] D. Honig and D. Mobius, *J. Phys. Chem.* 95, 4590 (1991).

[78] E. Teer and C. M. Knobler, *J. Chem. Phys.* 106, 1913 (1997).

[79] F. Wu, A. Gericke, C. R. Flach, T. R. Mealy, B. A. Seaton and R. Mendelson, *Biophys. J.* 74, 3273 (1998).

[80] R. K. Gupta, K. A. Suresh, S. Kumar, L. M. Lopatina, R. L. B. Selinger and J. V. Selinger, *Phys. Rev. E* 78, 041703 (2008).

[81] V . von Tscharner and H. M. McConnell, *Biophys. J.* 36, 409 (1981).

[82] M. Losche and H. Möhwald, *Coll. Surf. A* 10, 217 (1984).

[83] P. Kruger, M. Schalke, Z. Wang, R. H. Notter, R. A. Dluhy and M. Losche, *Biophys. J.* 77, 903 (1999).

[84] J. Teissie, *Chem. Phys. Lipids* 25, 357 (1979).

[85] H. Gruniger, D. Mobius and H. Meyer, *J. Chem. Phys.* 79, 3701 (1983).

[86] L. J. Kloeppner and R. S. Duran, *Langmuir* 14, 6734 (1988).

[87] T. Ubukata, T. Seki and K. Ichimura, *Macromol. Symp.* 137, 25 (1999).

[88] R. A. Dluhy, S. M. Stephens, S. Widayati and A. D. Williams, *Spectrochim. Acta* 51A, 1413 (1995).

[89] H. Sakai and J. Umemura, *Langmuir* 14, 6249 (1998).

[90] R. Mendenlson, J. W. Brauner and A. Gericke, *Annu. Rev. Phys. Chem.* 46, 305 (1995).

[91] R. Baekmark, T. Wiesenthal, P. Kuhn, T. M. Bayerl, O. Nuyken and R. Merkel, *Langmuir* 13, 5521 (1997).

[92] V. M. Kaganer, G. Brezesinski, M. Möhwald, P. B. Howes and K. Kjaer, *Phys. Rev. E* 59, 2141 (1999).

[93] I. Watanabe, H. Tanida and S. Kawauchi, *J. Am. Chem. Soc.* 119, 12018 (1997).

[94] Z. X. Li, C. D. Bain, R. K. Thomas, D. C. Duffy and J. Penfold, *J. Phys. Chem. B* 102, 9473 (1998).

[95] J. R. Lu, T. J. Su, Z. X. Li, R. K. Thomas, E. J. Staples, I. Tucker and J. Penfold, *J. Phys. Chem. B* 101, 10332 (1997).

[96] J. Majewski, T. L. Kuhl, K. Kjaer, M. C. Gerstenberg, J. Als-Nielsen, J. N. Israelachvili and G. S. Smith, *J. Am. Chem. Soc.* 120, 1469 (1998).

[97] C. F. Brooks, G. G. Fuller, C. W. Frank and C. R. Robertson, *Langmuir* 15, 2450 (1999).

[98] M. Thoma and H. Möhwald, *Coll. Surf. A* 95, 193 (1995).

[99] S. Cannan, J. Zhang, F. Grunfeld and P. R. Unwin, *Langmuir* 20, 701 (2004).

[100] T. Adalsteinsson and H. Yu, *Langmuir* 16, 9410 (2000).

[101] M. V. Meli, I. H. Lin and N. L. Abbott, *J. Am. Chem. Soc.* 130, 4326 (2008).

[102] H. Kraack, B. M. Ocko, P. S. Pershan, E. Sloutskin and M. Deutsch, *Science* 298, 1404 (2002).

[103] L. Tamam, B. M. Ocko and M. Deutsch, *Langmuir* 28, 15586 (2012).

[104] K. B. Blodgett, *J. Am. Chem. Soc.* 56, 495 (1934).

[105] K. B. Blodgett, *J. Am. Chem. Soc.* 57, 1007 (1935).

[106] J.A. Zasadzinski, R. Viswanathan, L. Madsen, J. Garnaes and D.K. Schwartz, *Science* 263, 1726 (1994).

[107] S. Watanabe, R. Okodu, R. Azumi, H. Sakai, M. Abe, and M. Matsumoto, *J. Coll. Inter. Sci.* 363, 379 (2011).

[108] H. D. Sikes and D. K. Schwartz, *Science* 278, 1604 (1997).

[109] M. K. Mukopadhyay, M. K. Sanyal, A. Datta, M. Mukherjee, Th. Geue, J. Grenzer and U. Pietsch, *Phys. Rev. B* 70, 245408 (2004).

[110] A. Gupta, P. Rajput, S. Bernstorff and H. Amenitsch, *Langmuir* 24, 7793 (2008).

[111] S. R. Naik and A. Gupta, *Coll. Surf. B: Biointerfaces*, 77, 150 (2010).

[112] T. Kato, *Science* 295, 2414 (2002).

[113] T. Kato, N. Mizoshita and K. Kishimoto, *Angew. Chem. Int. Ed*, 45, 38 (2006).

[114] C. D. Bain, E. B. Troughton, Y.-T. Tao, J. Evall, G. M. Whitesides and R. G. Nuzzo, *J. Am. Chem. Soc.* 111,321 (1989).

[115] S. H. Chen and C. W. Frank, *Langmuir* 5, 978 (1989).

[116] L. H. Dubois and R. G. Nuzzo, *Ann. Phys. Chem.* 43, 437 (1992).

[117] J. J. Hickman, D. Ofer, C. Zou, M. S. Wrighton, P. E. Laibinis, and G. M. Whitesides, *J. Am. Chem. Soc.* 113, 1128 (1991).

[118] R. G. Nuzzo, B. R. Zegarski and L. H. Dubois, *J. Am. Chem. Soc.* 109, 733 (1987).

[119] D. M. Jaffey and R. J. Madix, *J. Am. Chem. Soc.* 116, 3012 (1994).

[120] F. Schreiber, *J. Phys. Cond. Matter* 16, R881 (2004).

[121] G. E. Poirier, *Langmuir* 15, 1167 (1999).

[122] J. C. Love, L. A. Estroff, J. K. Kriebel, R. G. Nuzzo, G. M. Whitesides, *Chem. Rev.* 105, 1103 (2005).

[123] A. Dhirani, M. A. Hines, A. J. Fisher, O. Ismail, P. GuyotSionnest, *Langmuir* 11, 2609 (1995).

[124] P. Fenter, P. Eisenberger, J. Li, N. Chamillone, S. Bernasek, G. Scoles, T. A. Ramanarayanan and K. S. Liang, *Langmuir* 7, 2013 (1991).

[125] C. A. Alves, E. L. Smith and M. D. Porter, *J. Am. Chem. Soc.* 114, 1222 (1992).

[126] L. Strong and G. M. Whitesides, *Langmuir* 4, 546 (1988).

[127] C. E. D. Chidsey and D. N. Loiacono, *Langmuir* 6, 709 (1990).

[128] L. H. Dubois, B. R. Zegarski and R. G. Nuzzo, *J. Chem. Phys.* 98, 678 (1993).

[129] G. E. Poirier, M. J. Tarlov and H. E. Rushneier, *Langmuir* 10, 3383 (1994).

[130] P. E. Laibinis, G. M. Whitesides, D. L. Allara, Y. T. Tao, A. N. Parikh and R. G. Nuzzo, *J. Am. Chem. Soc.* 113, 7152 (1991).

[131] A. Ulman, *Chem. Rev.* 96, 1533 (1996).

[132] C. D. Bain and G. M. Whitesides, *Adv. Mater.* 1, 506 (1989).

[133] J. P. Folkers, J. A. Zerkowski, P. E. Laibinis, C. T. Seto and G. M. Whitesides, *Supramolecular Architecture* T. Bein Ed., *ACS Symposium Series* 499; American Chemical Society: Washington, DC, 1992; pp 10-23.

[134] T. R. Lee, P. E. Laibinis, J. P. Folkers and G. M. Whitesides, *Pure Appl. Chem.* 63, 821 (1991).

[135] G. M. Whitesides and G. S. Ferguson, *Chemtracts-Org. Chem.* 1, 171 (1988).

[136] H. Sellers, A. Ulman, Y. Shnidman and J. E. Eilers, *J. Am. Chem. Soc.* 115, 9389 (1993).

[137] A. Ihs, K. Uvdal and B. Liedberg, *Langmuir* 9, 733 (1993).

[138] C. E. D. Chidsey and D. N. Loiacono, *Langmuir* 6, 682 (1990).

[139] K. Dohlhofer, J. Figura and J.-H. Fuhrhop, *Langmuir* 8, 1811 (1992).

[140] R. Dahint, F. Bender and F. Morhard, *Anal. Chem.* 71, 3150 (1999).

[141] D. M. Collard and M. A. Fox, *Langmuir* 7, 1192 (1991).

[142] C. E. D. Chidsey, C. R. Bertozzi, T. M. Putvinski and A. M. Mujsce, *J. Am. Chem. Soc.* 112, 4301 (1990).

[143] K. Uosaki, Y. Sato and H. Kita, *Langmuir* 7, 1510 (1991).

[144] C. E. D. Chidsey, *Science* 251, 919 (1991).

[145] Y. Sato, B. L. Frey, R. M. Corn and K. Uosaki, *Bull. Chem. Soc. Jpn.* 67, 21 (1994).

[146] G. K. Rowe and S. E. Creager, *Langmuir* 10, 1186 (1994).

[147] L. Häussling, H. Ringsdorf, F.-J. Schmitt and W. Knoll, *Langmuir* 7, 1837 (1991).

[148] L. Häussling, B. Michel, H. Ringsdorf and H. Rohrer, *Angew. Chem., Int. Ed. Engl.* 30, 679 (1991).

[149] F.-J. Schmitt, L. Häussling, H. Ringsdorf and W. Knoll, *Thin Solid Films* 210/211, 815 (1992).

[150] J. Spinke, J. Liley, H.-J. Guder, L. Angermaier and W. Knoll, *Langmuir* 9, 1821 (1993).

[151] J. Spinke, M. Liley, F.-J. Schmitt, H.-J. Guder, L. Angermaier and W. Knoll, *J. Chem. Phys.* 99, 7012 (1993).

[152] S. Morgenthaler, S. Lee, S. Zuercher and N. D. Spencer, *Langmuir* 19, 10459 (2003).

[153] A. Ulman, S. D. Evans, Y. Shnidman, R. Sharma, J. E. Eilers and J. C. Chang, *J. Am. Chem. Soc.* 113, 1499 (1991).

[154] J. O. Olbris, A. Ulman and Y. Shnidman, *J. Chem. Phys.*, 102, 6865 (1995).

[155] Y. Xia, X.-M. Zhao and G. M. Whitesides, *Microelectron. Eng.* 32, 255 (1996).

[156] G. M. Whitesides and P. E. Laibinis, *Langmuir* 6, 87 (1990).

[157] R. Jr. Colorado and T. R. Lee, *Langmuir* 19, 3288 (2003).

[158] J. E. Houston and H. I. Kim, *Acc. Chem. Res.* 35, 547 (2002).

[159] G. K. Jennings, T.-H. Yong, J. C. Munro, and P. E. Laibinis, *J. Am. Chem. Soc.* 125, 2950 (2003).

[160] V. F. Petrenko and S. Peng, *Can. J. Phys.* 81, 387 (2003).

[161] Z.P. Yang, I. Engquist, J.-M. Kauffmann and B. Liedberg, *Langmuir* 12 1704 (1996).

[162] D. A. Newman, *Semiconductor Physics and Devices* (McGraw-Hill, Boston, 1997).

[163] G. Ertl, H. Knoezinger and J. Weitkamp, *Handbook of heterogeneous catalysis* (Wiley-VCH, Weinheim, 1997).

[164] C. R. Kagan, A. Afzali, R. Martel, L. M. Gignac, P. M. Solomon, A. G. Schrott and B. Ek, *Nano Lett.* 3, 119 (2003).

[165] J. -O.Lee, G. Lientschnig, F. Wiertz, M. Struijk, R. A. J. Janssen, R. Egberink, D. N. Reinhoudt, P. Hadley and C. Dekker, *Nano Lett.* 3, 113 (2003).

[166] K. Ichimura, S. -K. Oh and M. Nakagawa, *Science* 288, 1624 (2000).

[167] J. Lahann, S. Mitragotri, T. -N. Tran, H. Kaido, J. Sundaram, I. S. Choi, S. Hoffer, G. A. Somorjai and R. Langer, *Science* 299, 371 (2003).

[168] G. Binnig, H. Rohrer, C. Gerber and E. Weibel, *Phys. Rev. Lett.* 49, 57 (1982).

[169] S. Kalinin and A. Gruverman (eds), *Scanning Probe Microscopy: Electrical and electrochemical phenomena at the nanoscale*, Vol 1, (Springer, New York, 2007).

[170] G. Binnig, C. F. Quate and C. Gerber, *Phys. Rev. Lett.* 56, 930 (1986).

[171] J.I. Goldstein, C.E. Lyman, D.E. Newburry, E. Lifshin, P. Echlin, L. Sawyer, D.C. Joy and J.R. Michael, *Scanning Electron Microscopy and X-ray Microanalysis* (Kluwer Academic/Plenum Publishers, New York, 2003).

[172] G. Thomas and M. J. Goringe, *Transmission Electron Microscopy of Materials* (John Wiley and Sons, New York, 1979).

[173] P. Fenter, *Self-assembled monolayers of thiols*, in: A. Ulman (Ed.), Thin Films, Vol. 24, (Academic Press, San Diego, 1998).

[174] M. G. Samart, C. A. Brown and J. G. Gordon, *Langmuir* 9, 1082 (1993).

[175] R. Maoz, J. Sagiv, D. Degenhardt, H. Möhwald and *P. Quint, Supramol. Sci.* 2, 9 (1995).

[176] V. P. Tolstoy, I. V. Chernyshova and V. A. Skryshevsky (eds) *Handbook of Infrared Spectroscopy of Ultrathin Films* (John Wiley Sons, Inc., New Jersey, 2003).

[177] G. A. Beitel, A. Laskov, H. Oosterbeek and E. W. Kuipers, *J. Phys. Chem.* 100, 12494 (1996).

[178] R. G. Greenler, *J. Chem. Phys.* 44, 310 (1966).

[179] A. N. Parikh and D. L. Allara, *J. Chem. Phys.* 96, 927 (1992).

[180] J. E. Sipe, V. Mizrahi and G. I. Stegeman, *Phys. Rev. B* 35, 9091 (1987).

[181] P. Guyot-Sionnest and A. Tadjeddine, *J. Chem. Phys.* 92, 734 (1990).

[182] R. M. Corn and D.A. Higgins, *Chem. Rev.* 94, 107 (1994).

[183] M. K. Massey, C. Pearson, D. A. Zeze, B. G. Mendis and M. C. Petty, *Carbon* 49, 2424 (2011).

[184] J. Hamola, *Chem. Rev.* 108, 462 (2008).

[185] G. J. Ashwell, K. Skjonnemand, M. P. S. Roberts, D. W. Allen, X. Li, J. Sworakowski, A. Chyla and M. Bienkowski, *Coll. Surf. A* 155, 43 (1999).

[186] H. G. Tompkins and E. A. Irene (eds) *Handbook of Ellipsometry* (William Andrew Publishing Inc., New York, 2005).

[187] H. Fujiwara, *Spectroscopic Ellipsometry, Principles and Applications* (John Wiley and Sons, Ltd , Chichester, 2007).

[188] F. Eisert, O. Dannenberger and M. Buck, *Phys. Rev. B* 58, 10860 (1998).

[189] O. Dannenberger, M. Buck and M. Grunze, *J. Phys. Chem. B* 103, 2202 (1999).

[190] K. Bierbaum, M. Kinzler, Ch. Wöll, M. Grunze, G. Hähner, S. Heid and F. Effecberger, *Langmuir* 11, 512 (1995).

[191] S. R. Wasserman, G. M. Whitesides, I. M. Tidswell, B. M. Ocko, P. S. Pershan and J. D. Axe, *J. Am. Chem. Soc.* 111, 5852 (1989).

[192] K. Kneipp, M. Moskovits, and H. Kneipp (eds), *Surface - enhanced Raman scattering - Physics and applications, in Topics in Applied Physics*, vol. 103 (Springer, Berlin, 2006).

In: Comprehensive Guide for Nanocoatings Technology. Vol. 1 ISBN: 978-1-63482-447-7
Editor: Mahmood Aliofkhazraei

Chapter 11

ELECTRODEPOSITION OF INDIUM AND INDIUM ALLOYS

Ts. Dobrovolska, I. Krastev and M. Georgiev
Institute of Physical Chemistry,
Bulgarian Academy of Sciences, Sofia, Bulgaria

ABSTRACT

In this chapter, we attempt to summarize the available information about the electrodeposition of indium and indium alloys. Nowadays, indium is becoming an attractive alloying element enhancing the spectrum of properties of the basic metals. We focus our attention on the special features of electroplating of indium from aqueous solutions and its possible perspectives for practical application.

Indium can be electrodeposited from both acid and alkaline electrolytes. Precipitation of the electrolyte is possible by increasing of the pH value above 3.4. In order to prevent the creation of the hard-to-dissolve indium hydroxide under these conditions, ions of ammonia and sucrose (levulose and glucose) are used. Indium alloys have characteristics that depend on such specific to indium properties as low melting point, high plasticity, and corrosion resistance.

An improved corrosion resistance compared with nickel is observed in the nickel-indium alloy.The alloys of indium with bismuth, antimony, and tin are used for solders.

The co-deposited indium contributes to the significant decrease of the internal stress of the palladium and ruthenium coatings.

Decorative coatings of various colors are obtained by co-depositing of indium, palladium, and gold.

A number of separate intermetallic phases of the indium alloys with different metals demonstrate specific properties. For example, shape-memory effects have been registered in the indium-gold, cadmium-indium, and thallium-indium alloys. Intermetallic bismuth-indium compounds are used as zinc anodes in zinc-alkaline batteries; indium-antimonide is a photodetector for infrared and visible rays.

Self-organization phenomena are observed during electrodeposition of several indium alloys – with silver, cobalt and palladium. In this case, the increase in the content of the alloying element in electrodeposited alloys reflects the changes of their phase composition when the saturation limit of the lattice of the basic metal is reached.

INTRODUCTION

Indium *(In)* is a silver-bright, easily meltable soft metal, with an atomic weight of 114.82. It does not oxidize when exposed to air and it keeps its lustre for a long time. In its main chemical and crystal-chemical characteristics, indium is relative to zinc, tin, lead, copper and iron[1,2].

Usually, indium is obtained by extraction from zinc ores. In 2003, Alfantazi and Moskalyk [3] published a survey article devoted to the processing of indium. The authors stress that their survey serves to assimilate the major highlights of indium, which has both strategic importance and is well suited for electronic applications. The article is very informative with many cited literature sources. It also provides a readily accessible and affordable summary of the pertinent facts regarding the procurement of indium and its disposition.

The industrial usage of indium as a galvanic coating began in the 1930s. Indium was used to increase the quality of bearings in jewelry and dentistry [4-9]. After the 1960s the compounds of indium were widely used in electronics - in the production of semiconductors, high and low temperature transistors, infrared detectors, and etc. [10,11].

The electronic industry widely uses both the monocrystals of indium phosphate and the semi-conductor compounds of the $A^{III}B^{V}$ type like the copper-indium selenide (CIS ($CuInSe_2$)). The latter has become one of the most preferable semi-conductor materials due to its high absorbing factor, making it extremely suitable for solar cells.

Indium-tin oxide presents one of the major end-use products of indium metal, thanks to the combination of the important characteristics of electrical conductivity and optical transparency. The current application of indium in the coatings of various screens like liquid crystal display (LCD's) screens, flat panel displays (touch screens) is massive.

Some indium alloys, more precisely some of the alloy phases, show a shape–memory effect. This remarkable effect has become the reason why the current research and development activities have been directed towards advanced technologies [12,13].

ELECTRODEPOSITION OF INDIUM

One of the most detailed surveys of electrodeposition of indium is presented by R. Piercy and N. Hampson [14]. It encompasses all investigations up to 1975 that had been devoted to the applicable aspects of the electrodeposition of indium. They summarize the available data on the potential of the indium electrode, the mechanisms of the electrode reactions, and the reactions of hydrogen evolution on indium in different electrolytes.

Survey articles, devoted to the electrochemistry of indium have been published by R. Piercy and N.A. Hampson [14] F.C. Walsh and D.R. Gabe [15], and R. Walker and S.Y. Duncan [16-18]. S.P. Yatsenko's monography "Indium. Characteristics and application" [19] contains a survey chapter, devoted to the electrochemistry of the indium and indium alloys.

Indium is deposited from acidic as well as from alkaline electrolytes based on sulfate, sulfamate, chloride, perchlorate, bromide, iodide, tartaric, cyanide, and other compounds.

There are some important points about the indium aqueous electrolytes, which should be noted:

1. In water solutions, indium is found mainly in the three-valence state. The compounds of one- and two-valenced indium are instable. In acidic solutions In^{3+} coordinates with six molecules of water. Hydroxide starts precipitating in pH levels over 3.4 [19-21].
2. In alkaline solutions – those with excess of cyanide and hydroxide (sodium or potassium), as in the case of indium salts, a peptization of the hydroxide takes place with a creation of negatively charged colloid solution. The precipitation can be hindered by adding to the solution a large amount of sugars. When transferred to alkaline pH-values, these substances aid in the formation of negatively charged sol of the indium hydroxide [4,22-24].

The main composition and electrolysis conditions for most of the indium electrolytes are presented in the survey articles of R. Walker and S.Y. Duncan [16-18], as well as of F. Walsh and D.R. Gabe [15]. Therefore, we choose not to repeat these compositions and conditions.

Mainly copper, bronze, and brass are used as substrates for indium plating.

Perhaps due to the relative simplicity of preparing, the sulfate electrolytes are the ones most closely studied and widely applied electrolytes [25,26].

Indium Corporation of America suggested one of the most used sulfamate electrolytes for indium deposition. Indium sulfamate bath, according to J. Slattery [27,28] is the preferred indium plating bath for the following reasons:

1. The bath is characterized by a high throwing power. It is stable, easy to maintain and operate, and has a wide interval of operational parameters, which is appropriate for heavy industrial use.
2. The bath uses soluble indium substances.
3. The solution begins to turn milky white when the pH rises above 3.5, indicating the need to add sulfamic acid – the only additive required in the operation of the bath.

Chloride electrolytes and bromide electrolytes do not seem to have a realistic application.

The initial stages of In electrodeposition onto vitreous carbon have been studied in chloride solutions by Munoz et al. [29].The nucleation process has been analyzed according tothe atomistic theory of nucleation. A model based on the nucleation of active centers of vitreous carbon, involving adsorption of Cl^- on the initially adsorbed indium atoms, has been proposed.

Thiocyanate solution of indium has been determined by Koper and Sluyters as the “classic” indium/thiocyanate oscillator [30]. The authors investigated some multiperiodic “mixed-mode” oscillations and a periodic oscillations which appear near the system’s transition from steady state to oscillatory behavior. Abasova et al. [31] investigated the electrodeposition of indium from tartrate electrolytes on copper, pyrographite, and indium by cyclic voltammetry and rotating disk electrode techniques. The reaction of the three-valenced indium ion to the metallic indium is shown to occur with greater overvoltage onto the pyrographite cathode than that onto indium and copper.

Indium electrodeposition from acid tartrate, acetate, and sulfate electrolytes using asymmetric alternating current has been studied by Kireev et al. [32] The hydrogen evolution reaction in the indium ammoniacal-tartaric electrolyte onto three different substrates-platinum, silver, and indium has been investigated by Krastev et al. [33] Their studies have shown that the hydrogen overvoltage is the highest onto the indium substrate. Detailed information about cyanide electrolytes for the electrodeposition of indium is presented elsewhere [34].

ELECTRODEPOSITION OF INDIUM ALLOYS

In Brenner's monograph, a book considered classic on alloy electrodeposition until 1963, only two indium alloys are mentioned - Ag-In and In-Pb [35]. One of the few articles, devoted to the electrodeposition of indium alloys, is that of F.C. Walsh and D.R. Gabe [36] presentingthe electrolytes for acquiring some alloy systems such as In–Pb, In–Sn, In–Cd, In–Ag and In–Ni. Shortly after that, an article in 3 parts is published by R. Walker and S. J. Duncan. [16-18] They also described the electrolytes, suitable for the electrodeposition of indium and it's alloys. During the period 1973-1985 thorough research devoted to some indium alloy was done by Y. N. Sadana et al. [37-41]The known literature data about electrodeposition of indium alloys are summarized in Table 1.

Table 1.

Alloy	Electrolytes	Properties	Ref.
Ag-In	Cyanide	Antifriction properties, appearance of spatio-temporal structures onto the electrode surface	[34], [42]
	Thiocyanate	Electrochemical instabilities and oscillations	[44]
Au-In	EDTA, acetate	Coloured (blue) gold alloy	[33], [48-49]
Bi-In	Diethylenetriaminepenta acetate, trilonate	Solders	[40], [51-52]
Cd-In	Alkaline tartrate, sulphate	Shape-memory alloys	[56-57]
Cu-In (Ga, Se)	Sulfate, thiocyanate, citrate	CIGS, semiconductors	[59-60], [66]
Co-In	Citrate, tartrate	Improved magnetic properties, appearance of spatio-temporal structures onto the electrode surface	[38], [21]
Pd-In	Ammoniacal-EDTA	Decreased internal stress of the palladium coatings	[75-77]
Ni-In	Sulphate	Heterogeneous coatings	[91]
Sb-In	Citrate, tartrate, chloride	photodetector which detects infrared and visible rays	[93-98]
Sn-In	Citrate, sulphate	Material for the lead-free interconnection technology with low-melting temperature (120°C)	[102-104]
Tl-In	Sulphate	Shape-memory alloy	[39]

Silver-Indium

The first "crystallization" spirals on the surface of an electrodeposited alloy coating were observed by E. Raub and A. Schall during the electrodeposition of Ag-In alloys [42]. After almost 70 years, the phenomenon of formation of periodical spatio-temporal structures was rediscovered during the electrodeposition of the alloy from cyanide solutions (Figure 1).

Figure 1. Optical image of electrodeposited Ag-In alloy coating, width of sample -1 cm.

It was realized that the phenomenon presents a rich variety of waves, target patterns, left- and right-handed two-and three-arm spirals, and annihilating and reappearing waves. It was established by X-ray investigations with a thermal treatment of the alloys and anodic sweep linear voltammetry (ALSV) technique that the spatio-temporal structures were probably formed by the fcc α-phase of silver and the hexagonal γ-phase (Ag_3In ot Ag_4In) of the alloy [43]. More detailed information for electrodeposition of indium-silver alloys has been published by Dobrovolska and Krastev. [34]

Recently, the non-cyanide electrolytes (with the excess of thiocyanate and iodide complexing agents) for the silver-indium alloys deposition were investigated. It has been suggested that the observed oscillations of the potential during galvanostatic deposition of indium in the thiocyanate solution appear due to the formation/destruction of some passive layers on the electrode surface. [44]

Gold-Indium

According to the phase diagram, the alloy of gold-indium has a couple of intermetallic, differently colored phases, which makes the alloy gold-indium rather interesting for jewelers and is known as blue gold [45].

Morris et al. [46] reported that the intermetallic compound $AuIn_2$ shows shape-memory properties. This intermetallic compound appears with a content of about 46.3 wt.% Au and about 54.7 wt.% In. Such alloys containing $AuIn_2$ intermetallic compound can be formed into a gasket or seal, deformable at low stress levels while additionally having corrosion resistant properties.

As an electrodeposited alloy, the gold-indium plating alloy is used for producing the p-type of germanium transistors [47]. An electrolyte is proposed on the basis of EDTA-complex; in absence of EDTA, indium precipitates out of the bath in a few days [33].

Kuhn and Zilske (Degussa Company) [48] obtained by electrodeposition some light yellow, glossy and ductile gold-indium alloys with a good resistance to gradual corrosion by silver sulfide. The electrolytic bath contains 1 to 20 g/l of gold in the form of alkali or ammonium tetracyanoaurate (III), 0.5-50 g/l of indium in the form of water-soluble indium salt, a buffer or conducting salt and 0.5 to 10 mg/l of selenium and/or tellurium. Selenium and tellurium are mentioned as brighteners. An acid cyanide, gold-indium electroplating bath is described by Archibald and Sanderson [49]. Depending upon the plating conditions, the bath allows the deposition of a wide range of alloy compositions including one containing approximately 63 wt.% gold, which is white in appearance. Measurements of wear and corrosion resistance show that the properties of the Au-In deposits are superior to those of silver.

The electrodeposited gold-indium coatings, consisting of less than 60% gold are heterogeneous and represent a white coating with blue dots on it, which suggests a possibility for appearance of spatio-temporal structures.

Bismuth-Indium

The deposition of this alloy results in decorative, fusible and solderable coatings. Intermetallic bismuth-indium compounds are used in zinc anodes of zinc-alkaline batteries in order to avoidany gas formation during the oxidation of zinc [50].

Indium-bismuth alloys were obtained by Sadana and Wang in electrolytes with diethylenetriaminepentaacetate ions as complexing agents [40]. The authors produce coatings of good visual quality, containing 63-89 wt.% indium. The indium content and the current efficiency decrease with increasing of the current density. According to the phase diagram of the metallurgically obtained alloy, two intermediate phases could be present – InBi and In_2Bi. The analysis of the X-ray data of the electrodeposited alloy shows the existence of two intermediate phases InBi and In_3Bi. The latter is suggested to be a new intermediate phase.

Povetkin and Shibleva studied the electroplating of Bi-containing alloys, including indium from trilonate electrolytes [51,52]. The authors came to the conclusion that the very low corrosion rate for alloys containing at 10-15%. It is attributed to the excess phase of the intermetallic compound InBi in the coatings.

Canegallo et al. [53-55] investigated indium electrodeposition on bismuth cathodes at constant current density and room temperature. These conditions cause the formation of three intermetallic compounds - In_2Bi, In_5Bi_3, and InBi due to the indium diffusion and reaction inside the electrode. The average diffusion coefficient of indium in the three intermetallic compounds is calculated (ca. 10^{-15}–10^{-16} m^2 s^{-1}). Moreover, the indium chemical potentials within each compound made the authors expect the evolution of InBi [54].

Cadmium-Indium

Primarily, indium-cadmium was deposited for as coating for bearings [16,36]. Metallurgically obtained alloy with the same content, cooled to room temperature, show a shape recovery when heated to moderate temperatures [12]. A light and fine grained alloy with 22-25 wt.% of cadmium could be deposited from alkaline tartrate [56] and sulfate electrolytes [57], besides those, mentioned in other papers [16,17,36]. The coatings of cadmium-indium alloys are corrosion-resistant, particularly in an environment containing the oxidation products of lubricating oils and organic acids.

The electrochemical route for the preparation of thin films of Cd-In-Te without post - deposition annealing is reported by Jain et al. [58]. In this case, the deposition is based on the initial deposition of Te followed by co-deposition of Cd and In onto the cathode by an under potential deposition an aqueous acidic medium of a pH value about 2. The optical energy band gap for these films is presented to 1.2-1.1 eV.

Copper-Indium (Galium-, Selenium-)

The electrodeposition of Cu, Ga, In together with Se falls into a vast area – the electrochemistry of semiconductors. CIGS films are semi-conductor materials with very strong absorbing characteristics and are extremely suitable as materials for solar cells. An impressive review of the state of art of using electrodeposition to prepare chalcopyrite absorber layer in thin film solar cells is presented by D. Lincot et al. [59,60]. These survey articles cover the research of the electrodeposition of chalcogenides from the late 19th century up to 2004-2005.

Impressive are the following features of their research:

- A road map of the electrodeposition of the main inorganic semiconductors encompassing the period from 1865 to 2002);

From 2000 to 2005, about 3-4 papers had appeared annually on the subject of the electrodeposition of CIGS in highly impact journals. Later, from 2008 to 2012, the number of articles devoted to this subject increased to 12-15 annually.

- Flow chart or low cost strategy for chalcopyrite thin film solar cells.

The classical high cost direct deposition (PVD, CVD) onto Mo/glass substrate is compared with the low cost strategy, starting with a precursor preparation by electrodeposition or screen printing of a mixed chalcopyrite layer or stacked layers. A systemized chronicle of the knowledge is presented on the deposition of such films-primarily, the metals deposited separately from different baths with a following thermal treatment in atmosphere of Se, followed by a combination of 2 elements (indium and selenium for example) [61].

The film formation depends on the applied potential and the solution composition and follows the direct electrodeposition of CIS precursor films by using a unique bath containing the three elements (one step electrodeposition), as initiated by Bhattacharya [63-64]. A key point of the proposed deposition mechanism which is widely accepted is that the composition of the film is controlled by the ratio between selenium and copper fluxes (a) to the electrode (addition of partial currents under diffusion limitation) which in turn controls the insertion of indium by the Kroger's mechanism between excess selenium in the copper selenide film and indium ions [60]. The research of Bhattacharya, continues in the direction of one-step electrodeposited two-layer Cu–In–Ga–Se/In–Se thin films by applying dc voltage with superposition of ac voltage [62-64]. The electrodeposition of $CuInSe_2$ films by an alternating double potent iostatic method using nearly neutral, citrate electrolytes has been presented by Yang et al. [65].

According to M. Ganchev et al. [66], the Cu–In–Ga layers with various Ga-content on molybdenum substrates were produced using complex thiocyanate electrolytes. The use of a complex thiocyanate electrolyte allows a remarkable shift of the deposition potential of copper to the deposition potentials of indium and gallium.

J. Stickney et al. [67] invented the method for electrochemical deposition of semiconductor compounds, including In-containing layers, mostly with Se. The work in his group involves development of a method for the formation of semiconductor compounds in monoatomic layer steps. This method, called Atomic Layer Epitaxy (ALE), is based on the alternated deposition of atomic layers of the elements composing a compound using surface limited reactions - at under potential deposition conditions (UPD). ECALE is the combination of ALE and UPD. Solutions containing a single reactant are pumped to the substrate for a time sufficient to equilibrate. The substrate is exposed to the solution at a potential where the compound is formed at by UPD of the element [68,69].

Cobalt-Indium

The research on the cobalt-indium alloy electrodeposition is presented in a single article of Sadana et al. [38], where the authors describe the possibility of co-deposition of the two metals from citrate electrolyte. Later, a discovery had been made that coatings with a high indium content are heterogeneous. They are multiphase coatings with spatio-temporal structures - Figure 2 [21,70].

The structures are obtained for the first time from silver- and cyanide-free non-alkaline stable electrolytes of a relatively simple composition.Spatio-temporal structures could be observed also during electrodeposition under intensive hydrodynamic flow (electrodeposition in a Jet cell at current densities between 10 and 50 A dm^{-2}) and improved mass transport conditions (Figure 3) [70].

The spatio-temporal structures consist of different phases of the alloy - most probably In_3Co (In_2Co) and Co, which is confirmed not only by a conventional X-ray technique, but also by anodic linear sweep voltammetry [71]. Recently, the strong effect of addition of small indium amounts to some well-known and thoroughly studied permanent magnet materials like some cobalt alloys was reported [72]. The decrease of the sintering temperature and the high values of coercivity which could be reached without losing the magnet density have been associated with the indium-rich grain-boundary phases. High values of coercivity were also registered during our preliminary investigations on the magnetic properties of electrodeposited indium-rich cobalt alloys.

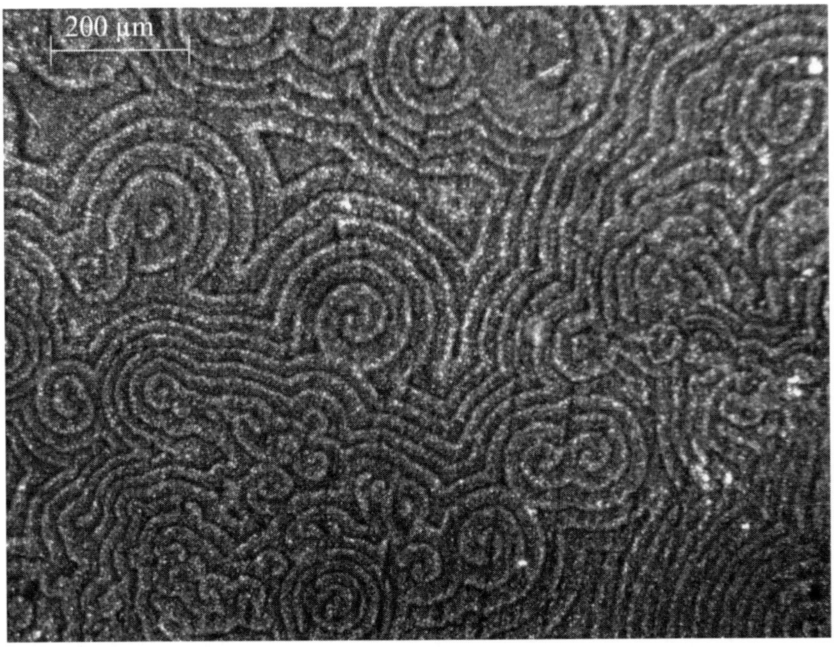

Figure 2. Optical image of electrodeposited Co-In coatings.

Figure 3. Optical image of electrodeposited in jet-cell Co-In coatings.

The morphology of the formed structures, obtained during the electrodeposition of Co-In alloys has been used as a benchmark for the validation of the proposed by Bozzini et al. [73] novel mathematical model for the formation of spatio-temporal patterns in electrodeposition.This model accounts for the coupling between surface morphology and surface composition important for the understanding the formation of morphological patterns found in electroplating.

Palladium-Indium

The main reason to alloy palladium with indium is two-fold: once, to decrease internal stress of electrodeposited pure palladium coatings and second, to obtain coloured coatings– according to metallurgical phase diagram some Pd-In phases are coloured-from yellow, to lilac [74].

The possibility to deposit palladium-indium alloys from ammoniacal-EDTA baths under direct and pulse currents was described in detail by the group of Russian scientists [75-77]. The investigations on the internal stress of the indium-palladium coatings, obtained from these electrolytes show, that coatings with a low content of indium (about 1.8 wt.%) are cracked. In the alloys with indium content more than 13 wt.% the internal stresses are much lower and the coatings are not cracked. It is supposed that the decrease of the internal stresses in these alloys with 20 wt.% of indium is due to the simultaneous action of two opposite effects: A decrease of the hydrogenation of the alloy and an increase of the number of defects in the crystallites during the formation of the saturated solid solution. The addition of saccharin to the electrolyte (2 g/l) leads to a significant decrease (two times) of the internal stress - according to the authors the reason is the more ordered structure of the crystallites.

The phase composition of the electrodeposited palladium-indium alloy is compared with the metallurgical ones in the paper of Reshetnikova et al. [78]. It proves that the chemical compound PdIn with a homogeneity region between 45 and 70 wt.% of indium (which is wider than the metallurgical one) has a pink-lilac colour. The paper also describes some properties of the intermetallic compound like increased hardness and low friction coefficient [79]. The patented electrolyte does not contain free cyanide and the resulting coating consists of an almost intermetallic compound with the formula PdIn. The color of the obtained coatings is pink-lilac and does not change with the thickness of the coatings (up to 30 μm). One next patent, from the same authors, is based on sodium nitril acetate where the obtained coatings are soft, white, and thin [80].

The effect of the electrolysis conditions and the composition of the electrolytes on the content and quality of the obtained palladium-indium coatings is shown by Tihonov et al. [81]. These authors investigated eight different electrolytes. They propose chloride electrolyte with sulfaminic acid as an appropriate electrolyte. The industrial application of the indium electrodeposition is directed toward plating of collectors for electric motors for direct current and for electrical contacts [82].

The ammoniacal–EDTA electrolyte for deposition of palladium–indium is suggested by Kadaner et al. [83]. The authors claimed that the addition of dysprosium chloride could avoid the formation of hydroxide layers in the vicinity of the cathode, which results in the precipitation of indium hydroxide.

Palladium-indium alloys were deposited from ammonia-containing electrolytes. At a low current densities (0.1–0.3 A dm-2) spatio-temporal structures were observed onto the surface of the electrode (Figure 4a, b).

Their composition is almost the same in both the light and dark zones (average content 85 wt.% Pd). The X-ray analysis shows the superposition and coincidence of the spectra of the phases Pd и Pd_3In.

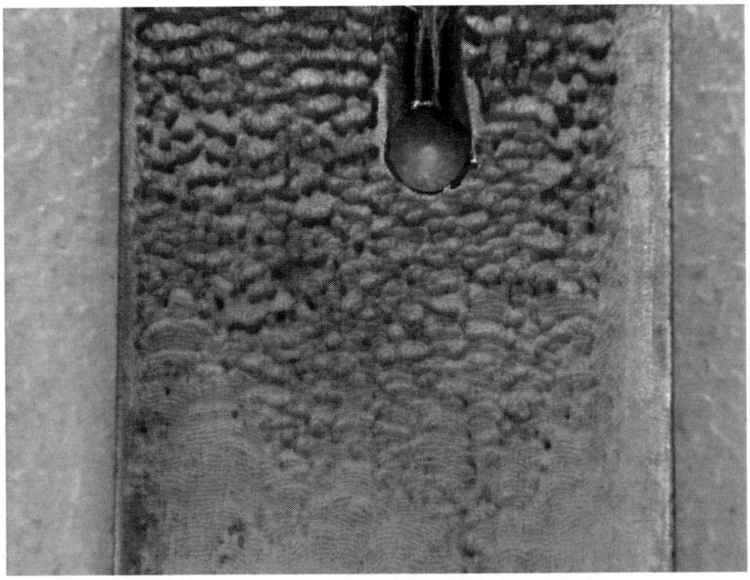

a

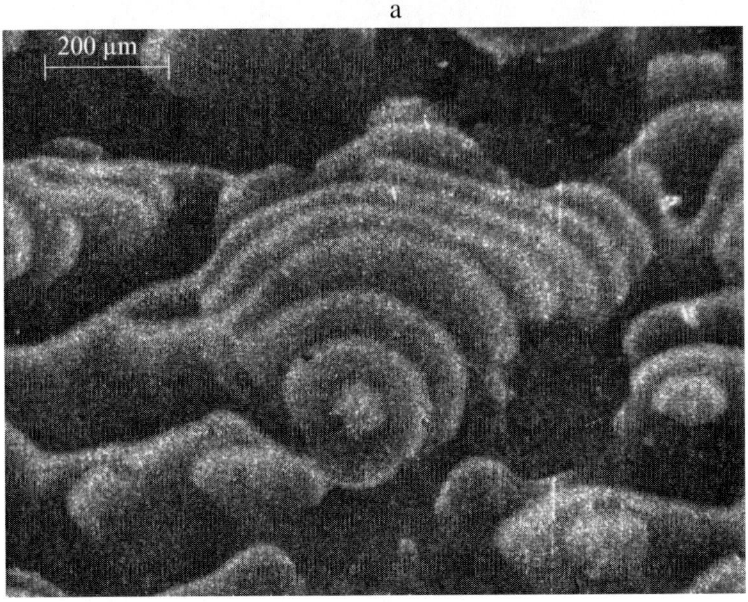

b

Figure 4. a, b. Optical images of electrodeposited Pd-In coating under different magnification a).Width of sample -1 cm.

Nickel-Indium

The metallurgically obtained Ni-In alloy is known as a solderable alloy layer. It is used in the manufacture of printed circuit boards [84] and for the improving of corrosion resistance [85]. A possibility to obtain semi-bright deposits of nickel-indium from acetate-chloride electrolyte is reported by Perelygin et al. [86,87]. Ionycheva et al. [88] reported obtaining similar semi-bright deposits from ammonia-tartrate electrolyte.

The microdistribution of Ni-In electrodeposits (In content up to 5%) has been studied in acid (sulfate, sulfamate, acatate-chloride) and alkaline (ammonia sulfate) baths. Uniform microdistribution and true leveling or negative leveling have been observed depending on bath composition and operating conditions. The conclusion has been reached that the indium content in the deposits is higher on micro-protrusions and lower in recesses of the cathode surface [89].

Nickel-indium galvanic coatings have been investigated as well. It was determined that annealing in a reducing atmosphere broadens the limits of indium content in electrolytically obtained solid solutions to 1 at.% while improved solderability and ductility of the coatings are retained [90].

Figure 5. Optical image of electrodeposited Ni-In coating, width of sample -1 cm.

Sadana [91] investigated the electrodeposition of indium-nickel alloys from sulphate electrolytes and established that in the nickel-rich region, deposits containing up to 29.5 per cent indium exhibit nickel structure. Also, they are homogeneous and present supersaturated solutions within this composition range. The indium-rich electrodeposits on the other hand were heterogeneous [91]. At the electrolysis conditions, shown by Sadana et al., the heterogeneity of the coatings was reproduced by us (Figure 5).

The obtained heterogeneity is quite close to periodicity and the appearance of spatio-temporal structures and is most probably only a problem of adjusting deposition conditions.

Antimony-Indium

According to the phase diagram, only one intermediate phase, SbIn, is present in this binary system. This phase, indium antimonide, is a photodetector which detects infrared and visible rays [92]. According to Y.N. Sadanaand J.P. Singh [93], bright, antimony-rich indium alloys could be deposited from acid citrate solutions. The InSb phase in the electrodeposited alloys exists in the composition range between 49.9 and 54.1 percent antimony, whereas in cast alloys it has been reported to exist at a fixed composition. The authors concluded that the solutions investigated for plating antimony-indium alloys produced bright, tarnish resistant, indium–rich deposits that are promising as decorative and protective coatings.

The electrodeposition of indium-antimony alloys from tartrate solution was also studied by Titov et al. [94]. The authors claim that during electrodeposition only 2 phases appear-antimony and indium (withoutany intermetallic phase). Galvanostatic indium deposition from aqueous solutions on antimony electrodes has been investigated. The formation of InSb has been detected thanks to the indium diffusion into the bulk of the cathode and to its reaction with antimony [95]. The influence of the morphology and structure of antimony substrates on the diffusion and reaction process of an electrodeposited indium overlayer was also investigated [96].

High quality deposits of indium-antimony alloys with a wide range of compositions can be deposited from indium and antimony chloride-tartrate solutions in glycerol, according to Kochegarov and Belitskaya [97] as well as from chloride solutions [98]. The electrodeposition of quasi–stoichiometrical polycrystalline indium antimonide was investigated in different non-aqueous solutions [99].

Tin-Indium

The In-Sn alloys are interesting as materials used in the lead-free interconnection technology with low-melting temperature (120°C). Dense, silvery, fine-crystal deposits of the indium alloy containing up to 16% of tincould be obtained from fluoride electrolytes [100]. The electrodeposition of indium-tin alloys also could be performed in alkaline solutions onto low carbon steel substrates [101].

Ozga et al. [102] investigated the indium-tin layers electrodeposited from complex citrate solutions. It was observed that the applied potential, hydrodynamic conditions, pH, composition of solution, and additional organic compounds have a strong effect on the chemical composition, the phase structure, and the texture of the electrodeposited layers. The X-ray structural investigations of the thermal stability of the deposits on the copper substrate were carried out within the temperature range from 25 to 400°C. The electrodeposition kinetics of a tin-indium alloy from a sulfate electrolyte in the simultaneous presence of synthanol DS-10, formalin, and 1,4-butynediol was studied by the Faraday impedance method [103]. Optimal organic additives providing surface microprofile leveling were found in a study of the electrodeposition of a tin-indium alloy from sulfate electrolytes [104].

Indium-Thalium

The indium-thallium alloys show shape-memory effects [13]. O'Keefe patented the acid electrolyte for the deposition of indium-thallium shape memory alloys [105]. Indium-thallium alloys in the range 15-38 at.% Tl were electrodeposited from a sulfate electrolyte by using pulsed current. The relationship between composition and phase structure of the alloys deposited was investigated with regard to the shape memory effect and the shape recovery for the electrodeposited alloys was also confirmed [106]. The results show that the temperature and the temperature interval for the transformation of the electrodeposited alloys are substantially different from those of the thermally prepared alloys.

Indium-thallium alloys containing up to 84.1% indium were deposited from aqueous sulphate solutions by Sadana [39]. In these solutions, thallium was nobler than indium and was preferentially deposited under the conditions applied. Thallium-rich deposits were powdery with poor adhesion and tarnished readily on exposure to air. Indium-rich deposits resembled indium in appearance. The indium content of the deposit increase with thecurrent density but decreased with increase in temperature. X-ray studies of the alloy deposits indicated the presence of the same phases as those reported in cast alloys, but the limits of the regions containing mixed phases were considerably extended in electrodeposited alloys, thus narrowing the regions of existence of the pure phases [39].

Miscellaneous

The dense uniform crystalline deposits of lead-indium alloy with an indium content of up to 50% and a thickness up to 100 μm were obtained from tartrate solution by Perelygin [107].

The experiments for the deposition of indium arsenide films by an electrodeposition process are described by Lajnef et al. [108]. According to the X-ray, studies of diffraction showed that the as-grown films are poorly crystallized and that the heat treatment improved the crystallinity of the InAs films.

Conclusion

Indium can be electrodeposited from both acid and alkaline electrolytes. Precipitation of the electrolyte is possible by increasing of thepH value above 3.4. In order to prevent the creation of the hard-to-dissolve indium hydroxide under these conditions, ions of ammonia and sucrose (levulose and glucose) are used. Indium alloys have characteristics that depend on such specific to indium properties as low melting point, high plasticity, and corrosion resistance. An improved corrosion resistance compared with nickel is observed in the nickel-indium alloy. The alloys of indium with bismuth, antimony, and tin are used for solders.

The co-deposited indium contributes to the significant decrease of the internal stress of the palladium and ruthenium coatings. Decorative coatings of various colors are obtained by co-depositing of indium, palladium, andgold. A number of separate intermetallic phases of the indium alloys with different metals demonstrate specific properties. For example, shape-memory effects have been registered in the indium-gold, cadmium-indium, and thallium-indium alloys. Intermetallic bismuth-indium compounds are used as zinc anodes in zinc-alkaline batteries; indium-antimonide is a photodetector for infrared and visible rays.

All investigated alloy coatings present average values of the properties of thephases presented in the coatings. Measurements of the properties of the separate phases are not easy to perform because of their fine distribution in the coating at the nano-level.

Self-organization phenomena are observed during electrodeposition of several indium alloys. In this case, the increase in the content of the alloying element in electrodeposited alloys reflects the changes of their phase composition when the saturation limit of the lattice of the basic metal is reached. At higher percentages, the excess amount of the alloying element forms one or more new, richer in this element phases. The coatings become multi-phase, heterogeneous and their physical-mechanical properties change.

Examples of self-organization phenomena during electrodeposition of different alloy systems, could be observed during electrodeposition of indium alloys with silver, palladium and cobalt. They are similar to the effects observed in the coatings of silver-antimony, silver-bismuth, silver-tin, silver-cadmium and copper-antimony alloys, resulting in pattern formation and formation of spatio-temporal structures on the surface of the obtained coatings are presented and compared.

One of the challenges of the electrodeposition of indium alloys is to obtain periodically structured coatings possessing new, unexpected properties, which could increase the possibilities for their application in the micro- and nanodomains.

REFERENCES

[1] Cotton, F. A.; Wilkinson, G. *Advanced Inorganic Chemistry (in russian);* Mir: Moscow, 1969; Vol. 2;.pp. 281.

[2] Gmelin, X. Indium. In *Handbook of Inorganic and organometallic chemistry*, Springer: Berlin, 1991; Vol. 37.

[3] Alfantazi, A. M.; Moskalyk, R. R. Processing of indium: A review. *Minerals Eng.* 2003, *16* (8), 687-694.

[4] Gray, D. Process of electrodepositing indium. US1849293, 1932.

[5] William, M.; Gray, D. Electroplating bath. US1965251, 1934.

[6] Smart, C. Electroplated article. US2755537, 1956.

[7] Indium Corporation of America Processes and compositions for the electrodeposition of indium alone or with other metals. GB602879, 1948.

[8] Bourne, A.; Knapton, A. Jewelry alloys. US4165983, 1979.

[9] MacCulloch, W. T. The effects of adding indium to dental amalgams. *British Dental Journal* 1967, *123* (11), 519-525.

[10] Chynoweth, A. G. High Technology Materials. *Science* 1976, *191*, 724-732.

[11] Mead, R. Battery cell construction with an indium anode and indium hydroxide in the electrolyte. US3857736, 1974.

[12] Fujita, Y.; Nittono, O. Measurements of electrical resistivity due to phase transformations in indium-rich alloys. *Jpn. J. Appl. Phys. Part 1* 1985, *24* (6), 652-655.

[13] Koyama, Y.; Nittono, O. Surface morphology associated with phase transformations in indium-rich solid solutions. *Trans. Jpn. Inst. Met.* 1982, *23* (12), 736-747.

[14] Piercy, R.; Hampson, N. A. The electrochemistry of indium. *J. Appl. Electrochem.* 1975, *5* (1), 1-15.

[15] Walsh, F. C.; Gabe, D. R. The electrodeposition of indium. *Surface Technology* 1979, *8* (2), 87-99.

[16] Walker, R.; Duncan, S. J. Indium and indium alloy deposition, part I. *Met. Finish* 1982, *80* (9), 21-25.

[17] Walker, R.; Duncan, S. J. Indium and indium alloy deposition, part II. *Met. Finish* 1982, *80* (10), 77-81.

[18] Walker, R.; Duncan, S. J. Indium and indium alloy deposition, part III. *Met. Finish* 1982, *80* (11), 59-62.

[19] Yatsenko, S. P. *Indium. Characteristics and applications;* Nauka: Moscow, 1987.

[20] Bleshinskij, S. V.; Abramova, V. F. *Chemistry of indium (in russian);* Akademija nauk KirgSSR: Frunze, 1958.pp. 372.

[21] Krastev, I.; Dobrovolska, T.; Lacnjevac, U.; Nineva, S. Pattern formation during electrodeposition of indium-cobalt alloys. *J. Solid State Electrochem.* 2012, *16* (11), 3449-3456.

[22] Dobrovolska, T.; Krastev, I.; Zielonka, A. Effect of the electrolyte composition on in and Ag-In alloy electrodeposition from cyanide electrolytes. *J Appl Electrochem* 2005, *35* (12), 1245-1251.

[23] Moeller, Th. Contributions to the Chemistry of Indium. IV. The Effects of Certain Poly-hydroxy Non-electrolytes upon the Precipitation of Hydrous Indium Hydroxide. *The journal of physical chemistry* 1941, 1235-1241.

[24] Dobrovolska, T.; Krastev, I.; Zielonka, A. Electrodeposition of silver-indium alloy from cyanide-hydroxide electrolytes. *Russ. J. Electrochem* 2008, *44* (6), 676-682.

[25] Walsh, F. C.; Gabe, D. R. Electrode reactions during the electrodeposition of indium from acid sulphate solutions. *Surface Technology* 1978, *6* (6), 425-436.

[26] Szocs, E.; Schwager, F.; Toben, M.; Brese, N. High-speed indium electrodeposition: Efficient, reliable TIM technology. In *2008 2nd Electronics Systemintegration Technology Conference, ESTC*; Greenwich, 2008; pp 347-350.

[27] Slattery, J. A. Overview of plating with indium. *Electrionics* 1986, *32* (9), 16-18.

[28] Slattery, J. A. Indium plating. *Met. Finish.* 2007, *105* (10), 216-218.

[29] Munoz, A. G.; Saidman, S. B.; Bessone, J. B. Electrodeposition of indium onto vitreous carbon from acid chloride solutions. *J. Electrochem. Soc.* 1999, *146* (6), 2123-2130.

[30] Koper, M. T. M.; Sluyters, J. H. Electrochemical oscillators: an experimental study of the indium/thiocyanate oscillator. *J. Electroanal. Chem.* 1991, *303* (1-2), 65-72.

[31] Abasova, N. K.; Tarasevich, M. R.; Bogdanovskaya, V. A. Electroreduction of indium(III) at various electrodes in tartrate electrolytes. *Russ. J. Electrochem* 1996, *32* (10), 1084-1088.

[32] Kireev, S. Yu.; Peristaya, G. A.; Perelygin, Y. Electrodeposition of Indium from acid electrolytes with the use of asymmetric alternating current. *Russ. J. Appl. Chem.* 2000, *73* (9), 1641-1642.

[33] Krastev, I.; Baumgärtner, M. E.; Raub, C. Zyklisch-voltammetrische Untersuchungen der Silber-Indium-Legierungsabscheidung aus Thiocyanat-Tartrat-Elektrolyten. *Galvanotechnik* 1997, *88* (6), 1882-1891.

[34] Dobrovolska, Ts.; Krastev, I. Electrodeposition of silver-indium alloys. In *Electrolysis: Theory, Types and Applications*, Shing Kuai, Ji Meng ed.; Nova Science Publishers Inc: New York, 2010; pp 303-326.

[35] Brenner, A. *Electrodeposition of Alloys. Principles and Practice;* Academic Press: New York, London, 1963; Vol. 2;.pp. 158-168.

[36] Walsh, F. C.; Gabe, D. R. The electrodeposition of indium alloys. *Surface Technology* 1981, *13* (4), 305-314.

[37] Sadana, Y. N.; Singh, J. P. Electrodeposition and X-ray structure of antimony-indium alloys. *Plat. Surf. Finish* 1985, *72* (12), 64-70.

[38] Sadana, Y. N.; Keskinen, A. E.; Guindon, M. Electrodeposition of alloys III. Electrodeposition and x-ray structure of cobalt-indium alloys (initial studies). *Electrodeposition and Surface Treatments* 1975, *3* (3), 149-157.

[39] Sadana, Y. N. Electrodeposition of alloys VI. Electrodeposition and x-ray structure of indium- thallium alloys from aqueous sulphate solutions. *Surface Technology* 1978, *6* (5), 369-378.

[40] Sadana, Y. N.; Wang, Z. Z. Electrodeposition of alloys XV: Electrodeposition and x-ray structure of Bi-In alloys. *Surface Technology* 1985, *25* (1), 17-29.

[41] Sadana, Y. H. Electrodepositions of alloys - 2. X-ray structure of indium-nickel electrodeposits. *Plating* 1973, *60* (2), 146-148.

[42] Raub, E.; Schall, A. *Zeitschrift fur Metallkunde* 1938, *30* (5), 149-151.

[43] Dobrovolska, T.; Jovic, V. D.; Jovic, B. M.; Krastev, I. Phase identification in electrodeposited Ag-In alloys by ALSV technique. *J Electroanal. Chem.* 2007, *611* (1-2), 232-240.

[44] Dimitrova, N.; Dobrovolska, T.; Krastev, I. Electrodeposition of silver-indium alloys from non-cyanide electrolytes. *Arch. Metall. Mater.* 2013, *58* (1), 255-260.

[45] Klotz, U. E. Metallurgy and processing of coloured gold intermetallics - Part I: Properties and surface processing. *Gold Bull* 2010, *43* (1), 4-10.

[46] Morris, R.; Narasimhan, D.; Hasan R. Gold-indium intermetallic compound, shape memory alloys formed therefrom and resulting articles. US2001026770, 2001.

[47] Brenner, A. Electrodeposition of alloys. Principles and Practice. Academic Press: New York and London, 1963; pp 495-541.

[48] Kuhn, W.; Zilske Wolfgang. Bath and process for the electrolytic deposition of gold-indium alloys. US4617096, 1986.

[49] Archibald, L. C.; Sanderson, G. Electrodeposition of a white gold-indium alloy from an acid cyanide electrolyte. *Trans Institute Metal Finish* 1978, *55* (pt 4), 149-154.

[50] Hymer, T.; James, S. Zinc alloy containing a biamuth-indium intermetallic compound for use in alkaline batteries. WO2004021483, 2004.

[51] Povetkin, V. V.; Shibleva, T. G. Electrodeposition and properties of bismuth alloys with low-melting metals. *Prot. Met.* 2006, *42* (5), 516-519.

[52] Povetkin, V. V.; Shibleva, T. G. Electrodeposition of Bi-In alloys from trilonate electrolyte. Properties and structure. *Zashch. Met.* 1993, *29* (3), 518-530.

[53] Canegallo, S.; Demeneopoulos, V.; Peraldo Bicelli, L.; Serravalle, G. Intermetallic compounds formed by electrodeposition of indium on bismuth. *J. Alloys Compd.* 1994, *216* (1), 149-154.

[54] Canegallo, S.; Demeneopoulos, V.; Peraldo Bicelli, L.; Serravalle, G. Mechanism of indium electrodeposition on bismuth cathodes and time evolution of the deposits. *J. Alloys Compd.* 1995, *228* (1), 23-30.

[55] Canegallo, S.; Agrigento, V.; Moraitou, C.; Toussimi, A.; Bicelli, L. P.; Serravalle, G. Indium diffusion inside InBi during and after electrodeposition at various temperatures. *J. Alloys Compd.* 1996, *234* (2), 211-217.

[56] Perelygin, Y. Electrodeposition of an indium-cadmium alloy. *Prot. Met.* 1989, *24* (5), 663-664.

[57] Peristaya, G. A.; Perelygin, Y. The electrodeposition of indium-cadmium alloy. *Prot Met*2000, *36* (3), 294-295.

[58] Jain, K.; Sharma, R. K.; Kohli, S.; Sood, K. N.; Rastogi, A. C. Electrochemical deposition and characterization of cadmium indium telluride thin films for photovoltaic application. *Curr Appl Phys* 2003, *3* (2-3), 251-256.

[59] Lincot, D. Electrodeposition of semiconductors. *Thin Solid Films* 2005, *487* (1-2), 40-48.

[60] Lincot, D.; Guillemoles, J. F.; Taunier, S.; Guimard, D.; Sicx-Kurdi, J.; Chaumont, A.; Roussel, O.; Ramdani, O.; Hubert, C.; Fauvarque, J. P.; Bodereau, N.; Parissi, L.; Panheleux, P.; Fanouillere, P.; Naghavi, N.; Grand, P. P.; Benfarah, M.; Mogensen, P.; Kerrec, O. Chalcopyrite thin film solar cells by electrodeposition. *Sol Energy* 2004, *77* (6), 725-737.

[61] Massaccesi, S.; Sanchez, S.; Vedel, J. Electrodeposition of indium selenide In2Se3. *J. Electroanal. Chem.* 1996, *412* (1-2), 95-101.

[62] Bhattacharya, R. N. Electrodeposited Cu-In-Ga-Se thin films for CIGS-based solar cells. In *2009 MRS Spring Meeting*; San Francisco, CA, 2010; pp 37-40.

[63] Bhattacharya, R. N.; Oh, M. K.; Kim, Y. CIGS-based solar cells prepared from electrodeposited precursor films. *Sol. Energ. Mater. Sol. Cells* 2012, *98*, 198-202.

[64] Bhattacharya, R. N. CIGS-based solar Cells prepared from electrodeposited stacked Cu/In/Ga Layers. *Sol. Energ. Mater. Sol. Cells* 2013, *113*, 96-99.

[65] Yang, J.; Jin, Z.; Li, C.; Wang, W.; Chai, Y. Electrodeposition of CuInSe₂ films by an alternating double-potentiostatic method using nearly neutral electrolytes. *Electrochem. Commun.* 2009, *11* (3), 711-714.

[66] Ganchev, M.; Kois, J.; Kaelin, M.; Bereznev, S.; Tzvetkova, E.; Volobujeva, O.; Stratieva, N.; Tiwari, A. Preparation of Cu(In,Ga)Se_2layers by selenization of electrodeposited Cu-In-Ga precursors. *Thin Solid Films* 2006, *511-512*, 325-327.

[67] Stickney, J.; Gregory, B.; Villegas, I. Method to electrochemically deposit compound semiconductors. US5320736, 1994.

[68] Vaidyanathan, R.; Stickney, J. L.; Cox, S. M.; Compton, S. P.; Happek, U. Formation of In2Se3 thin films and nanostructures using electrochemical atomic layer epitaxy. *J. Electroanal. Chem.* 2003, *559*, 55-61.

[69] Kim, J. Y.; Stickney, J. L. Ultrahigh vacuum surface studies of the electrochemical atomic layer deposition of indium telluride on n-type GaAs(100). *J. Phys. Chem. C* 2008, *112* (15), 5966-5971.

[70] Krastev, I.; Dobrovolska, T. Pattern formation during electrodeposition of alloys. *J. Solid State Electrochem.* 2013, *17* (2), 481-488.

[71] Krastev, I.; Dobrovolska, Ts. Self-organized structure formation and phase identification in electrodeposited silver-cadmium, silver-indium and cobalt-indium alloys. *Journal of Engineering&Processing management. An International Journal* 2010, *2* (2), 99-105.

[72] Gabay, A. M.; Hadjipanayis, G. C. Phases and phase equilibria in cobalt-rich Pr-Co-In alloys for permanent magnets. *J Alloys Compd* 2010, *500* (2), 161-166.

[73] Bozzini, B.; Lacitignola, D.; Sgura, I. Spatio-temporal organization in alloy electrodeposition: A morphochemical mathematical model and its experimental validation. *J. Solid State Electrochem.* 2013, *17* (2), 467-479.

[74] *Blagorodnye metally;* Metallurgia: Moscow, 1984.

[75] Vinogradov, S. N.; Perelygin, Y. Electrodeposition of palladium-indium alloy. *Prot. Met.* 1980, *16* (4), 409-410.

[76] Vinogradov, S. N.; Perelygin, Y.; Efimov, E. A. Electrodeposition of palladium-indium alloy from ammoniacal citrate electrolyte. *Prot. Met.* 1985, *21* (1), 126-128.

[77] Vinogradov, S. N.; Perelygin, Y.; Efimov, E. A. Kinetic investigation of indium-palladium alloy electrodeposition. *Sov. Electrochem.* 1987, *23* (7), 909-911.

[78] Reshetnikova, N. F.; Pedan, K. S. Electrodeposition of palladium-indium alloys. *Zhurnal prikladnoj khimii* 1982, (2), 1996-1999.

[79] Pedan, K. S.; Reshetnikova, N. F. Electrolyte for depositing palladium-indium alloy coatings. SU917570, 1985.

[80] Pedan, K. S.; Reshtenikova, N. F. Electrolyte for depositing coatings of palladium-indium alloy. SU931812, 1982.

[81] Tihonov, A. A.; Vjacheslavov, P. M.; Burkat, G. K. Investigation of the electrodeposition of palladium-indium alloys from ammoniacal-trilonate and sulfamic electrolytes. *Zhurnal prikladnoj khimii* 1981, (2), 364-368.

[82] Vjacheslavov, P. M.; Burkat, G. K.; Tihonov, A. A.; Dzhandubaeva, F. M. Electrodeposition of palladium-indium alloys. *Zhurnal prikladnoj khimii* 1990, (2), -by.

[83] Kadaner, L.; Slyusarskaya, T.; Borodaj, Z.; Reznichenko, V. Electrolyte for depositing palladium-indium alloy coatings. SU834263, 1981.

[84] Samuels, G. J.; Garman, J. M. Novel nickel/indium alloy and method of using same in the manufacture of printed circuit boards. EP0163059 (A2), 1985.

[85] Kravtsov, Ev.; Kondratenko, T.; Filippova, E.; Kuzmin, A. Electrolyte for depositing indium-nickel coatings. SU1615232 , 1990.

[86] Vinogradov, S. N.; Perelygin, Y. Electrodeposition of a nickel-indium alloy from an ammonium citrate electrolyte. *Prot. Met.* 1988, *23* (5), 642-644.

[87] Perelygin, Y. Electrodeposition of nickel-indium alloy from an acetate electrolyte. *Prot. Met.* 1989, *25* (2), 243-244.

[88] Ionycheva, L. S.; Krapivnaya, E. D.; Dunaevskaya, F. M.; Krasnozhon, L. N.; Loboda, L. N. Comparative characteristics of electrodeposition of indium alloys. *Prot. Met.* 1981, *17* (1), 110-112.

[89] Kruglikov, S. S.; Novozhilova, R. A.; Rusanova, E. A. Microdistribution of nickel-indium alloys. In *Proceedings of the 80th AESF Annual Technical Conference*; Anon, Ed.; Publ by American Electroplaters & Surface Finishers Soc Inc: Anaheim, CA, USA, 1993; pp 747-753.

[90] Novikov, L. G.; Kurnoskin, G. A.; Flerov, V. N.; Shul'pin, G. P. Physicomechanical characteristics of galvanic nickel-indium coatings. *Prot Met* 1989, *25* (2), 253-254.

[91] Sadana, Y. H. Electrodepostion of alloys - 2. X-ray structure of indium-nickel electrodeposits. *Plating* 1973, *60* (2), 146-148.

[92] Ichiroo, K.; Haabaato, E.; Suteibun, E. Indium antimony photodetector. JPH05206499, 1993.

[93] Sadana, Y. N.; Singh, J. P. Electrodeposition and x-ray structure of antimony-indium alloys. *Plating and Surface Finishing* 1985, *72* (12), 64-70.

[94] Titov, P. S.; Anisimova, I. V.; Afanasieva, V. S. Elektroosazhdemie splava indij - surma iy tartratnogo elektrolita. *Zashch. Met.* 1968, *4* (5), 595-596.

[95] Kozlov, V. M.; Agrigento, V.; Bontempi, D.; Canegallo, S.; Moraitou, C.; Toussimi, A.; Bicelli, L. P.; Serravalle, G. Intermetallic compound formed by electrodeposition of indium on antimony. *J. Alloys Compd.* 1997, *259* (1-2), 234-240.

[96] Kozlov, V. M.; Agrigento, V.; Mussati, G.; Bicelli, L. P. Influence of the structure of the electrodeposited antimony substrate on indium diffusion. *J. Alloys Compd.* 1999, *288* (1-2), 255-261.

[97] Kochegarov, V. M.; Belitskaya, T. B. Electrodeposition of indium- antimony alloys from chloride tartrate solutions in glycerol. *Zh Prikl Khim* 1971, *44* (2), 452-454.

[98] Shibleva, T.; Povetkin, V.; Shindler, R. Electrolyte for deposition of indium-antimony alloy. RU2359073, 2009.

[99] Carpenter, M. K.; Verbrugge, M. W. Electrochemical codeposition of indium and antimony from a chloroindate molten salt. *J. Mater. Res.* 1994, *9* (10), 2584-2591.

[100] Mogilev, V. M.; Falicheva, A. I. Electrodeposition of indium-tin alloys. *Prot Met* 1974, *10* (2), 182-183.

[101] Abd El-Rehim, S. S.; Awad, A.; El-Sayed, A. Electroplating of indium and indium-tin alloys from alkaline baths. *Hung. J. Ind. Chem.* 1988, *16* (4), 419-426.

[102] Ozga, P.; Uwianctek, Z.; Michalec, M.; Onderka, B.; Bonarski, J. Phase structure and texture of electrodeposited InSn alloys on copper substrate. *Arch. Metall. Mater.* 2008, *53* (1), 307-315.

[103] Medvedev, G. I.; Rybin, A. A.; Makrushin, N. A. A study of the electrodeposition kinetics of a tin-indium alloy from a sulfate electrolyte with organic additives. *Russ. J. Appl. Chem.* 2012, *85* (8), 1163-1166.

[104] Medvedev, G. I.; Rybin, A. A.; Makrushin, N. A. Leveling power of sulfate electrolytes with organic additives in electrodeposition of tin-indium alloy. *Russ. J. Appl. Chem.* 2013, *86* (3), 351-354.

[105] O'Keefe, Th. Electrodeposition of indium-thallium shape memory alloys . US5203931, 1993.

[106] Chong, H. S.; O'Keefe, T. J. Characterization of phase transformation behavior in electrolytically produced indium-thallium shape memory alloy films. *Mater. Charact.* 1994, *33* (4), 311-319.

[107] Perelygin, Y. Electrodeposition of a lead-indium alloy from an alkaline-tartrate electrolyte. *Prot. Met.* 1991, *26* (4), 532-533.

[108] Lajnef, M.; Ezzaouia, H.; Chtourou, R. Optical characterization of InAs film grown on SnO_2 substrate by the electrodeposition technique. *J. Phys. D* 2008, *41* (12).

In: Comprehensive Guide for Nanocoatings Technology. Vol. 1 ISBN: 978-1-63482-447-7
Editor: Mahmood Aliofkhazraei

Chapter 12

ELECTROCHEMISTRY ROLE IN PRODUCTION OF NANO-COATINGS

***Katarzyna Krukiewicz** and Jerzy K. Zak*†**
Faculty of Chemistry, Silesian University of Technology, Gliwice, Poland

ABSTRACT

Out of numerous synthesis methods of nano-coatings, electrochemical techniques have gained much interest, mainly because of their versatility. By means of advanced electrochemical strategies it is possible to obtain nano-coatings of diverse forms. Rapidity of the process, purity of products as well as production of coatings on widely varying surfaces are the most noticeable advantages of electrochemical synthetic methods. In this chapter we will present two principal electrochemical techniques: electrochemical deposition and electrophoretic deposition, together with less popular alternatives, such as plasma electrolytic deposition, laser-assisted electrodeposition and alternating current electrophoretic deposition. The fundamentals of each technique will be followed by specific examples illustrating details of the synthetic methods.

INTRODUCTION

Electrochemistry is considered to be one of the most advantageous methods of the production of nano-coatings [1]. Out of numerous synthesis methods, these are electrochemical techniques that provide high versatility together with the precise control of the properties of obtained materials [2-4]. Variety of electrochemical strategies enables the formation of diversified nano-coatings composed of nanoparticles, nanocomposites, nanotubes, nanostructured fibers, nanospheres or nanoclusters [5-9]. By changing the process parameters, it is possible to tailor the crucial properties of nano-coatings such as their thickness and morphology [3-4,7]. Besides, electrochemistry as the synthesis method seems

† jerzy.zak@polsl.pl
* katarzyna.krukiewicz@polsl.pl

to be of considerable economic importance because of its low costs, reduced amount of waste materials and the lack of need of further product purification [1,10-11]. These advantages, together with the high deposition rates, ability of continuous processing, and wide availability of processable substrates, place electrochemistry in the centre of attention for the researchers' requirements in the technology of nano-coatings.

Electrochemical techniques can be divided into two principal groups, namely electrochemical deposition and electrophoretic deposition (Figure 1). Nonetheless, with the development of science these two basic techniques evolved in the large number of alternative processes, such as electrochemical polymerization, ultrasound assisted electrodeposition, laser-assisted electrodeposition, pulse current electrochemical deposition, localized electrochemical deposition, electrophoretic deposition in magnetic field, alternating current electrophoretic deposition, pulsed direct current electrophoretic deposition and plasma electrolytic deposition. The aim of this chapter is to provide the summary of given electrochemical techniques of the production of nano-coatings followed by specific examples illustrating details of the synthetic methods.

ELECTROCHEMICAL DEPOSITION

Electrochemical deposition (ECD), also known as electrodeposition, electroplating or electrolytic deposition, is a process of surface modification in which a thin and tightly adherent coating of metal, oxide, salt or macromolecule is deposited onto the conducting substrate by simple electrolysis of a solution containing a desired substance [12]. The history of electrodeposition (Figure 2) started in 1800 with the invention of electrical battery by Alessandro Volta [13]. Significant contributions to electrochemical deposition were made by Faraday (Faraday's laws of electrolysis, 1833), Tafel (the concept of overpotential, 1905), Erdey-Gruz and Volmer (current–potential relationship, 1930), Eyring and Wynne-Jones (formulation of absolute rate theory, 1935), Lorenz (consideration of rate-determining surface diffusion of adions, 1954), Conway and Bockris (calculation of probabilities of charge transfer to different sites on the metal surface, 1958) and Dickson (study of nucleation and growth of electrodeposited gold on surfaces of silver, 1965) [14].

The utilization of electrochemical deposition in the production of nano-coatings has emerged as an area of great interest and advanced the synthesis of nano-materials from a laboratory scale to practical industrial use [1]. The advantages of this century-old process include rapidity, low costs, high purity of products, high deposition rates, homogenous distribution of particles, the ease of process control, variability of the coating composition, sustainability for complex geometries of substrate, reduced waste materials and ability of continuous processing [1,10-11]. What is more, by controlling ECD conditions such as initial pH, electrolyte composition, current density, deposition time and reaction temperature it is possible to produce coatings with tailored morphology and structures [3-4]. ECD can be successfully applied in the production of coatings with a varying degree of crystallinity under milder conditions and shorter reaction times than any other method of production [15]. However, the application of ECD method is limited to electrically conducting materials such as metals, alloys, semiconductors, electrically conductive polymers and oxides [16].

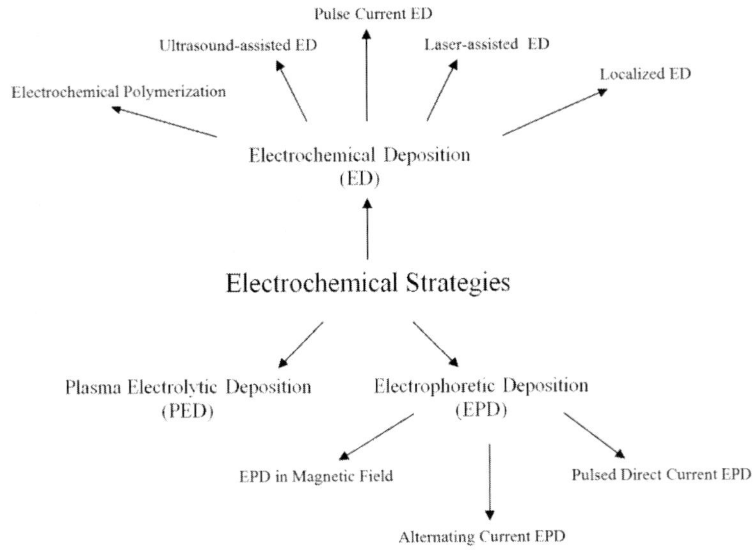

Figure 1. The classification of electrochemical techniques.

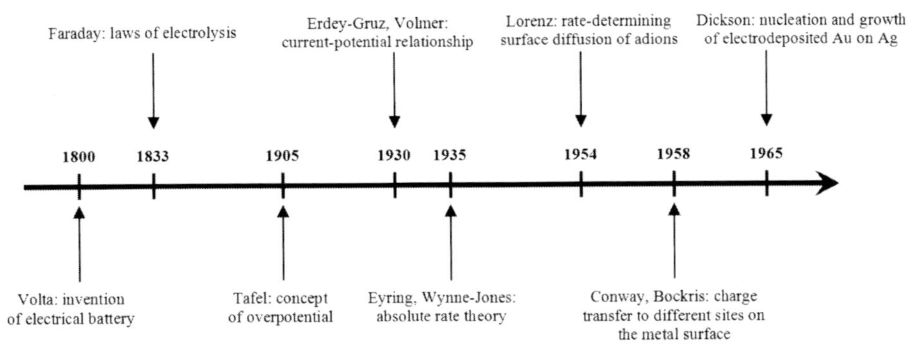

Figure 2. The history of electrochemical deposition.

The typical application of ECD is the research and development of magnetic thin films [17]. The effectiveness of this technique relies on its low costs, high throughput, and high quality of the deposits – these properties are beneficial in the magnetic recording industry, for instance to form Permalloy films [18]. The main area of the application of magnetic films includes microelectrochemical systems (MEMS) and devices such as sensors, microactuators, micromotors and frictionless microgears [19]. These materials are successfully applied in the production of magnetic recording head devices; the ease of production of precise structures by ECD makes the areal recording density constantly increasing [17].

ECD is a method particularly important for the production of precious metal coatings. This kind of deposit has many applications including anti-microbially active coatings [20], precious metal catalysts [21] and corrosion protective layers [22]. High controllability of ECD process enables the production of different kinds of deposit structures with defined parameters by tailoring operating parameters such as temperature, pH and current density [23]. The mechanism of reduction of a simple solvated metal salt is shown in Figure 3. Current from external source passing through an electrochemical cell, together with diffusion and convection phenomena, results in the migration of solvated ion present in the electrolyte (a) to the cathode, where it enters the diffusion layer (b). The electrical field in diffusion layer is strong enough to align solvating water molecules but too weak to liberate free metal ion. When solvated metal ion reaches Helmholtz layer (c), the solvating water molecules are removed leaving free metal ion. Finally, the metal ion is reduced on the surface of cathode and the metal deposit is formed [24].

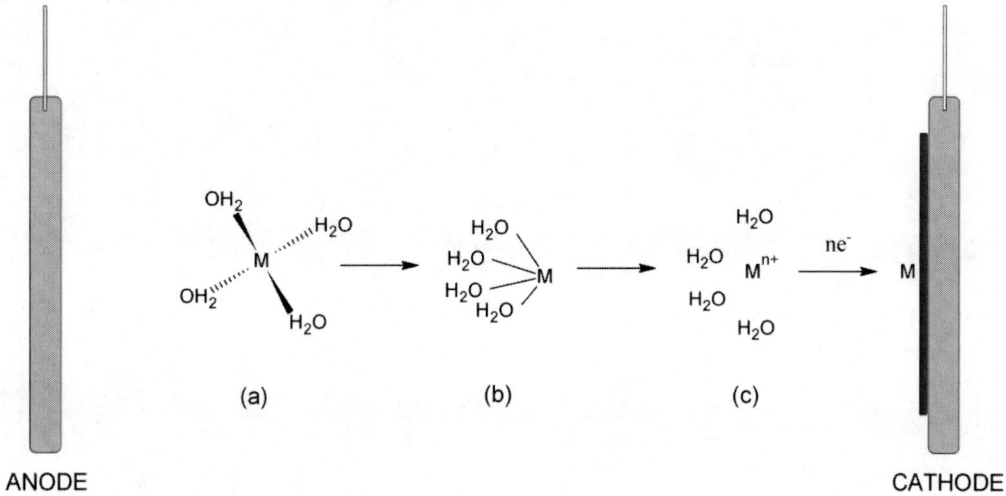

Figure 3. The mechanism of electrochemical reduction of a simple solvated metal ion in three regions: a) bulk electrolyte, b) diffusion layer, c) Helmholtz layer.

The special case of ECD is **electrochemical polymerization**, the technique resulting in the formation of conducting polymer coatings. During the process of electropolymerization, the anodic oxidation of monomer present in electrolytic solution takes place at the surface of the working electrode [25]. The mechanism of this process involves the formation of a radical cation (polaron) in the first step and chemical pairing reaction between two radical cations in the second step, resulting in the formation of dication (bipolaron). The process is repeating until the resulting oligomer becomes insoluble in solvent and precipitates on the surface of the anode [26]. The schematic representation of the electrochemical polymerization of pyrrole is shown in Figure 4.

The advantages of electrochemical polymerization include the easiness of the control of film thickness, direct grafting of polymer onto the electrode surface as well as possibility to perform in situ characterization of the growing polymer with the use of electrochemical and/or spectroscopic methods [27]. Conducting polymers are versatile materials that are successfully used in a variety of applications. Polyaniline coatings are found to exhibit excellent corrosion resistance, even in aggressive environments [28-30], polypyrrole and its derivatives are applied as biocompatible coatings in biomedical engineering [31-33], while poly(3,4-ethylenedioxythiophene) can be utilized in photovoltaic [34] and electrochromic devices [35-36].

Apart from "classical" electrodeposition process, there are numerous alternatives of this technique appearing in the literature. One of them is **ultrasound assisted electrodeposition**. Ultrasounds are recognized to induce strong physical, chemical and mechanical changes that are able to activate chemical and electrochemical reactions [37]. That is why extensive research has been conducted in order to couple an ultrasound field with an electric field [38]. This technique was successfully applied in a process of fabrication of TiO_2 nanotube array/CdSe nanoparticle/TiO_2 composite, making CdSe nanoparticles prone to disperse into the nanotube arrays [5].

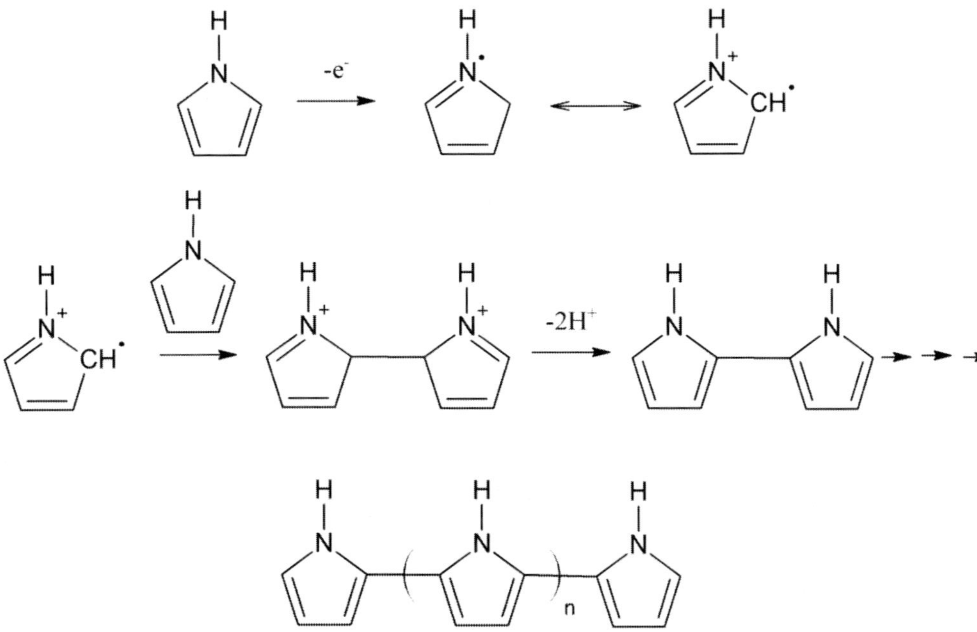

Figure 4. The mechanism of electrochemical polymerization of pyrrole.

The other alternative to classical electrodeposition method is a **laser-assisted electrodeposition**. For the first time it appeared in literature in 1979 when von Gutfeld achieved maskless plating with the use of pulsed laser focused onto an electrode in an electroplating bath [39]. Under laser irradiation at intensities near 10 kW/cm^2 (Ar or Kr laser of 1.5 W or 100 mW output power, respectively) it was possible to enhance the process of electrodeposition up to 1000 times. In the laser assisted method, improvement of the deposition is associated with the local activation of both irradiated electrode surface and solved species [40]. This type of deposition is of a particular importance for creating zinc electrodeposits because of their improved morphology, hence higher adherence and compactness of obtained material comparing with other methods [41].

One of the process operational variables that has a tremendous effect on the composition and resultant properties of coating, is the applied current. It was found that the consecutive changes in charge and mass transfer processes can influence the chemical composition, properties and morphology of obtained coatings – that is why **pulse current (PC) electrochemical deposition** has been introduced as an interesting alternative to a direct current (DC) method [42]. In PC ECD, the value of one of two parameters, the potential or current density, is rapidly changed by adjusting a series of pulses with equal amplitude, polarity and duration, separated by periods of zero current [43]. The schematic representation of pulse current is shown in Figure 5. PC electrodeposition can adopt higher current densities resulting in the production of nanocrystalline coatings with more compact structure and finer grains than DC deposits [6]. The clear proof of the effectiveness and versatility of PC electrodeposition is the wide range of its application in the production of nano-coatings. Pulse current electrodeposition has been used to fabricate fine-grained and smooth nanocrystalline nickel coatings [44], nickel hexacyanoferrate/chitosan/carbon nanotubes nanocomposite films with controllable structures [45], bismuth telluride films [46], binary Co–Sn alloys for lithium secondary batteries [47] and Ni-W-B composites containing CeO_2 nano-particle [48].

Modern applications of coatings, such as nano-electronics and photonics, require of the development of new technologies for the fabrication of structures at the nanometer scale [49]. The electrochemical method enabling the realization of this goal is called **localized electrochemical deposition** (LECD), which was introduced in 1996 by Madden and Hunter [50]. In LECD, the current density is confined by limiting the surface area of counter electrode (CE) which is accomplished by using microtips as CE; the three-dimensional nano-features are obtained by moving CE along a chosen trajectory with respect to the substrate [51]. LECD technique has numerous advantages: it is applicable to many materials without limitation on the type of substrate or thickness of electrodeposit, it works in ambient conditions and enables tailoring of the properties of coating by modifying electrochemical parameters [51-52]. This method is believed to be more economic and effective than the conventional lithography, hence, it is widely used in semiconductor industry [53]. So far, LECD has been applied in the production of nickel microstructures [54], copper electrical interconnects [55] as well as titania-based microwires [49].

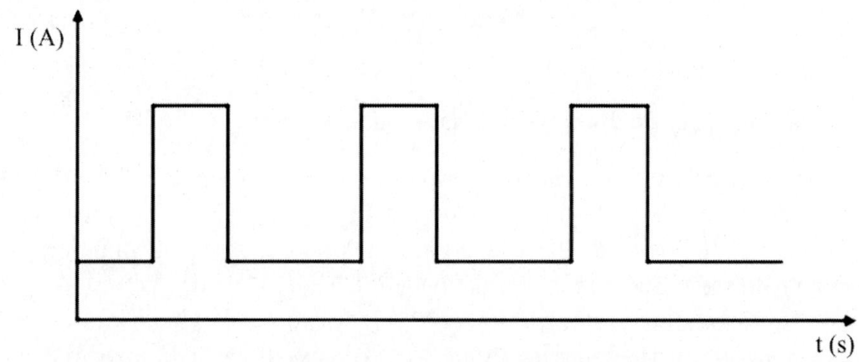

Figure 5. The schematic representation of pulse current.

Electrophoretic Deposition

Electrophoretic deposition (EPD) is a process consisting of two steps: the movement of particles suspended in liquid toward electrode by applying electrical field (electrophoresis) and the collection of these particles on the surface of electrode followed by the formation of a coherent deposit [56]. The phenomenon of electrophoretic deposition was initially observed in 1808 by the Russian scientist Ruess (Figure 6), however, it was 1933 when EPD technique was used for the first time practically for the deposition of thoria particles on platinum electrode as an emitter for electron tube application patented in USA [57]. The significant contribution to understanding EPD process was made by Hamaker who was the first researcher specialized in EPD of ceramics [58]. In late 1950s, EPD focused the centre of attention of Dr. George E. F. Brewer's research group from Ford Motor Company; through their studies EPD began to take its current shape [59].

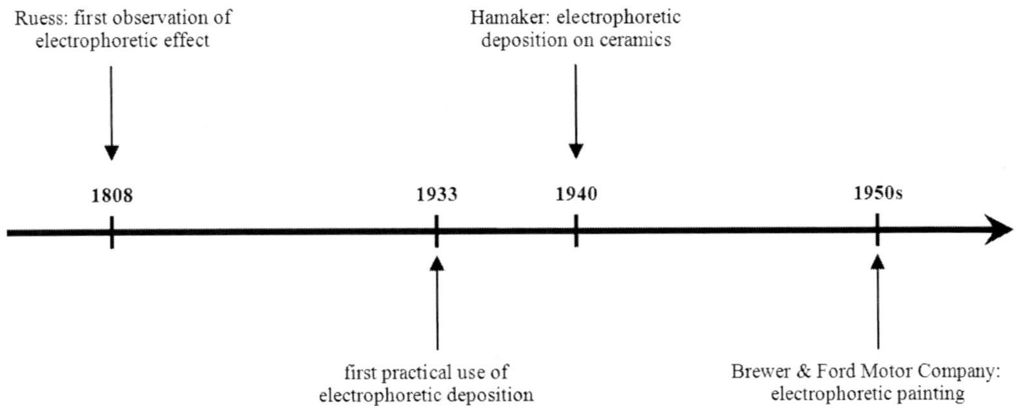

Figure 6. The history of electrophoretic deposition.

EPD can be divided into two types according to the charge of deposited particles and electrode on the surface of which the deposit is formed [60]. When the particles are positively charged, the deposit is formed on the cathode – this process is called cathodic electrophoretic deposition. When the particles are negatively charged, the process of deposition occurs at the surface of anode and is called anodic electrophoretic deposition. These two processes are schematically shown in Figure 7.

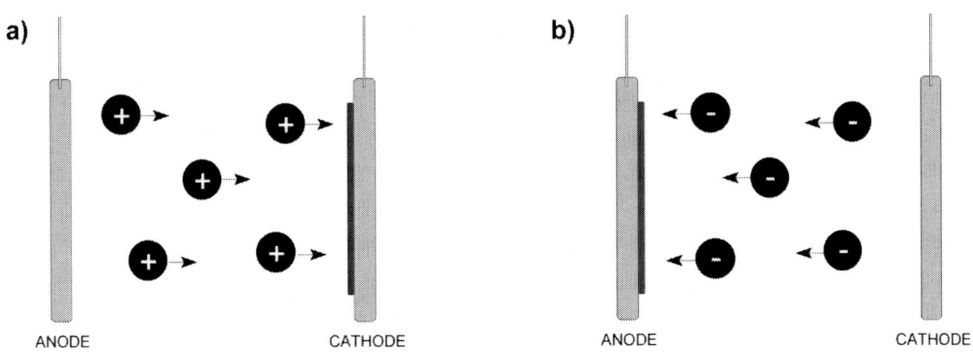

Figure 7. The schematic representation of two types of electrophoretic deposition: a) cathodic EPD, b) anodic EPD.

EPD is a complex process that cannot be described with a single mechanism [61]. The examples of mechanisms that have been so far described in the literature include deposition *via* suspension densification [61], direct action of electrostatic force on individual particles

[62], increase of voltage gradient due to ion depletion at the electrode [63], the effect of salting out [64], etc.

The process of electrophoretic deposition can be performer either in aqueous medium or in organic solvents. The use of water as a solvent is preferred because of low costs, high efficiency and low impact on the environment [65]. The drawback of aqueous EPD is the limitation of DC voltage used in order to avoid water decomposition and gas evolution – this hindrance can be, however, omitted by working with low voltage, addition of a co-solvent, application of pulse potential or AC-electric field [66]. Among organic solvents, the most commonly used are acetone, isobutylmethylketone, propanol and butanol [67]. By using solvents differing in molecular structures, viscosities and dielectric constants it is possible to obtain coatings with various microstructures [68]. For instance, Panigrahi *et al* studied the effect of electrophoretic deposition of doped ceria in various non-aqueous solvents (ethanol, butanol, acetyl acetone) [69]. It was found that the choice of solvent had the effect on deposition yield and the quality of deposit, its morphology and porosity.

EPD is not only a highly versatile method enabling the utilization of any powdered solid able to form a stable suspension, but it is also a cost-effective technique that does not require developed processing equipment nor infrastructure [2]. Deposits obtained by means of EPD are characterized by high microstructural homogeneity as well as highly controlled thickness and, what is more, they can be arranged in advanced nanostructure coatings and films with enhanced properties [7]. EPD is commonly employed in processing of ceramics [70-72], biomedical engineering [73-75] and corrosion protection [76-78].

EPD has been also successfully applied in the fabrication of carbon nanotube (CNT)-containing structures potentially useful as components of field emission devices, fuel cells or supercapacitors [79]. The major advantage of this method is the avoidance of the problem of agglomeration of CNT and the possibility to form uniform composites with high level of homogeneity [80]. EPD was applied to produce ultra-fine (20 nm) hydroxyapatite bioactive coatings reinforced with CNT on the surface of Ti-6Al-4V alloys, with the effect of increased hardness, elastic modulus as well as inter-laminar shear strength [81]. This technique was also used to fabricate MWCNT/hexamethylene diisocyanate composite coatings with excellent microstructural homogeneity for the protection of copper substrate against oxidation and corrosion [82].

A promising alternative to classical EPD is **electrophoretic deposition in magnetic field**. This technique utilizes the anisotropic magnetic susceptibilities exhibited by many materials having asymmetric crystalline structures [83]. Such crystals placed in a magnetic field start rotating and become aligned in the direction determined by the field. This results in the high control of crystalline orientation – the property that has never been controlled in the conventional process of EPD [84]. Highly oriented crystals are materials of particular interest for ceramic industry because they are believed to possess improved mechanical, electrical, piezoelectric as well as optical properties influencing their structural and functional applications [85]. Electrophoretic deposition in magnetic field has been successfully applied in the production of textured monoliths and laminate composites of α-alumina [85], alumina/alumina laminar composites with alternate crystalline oriented layers [86], highly textured TiO_2 thick films [87-88], high temperature superconducting, grain-aligned $YBa_2Cu_3O_{7-x}$ films [84] and *c*-axis oriented zinc oxide in an innovative technique utilizing rotating magnetic field [89].

Although it is direct current (DC) which is the most commonly applied in a process of EPD, **alternating current (AC) electrophoretic deposition** seems to be an interesting alternative due to the considerable advantages that this technique offers, among which the suppression of hydrolysis at high voltages is the most prominent [90-91]. The fundamental concept of AC EPD is the periodically reversed direction of electric field resulting in oscillations and migration of suspended particles between electrodes [92]. The major parameters influencing oscillating migration, and therefore the morphology of deposit, are frequency and asymmetry of the wave applied – the example of one period of asymmetric AC signal is presented in Figure 8.

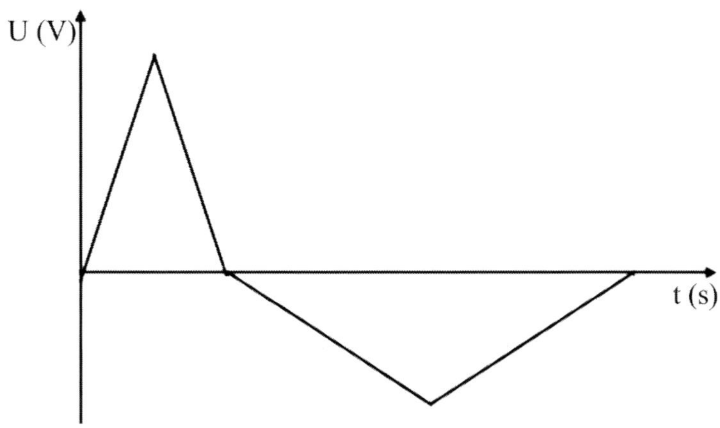

Figure 8. The example of one period of asymmetric AC signal.

Deposits formed by AC EDP technique are usually denser and less cracked than coatings obtained by DC EPD at similar thickness [92]. What is more, the comparison between these two types of EDP showed that AC EDP, because of the possibility to control the migration of particles, is a better choice for depositing powders with broad particle size distribution. This technique has been applied mainly in sensor and biosensor fabrication, e.g., to align and deposit In_2O_3 nanoribbons on the interdigitated electrode surface of NO_2 sensor [8], to deposit glucose oxidase, glutamate oxidase and β-galactosidase on platinum electrodes for glucose, glutamate and lactose biosensors [93-95] and to deposit highly active layers of catalase used for selective and sensitive detection of H_2O_2 [96]. AC EDP can be also used to deposit nanoparticles alternatively on both electrodes, as it was described on the example of SnO_2 nano-layer in the process of low frequency AC electrophoresis [9].

Another EPD alternative applied in order to suppress the evolution of gas bubbles at the surface of the electrodes and thus enabling the formation of homogeneous deposits from aqueous suspensions is **pulsed direct current electrophoretic deposition** [97]. This technique relies on the fact that the electrolysis of water has a slow kinetics that prevents formation of reaction products when short DC pulses are applied [98]. The nucleation of bubbles is further prevented by extending time intervals between subsequent pulses to enable the products or half products of electrolysis to diffuse away from the electrodes. It follows that for each process there exists a narrow window in the plot of applied current vs. pulse width within which it is possible to obtain bubble-free deposits by continuous current pulsed EPD [99]. The working window is apparently wider and lies at a higher value of pulse for lower applied potentials; when the potential is increased, the upper and lower limits of pulse tend to merge with each other [100]. That is why pulsed EPD is usually carried out at lower potential applied, even though it leads to the formation of thinner deposit layers. This method is also found to generate low changes of pH at the surface of electrodes what makes it convenient for the formation of biologically active coatings out of biochemical and biological species [101]. Pulsed DC EPD has turned out to be a convenient and controllable method of zeolite deposition (from submonolayer to multilayers) [102], homogeneous and uniform CNT/Ag/SiO_2 films [103] as well as uniform deposition of single TiO_2 particles [104].

Finally, the combination of both AC EPD and pulsed DC EPD techniques has been described recently in the study of the deposition of Al_2O_3 particles from aqueous suspension [105]. The utilization of unbalanced pulses of alternating current resulted in the reduction of electrolysis of water to a greater extend and formation of bubble-free local Al_2O_3 deposits. The combination of these two methods is believed to be of interest as a cost effecting alternative to current techniques in applications where small local deposition of nanomaterials is crucial, e.g., in the sector of dental biomaterials [106].

PLASMA ELECTROLYTIC DEPOSITION

A novel technique of electrochemical surface modification that has been gaining growing interest in surface engineering is plasma electrolytic deposition (PED), also known as micro-arc electrodeposition or plasma electrolytic oxidation [107]. This technique relies on the appearance of large potential drop across the insulating barrier formed on the surface of anode in a suitable electrolyte; when the potential drop is high enough, the electrical breakdown occurs and plasma process appears on the surface of electrode [108].

The combination of discharge process and electrolysis was discovered by Sluginov (1880) and studied by Günterschultze and Betz (1937), however, it was McNiell and Gruss who for the first time employed PED for practical application (cadmium niobate deposition on a cadmium anode in Nb-containing electrolyte, 1958) [109]. Significant contributions to electrochemical deposition were made by Markov (1976), Snezhko (1980), Kurze (1986), Fyedorov (1988) and Gordienko (1989) [109].

PED is a method allowing the in-situ growth of ceramic coatings of layered structures (porous top layer, dense intermediate layer and thin inner layer) with fine properties: improved corrosion resistance, wear resistance, interfacial adhesion, anti-abrasion and decorative properties on the surface of Al, Ti, Mg, Nb, etc. [110-111].

Through the change of the electrolyte composition and process parameters, such as time of treatment or volt-ampere mode, it is possible to tailor thickness and morphology of obtained coatings [112]. What is more, coatings can be produced quickly and economically on the surfaces of any size and shape [113]. As a low-temperature surface treatment method, PED has no adverse effect on the bulk properties of specimen [114].

PED has been applied to fabricate various nanostructures such as nano-crystalline graphite films, diamond-like carbon films, ceramic coatings on the surface of metals and alloys [115].

Table 1a. Electrochemical strategies to production of nano-coatings: Electrochemical Deposition (ED)

Technique	Electrochemical Deposition (ED)					
	Classical ED	Electrochemical Polymerization	Ultrasound-assisted ED	Laser-assisted ED	Pulse Current ED	Localized ED
Advantages	rapid, low costs, high purity of products, high deposition rates, homogeneous distribution of particles, easy process control, variable composition of coatings, sustainable for complex geometries of substrate, reduces waste material [1] [10-11]	easy control of film thickness, direct grafting of polymer onto the surface of electrode, enables to perform in situ characterization of the growing polymer with the use of electrochemical and/or spectroscopic methods [27]	stimulates chemical and electrochemical reactions [37]	enhances the process of electrodeposition up to 1000 times [39-40], improves morphology, higher adherence and compactness [41]	higher current densities, production of nanocrystalline coatings with more compact structure and finer grains than DC deposits [6]	applicable to many materials without limitation on the type of substrate or thickness of electrodeposit, ambient process conditions, easy tailoring of the properties of coating by modifying electrochemical parameters [51-52]
Typical applications	magnetic thin films [17], Permalloy films [18], anti-microbial coatings [20], precious metal catalysts [21], corrosion protective layers [22]	corrosion protection coatings [28-30], biocompatible coatings for biomedical engineering [31-33], photovoltaics [34], electrochromic devices [35-36]	TiO_2 nanotube array/CdSe nanoparticle/TiO_2 composites [5]	zinc electrodeposits [41]	nanocrystalline nickel coatings [44], nickel hexacyanoferrate/chitosan/carbon nanotubes films [45], bismuth telluride films [46], binary Co–Sn alloys for lithium secondary batteries [47], Ni-W-B composites containing CeO_2 nano-particle [48]	nickel microstructures [54], copper electrical interconnects [55], titania-based microwires [49]

Table 1b. Electrochemical strategies to production of nano-coatings: Electrophoretic Deposition (EPD) and Plasma Electrolytic Deposition (PED)

Technique	Electrophoretic Deposition (EPD)				Plasma Electrolytic Deposition (PED)
	Classical EPD	EPD in Magnetic Field	Alternating Current EPD	Pulsed Direct Current EPD	
Advantages	versatile, enables utilization of any powdered solid able to form a stable suspension, cost-effective, high microstructural homogeneity of composites, highly controlled thickness [2,7]	high control of crystalline orientation [84], improved mechanical, electrical, piezoelectric and optical properties of coatings [85]	suppresses hydrolysis at high voltages [90-91]	suppresses evolution of gas bubbles at the surface of the electrodes [97], wide working window [100], low pH changes at the surface of electrodes [101]	allows in-situ growth of ceramic coatings of layered structures with improved corrosion resistance, wear resistance, interfacial adhesion, anti-abrasion and decorative properties [110-111]
Typical applications	ceramics processing [70-72], biomedical engineering [73-75], corrosion protection [76-78],	alumina/alumina laminar composites [85-86], highly textured TiO_2 films [87-88], high temperature superconducting, grain-aligned $YBa_2Cu_3O_{7-x}$ films [84], *c*-axis oriented zinc oxides [89]	NO_2 sensor [8], glucose, glutamate and lactose biosensors [93-95] selective and sensitive detection of H_2O_2 [96]	zeolite deposition [102], homogeneous and uniform CNT/Ag/SiO_2 films [103], uniform deposition of single TiO_2 particles [104]	biocompatible coatings [114], nano-crystalline graphite films, diamond-like carbon films, ceramic coatings [115]

PED is also successfully applied for the improvement of the corrosion resistance of AZ31 magnesium alloy by the deposition of crack-free TiO_2 layer composed of randomly oriented rutile TiO_2 nanocrystallites (< 10 nm) [116]. The crystalline TiO_2 (rutile) oxide coatings containing Mn_2O_3 and Mn_3O_4 formed in a process of PED are found to have catalytic activity in $CO \rightarrow CO_2$ oxidation reactions dependent on the morphology, namely PED process parameters [117]. It is also possible to synthesize nanostructured titania coatings with a large range of different morphologies (nanorods, nanofibers, nanoparticles) and crystallinities by tailoring applied voltage, titanium initial concentration as well as the time of the process [118]. The possibility of low-temperature treatment is particularly important in case of Nitinol, a nickel-titanium alloy used in a wide range of bio-applications, because of its high sensitivity to heat input and deterioration of properties in high temperatures [114].

That is why PED has been chosen as a method for fabrication of anatase titanium oxide on the surface of Nitinol resulting in rough and porous coating with safer surface for implant applications [119] as well as Al_2O_3 coating improving osteoblast growth behavior [120]. Recently, PED has been combined with ion vapor deposition (IVD) to improve the corrosion properties of Al coatings for airframe components, aircraft landing gear and fasteners made from high strength steel and Ti alloy [121]. This duplex technique turns out to be a cost effective process providing unique improvements in the corrosion resistance of IVD Al alloy coating: more positive corrosion potential, lower corrosion current and increased polarization resistance compared with traditional treatments [122].

Even though plasma electrolytic deposition is an advantageous technique of electrochemical surface modification with wide range of applications, it is mainly used to obtain coatings with thicknesses in the range of several micrometers.

Nevertheless, the major contribution of PED to the production of nano-coatings is the fact that this technique enables formation of nano-structured coatings composed of nano-sized particles, nanocrystallites, nanorods and nanofibers. Consequently, PED may be thought to be an effective tool to tailor the nano-morphology of the surface of coatings.

Conclusion

Electrochemical techniques proved to be advantageous methods in production of nano-coatings. They have been successfully utilized in numerous applications including corrosion protection, biomedical engineering, fuel cells, supercapacitors as well as photovoltaic and electrochromic devices. The continuous research in developing novel strategies alternative to well-known processes enables the precise control of the properties of the obtained coatings making electrochemistry an irreplaceable method for the synthesis of numerous advanced materials.

Acknowledgments

Jerzy K. Zak and Katarzyna Krukiewicz are grateful to the National Science Centre in Poland for financing the research (2011/01/B/ST5/03794). Katarzyna Krukiewicz is greatly indebted for financing the research in the framework of PRELUDIUM (2012/07/N/ST5/01878).

REFERENCES

[1] Gurrappa, I.; Binder, L. Electrodeposition of nanostructured coatings and their characterization – a review. *Sci. Technol. Adv. Mater.* 2008, 9, 043001–043011.

[2] Boccaccini, A.R.; Keim, S.; Ma, R.; L, Y.; Zhitomirsky, I. Electrophoretic deposition of biomaterials. *J. R. Soc. Interface* 2010, 7, S581-613.

[3] Li, H.; Zhao, X.; Cao, S.; Li, K.; Chen, M.; Xu, Z.; Lu, J.; Zhang, L. Na-doped hydroxyapatite coating on carbon/carbon composites: Preparation, in vitro bioactivity and biocompatibility. *Appl. Surf. Sci.* 2012, 263, 163-173.

[4] Thiemig, D.; Cantaragiu, A.M.; Schachschal, S.; Bund, A.; Pich, A.; Carac, G.; Gheorghies, C. Electrocodeposition of hydroxyapatite nanoparticles with zinc–iron alloys. *Surf. Coat. Technol.* 2009, 203, 1488-1493.

[5] Ouyang, J.; Chang, M.; Zhang, Li, X. CdSe-sensitized TiO_2 nanotube array film fabricated by ultrasonic-assisted electrochemical deposition and subsequently wrapped with TiO_2 thin layer for the visible light photoelectrocatalysis. *Thin Solid Films* 2012, 520, 2994-2999.

[6] Lajevardi, S.A.; Shahrabi, T. Effects of pulse electrodeposition parameters on the properties of Ni–TiO_2 nanocomposite coatings. *Appl. Surf. Sci.* 2010, 256, 6775-6781.

[7] Chavez-Valdez, A.; Shaffer, M.S.; Boccaccini, A.R. Applications of Graphene Electrophoretic Deposition. A Review. *J. Phys. Chem. B* 2013, 117, 1502-1515.

[8] Sowti Khiabani, P.; Hosseinmardi, A.; Marzbanrad, E.; Ghashghaie, S.; Zamani, C.; Keyanpour-Rad, M.; Raissi, B. NO_2 gas sensor fabrication through AC electrophoretic deposition from electrospun In_2O_3 nanoribbons. *Sens. Actuators, B* 2012, 162, 102-107.

[9] Gardeshzadeh, A.R.; Raissi, B.; Marzbanrad, E. Electrophoretic deposition of SnO_2 nanoparticles using low frequency AC electric fields. *Mater. Lett.* 2008, 62, 1697-1699.

[10] Lajevardi, S.A.;, Shahrabi, T.; Szpunar, J.A. Synthesis of functionally graded nano Al_2O_3–Ni composite coating by pulse electrodeposition. *Appl. Surf. Sci.* 2013, 279, 180-188.

[11] Narayanan, R.; Seshadri, S.K.; Kwon, T.Y.; Kim, K.H. Electrochemical nano-grained calcium phosphate coatings on Ti–6Al–4V for biomaterial applications. *Scripta Mater.* 2007, 56, 229-232.

[12] Electrochemical deposition. *Encyclopedia of Tribology*; Springer-Verlag: New York, 2013.

[13] Schwarzacher, W. Electrodeposition: a technology for the future. *Electrochem. Soc. Interface* 2000, 15, 32–35.

[14] Paunovic, M.; Schlesinger, M. *Fundamentals of electrochemical deposition*, 2nd ed.; Wiley: New York, 2006.

[15] Narayanan, R.; Kwon, T.Y.; Kim, K.H. Preparation and Characteristics of Nano-Grained Calcium Phosphate Coatings on Titanium From Ultrasonated Bath at Acidic pH. *J. Biomed. Mater. Res. B: Appl. Biomater.* 2008, 85, 231-239.

[16] Cao, G.; Liu, D. Template-based synthesis of nanorod, nanowire, and nanotube arrays. *Adv. Colloid Interface Sci.* 2008, 136, 45-64.

[17] Osaka, T.; Takano, N.; Yokoshima, T. Microfabrication of electro- and electroless-deposition and its application in the electronic field. *Surf. Coat. Technol.* 2003, 169-170, 1-7.

[18] Munford, M.L.; Seligman, L.; Sartorelli, M.L.; Voltolini, E.; Martins, L.F.O.; Schwarzacher, W.; Pasa A.A. Electrodeposition of magnetic thin films of cobalt on silicon. *J. Magn. Magn. Mater.* 2001, 226, 1613-1615.

[19] Park, D.Y.; Yoo, B.Y.; Kelcher, S.; Myung, N.V. Electrodeposition of low-stress high magnetic moment Fe-rich FeCoNi thin films. *Electrochim. Acta* 2006, 51, 2523-2530.

[20] Scholz, J.; Nocke, G.; Hollstein, F.; Weissbach, A. Investigations on fabrics coated with precious metals using the magnetron sputter technique with regard to their anti-microbial properties. *Surf. Coat. Technol.* 2005, 192, 252-256.

[21] Ronkkonen, H.; Simell, P.; Reinikainen, M.; Niemela, M.; Krause, O. Precious metal catalysts in the clean-up of biomass gasification gas Part 1: Monometallic catalysts and their impact on gasification gas composition. *Fuel Process. Technol.* 2011, 92, 1457-1465.

[22] Leyens, C.; Pint, B.A.; Wright, I.G. Effect of composition on the oxidation and hot corrosion resistance of NiAl doped with precious metals. *Surf. Coat. Technol.* 2000, 133-134, 15-22

[23] Rostami, M.; Fahami, A.; Nasiri-Tabrizi, B.; Ebrahimi-Kahrizsangi, R.; Saatchi, A. Characterization of electrodeposited Ni–SiC–Cg nanocomposite coating. *Appl. Surf. Sci.* 2013, 265, 369-374.

[24] Raub, E.; Muller, K. *Fundamentals of Electrodeposition*; Elsevier: The Netherlands, 1967, p. 60.

[25] Heinze, J.; Frontana-Uribe, B.A.; Ludwigs, S. Electrochemistry of Conducting Polymers Persistent Models and New Concepts. Chem. Rev. 2010, 110, 4724–4771.

[26] Guimard, N.K.; Gomez, N.; Schmidt, C.E. Conducting polymers in biomedical engineering. *Prog. Polym. Sci.* 2007, 32, 876-921.

[27] Roncali, J. Conjugated Poiy(thiophenes): Synthesis, Functionalizatlon, and Applications. *Chem. Rev.* 1992, 92, 711-738.

[28] Radhakrishnan, S.; Siju, C.R.; Mahanta, D.; Patil, S.; Madras, G. Conducting polyaniline–nano-TiO_2 composites for smart corrosion resistant coatings. *Electrochim. Acta.* 2009, 54, 1249-1254.

[29] Patil, R.C.; Radhakrishnan, S. Conducting polymer based hybrid nano-composites for enhanced corrosion protective coatings. *Prog. Org. Coat.* 2006, 57, 332-336.

[30] Baldissera, A.F.; Ferreira C.A. Coatings based on electronic conducting polymers for corrosion protection of metals. *Prog. Org. Coat.* 2012, 75, 241-247.

[31] Wadhwa, R.; Lagenaur, C.F.; Cui, X.T. Electrochemically controlled release of dexamethasone from conducting polymer polypyrrole coated electrode. *J. Controlled Release* 2006, 110, 531-541.

[32] Ravichandran, R.; Sundarrajan, S.; Venugopal, J.R.; Mukherjee, S.; Ramakrishna, S. Applications of conducting polymers and their issues in biomedical engineering. *J. R. Soc. Interface* 2010, 7, 559-579.

[33] Svennersten, K.; Larsson, K.C.; Berggren, M.; Richter-Dahlfors, A. Organic bioelectronics in nanomedicine. *Biochim. Biophys. Acta* 2011, 1810, 276–285.

[34] Yan, J.; Sun, C.; Tan, F.; Hu, X.; Chen, P.; Qu, S.; Zhou, S.; Xu, J. Electropolymerized poly(3,4-ethylenedioxythiophene):poly(styrene sulfonate) (PEDOT:PSS) film on ITO glass and its application in photovoltaic device. *Sol. Energy Mater. Sol. Cells* 2010, 94, 390-394.

[35] Abaci, U.; Guney, H.Y.; Kadiroglu, U. Morphological and electrochemical properties of PPy, PAni bilayer films and enhanced stability of their electrochromic devices (PPy/PAni–PEDOT, PAni/PPy–PEDOT). *Electrochim. Acta* 2013, 96, 214-224.

[36] Ma, L.J.; Li, Y.X.; Yu, X.F.; Yang, Q.B.; Noh, C.H. Using room temperature ionic liquid to fabricate PEDOT/TiO_2 nanocomposite electrode-based electrochromic devices with enhanced long-term stability. *Sol. Energy Mater. Sol. Cells* 2008, 92, 1253-1259.

[37] Kang, J.; Shin, Y.; Tak, Y. Growth of etch pits formed during sonoelectrochemical etching of aluminum. *Electrochim. Acta* 2005, 51, 1012-1016.

[38] Zhao, X.; Li, H.; Chen, M.; Li, K.; Wang, B.; Xu, Z.; Cao, S.; Zhang, L.; Deng, H.; Lu, J. Strong-bonding calcium phosphate coatings on carbon/carbon composites by ultrasound-assisted anodic oxidation treatment and electrochemical deposition. *Appl. Surf. Sci.* 2012, 258, 5117-5125.

[39] Gutfeld, R.J.; Tynan, E.E.; Melcher, R.L.; Blum, S.E. Laser enhanced electroplating and maskless pattern generation. *Appl. Phys. Lett.* 1979, 35, 651-653.

[40] Zouari, I.; Lapicque, F.; Calvo, M.; Cabrera, M. Laser assisted metal electrodeposition: comprehensive investigation of zinc deposition. *Chem. Eng. Sci.* 1990, 45, 2467-2474.

[41] Zouari, I.; Calvo, M.; Lapicque, F.; Cabrera, M. Influence of laser assistance on the nucleation and growth processes. *Appl. Surf. Sci.* 1992, 54, 311-316.

[42] Adelkhani, H.; Arshadi, M.R. Properties of Fe–Ni–Cr alloy coatings by using direct and pulse current electrodeposition. *J. Alloys Compd.* 2009, 476, 234-237.

[43] Ghaemi, M.; Binder, L. Effects of direct and pulse current on electrodeposition of manganese oxide. *J. Power Sources* 2002, 111, 248-254.

[44] Tian, Z.J.; Wang, D.S.; Wang, G.F.; Shen, L.D.; Liu, Z.D.; Huang, Y.H. Microstructure and properties of nanocrystalline nickel coatings prepared by pulse jet electrodeposition. *T. Nonferr. Metal Soc.* 2010, 20, 1037-1042.

[45] Wang, Z.; Hao, X.; Zhang, Z.; Liu, S.; Liang, Z.; Guan, G. One-step unipolar pulse electrodeposition of nickel hexacyanoferrate/chitosan /carbon nanotubes film and its application in hydrogen peroxide sensor. *Sens. Actuators B* 2012, 162, 353-360.

[46] Richoux, V.; Diliberto, S.; Boulanger, C.; Lecuire, J.M. Pulsed electrodeposition of bismuth telluride films: Influence of pulse parameters over nucleation and morphology. *Electrochim. Acta* 2007, 52, 3053-3060.

[47] Ui, K.; Kikuchi, S.; Jimba, Y.; Kumagai, N. Preparation of Co–Sn alloy film as negative electrode for lithium secondary batteries by pulse electrodeposition method. *J. Power Sources* 2011, 196, 3916-3920.

[48] Wang, J.; Xu, R.; Zhang, Y. Study on characteristics of Ni-W-B composites containing CeO_2nano-particles prepared by pulse electrodeposition. *J. Rare Earth.* 2012, 30, 43-47.

[49] Reveron, H.; Elissalde, C.; Lecoutre, C.; Garrabos, Y. Synthesis of titania-based microwires by localized electrolytic deposition using a scanning electrochemical microscope (SECM). *Mater. Lett.* 2008, 62, 1067-1070.

[50] Madden, J.D. Three-dimensional microfabrication by localized electrochemical deposition. *J. Microelectromech. Syst.* 1996, 5, 24-32.

[51] Pane, S.; Panagiotopoulou, V.; Fusco, S.; Pellicer, E.; Sort, J.; Mochnacki, D.; Sivaraman, K.M.; Kratochvil, B.E.; Baro, M.D.; Nelson B.J. The effect of saccharine on the localized electrochemical deposition of Cu-rich Cu–Ni microcolumns. *Electrochem. Commun* 2011, 13, 973-976.

[52] Said, R.A. Microfabrication by localized electrochemical deposition: experimental investigation and theoretical modeling. *Nanotechnology* 2003, 14, 523-531.

[53] Lin, J.C.; Chang, T.K.; Yang, J.H.; Chen, Y.S., Chuang, C.L. Localized electrochemical deposition of micrometer copper columns by pulse plating. *Electrochim. Acta* 2010, 55, 1888-1894.

[54] Jansson, A.; Thornell, G.; Johansson, S. High resolution 3D microstructures made by localized electrodeposition of nickel. *J. Electrochem. Soc.* 2000, 147, 1810-1817.

[55] El-Giar, E.M.; Thomson, D.J. Localized electrochemical plating of interconnectors for microelectronics. WESCANEX 97: Communications, Power and Computing. Conference Proceedings., IEEE, Winnipeg, Man., May 22-23 1997, 327-332.

[56] Van der Biest, O.O.; Vandeperre, L.J. Electrophoretic deposition of materials. *Annu. Rev. Mater. Sci.* 1999, 29, 327–352.

[57] Besra, L.; Liu, M. A review on fundamentals and applications of electrophoretic deposition (EPD). *Prog. Mater. Sci.* 2007, 52, 1-61.

[58] Hamaker, H.C. Formation of deposition by electrophoresis. *Trans. Farad. Soc.* 1940, 36, 279–83.

[59] Brewer, G.E.F. Electrophoretic painting. *J.Appl. Electrochem.* 1983, 13, 269-275.

[60] Hu, R.G.; Zhang, S., Bu, J.F.; Lin, C.J.; Song, G.L. Recent progress in corrosion protection of magnesium alloys by organic coatings. *Prog. Org. Coat.* 2012, 73, 129-141.

[61] Van Tassel, J.J.; Randall, C.A. Mechanisms of Electrophoretic Deposition. *Key Eng. Mater.* 2006, 314, 167-174.

[62] Van Tassel, J.; Randall, C.A. Potential for integration of electrophoretic deposition into electronic device manufacture; demonstrations using silver/palladium. *J. Mater. Sci.* 2004, 39, 867-879.

[63] Sarkar, P.; Huang, X.; Nicholson, P.S. Electrophoretic Deposition and Its use to Synthesize YSZ/A1203 Micro'Laminate Ceramic/Ceramic Composites. *Ceram. Eng. Sci. Proc.* 1993, 14, 707-726.

[64] Koelmans, H.; Overbeek, J.T.G. Stability and Electrophoretic Deposition of Suspensions in Non-Aqueous Media. *Discuss. Faraday Soc.* 1954, 18, 52-63.

[65] Zhao, J,; Wang, X.; Li, L. Electrophoretic deposition of $BaTiO_3$ films from aqueous suspensions. *Mater. Chem. Phys.* 2006, 99, 350-353.

[66] Benehkohal, N.P.; Gomez, M.A.; Gauvin, R.; Demopoulos, G.P. Enabling aqueous electrophoretic growth of adherent nanotitania mesoporous films via intrafilm cathodic deposition of hydrous zinc oxide. *Electrochim. Acta* 2013, 87, 169-179.

[67] Dusoulier, L.; Cloots, R.; Vertruyen, B.; Moreno, R.; Burgos-Montes, O.; Ferrari, B. $YBa_2Cu_3O_{7-x}$ dispersion in iodine acetone for electrophoretic deposition: Surface charging mechanism in a halogenated organic media. *J. Eur. Ceram. Soc.* 2011, 31, 1075-1086.

[68] Sadeghi, A.A.; Ebadzadeh, T.; Raissi, B.; Ghashghaie, S. Electrophoretic deposition of TiO_2 nanoparticles in viscous alcoholic media. *Ceram. Int.* 2013, 39, 7433-7438.

[69] Panigrahi, S.; Bhattacharjee, S.; Besra, L.; Singh, B.P.; Sinha, S.P. Electrophoretic deposition of doped ceria: Effect of solvents on deposition microstructure. *J. Eur. Ceram. Soc.* 2010, 30, 1097-1103.

[70] Kaya, C.; Kaya, F.; Su, B.; Thomas, B.; Boccaccini, A.R. Structural and functional thick ceramic coatings by electrophoretic deposition. *Surf. Coat. Technol.* 2005, 191, 303-310.

[71] Boccaccini, A.R.; Schindler, U.; Kruger, H.G. Ceramic coatings on carbon and metallic fibres by electrophoretic deposition. *Mater. Lett.* 2001, 51, 225-230.

[72] Wang, Z.; Shemilt, J; Xiao, P. Fabrication of ceramic composite coatings using electrophoretic deposition, reaction bonding and low temperature sintering. *J. Eur. Ceram. Soc.* 2002, 22, 183-189.

[73] Zhatomirsky, D.; Roether, J.A.; Boccaccini, A.R.; Zhitomirsky, I. Electrophoretic deposition of bioactive glass/polymer composite coatings with and without HA nanoparticle inclusions for biomedical applications. *J. Mater. Process. Technol.* 2009, 209, 1853-1860.

[74] Dondapati, S.K.; Lozano-Sanchez, P.; Katakis, I. Controlled electrophoretic deposition of multifunctional nanomodules for bioelectrochemical applications. *Biosens. Bioelectron.* 2008, 24, 55-59.

[75] Mohan, L.; Durgalakshmi, D.; Geetha, M.; Sankara Narayanan, T.S.N.; Asokamani, R. Electrophoretic deposition of nanocomposite (HAp + TiO_2) on titanium alloy for biomedical applications. *Ceram. Int.* 2012, 38, 3435-3443.

[76] Zhang, J.; Wu, C. Corrosion protection behavior of AZ31 magnesium alloy with cathodic electrophoretic coating pretreated by silane. *Prog. Org. Coat.* 2009, 66, 387-392.

[77] He, W.; Zhu, L.; Chen, H.; Nan, H.; Li, W.; Liu, H.; Wang, Y. Electrophoretic deposition of graphene oxide as a corrosion inhibitor for sintered NdFeB. *Appl. Surf. Sci.* 2013, 279, 416-423.

[78] Wu, L.K.; Zhang, J.T.; Hu, J.M.; Zhang, J.Q. Improved corrosion performance of electrophoretic coatings by silane addition. *Corros. Sci.* 2012, 56, 58-66.

[79] Boccaccini, A.R.; Cho, J.; Roether, J.A.; Thomas, B.J.C.; Minay, E.J.; Shaffer, M.S.P. Electrophoretic deposition of carbon nanotubes. *Carbon* 2006, 44, 3149-3160.

[80] Schausten, M.C.; Meng, D.; Telle, R.; Boccaccini, A.R. Electrophoretic deposition of carbon nanotubes and bioactive glass particles for bioactive composite coatings. *Ceram. Int.* 2010, 36, 307-312.

[81] Kaya, C. Electrophoretic deposition of carbon nanotube-reinforced hydroxyapatite bioactive layers on Ti–6Al–4V alloys for biomedical applications. *Ceram. Int.* 2008, 34, 1843-1847.

[82] Castro, R.H.R.; Hidalgo, P.; Diniz, E.C. Enhanced electrical conduction in aluminum wires coated with carbon nanotubes. *Mater. Lett.* 2011, 65, 271-274.

[83] Uchikoshi, T.; Suzuki, T.H.; Iimura, S.; Tang, F.; Sakka, Y. Control of crystalline texture in polycrystalline TiO_2 (Anatase) by electrophoretic deposition in a strong magnetic field. *J. Eur. Ceram. Soc.* 2006, 26, 559-563.

[84] Kawachi, M.; Sato, N.; Noto, K.; Yoshizawa, M. Fabrication of grain-aligned YBCO films using electrophoretic deposition in a magnetic field. Phys. C 2002, *Physica C* 2002, 372-376, 802-805.

[85] Uchikoshi,T.; Suzuki, T.H.; Okuyama, H.; Sakka, Y.; Nicholson, P.S. Electrophoretic deposition of alumina suspension in a strong magnetic field. *J. Eur. Ceram. Soc.* 2004, 24, 225-229.

[86] Suzuki, T.H.; Uchikoshi, T.; Okuyama, H.; Sakka, Y.; Hiraga K. Mechanical properties of textured, multilayered alumina produced using electrophoretic deposition in a strong magnetic field. *J. Eur. Ceram. Soc.* 2006, 26, 661-665.

[87] Tang, F.; Uchikoshi, T.; Suzuki, T.H.; Sakka, Y. Alignment of TiO_2 particles by electrophoretic deposition in a high magnetic field. *Mater. Ress. Bull.* 2004, 39, 2155-2161.

[88] Uchikoshi, T.; Suzuki, T.S.; Tang, F.; Okuyama, H.; Sakka, Y. Crystalline-oriented TiO_2 fabricated by the electrophoretic deposition in a strong magnetic field. *Ceram. Int.* 2004, 30, 1975-1978.

[89] Uchikoshi, T.; Suzuki, T.H.; Sakka, Y. Fabrication of c-axis oriented zinc oxide by electrophoretic deposition in a rotating magnetic field. *J. Eur. Ceram. Soc.* 2010, 30, 1171-1175.

[90] Yoshioka, T.; Chavez-Valdez, A.; Roether, J.A.; Schubert, D.W.; Boccaccini, A.R. AC electrophoretic deposition of organic–inorganic composite coatings. *J. Colloid Interface Sci.* 2013, 392, 167-171.

[91] Neirinck, B.; Fransaer, J.; Van ser Biest, O.; Vleugels, A. Aqueous electrophoretic deposition in asymmetric AC electric fields (AC–EPD). *Electrochem. Commun.* 2009, 11, 57-60.

[92] Ozhukil Kollath, V.; Chen, Q.; Closset, R.; Luyten, J.; Traina, K.; Mullens, S.; Boccaccini, A.R.; Cloots, R. AC vs. DC electrophoretic deposition of hydroxyapatite on titanium. *J. Eur. Ceram. Soc.* 2013, 33, 2715-2721.

[93] Ammam, M.; Fransaer, J. Glucose microbiosensor based on glucose oxidase immobilized by AC-EPD: characteristics and performance in human serum and in blood of critically ill rabbits. *Sens. Actuators, B* 2010, 145, 46-53.

[94] Ammam, M.; Fransaer, J. Highly sensitive and selective glutamate microbiosensor based on cast polyurethane/AC-electrophoresis deposited multiwalled carbon nanotubes and then glutamate oxidase/electro-synthesized polypyrrole/Pt electrode. *Biosens. Bioelectron.* 2010, 25, 1597-1602.

[95] Ammam, M.; Fransaer, J. Two-enzyme lactose biosensor based on β-galactosidase and glucose oxidase deposited by AC-electrophoresis: characteristics and performance for lactose determination in milk. *Sens. Actuators, B* 2010, 148, 583-589.

[96] Ammam, M.; Fransaer, J. AC-electrophoretic deposition of metalloenzymes: Catalase as a case study for the sensitive and selective detection of H_2O_2. *Sens. Actuators, B* 2011, 160, 1063-1069.

[97] Besra, L.; Uchikoshi, T.; Suzuki, T.S.; Sakka, Y. Experimental verification of pH localization mechanism of particle consolidation at the electrode/solution interface and its application to pulsed DC electrophoretic deposition (EPD). *J. Eur. Ceram. Soc.* 2010, 30, 1187-1193.

[98] Neirinck, B.; Van der Biest, O.; Vleugels, J. A Current Opinion on Electrophoretic Deposition in Pulsed and Alternating Fields. *J. Phys. Chem. B* 2013, 117, 1516-1526.

[99] Besra, L.; Uchikoshi, T.; Suzuki, T.S.; Sakka, Y. Application of constant current pulse to suppress bubble incorporation and control deposit morphology during aqueous electrophoretic deposition (EPD). *J. Eur. Ceram. Soc.* 2009, 29, 1837-1845.

[100] Besra, L.; Uchikoshi, T.; Suzuki, T.S.; Sakka, Y. Bubble-Free Aqueous Electrophoretic Deposition (EPD) by Pulse-Potential Application. *J. Am. Ceram. Soc.* 2008, 91, 3154-3159.

[101] Amman, M. Electrophoretic deposition under modulated electric fields: a review. *RSC Adv.* 2012, 2, 7633-7646.

[102] Yu, B.; Khoo, S.B. Controllable zeolite films on electrodes – comparing dc voltage electrophoretic deposition and a novel pulsed voltage method. Electrochem. Commun. 2002, 4, 737-742.

[103] Wang, S.C.; Huang, B.C. Field emission properties of Ag/SiO_2/carbon nanotube films by pulsed voltage co-electrophoretic deposition. *Thin Solid Films* 2008, 517, 1245-1250.

[104] Nazli Naim, M.; Iijima, M.; Kamiya, H.; Wuled Lenggoro, I. Electrophoretic packing structure from aqueous nanoparticle suspension in pulse DC charging. *Colloids Surf. A* 2010, 360, 13-19.

[105] Nold, A.; Clasen, R. Bubble-free electrophoretic shaping from aqueous suspension with micro point-electrode. *J. Eur. Ceram. Soc.* 2010, 30, 2971-2975.

[106] Chavez-Valdes, A.; Boccaccini, A.R. Innovations in electrophoretic deposition: Alternating current and pulsed direct current methods. *Electrochim. Acta* 2012, 65, 70-89.

[107] Wang. Y.; Jiang, Z.; Liu, X.; Yao, Z. Influence of treating frequency on microstructure and properties of Al_2O_3 coating on 304 stainless steel by cathodic plasma electrolytic deposition. *Appl. Surf.* 2009, 255, 8836-8840.

[108] Wirtz, G.P.; Brown, S.D.; Kriven, W.M. Ceramic coatings by anodic spark deposition. *Mater. Manuf. Processes* 1991, 6, 87-115.

[109] Yerokhin, A.L.; Nie, X.; Leyland, A.; Matthews, A.; Dowey, S.J. Plasma electrolysis for surface engineering. *Surf. Coat. Technol.* 1999, 122, 73-93.

[110] Yao, Z.; Jiang, Y.; Jiang, Z.; Wang, F.; Wu, Z. Preparation and structure of ceramic coatings containing zirconium oxide on Ti alloy by plasma electrolytic oxidation. *J. Mater. Proc. Technol.* 2008, 205, 303-307.

[111] Gu, W.; Shen, D.; Wang, Y.; Chen, G.; Feng, W.; Zhang, G.; Fan, S.; Liu, C.; Yang, S. Deposition of duplex Al_2O_3/aluminum coatings on steel using a combined technique of arc spraying and plasma electrolytic oxidation. *Appl. Surf. Sci.* 2006, 252, 2927-2932.

[112] Sandhyarani, M.; Rameshbabu, N.; Venkateswarlu, K.; Sreekanth, D.; Subrahmanyam, C. Surface morphology, corrosion resistance and in vitro bioactivity of P containing ZrO_2 films formed on Zr by plasma electrolytic oxidation. *J. Alloys Compd.* 2013, 553, 324-332.

[113] Curran, A.; Clyne, T.W. Thermo-physical properties of plasma electrolytic oxide coatingson aluminium. *Surf. Coat. Technol.* 2005, 199, 168-172.

[114] Yao, Z.; Jiang, Y.; Jia, F.; Jiang, Z.; Wang, F. Growth characteristics of plasma electrolytic oxidation ceramic coatings on Ti–6Al–4V alloy. *Appl. Surf. Sci.* 2008, 254, 4084-4091.

[115] Liu, P.; Pan, X.; Yang, W.; Cai, K.; Chen, Y. Al_2O_3–ZrO_2 ceramic coatings fabricated on WE43 magnesium alloy by cathodic plasma electrolytic deposition. *Mater. Lett.* 2012, 70, 16-18.

[116] Ryu, H.S.; Rark, D.S.; Hong, S.H. Improved corrosion protection of AZ31 magnesium alloy through plasma electrolytic oxidation and aerosol deposition duplex treatment. *Surf. Coat. Technol.* 2013, 219, 82-87.

[117] Rudnev, V.S.; Vasilyeva, M.S.; Kondrikov, N.B.; Tyrina, L.M. Plasma-electrolytic formation, composition and catalytic activity of manganese oxide containing structures on titanium. *Appl. Surf. Sci.* 2005, 252, 1211-1220.

[118] Paulmier, T.; Bell, J.M.; Fredericks, P.M. Plasma electrolytic deposition of titanium dioxide nanorods and nano-particles. *J. Mater. Process. Technol.* 2008, 208, 117-123.

[119] Siu, H.T.; Man, H.C. Fabrication of bioactive titania coating on nitinol by plasma electrolytic oxidation. *Appl. Surf. Sci.* 2013, 274, 181-187.

[120] Yan, Y.; Cai, K.; Yang, W.; Liu, P. Surface Modification of NiTi Alloy via Cathodic Plasma Electrolytic Deposition and its Effect on Ni Ion Release and Osteoblast Behaviors. *Plasma Sci. Technol.* 2013, 15, 648-654.

[121] Yu, Q.S.; Deffeyes, J.E.; Yasuda, H.K. Corrosion protection of ion vapor deposition (IVD) Al-coated Al alloys by low-temperature plasma interface engineering: Part I. DC cathodic polymerization with anode magnetron enhancement. *Prog. Org. Coat.* 2001, 42, 100-109.

[122] Alanazi, N.M.; Leyland, A.; Yerokhin, A.L.; Matthews, A. Substitution of hexavalent chromate conversion treatment with a plasma electrolytic oxidation process to improve the corrosion properties of ion vapour deposited AlMg coatings. *Surf. Coat. Technol.* 2010, 205, 1750-1756.

In: Comprehensive Guide for Nanocoatings Technology. Vol. 1 ISBN: 978-1-63482-447-7
Editor: Mahmood Aliofkhazraei

Chapter 13

SYNTHESIS OF GRAPHENE BY PLASMA CHEMICAL VAPOR DEPOSITION

Golap Kalita[1,2,*], Koichi Wakita[3], Masayoshi Umeno[3] and Masaki Tanemura[2]

[1]Center for Fostering Young and Innovative Researchers, Nagoya Institute of Technology, Gokiso-cho, Nagoya, Japan

[2]Department of Frontier Materials, Nagoya Institute of Technology, Gokiso-cho, Nagoya, Japan

[3]Department of Electronics and Information Engineering, Chubu University, Kasugai, Japan

ABSTRACT

The discovery of graphene, a two-dimensional (2D) sp^2 hybridized carbon atoms arranged in a honeycomb structure brought tremendous attention as next generation material for various applications. Recent significant development on synthesis of large-area continuous graphene by the chemical vapor deposition (CVD) process opens up new possibilities of application. The CVD process can be a reliable technique for control synthesis of high quality and large-area graphene on metal substrates. Among CVD technique, plasma CVD has evolved as a novel technique for deposition of graphene film at a comparatively lower temperature, than that of thermal CVD process. A low-temperature deposition technique can be more feasible for industrial application and can pave the way for direct deposition on any arbitrary substrates. In a plasma CVD process, carbon radicals are formed with hydrogen termination from a hydrocarbon, which absorb continuously on the substrate surface to form sp^2 carbon and eventually graphene structure. Recent development of the plasma CVD process for synthesis of graphene based materials is elaborately discussed.

* Corresponding author: Golap Kalita. Center for Fostering Young and Innovative Researchers, Nagoya Institute of Technology, Gokiso-cho, Nagoya 466-8555, Japan. Department of Frontier Materials, Nagoya Institute of Technology, Gokiso-cho, Nagoya 466-8555, Japan. E-mail: kalita.golap@nitech.ac.jp, phone/fax: +81-527355216.

1. INTRODUCTION

Graphene is a perfect two-dimensional (2D) sp^2 hybridized carbon sheet arranged in a honeycomb lattice configuration. It is the main building block of many other carbon allotropes such as, graphite, fullerene, and carbon nanotube. The first isolation of graphene was demonstrated by Andre Geim and Konstantin Novoselov in 2004 [1], since then significant importance has been given for graphene related materials owing to fascinating electrical, mechanical, thermal and chemical properties [2-8].

Although, graphene has been isolated only lately, the other related carbon allotropes have been discovered previously or existed in nature. Graphite is one of the *naturally occurring* carbon allotrope can be considered as a three dimensional (3D) stack of graphene layers held together by van der Waals interactions. Similarly, Fullerene (C_{60}), a rolled sheet of graphene to from a sphere like structure with zero dimension (0D) has been discover in 1986 by Kroto et al. [9].

Again, the one dimensional (1D) carbon nanotube (CNT) is another exciting form of carbon, which brought significant research interests to develop all kinds of nanostructure materials in the last two decades [10, 11].

Basically, a carbon nanotube can be considered as rolled sheet of graphene forming 1D tubular structure. The discovery and development of various sp^2 hybridized carbon materials is summarized in the following table.

2. PROPERTIES OF GRAPHENE

Graphene is a semiconductor material with zero band gap, where the electrons can only move between carbon atoms in the 2D crystal lattice.

The energy band structure of graphene involves π electrons of sp^2 hybridized carbon atoms [18]. The quantum confinement of the electrons in the 2D system introduces various novel properties in graphene. At low temperatures, electrons in graphene behave as massless relativistic fermions, a property unusual for a condensed-matter system [19]. Various properties of graphene are summarized in the following table to have a better understanding.

Table 1. Development of graphitic or sp2 hybridized carbon related materials

Years	Carbon related materials	References
1789	Graphite	Named by Abraham Gottlob Werner (Wikipedia) [9]
1958	Graphene oxide	Hummers et al. [10]
1963	Glassy carbon	Lewis et al. [11]
1986	Buckyball or fullerene	Kroto et al. [12]
1991	Carbon nanotubes	Iijima et al. [13]
2004	Graphene sheet	Novoselov et al. [1]
2009	Graphene nanoribbons	Kosynkin et al. [14]; Fujita et al. [15]
2009	Graphene quantum dot	Yang et al. [16]
2013	Grossly warped 'nanographene'	Kawasumi et al. [17]

3. DERIVING OR SYNTHESIS PROCESS OF GRAPHENE

Synthesis of graphene has been achieved by various physical and chemical synthesis approaches. Among these, micromechanical cleavage of highly ordered pyrolytic graphite (HOPG) [25, 26], reduction of chemically exfoliated graphene oxide (GO) [27-30], epitaxial growth on silicon carbide [31] and chemical vapor deposition (CVD) [32-41] are significantly explored. Mechanical exfoliation also known as scotch tape method has been widely studied to investigate the properties of graphene. In this approach, graphene is derived by placing a sticky common adhesive tape on bulk pyrolytic graphite. The tape is pressed on a desired substrate and subsequently peeled away.

Repeating the peel off process graphene flakes can be obtained on the desired substrate, along with some graphitic parts and residues of the adhesive tape. However, the mechanical exfoliation process depends on trial and error approach with difficulties to locate an appropriate flake consistently.

Again, the exfoliated graphene flakes are of the order of only few microns scattered randomly on the substrate, which make it difficult to identify and hinder device applications.

Reduction of chemically exfoliated GO is another approach, which generated lot of interest for room temperature bulk synthesis of graphene flakes. GO has been synthesized by a modified Hummer method that involves rigorous oxidation of pure graphitic materials [10, 42]. GO is an insulating graphene sheet with oxygen containing functional groups on the basal plane and at the edges [43, 44]. In a GO sheet both the conducting π state from sp^2 carbon and the σ state of sp^3 bonded carbon with oxygen can be available. Hence, the optical and electrical properties of GO can be controlled by optimizing sp^2 and sp^3 hybridized carbon ratio with oxidation and reduction process [45]. However, high temperature annealing and chemical treatments of GO are unavoidable to obtain conducting graphene sheet.

Significant structural defects remain in the graphene sheet after the reduction process of GO, which considerably effect the mechanical and electrical properties.

Table 2. Properties of graphene 2D crystals

Properties	Graphene	References
Electrical	1. Band structure for Graphite	By P.R. Wallace in 1947 [18]
	2. Carrier mobility~200,000 $cm^2\ V^{-1}s^{-1}$	Morozov et al. [20]
	3. Unusual (relativistic) quantum Hall effect	Zhang et al. [2]
	4. Excellent conductor (Can sustain 6 orders higher current than Cu)	Geim et al. [21]
Optical	1. Graphene absorb 2.3±0.1% of light (From Fine Structure Constant), Independent of wavelength	Nair et al. [5]
	2. Saturated absorber	Bao et al. [22]
	3. Photoluminescence of graphene sheet	Lui et al. [23]
Thermal	Thermal conductivity~4.84×10^3 to 5.3×10^3 W/mK	Balandin et al. [8]
Mechanical	Breaking strength of ~40 N/m and Young's modulus of ~1.0 TPa	Lee et al. [4]
Chemical	1. Functionalization with various functional groups	Loh et al. [45]
	2. Impermeable atomic membrane	Bunch et al. [24]

Similarly, graphene has been synthesized on silicon carbide surface by high temperature (~1500°C) thermal decomposition approach. The graphitization of hexagonal SiC crystal with annealing at a high temperate was first reported by Badami in 1961 [46]. In the annealing process, top layer of SiC undergoes thermal decomposition, where Si atoms desorb and carbon atoms remain on the surface rearranging and bonding to form graphene structure [30, 47]. The main drawback of graphene growth on SiC is small domain structures with presence of steps and terrace edges and difficult transfer process [48-50].

As discussed above there are several routes to obtain monolayer or few-layer graphene structures, however all these processes have their own limitations. The most promising and simple approach to synthesis large-area continuous graphene is CVD technique on various metal substrates. The better understanding and scalability of CVD process to synthesis high-quality and single crystal graphene synthesis on metal substrates opens up new possibilities for applications [35-40, 51-53]. Presently, CVD technique for graphene synthesis can be divided in two different parts as thermal and plasma CVD process.

In the following table we summarized the different synthesis or deriving process of graphene. In this chapter, we will emphasis and discuss about the plasma CVD approach to synthesis graphene at a comparatively lower temperature than that of thermal CVD.

4. Graphene Synthesis on Transition Metal by Plasma CVD

Recent progress in thermal CVD technique makes it possible to synthesize large-area continuous graphene film, however the trends of CVD process involve very high growth temperature, typically around 1000ºC. A low-temperature deposition technique can be more practicable for industrial development and can pave the way for direct deposition on transparent flexible substrates. The limitations of a thermal CVD process are inefficient decomposition and dehydrogenation of hydrocarbon gaseous source at a low temperature (<600ºC) for graphene growth. This is due to insufficient energy at low temperature for hydrogen detachment from the hydrocarbon to create active carbon atoms for graphene growth [54]. In this contrast, carbon radical for graphene growth can produced at a significantly low temperature with decomposition of hydrocarbon source gas in a plasma CVD.

4.1. Microwave Plasma CVD Process

Microwave assisted plasma CVD technique has become a well establish technique for low temperature graphene synthesis on transition metal surface.

The technique has special advantage as it can easily produce meter scale uniform plasma, which can be suitable for large area synthesis of graphene film.

Again, microwave assisted surface wave plasma (SW-MWP) is one of the most prominent technique with highly stable large-area plasma. The technique has been also used for deposition of amorphous carbon films and diamond coating on various substrates at near room temperature [55-58].

Table 3. Development of graphene synthesis process

Approach	Graphene deriving/synthesis	References
Mechanical exfoliation	1. Scotch tape method (Adhesive tape)	Novoselov et al. [25]
	2. Wedge based mechanical exfoliation	Jayasena et al. [26]
Chemical exfoliation	1. Reduction of graphite oxide obtained by Hummers' method	Stankovich et al. [27]
	2. Exfoliation of intercalated graphite	Su et al. [28]
	3. Electrochemical exfoliation	Xia et al. [29]
	4. Sonication assisted liquid-phase exfoliation	Ciesielski et al. [30]
Epitaxial growth on silicon carbide (SiC)	Graphene growth by Heating SiC at high temperature	Berger et al. [31] Badami [46]
Chemical vapor deposition (CVD)	1. Thermal CVD approach (Transition metal)	Somani et al. [32]; Kim et al. [36], Li et al. [37]
	2. Thermal CVD approach (Non-transition metal)	Lee et al. [39]
	3. Plasma CVD on metal substrate	Kim et al. [66]
	4. Nanographene by plasma CVD	Kalita et al. [75]
Solid phase reaction	Graphene growth by graphitization of solid materials	Hirano et al. [65]

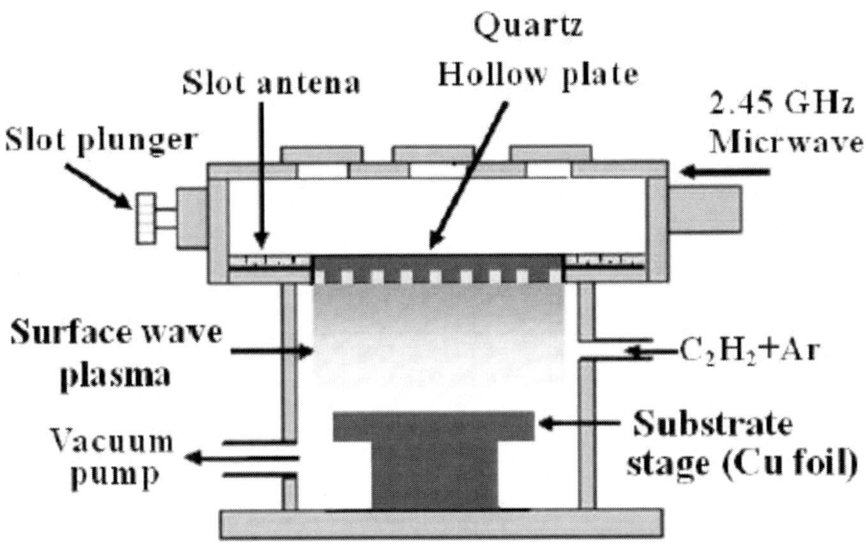

Reference [61] with permission from The Royal Society of Chemistry.

Figure 1. Schematic diagram of a MW-SWP CVD system used for low temperature synthesis of graphene films on transition metal foil.

Figure 1 presents a schematic diagram of MW-SWP CVD system used for graphene synthesis. Dense plasma can be produced even at low gas pressure, with a density higher than the cut-off density (7.4×10^{16} m^{-3} for 2.45 GHz). The plasma can be enlarged by employing a large dielectric window and multi-slot antenna without any magnetic field [59-61].

The MW-SWP has very low electron temperature and low plasma space potentials. Carbon radicals can form with hydrogen termination in the plasma, which absorb onto metal substrate by diffusion process to form sp^2 carbon, thereby achieving graphene growth at much lower substrate temperature [61].

Synthesis of graphene by a CVD process has been studied on various transition metals such as ruthenium (Ru) iridium (Ir), cobalt (Co), nickel (Ni), copper (Cu) and germanium (Ge) [32-40, 62, 63-65]. Among these metals, Cu and Ni have been significantly investigated for large area graphene growth in the thermal and plasma CVD growth process [36, 37]. Synthesis of graphene film on metal foils has significant importance in a plasma CVD process for large-area roll-to-roll production. In the following, we discuss in details about graphene growth on Cu and Ni foil using microw--ave assisted plasma process. Kim et al. has demonstrated synthesis of graphene in a SWP CVD process at a low-pressure such that substrate temperature is lower with low neutral gas temperature [66]. SWP generated at a low gas pressure exhibits outstanding properties for low temperature and rapid growth of graphene film. Graphene has been grown on Cu foil at a relatively low temperature (~400 °C) using a gas mixture of CH_4 and Ar. As shown in Figure 2a-c, large-area uniform graphene film can be synthesized on Cu foil as well as transferable to a flexible plastic sheet. The transferred graphene film remains continuous without any structural damage and can be excellent materials as a conducting optical window for various electronic devices.

In the demonstrated process, large-area graphene film with a sheet resistance and optical transmittance of 2.2 to 45 Ωk/sq and 78% to 94%, respectively has been achieved [66]. Raman studies shows presence of second order 2D Raman peak (2657 cm^{-1}), indicating graphene formation (Figure 2d). However, a strong defect induced D peak at 1326 cm^{-1}, indicate smaller graphene domain structure. Similarly, graphene film can be synthesized at a temperature as low as 240 °C, while the quality of graphene film varies with growth temperature (Figure 2e) [61]. Most importantly the low temperature deposited graphene films also remain intact transferring to an arbitrary substrate and can be used as coating and flexible transparent conductors.

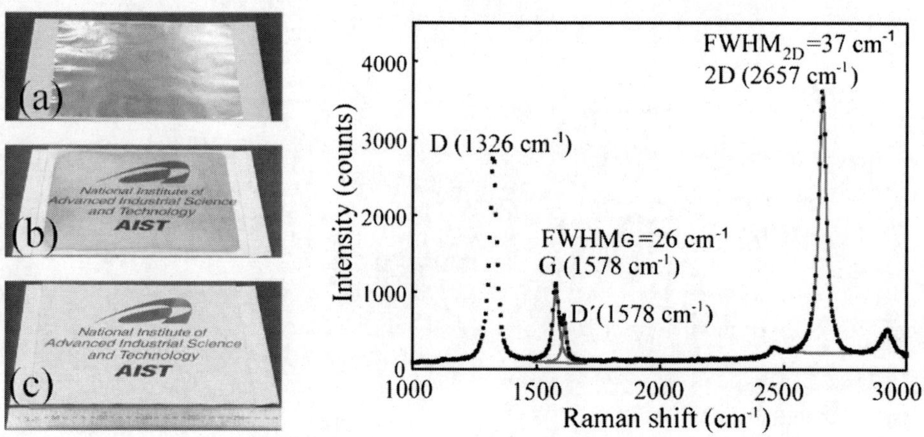

d

Figure 2. (Continued).

mote Plasma Enhanced Chemical Vapor Deposition

nilarly, graphene has been synthesized on Cu foil using a Remote plasma enhanced al vapor deposition (RPECVD) technique [71]. In this plasma CVD process, enzene group's hydrocarbons have been used as a precursor, where most of the plasma process used carbon gas sources. The growth mechanism can be explained as that of e wave plasma CVD process, where activated carbon radicals were generated in the a be decomposing cyclobenzene precursor.

aman study shows that graphene can be synthesized at a temperature as low as 300°C, h is an important aspect of the developed CVD process. The quality of the graphene film nced with increase in growth temperature. Above 600°C, sharp G and 2D can be rved without any defect related D peak. Similarly, the I_{2D}/I_G increases from 0.8 to 2.9 n FWHM of the 2D peak reducing from 54 to 33 as the growth temperature increases from) to 800°C [71]. The control in graphene nucleation density plays an important role in the thesis of uniform graphene film.

It has been observed that the densities of graphene domains increase along with the crease in heating temperature of the naphthalene precursor. Raman mapping studies of the , G and 2D bands for a graphene film deposited at 400°C indicates uniform and good uality graphene formation. Graphene growth by the RPECVD technique also can be xplained by a three step process of nucleation of domain, interconnection of domain and ormation of a continuous film [71].

4.3. Plasma-Assisted Thermal Chemical Vapor Deposition

It has been also demonstrated plasma process can be integrated with a thermal CVD process to synthesis high quality graphene at a low temperature. Figure 5a presents a schematic diagram of the developed plasma assisted thermal CVD process. In this approach, decomposition of hydrocarbon gas can be achieved by the generated plasma, rather than using high thermal energy as shown in figure 5b. Dehydrogenation of hydrocarbon gases is considered to be the critical steps for the graphene formation, which require relatively high growth temperature in a thermal CVD process [72].

On the other hand, the Dehydrogenation of hydrocarbon occurs in the plasma and creating carbon radical, which facilitate low temperature growth on transition metal substrate. It has been reported that by the plasma assisted thermal CVD process monolayer graphene film can be synthesized at a temperature as low as 600 °C using H_2 and CH_4 gaseous mixture [72]. From Raman studies, it has been obtained that the defect related D peak reduced and second ordered 2D Raman peak enhanced with increase H_2 flow rate (Figure 5c-d). This can be explained from the H_2 reaction to activate carbon radical on the Cu surface for effective graphene nucleation and growth.

In-situ plasma emission spectral studies showed that a distance must be maintained between plasma initial stage and deposition stage to allow the plasma to diffuse to substrate. The demonstrated plasma-assisted thermal CVD technique also can be significant for low temperature and cost effective approach for synthesis of high-quality graphene films.

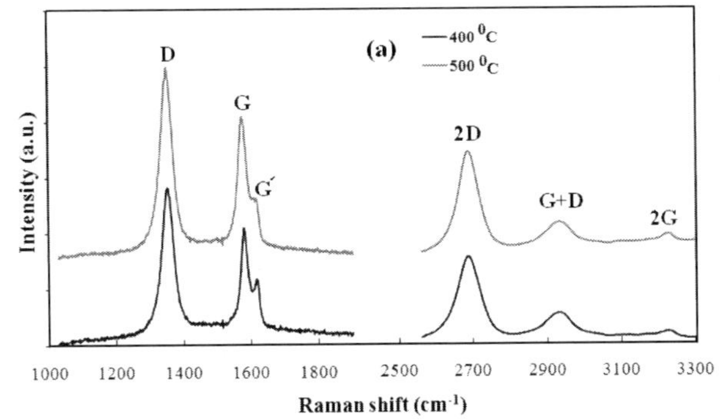

e

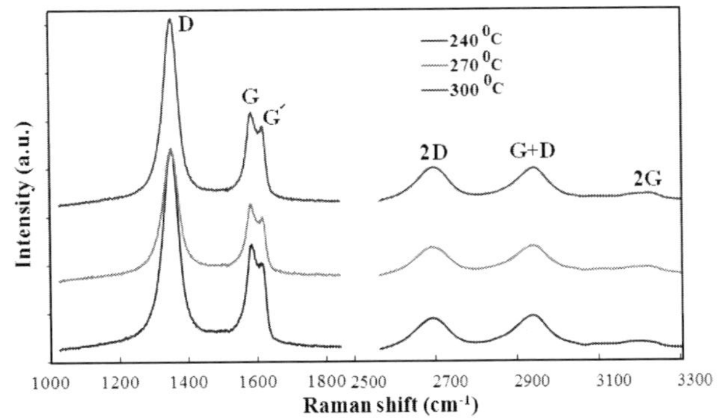

f

Reference [66] with permission from The Royal Society of Chemistry (a-d).
Reference [61] with permission from The Royal Society of Chemistry (e, f).

Figure 2. (a,b and c) synthesized graphene film on Cu foil by the SWP-CVD process and transferred graphene film with different thickness varying the growth durations on Cu foil. (d) Raman spectra of the synthesized graphene-based film by SWP-CVD. (e) and (f) Raman spectra of graphene film deposited by similar SWP-CVD technique at various growth temperature.

The SWP-CVD technique has been also extended for solid carbon source (camphor) instead of using a hydrocarbon gas source for graphene synthesis.

Similarly, Ni foil has been used for graphene growth by a MWP-CVD process at a comparatively low temperature. Monolayer and few-layer graphene have been synthesized on Ni foil by MWP-CVD process in a temperature range of 450 to 750°C [67]. Raman analysis of the synthesized graphene film presents reduce defect induced D peak with increase in growth temperature and increase in 2D peak intensity than that G peak (Figure 3a).

Figure 3b gives much better information about the quality of graphene with respect to growth temperature. The increase in quality of graphene with growth temperature can be explained from better carbon solubilitiy on Ni surface at an elevated temperature.

It has been observed that for same growth duration the overall coverage of monolayer graphene is around 80% regardless of deposition temperatures. The transferred graphene films showed transmittance of around 89% at 550 nm wavelength [67]. It has been observed that the sheet resistance increased with a lower graphene growth temperature (Figure 3b).

This is due to the fact that the quality of the graphene is poor at lower growth temperature.

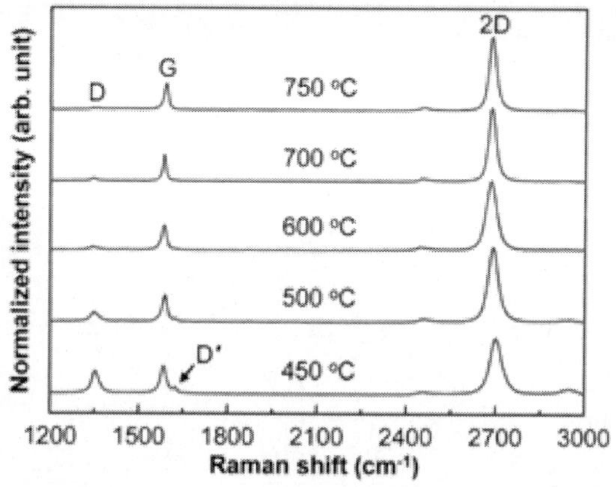

a

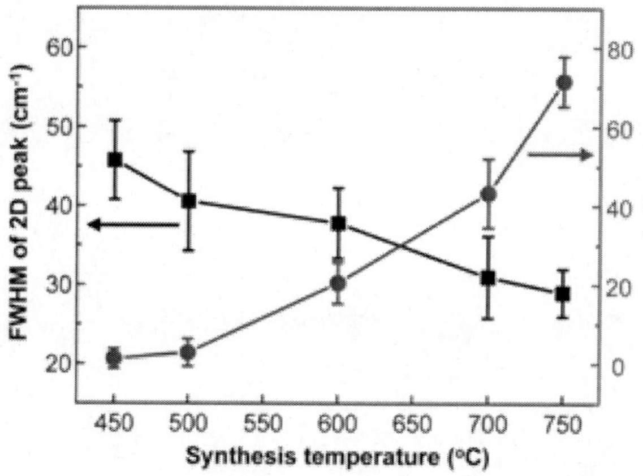

b

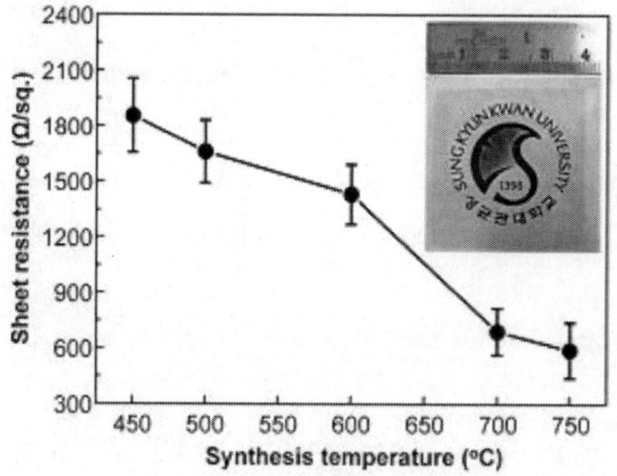

c

Figure 3. (a) Raman spectra of the graphene film synthesized at various substrate temperatures on Ni foil. (b) FWHM of 2D peak and intensity ratio of D to G peak as a function of growth temperature. (c) Sheet resistance of transferred graphene film as a function of growth temperature. Inset of the figure presnets a photograph of transferred graphene onto a PET substrate.

The results of graphene growth on Cu and Ni foil by a M plasma technology can be effectively used for graphene growth thereby paving the way for effective industrial production. Again, that a few-layer graphene film deposited by SWP-CVD proces oxidation resistive barrier of metal surface (Figure 4a-b).

Deposition of graphene layer on Cu foil as oxidation barrier dep temperature as low as 450°C. It has been observed that heating (~150 Cu foil in atmospheric air, a strong oxidation resistance performance Synthesized graphene film effectively passivated the Cu surface to res The oxidation barrier characteristics of a low temperature deposited few also similar to that of high temperature deposited thermal CVD graphene

The demonstrated process can be exploited for coating various graphene growth is weakly related to catalytic ability of metal surface. The technology can be significantly viable for roll-to-roll deposition and rea transfer for industrial production [69, 70].

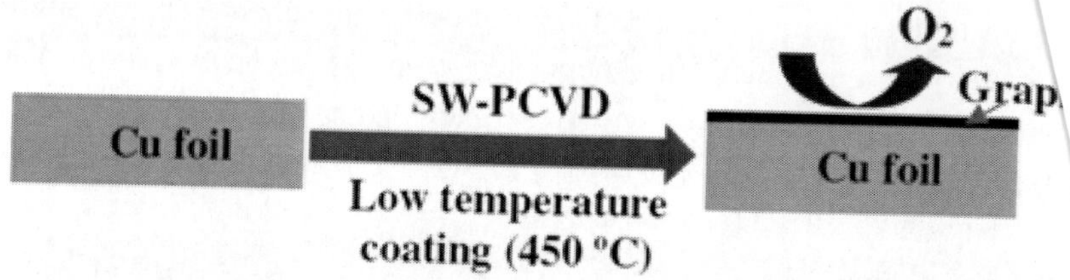

a

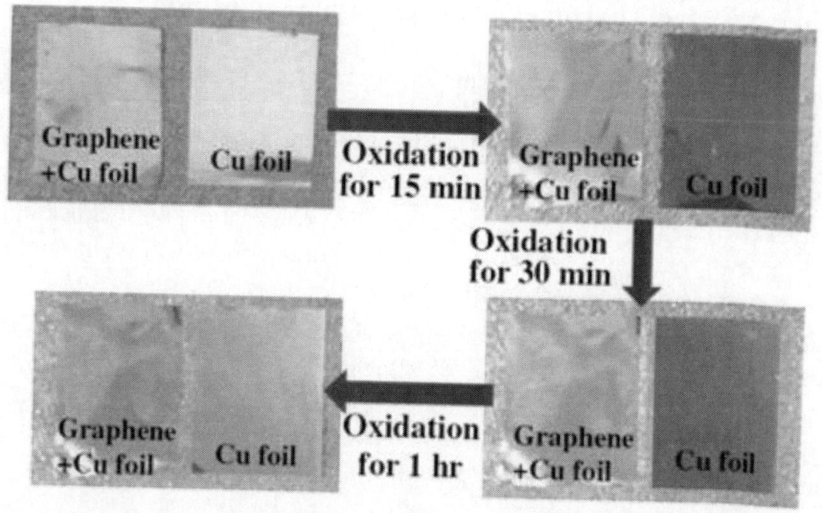

b

Figure 4. (a) Schematic diagram of few-layer graphene coating on Cu foil by SWP CVD as an oxidation resistive barrier. (b) Photographs of oxidized graphene-coated and bare Cu foil at various duration heating at 150 °C in atmospheric conditions.

4.2. Re

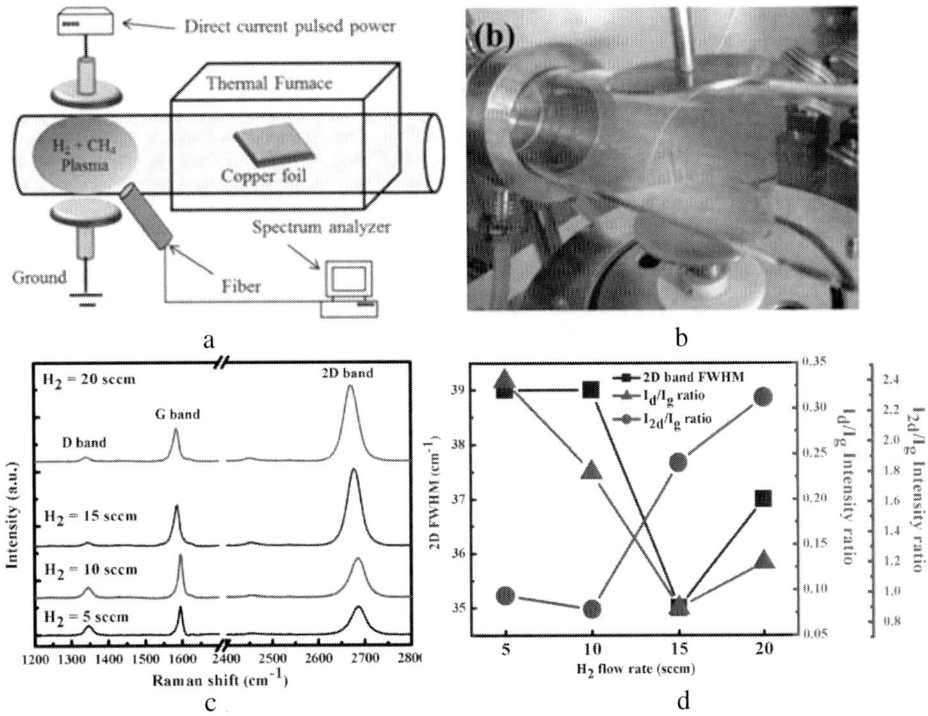

Reprinted from Chan et al. Reference [72].

Figure 5. (a) schematic diagram of the plasma assisted thermal CVD process. (b) Plasma generation in the quartz tube using a gas mixture. (c) Raman spectra, (d) FWHM and I_D/I_G ratio of graphene films synthesized with different H_2 flow rates.

5. Direct Growth of Graphene on Arbitrary Substrate by Plasma CVD

Another important aspect of plasma CVD process is synthesizing graphene like structures on non-catalytic arbitrary substrates. A catalyst-free direct growth process at low temperature can have much advantages for integration in electronic devices. This will allow us direct integration of graphene film in electronic devices, creating hybridized heterostructure as well as coating on various type of substrates. In the following, we discuss various development of plasma CVD process for direct growth of graphene like materials [74].

5.1. Free-Standing Graphene by Microwave Plasma CVD

Microwave plasma CVD technique has been demonstrated for synthesis of high quality, crystalline, freestanding few-layer graphene sheet on an arbitrary substrate. Various substrate materials, such as quartz, Si, stainless steel and transition metals (Ni Pt, Ge, Ti, W, Mo etc.) have been used in this synthesis process. The scanning electron microscope (SEM) study presents few-layer graphene growth vertically aligned to the substrate surface (figure 6a-b).

Copyright permission from Malesevic et al. Reference [74].

Figure 6. (a) Cross-sectional SEM image of the free-standing graphene sheet synthesized by the plasma CVD process. (b) Top view SEM image presenting highly dense vertically aligned graphene sheets. (Scale bars~1μm). (c) and (d) TEM images presenting average thickness of the graphene flakes as ~4-6 layers.

This type of few-layer graphene structure can be directly used for field emission and energy storage devices.

The developed process is a straightforward and highly reproducible technique as it does not depend on catalyst nature of the substrate. TEM studies also confirm formation of high quality graphene sheets as shown in figure 6c-d [74].

The graphene growth in this CVD technique also can be explained by a three step process. In a first step, a base graphite layer is formed on the substrate surface which is not continuous and presence irregular cracks. The density of the cracks increases with increasing growth time. Subsequently, in the second step, freestanding few-layer graphene nucleates from these upward curling crack edges, changing their growth direction from parallel to vertical to the substrate surface.

In the third and final step comprises the accumulation and incorporation of carbon radicals at the edges of the freestanding graphene structure and further growth occurs to a larger sheet [74]. The growth of vertical few-layer graphene structure on arbitrary substrates independent of catalytic nature can have significant importance for several practical device applications.

5.2. Synthesis of Nanographene on Arbitrary Substrate

5.2.1. Surface Wave Microwave Plasma CVD

Nanographene films have been deposited on Si and SiO_2 substrates by using the MW-SWP-CVD technique similar to that of as shown in figure 1. The formation of nanographene structure on a non-catalytic substrate can be explained from the decomposition hydrocarbon and formation of carbon radical in the generated plasma. The deposited carbon material as characterized by Raman study (Figure 7a-b) presents a graphitic G and seconder ordered Raman 2D band at 1568 and 2669 cm^{-1}, respectively, presenting crystalline sp^2 structure. However, hightly disordered induced D peak at around 1339 cm^{-1} was observed, this due to fact that low crystallite sizes or an abundance of edges in the nanographene structures. In the SW-PCVD process growth of nanographene films on an arbitrary substrate was achieved in the temperature range of 400-560°C. TEM study also confirmed the formation of few-layer graphene as shown in figure 7c-d. The presented graphene layers are of individual nanographene domain, which are overlapped each other in the deposited films. Atomic force microscopy (AFM) analysis showed that the graphene film deposited on Si and SiO_2 substrate form triangle shape nanographene with a side of the order of 80~100 nm [75]. This study is an important contribution in the development of direct growth of grahene based material by a plasma CVD technique. In the growth mechanism, carbon radical formed with hydrogen termination in the microwave plasma and plays the key role in the nucleation and growth process in the metal catalyst-free substrate. The carbon radicals from the plasma absorb onto the substrate and link each other by diffusion process to form sp^2 hybridized carbon and eventually the graphene structure [75].

The synthesized nanographene films on SiO_2 and Si substrate were adequately thin to be transparent for long wavelength ranges (0.3-2μm). The nanographene films deposited for different durations are highly transparent comparing to a standard indium tin oxide (ITO) film (Figure 8a).

Transmittance of a graphene film deposited for shorter duration was achieved as high as 82% at 550 nm wavelength (visible light). Sheet resistance of a graphene film with transmittance of 82% at 550 nm wavelength was obtained as 6.8 KΩ/sq (Figure 8b), which shows better conductivity than that some of reduced GO film.

At the same time, lateral conductivity of synthesized nanographene film will be significantly high as the thickness is only few nanometers [75, 76].

Fabrication of Schottky junction photovoltaic device has been demonstrated with the direct grown nanographene film on n-Si substrate.

A schematic diagram of the fabricated nanographene/n-Si schottky junction device is presented in figure 9a.

The nanographene film has metallic conductivity and the work function can be considered as that of graphite~4.6 eV. As explained before, the thickness of the synthesized nanographene film is significantly less, which enable transmission of incident light to the graphene/Si interface. Figure 10b presents an energy band diagram of such type of Schottky junction device in thermal equilibrium [75]. It has been observed that a graphene/Si Schottky junction can create a relatively large built-in filed (V_{bi}, 0.55–0.75 V) [77, 78].

In presence of a built-in field in the fabricated graphene/n-Si heterojunction, there can be efficient charge separation and transportation. Figure 9c presents a log scale plot of the dark current (J)-voltage (V) characteristic of the fabricated heterojunction device.

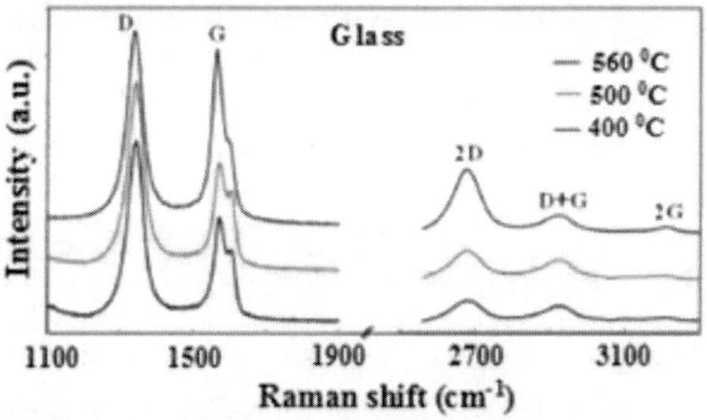

a

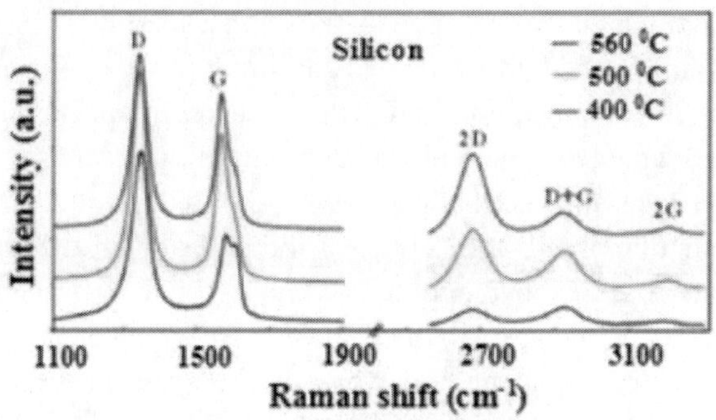

b

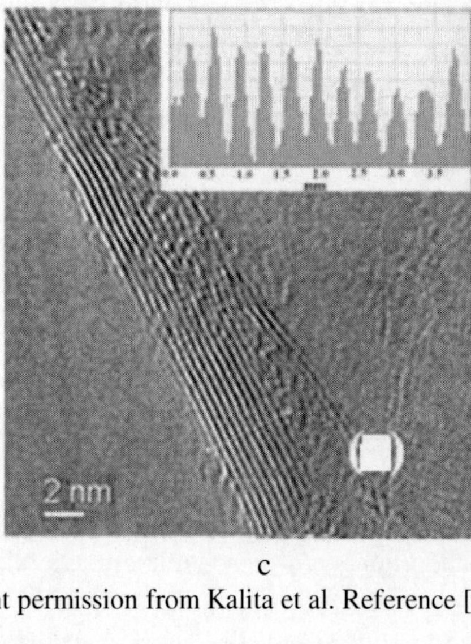

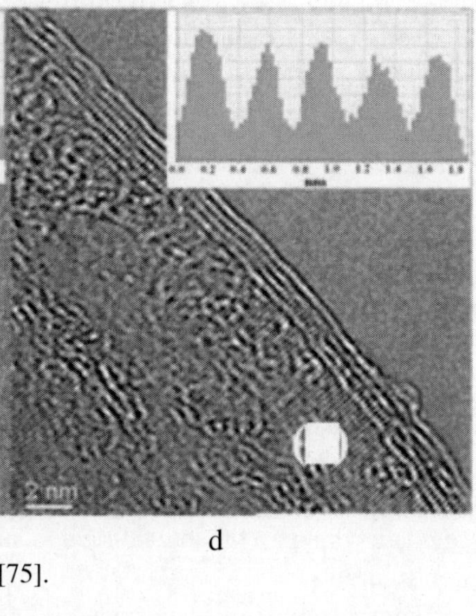

c d

Figure 7. Raman spectra of the nanographene film deposited at 400, 500 and 560 ºC on (a) glass (SiO_2) and (b) Si substrates. (c) and (d) TEM images of nanographene edges deposited at 560 ºC presenting few-layers graphene structure (inset of the figure, intensity pattern of the few-layer graphene).

The dark characteristic indicates rectifying diode characteristic with a rectification ratio of 10^4, which is similar to other Schottky diodes.

The fitting plot of ln(I)-V in forward bias shows linear characteristic in the range of 0.1 ≈ 0.37 V having a diode ideality factor (n) of 2.44 [75].

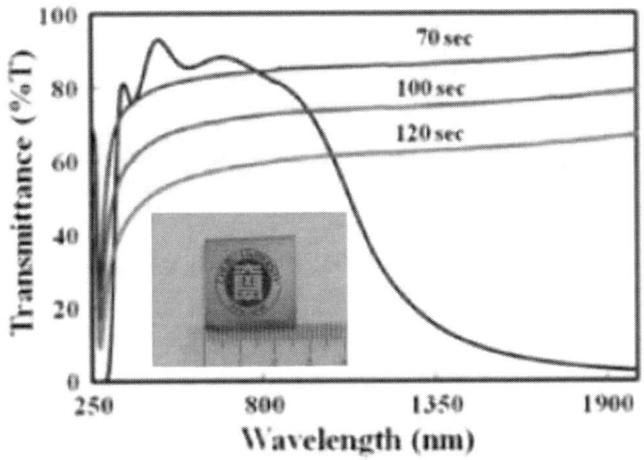

a

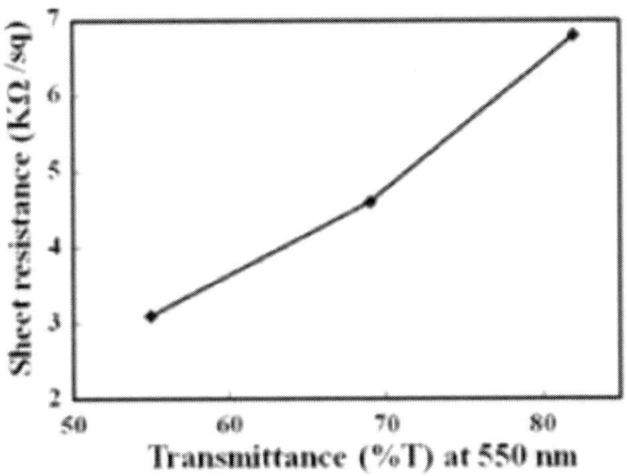

b

Copyright permission from Kalita et al. Reference [75].

Figure 8. (a) Transmittance of the nanographene films deposited at 560 ºC for different durations in comparison to indium tin oxide (inset of the figure, a photograph of nanographene deposited SiO_2 substrate). (b) Sheet resistance of the graphene films with variation of transmittance at 550 nm wavelength.

A photovoltaic action with open circuit voltage (V_{oc}) 0.41 V, short circuit current density (J_{sc}) 15.6 mA/cm^2 and fill factor (FF) of 33% and a conversion efficiency (η) of 2.11% have been achieved in the fabricated device (Figure 9d). V_{oc} of the photovoltaic device with direct growth nanographene film is consistent with the fabricated device using transferred CVD graphene film.

The observation of photovoltaic action with the plasma CVD deposited graphene film on Si substrate can be really significant to extend it for other optoelectronic device application.

Improving the quality and domain size of the graphene structure, we can expect highly efficient photodiode, photovoltaic device and transparent conductor for display electronics. Similarly, other plasma CVD techniques have been also explored with the possibility of direct growth of graphene film on various kinds of substrates [79-82].

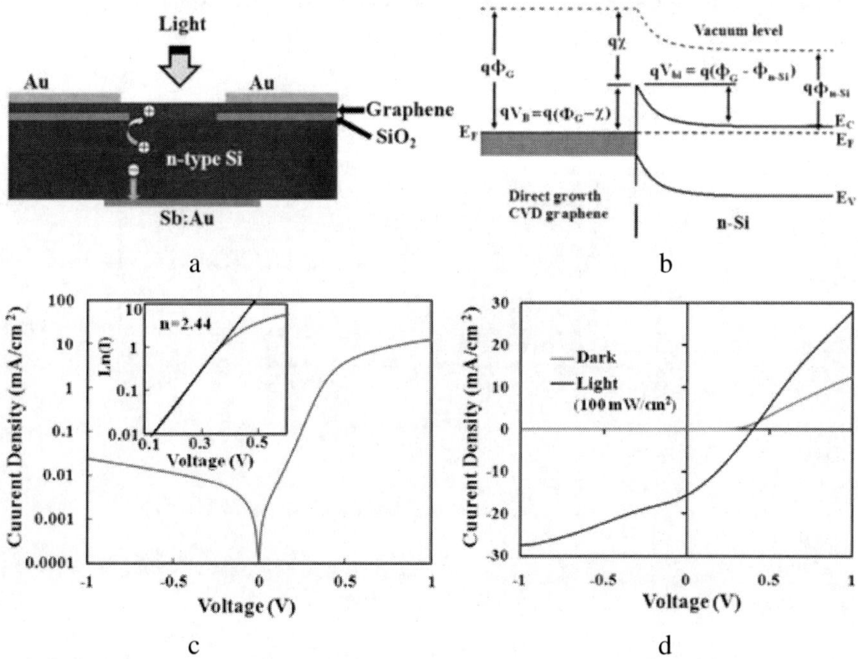

Figure 9. (a) schematic diagram of fabricated nanographene/Si Schottky junction. (b) Corresponding energy-band diagram of the device; presenting formation a potential barrier at the interface. (c) J-V curve in log plot in dark condition (inset of the figure, a fitting plot of ln(I)-V in forward bias). (d) J-V characteristic under dark and illumination of 100 mW/cm^2.

5.2.2. Plasma Enhancement CVD

Remote plasma CVD has been also demonstrated as that of SW-MWP-CVD process for direct synthesis of graphene film on an arbitrary substrate. In this process, nanocrystalline graphene with a size about tens of nanometers has been directly grown on various substrates at a temperature as low as 550°C.

In the growth process, nanographene are nucleated all over the SiO_2 surface. The larger graphene islands are ~1.2–1.5 nm thick, consisting of 2 or 3 layers. Constant growth with increase in growth duration yields more nucleation sites on the substrate surface, and thereby forming a continuous and uniform nanographene film. Again, XPS studies confirmed that the nanographene film contains high amount of sp^2 hybridized carbon atoms as shown in figure 11f, where presence of carboxyl and hydroxyl groups are negligible. The synthesized nanographene film showed ohmic conductivity, attributing to metallic characteristics [79].

Similarly, high quality graphene has been synthesized by plasma enhanced CVD without using any meta-catalyst as shown in figure 10 [80]. Graphene growth has been achieved at a temperature as low as 600°C. It has been demonstrated that the synthesized graphene can be directly integrated for fabrication of light emitting diode (LED).

The synthesized graphene provides an in situ ohmic-contact formation between graphene and p-GaN, thereby achieving significant improvement in LED performance. The main advantages of the direct growth process without using a metal catalytic layer are low material and process cost, avoiding metal impurities, depositing on a substrate of choice for application.

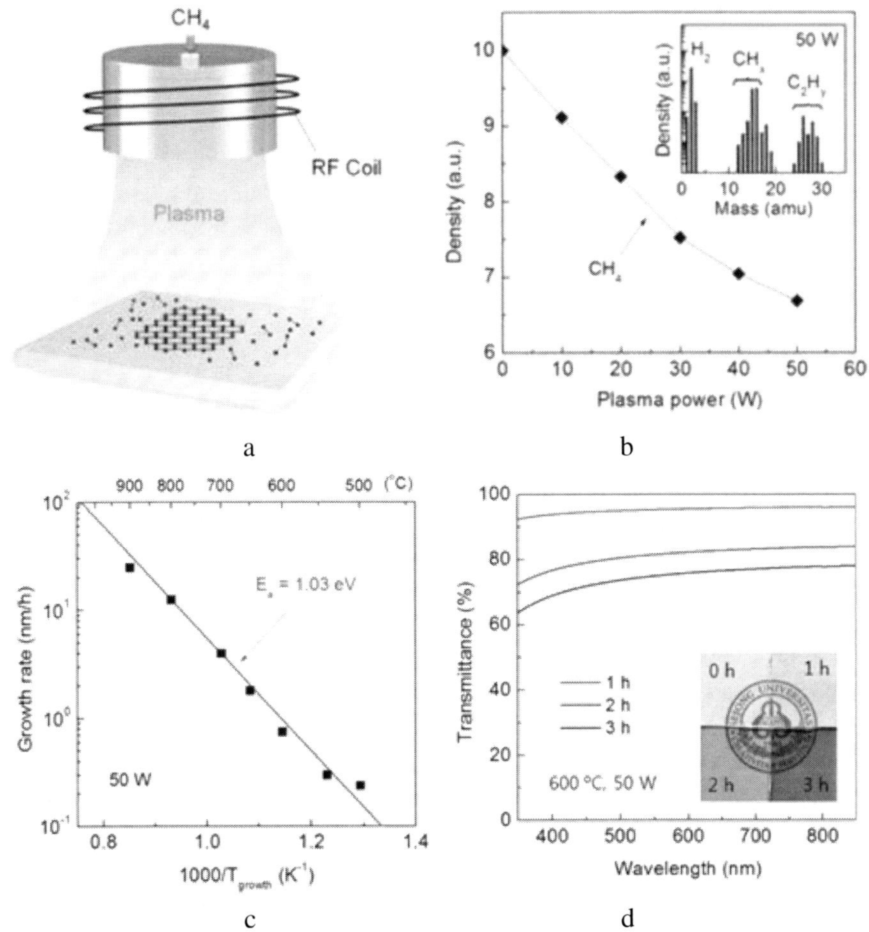

Reprinted with permission from Kim et al. Reference [80].

Figure 10. Schematic diagram of the graphene synthesis process using PECVD technique. (b) Density of CH_4 as a function of plasma power. (c) Arrhenius plot of growth rate as a function of reciprocal substrate temperature. (d) Time dependent thickness variation and transparency for the as-synthesized graphene film.

However, quality of the graphene structure is still poor than that of catalytic growth process, which remain a challenges ahead in this area. In general, the plasma CVD process has significant potential for graphene growth by a catalytic or non-catalytic approach to complete with the thermal CVD process [36-38, 61, 66, 67].

CONCLUSION

In the very beginning of the chapter, we presented a brief introduction of 2D crystalline materials. Significant development has been achieved in the last few years of two 2D material science and their practical device applications. Recent significant development on synthesis of large-area continuous graphene film by CVD process opens up new possibilities of applications.

Among CVD techniques, plasma CVD has evolved as a novel technique to deposit graphene film at a comparatively lower temperature than that of thermal CVD process. A low temperature deposition technique can be more feasible for industrial application and can pave the way for direct deposition on any arbitrary substrates. We discussed various plasma CVD techniques and their potential for large-area graphene growth on metal catalytic layer. Quality of graphene synthesized by a plasma CVD process has been achieved as high as that of thermal CVD process. Furthermore, non-catalytic direct growth of nanographene film on an arbitrary substrate has been achieved by the plasma CVD process.

As-grown nanographene film can be integrated in various applications as well as transparent optical window for electronic devices. The graphene growth mechanism in a plasma CVD process involves with hydrocarbon dissociated by hydrogen termination and creating carbon radicals. The carbon radicals absorb continuously on the substrate surface to form sp^2 carbon and eventually graphene structure. The plasma CVD process has significant potential and can be a key technology for deposition of high quality graphene film.

Acknowledgments

The work is supported by the funds for the development of human resources in science and technology, Japan.

References

[1] K. S. Novoselov, A. K. Geim, S. V. Morozov, D. Jiang, Y. Zhang, S. V. Dubonos, I. V. Grigorieva, and A. A. Firsov, *Science* 306, 666, (2004).

[2] Y. Zhang, Y. W. Tan, H. L. Stormer, and P. Kim, *Nature* 438, 201, (2005).

[3] K. S. Novoselov, A. K. Geim, S. V. Morozov, D. Jiang, M. I. Katsnelson, I. V. Grigorieva, S. V. Dubonos, and A. A. Firsov, *Nature* 438, 197, (2005).

[4] C. Lee, X. Wei and J. W. Kysar and J. Hone, *Science* 321, 385, (2008).

[5] R. R. Nair, P. Blake and A. N. Grigorenko, K. S. Novoselov, T. J. Booth, T. Stauber, N. M. R. Peres, and A. K. Geim, *Science* 320, 1308, (2008).

[6] A. K. Geim and K. S. Novoselov, *Nature mater*. 6, 183, (2007).

[7] C. Berger, Z. Song, X. Li, X. Wu, N. Brown, C. Naud, D. Mayou, T. Li, J. Hass, A. N. Marchenkov, E. H. Conrad, P. N. First, W. A. de Heer, *Science* 312, 1191, (2006).

[8] A. A. Balandin, S. Ghosh, W. Bao, I. Calizo, D. Teweldebrhan, F. Miao, and C. Ning Lau, *Nano Lett.* 8, 902, (2008).

[9] *Wikipedia, the free encyclopedia*, http://en.wikipedia.org/wiki/Graphite.

[10] W. S. Hummers and R. E. Offeman, *J. Am. Chem. Soc*., 80, 1339, (1958).

[11] J. C. Lewis, B. Redfern and F. C. Cowlard, *Solid-State Electronics* 6(3), 251, (1963).

[12] H. W. Kroto, J. R. Heath, S. C. Obrien, R. F. Curl, and R. E. Smalley, *Nature* 318, 162, (1985).

[13] S. Iijima, *Nature* 354, 56, (1991).

[14] D. V. Kosynkin, A. L. Higginbotham and A. Sinitskii, A. Sinitskii, J. R. Lomeda, A. Dimiev, B. K. Price, and J. M. Tour, *Nature* 458, 872, (2009).

[15] M. Fujita, K. Wakabayashi, K. Nakada, and K. Kusakabe, *Journal of the Physics Society Japan* 65 (7), 1920 (1996).
[16] S. T. Yang, L. Cao, P. G. Luo, F. S. Lu, X. Wang, H. F. Wang, M. J. Meziani, Y. F. Liu, G. Qi, and Y. P. Sun, *J. Am. Chem. Soc.* 131, 11308, (2009).
[17] K. Kawasumi, Q. Zhang, Y. Segawa, L. T. Scott, and K. Itami, *Nature Chemistry* 5, 739, (2013).
[18] P. R. Wallace, *Phys. Rev.* 71, 622, (1947).
[19] P. Avouris and C. Dimitrakopoulos, *Materials Today* 15, 86, (2012).
[20] S. V. Morozov, K. S. Novoselov and M. I. Katsnelson, F. Schedin, D. C. Elias, J. A. Jaszczak, and A. K. Geim, *Phys. Rev. Lett.* 100, 016602, (2008).
[21] A. K. Geim, *Science* 324, 1530, (2009).
[22] Q. Bao, H. Zhang, Z. Ni, Y. Wang, L. Polavarapu, Z. Shen, Q. H. Xu, D. Y. Tang, and K. P. Loh, *Nano Research* 4, 297, (2011).
[23] C. H. Lui, K. F. Mak, J. Shan, and T. F. Heinz, *Phys. Rev. Lett.* 105, 127404, (2010).
[24] J. S. Bunch, S. S. Verbridge, J. S. Alden, A. M. van der Zande, J. M. Parpia, H. G. Craighead, and P. L. *McEuen Nano Lett.*, 8, 2458, (2008).
[25] K. S. Novoselov, D. Jiang and F. Schedin, D. Jiang, F. Schedin, T. J. Booth, V. V. Khotkevich, S. V. Morozov, and A. K. Geim, *Proc. Natl. Acad. Sci.* 102, 10451, (2005).
[26] B. Jayasena and S. Subbiah, *Nanoscale Research Letters*, 6, 95, (2011).
[27] S. Stankovich, D. A. Dikin and G. H. B. Dommett, K. M. Kohlhaas, E. J. Zimney, E. A. Stach, R. D. Piner, S. T. Nguyen, R. S. Ruoff, *Nature* 442, 282, (2006).
[28] C. Y. Su, A. Y. Lu, Y. P. Xu, F. R. Chen, A. N. Khlobystov, L. J. Li, *ACS Nano*, 5, 2332, (2011).
[29] Z. Y. Xia, S. Pezzini, E. Treossi, G. Giambastiani, F. Corticelli, V. Morandi, A. Zanelli, V. Bellani, and V. Palermo, *Adv. Funct. Mater.* 23, 4684, (2013).
[30] A. Ciesielski and P. Samorı, *Chem. Soc. Rev.*, DOI: 10.1039/c3cs60217f.
[31] C. Berger, Z. Song and T. Li, C. Berger, Z. Song, T. Li, X. Li, A. Y. Ogbazghi, R. Feng, Z. Dai, A. N. Marchenkov, E. H. Conrad, P. N. First, and W. A. de Heer, *J. Phys. Chem.* B 108, 19912, (2004).
[32] P. R. Somani, S. P. Somani and M. Umeno, *Chem. Phys. Lett.* 430, 56, (2006).
[33] Q. Yu, J. Lian and S. Siriponglert, H. Li, Y. P. Chen and S. S. Pei, *Appl. Phys. Lett.* 93, 113103, (2008).
[34] G. Kalita, M. Masahiro, H. Uchida, K. Wakita, and M. Umeno, *Mater. Lett.* 64, 2180, (2010).
[35] A. Reina, X. Jia and H. John, D. Nezich, H. Son, V. Bulovic, M. S. Dresselhaus, and J. Kong, *Nano Lett.* 9, 30, (2009).
[36] K. S. Kim, Y. Zhao and H. Jang, S. Y. Lee, J. M. Kim, K. S. Kim, J. Ahn, H. P. Kim, J. Y. Choi, and B. H. Hong, *Nature* 457, 706, (2009).
[37] X. S. Li, W. W. Cai and J. H. An, S. Kim, J. Nah, D. X. Yang, R. Piner, A. Velamakanni, I. Jung, E. Tutuc, S. K. Banerjee, L. Colombo, and R. S. Ruoff, *Science* 324, 1312, (2009).
[38] S. Bae, H. K. Kim, Y. Lee, X. Xu, J. Park, Y. Zheng, J. Balakrishnan, D. Im, T. Lei, Y. Song, Y. Kim, K. Kim, B. Ozyimaz, J. Ahn, B. Hong, and S. Iijima, *Nat. Nanotechnol.* 5, 574, (2010).

[39] J. H. Lee, E. K. Lee, W. J. Joo, Y. Jang, B. S. Kim, J. Y. Lim, S. H. Choi, S. J. Ahn, J. R. Ahn, M. H. Park, C. W. Yang, B. L. Choi, S. W. Hwang, and D. Whang, *Science* 344, 286, (2014).

[40] M. E. Ayhan, G. Kalita, S. Sharma, and M. Tanemura, *Phys. Stat. Sol.-RRL* 7, 1076, (2013).

[41] X. Zhang, L. Wang, J. Xin, B. I. Yakobson, and F. Ding, *J. Am. Chem. Soc*. 136, 3040, (2014).

[42] G. Kalita, S. Sharma and K. Wakita, M. Umeno, M. Tanemura, *Phys. Chem. Chem. Phys.* 15, 1271, (2013).

[43] S. Gilje, S. Han, M. Wang, K. L. Wang, and R. B. Kaner, *Nano Lett.* 7, 3394, (2007).

[44] G. Eda, G. Fanchini and M. Chhowalla, *Nature Nanotechnol*. 3, 270, (2008).

[45] K. P. Loh, Q. Bao and G. Eda and M. Chhowalla, *Nature Chem*. 2, 1015, (2010).

[46] D. V. Badami, *Nature*, 193, 569, (1962).

[47] I. Forbeaux, J. M. Themlin, A. Charrier, F Thibaudau, and J. M. Debever, *Appl. Surf. Sci*. 162, 406, (2000).

[48] S. Unarunotai, Y. Murata, C. E. Chialvo, H. sik Kim, S. MacLaren, N. Mason, I. Petrov, and J. A. Rogers, *Appl. Phys. Lett*. 95, 202101, (2009).

[49] H. Hibino, S. Mizuno, H. Kageshima, M. Nagase, and H. Yamaguchi, *Phys. Rev.* B 80, 085406 (2009).

[50] C. Virojanadara, R. Yakimova, A. A. Zakharov, and L. I. Johansson, *J. Phys. D: Appl. Phys.* 43, 374010, (2010).

[51] Y. Wang, S. W. Tong and X. F. Xu, B. Özyilmaz and K. P. Loh, *Adv. Mater*. 23, 1514, (2011).

[52] Z. Liu, J. Li, Z. H. Sun, G. Tai, S. P. Lau, and F. Yan, *ACS Nano* 6, 810, (2012).

[53] Arco de, L. G., Zhang Y. and C. W. Schlenker, K. Ryu, M. E. Thompson and C. Zhou, *ACS Nano* 4, 2865, (2010).

[54] Z. Li, P. Wu, C. Wang, X. Fan, W. Zhang, X. Zhai, C. Zeng, Z. Li, J. Yang, and J. Hou, *ACS Nano* 5(4), 3385, (2011).

[55] M. Rusop, S. Adhikari, A. M. M. Omer, S. Adhikary, H. Uchida, T. Jimbo, T. Soga, and M. Umeno, *Mod. Phys. Lett.* B 18, 987, (2004).

[56] G. Kalita, S. Adhikari, H. R. Aryal, R. Afre, T. Soga, M. Sharon, and M. Umeno, *Diam. Relat. Mater*. 17, 1697, (2008).

[57] J. Kim, K. Tsugawa, M. Ishihara, Y. Koga, and M. Hasegawa, *Plasma Sources Sci. Technol.* 19, 015003, (2010).

[58] V. T. Airoldi, C. F. M. Borges, M. Moisan, and D. Guay, *Applied Optics,* 36(19), 4400, (1997).

[59] G. Zhong, M. Tachiki, H. Umezawa, T. Fujisaki, H. Kawarada, and I. Ohdomari, *Chem. Vap. Deposition* 10, 125, (2004).

[60] S. Nakao and H. Sugai, *Jpn. J. Appl. Phys*. 46, 1039, (2007).

[61] G. Kalita, K. Wakita and M. Umeno, ***RSC Adv*. 2**, 2815, (2012).

[62] P. W. Sutter, J. I. Flege and E. A. Sutter, *Nature Mater*. 7, 406, (2008).

[63] J. Coraux, A. T. N'Diaye, C. Busse, and T. Michely, *Nano Lett*. 8, 565, (2008).

[64] G. Kalita, M. Masahiro, H. Uchida, K. Wakita, and M. Umeno, *J. Mater. Chem.* 20, 9713, (2010).

[65] R. Hirano, K. Matsubara, G. Kalita, Y. Hayashia, and Masaki Tanemura, *Nanoscale* 4, 7791, (2012).

[66] J. Kim, M. Ishihara, Y. Koga, K. Tsugawa, M. Hasegawa, and S. Iijima, *Appl. Phys. Lett.* 98, 091502, (2011).

[67] Y. Kim, W. Song, S. Y. Lee, C. Jeon, W. Jung, M. Kim, and C. Y. Park, *App. Phys. Lett.* 98, 263106, (2011).

[68] G. Kalita, M. E. Ayhan, S. Sharma, S. M. Shinde, D. Ghimire, K. Wakita, M. Umeno, and M. Tanemura, *Corros. Sci.* 78, 183-187, (2014).

[69] T. Yamada, M. Ishihara, J. Kim, M. Hasegawa, and S. Iijima, *Carbon* 50, 2615, (2012).

[70] T. Yamada, J. Kim, M. Ishihara, and M. Hasegawa, *J. Phys. D: Appl. Phys.* 46, 063001, (2013).

[71] T. Wu, H. Shen, L. Sun, J. You, and Z. Yue, *RSC Adv.* 3, 9544, (2013).

[72] S. H. Chan, S. H. Chen, W. T. Lin, M. C. Li, Y. C. Lin, and C. C. Kuo, *Nanoscale Res. Lett.* 8(1), 285, (2013).

[73] S. Wang, L. Qiao, C. Zhao, X. Zhang, J. Chen, H. Tian, W. Zheng, and Z. Han, *New J. Chem.* 37, 1616, (2013).

[74] A. Malesevic, R. Vitchev, K. Schouteden, A. Volodin, L. Zhang, G. V. Tendeloo, A. Vanhulsel, and C. V. Haesendonck, *Nanotechnology* 19, 305604, (2008).

[75] G. Kalita, M. S. Kayastha, H. Uchida, K. Wakita, and M. Umeno, *RSC Adv.* 2, 3225, (2012).

[76] J. Kwak, J. H. Chu, J. K. Choi, S. D. Park, H. Go, S. Y. Kim, K. Park, S. D. Kim, Y. W. Kim, E. Yoon, S. Kodambaka, S. Y. Kwon, *Nat. Commun.,* 3, 645, (2012).

[77] G. Kalita, R. Hirano, M. E. Ayhan, and M. Tanemura, *J. Phys. D: Appl. Phys.* 46, 455103, (2013).

[78] X. Li, H. Zhu, K. Wang, A. Cao, J. Wei, C. Li, Y. Jia, Z. Li, X. Li, and D. Wu, *Adv. Mater.* 22, 2743, (2010).

[79] L. Zhang, Z. Shi, Y. Wang, R. Yang, D. Shi, and G. Zhang, *Nano Res.* 4 (3), 315, (2011).

[80] Y. S. Kim, K. Joo, S. K. Jerng, J. H. Lee, D. Moon, J. Kim, E. Yoon, S. H. Chun, *ACS Nano* 8, 2230, (2014).

[81] W. Yang, C. He, L. Zhang, Y. Wang, Z. Shi, M. Cheng, G. Xie, D. Wang, R. Yang, D. Shi, and G. Zhang, *Small* 8(9), 1429, (2012).

[82] C. S. Lee, C. S. Cojocaru, W. Moujahid, B. Lebental, M. Chaigneau, M. Châtelet, F. L. Normand, and J. L. Maurice, *Nanotechnology* 23, 265603, (2012).

In: Comprehensive Guide for Nanocoatings Technology. Vol. 1 ISBN: 978-1-63482-447-7
Editor: Mahmood Aliofkhazraei

Chapter 14

PRODUCING NANOGRAIN SURFACE LAYERS BY FRICTION STIR PROCESSING

Milena Folea[1], Adriana Roman[2], Cecile Langlade[2], Daniel Schlegel[3], Eric Gete[2], and Dominique Chamoret[4]

[1]Transilvania University of Brasov, Romania
[2]University of Technology of Belfort-Montbéliard, France, Laboratoire IRTES-LERMPS,
[3]School of Business and Engineering of Belfort, France
[4]University of Technology of Belfort-Montbéliard, France, Laboratoire IRTES-M3M

ABSTRACT

Friction stir processing (FSP) is a recent surface engineering method that allows obtaining in-situ ultrafine grains in surface layers. FSP is developed from friction stir welding, a thermo-mechanical welding process, now widely used due to technical, economic and environmental advantages. Friction stir basic principle is very simple and consists in moving a rotating tool along the workpiece surface while applying an axial load to increase generated heat.

Actually many studies are focused on friction stir processing as a surface modification method. The temperature and mechanical aspects must be balanced and optimal conditions need to be found in order to obtain the desired nanograin structure when processing different materials.

This chapter will present state-of-the art and future challenges in friction on stir as a method to obtain significantly refined nanograin structures at workpiece surface. The following topics are emphasized: friction stir processing - innovative surface engineering method, processing nanograin layers by friction stir processing of various materials and numerical control of friction stir processing.

1. FRICTION STIR PROCESSING - INNOVATIVE SURFACE ENGINEERING METHOD

Surface engineering includes a wide range of processing methods that aim to improve surface properties of metallic and nonmetallic components.

Generally, surface engineering methods are divided in two classes: surface modification and surface coating (Kenedy et al., 2005). By surface coating methods (e.g., Physical Vapor Deposition, Chemical Vapour Deposition, plasma and thermal spraying, cladding, electroplating etc.), thin films of additive material are deposited on a substrate. Surface modifications usually involve microstructure change.

Theyare obtained by thermal processes (e.g., hardening, laser or electron beam), mechanical processes (e.g., cold working). The chemical composition of the surface layer may also be altered in order to ameliorate its characteristics by diffusion treatments (e.g., nitriding, chromizing, boriding, carburizing). In Figure 1 are illustrated most frequently used fine grained techniques (Ma, 2008).

Friction stir, originally designed as a welding method, recently became used as a surface engineering method that allows obtaining in-situ ultrafine grains in surface layers.

Friction stir is a thermo-mechanical process with a very simple basic principle: a rotating tool is moved along the workpiece surface while applying an axial load to increase generated heat (Figure 2).

Friction stir welding was patented in 1991 by British scientists of The Welding Institute (TWI) as an assembly method, but lately it has also been developed as a surface modification method due to its technical, economic and environmental advantages.

In recent years, TWI has collected information and created a database with patents and patent applications referring to friction stir that is available on the company website (http://www.twi.co.uk/services/intellectual-property-licensing/friction-stir-welding/intellectual-property-rights). A quick look at this list of patents shows that important companies in automotive, aerospace and manufacturing industries were interested in developing Friction Stir Welding and Friction Stir Processing. Hundreds of patents have been obtained for friction stir process, for specific tools used during friction stir and for specialized machines and apparatuses. Regarding friction stir processing as surface modification method, there is still a long way to go before it will become a common method in manufacturing.

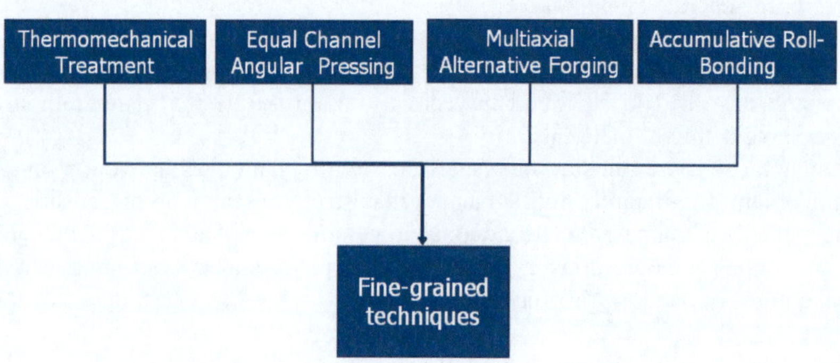

Figure 1. Fine grained techniques.

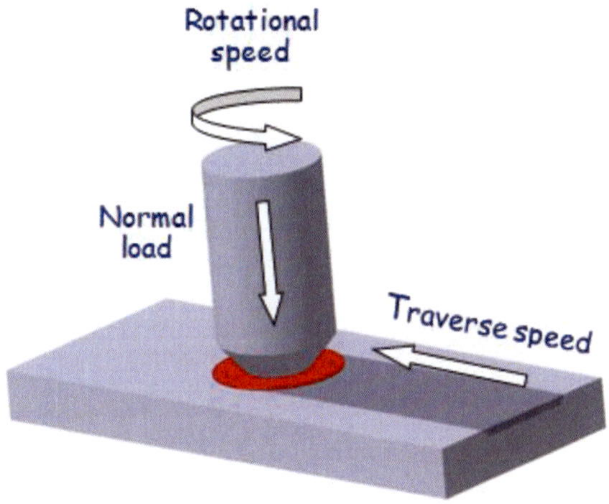

Figure 2. Basic principles of friction stir processing.

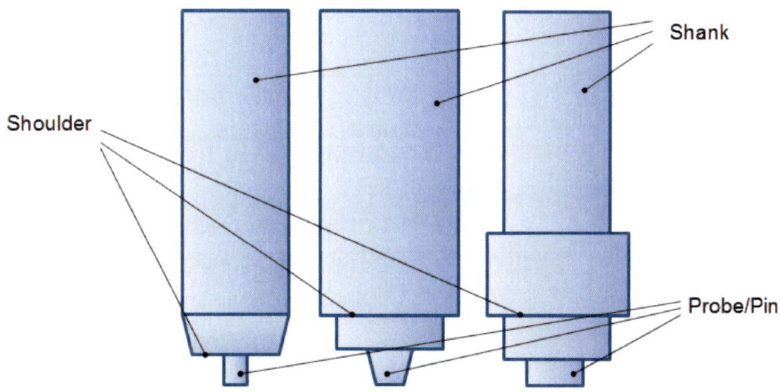

Figure 3. General appearance of friction stir tools.

Concerning microstructure evolution, current studies on the FSW or FSP layers are conducted to address several issues as:

- understanding the grain refinement mechanism
- analysis of structural changes in relation to the process parameters
- thermo-mechanical numerical simulation

Although there is still a controversy about the grain-refinement mechanism in the stirred zones, it is generally accepted that the grain refinement is due to dynamic recrystallization (Ma, 2008). Pilchak and Williams (2010) explain that adding the adjective "dynamic" in front of recrystallization implies that the process occurs concurrently with straining and new grains with high-angle boundaries form homogeneously throughout the microstructure, while as in conventional recrystallization, transformation is initially located in certain areas in the microstructure and gradually increases in magnitude.

For friction stir processing are generally used the same tools as for friction stir welding. The shank is cylindrical or conical to fit a standard tool holder and the active part consists of a shoulder and a probe (Figure 3).

Table 1. Synthesis of friction stir tool types

Criterion		Type	Remarks
Material		Monobloc tools	• higher mechanical resistance and rigidity
		Different materials for shank, shoulder or/and probe	• replaceable active part or probe when worn out
Adjustability		Fixed tool	• single piece tool • the length of the probe cannot be adjusted
		Adjustable probe	• probe is a separate piece with adjustable length
		Self-reacting tool	• bobbin type tool with the probe between two parts of the shoulder • tool must be positioned normally to the workpiece surface • only used for friction stir welding
Shoulder	End surface shape	Flat end surface	• simple shape most frequently used • sometimes flashing occurs
		Concave end surface	• most adequate shape retains • workpiece material underneath tool • needs to be slightly tilted
		Convex end surface	• usable for welding thin metal sheets with small diameter tools • full contact shoulder – workpiece • workpiece material tends to be displaced outside the edge of the shoulder
	End surface feature	Featureless end surface	• most common shoulder end surface style
		End surface with special features (e.g., scrolls, grooves)	• increase friction • mixing process
Probe (Pin)	Outer surface type	Cylindrical probe	• tools are easy to manufacture
		Conical probe	
		Prismatic shape	• flat lateral surfaces act as the cutting edge of a cutter. • effective mixing
		Pyramidal shape	
	Outer surface profile	Smooth	• usual shape of the probe
		Fluted	• easier flow of the plasticized material
		Threaded	• better material stirring

The amount of heat produced by the rotating tool and the material flow strongly depends on the shape of shoulder and probe. In Table 1 are presented the most common tool types (Mishra, 2005, Ma, 2008 and Zhang, 2012):

Due to its undeniable advantages, friction stir initially applied to aluminum and aluminum alloys (Mishra et al., 2005) extended lately to other materials including copper and copper alloys (Nakata, 2005 and Menon et al., 2012), magnesium alloys (Darras et al., 2007, Zhang et al., 2005), steels (Chabok and Dehghani, 2010, Mehranfar and Delgani, 2011, Dehghani and Chabok, 2011, Wei and Nelson, 2012) or titanium alloys (Reshad Seighalani et al., 2010, Zhang et al., 2008, Jata et al., 2004, Lauro, 2012).

Studies on various materials demonstrated that the quality of friction stir welds and the quality of surface layer processed by friction stir generally depend on:

- base material mechanical and thermal properties
- tool material, shape and dimensions (Leal et al., 2008, Elangovan and Balasubramanian, 2008, Hattingh et al., 2008)
- process parameters (rotational speed, traverse speed, axial force, tilt angle, cooling system) (Mishra et al., 2005, Jayaraman, 2009, Karthikeyan et al., 2005, Zhang, 2009)

In order to ensure proper surface quality with uniformly refined nanograined structure and homogeneous characteristics it is essential to use high-performance equipment. A very good control of process parameters can be obtained either on purpose-built machines, or on conventional CNC machines specially adapted for friction stir. During friction stir process tool wear may cause excessive increase of temperature in the tool-workpiece contact zone. Friction stir is a solid state process, therefore temperatures reached by the workpiece during friction stir are near the solidus temperature, but should not exceed it.

Overheating may be prevented by adjusting the rotating speed and the normal load/ penetration depth of the tool, since the amount of heat produced by friction depends mostly on these parameters.

Cooling rate is also an important parameter that influences recrystallization process and consequently determines the size of the grains in processed surface of the workpiece. An increased traverse speed decreases the amount of heat input resulting in a finer grain structure. Because of this, traverse speed is a very important parameter when a very fine structure is aimed to be obtained by friction stir. According to Babb et al. (2011) friction stir machines generally use one or two of the following input parameters for dynamic control: tool rotational speed, traverse speed, Z-axis load (normal load) and Z-axis position.

Numerous patents have been obtained for specialized friction stir control methods and devices. One of the most simple and efficient ways to implement dynamic control during friction stir is to monitor the temperature and modify input parameters in case measured temperature reaches a preset maximum limit as shown in Figure 4. A complex system with multiple functions has been developed by Cruz et al. (2012). Their system is monitoring and controlling several parameters: down force, rotational speed, traverse speed, acceleration speed and tool path.

The last one is controlled using a laser system and an alarm system to prevent collision between tool and workpiece backing surface.

Until recently, most studies have been conducted on friction stir welding and rather few have focused on friction stir processing as a surface modification method. Friction stir is a solid state thermo-mechanical process that may significantly improve the surface layer of the workpiece.

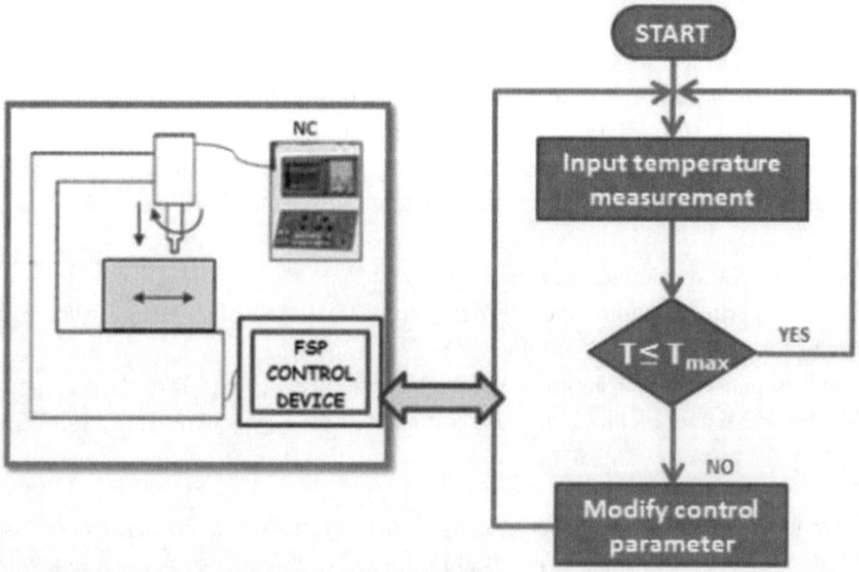

Figure 4. Dynamic control during friction stir processing.

In recrystallized nanograined layer, most important ameliorated characteristics, without being limited to, are: refined microstructure, higher strength and hardness, better fatigue resistance, abrasion resistance and corrosion resistance. Final microstructure and properties of the processed zone strongly depend on the FSP parameters.

The temperature and mechanical aspects must be balanced and optimal conditions need to be found in order to obtain desired nanograin structure when friction stir processing different materials. Reinforcement particles (e.g., ceramic materials) may be added on a metallic substrate before friction stir processing, the result being a composite superficial layer with enhanced mechanical and tribological properties.

2. Processing Nanograined Layers by Friction Stir Processing of Various Metal and Alloys

A. FSP of Light and Low Melting Point Metals and Alloys

Friction stir process is very suitable to light weight and non-ferrous materials such as aluminum, magnesium, copper and their alloys. Low melting temperature, high flow and ductility corroborated with very good heat conductivity allow smooth processing of surface layers by friction stir. Frequent defects of cast alloys are porosity and cold flakes. By friction stir these defects are eliminated from the surface of the workpiece, while tensile strength and hardness increase due to grain refinement.

Hoffman and Vecchio (2005) demonstrated that friction stir processing done underwater leads to a much finer microstructure of aluminum alloy Al-6061–T6 than friction stir processing in air. After submerged friction stir processing the nanolayer created at the workpiece surface had small sized grains under 200nm.

Compared to the initial size of the grains (approximately 50 μm), an important dimensional decrease was obtained considering that, generally, very fine grains are difficult to obtain by friction stir in aluminum alloys due to fast recrystallization. The authors of this study demonstrated that in the particular case of submerged friction stir processing, the faster cooling rate can retard grain growth and produce very fine microstructure.

Significant improvement of microstructural, tribological and mechanical properties was observed after friction stir processing of as-cast A413 aluminum alloy (Mahmoud et al., 2012). Along with a structurally refined superficial layer, reduction of pores and cavities contributed to ameliorate wear resistance. Finer grains, higher wear resistance and hardness were obtained when the traverse speed was decreased and the rotational speed was increased. Ultrafine-grained microstructure with an average grain size of 100nm was produced on aluminum alloy 7075Al-T651by using water, ice and methanol as coolant immediately after friction stir (Su et al., 2003).

Mishra et al. (2005) noted that recrystallized grain size can be reduced by decreasing the ratio of tool rotational speed/traverse speed. They obtained very good results on cast Al–Zn–Mg–Sc with a threaded pin tool, producing fine grains of with minimum size of 680 nm. By friction stir processing using a tool with a conical pin, low rotational speed and high traverse speed, Kwon et al. (2002, 2003) managed to produce nanolayers in aluminum alloys with grain size less than 1μm. At a rotational speed of 560 rpm and traverse speed of 155 mm/min, ultra-fine grains of 500nm were produced.

Studies on the effect of cooling rate on the grain size of the stirred zone after processing aluminum alloy Al5083, showed that the grain size of the stirred zone can be maintained in the nanometer range by immediate rapid cooling rate (Yazdipoura et al., 2009). A numerical model was elaborated and it revealed that decreasing the rotational speed and increasing the traverse speed decreases the grain size of stirred zone.

The effect of process parameters on the grain size during friction stir of cast Al–7Si–0.6Mg alloy was studied by Jana et al. (2010). They tempted to limit grain growth during recrystallization phase using different combinations of rotational speeds and traverse speeds during single pass and multi-pass friction stir runs. The team failed to produce grains under 1μm, but it was demonstrated that multi-pass runs were more resistant to abnormal grain growth.

Multi-pass friction stir processing determined significant improvement of surface layer's mechanical properties of ADC12 aluminum die cast alloy (Nakata et al., 2006). Even if microstructure could not be refined down to a nanoscale (1-2 μm average grain size), tensile strength and hardness almost doubled after friction stir processing. Moreover, experimental data confirmed the Hall-Petch relationship (Equation 1) between the grain size d and hardness H_v in the stirred zone:

$$H_v = H_0 + k_H \cdot d^{-1/2} \tag{1}$$

where H_0 and k_H are constants.

Even if until now most studies about FSP have been carried out on aluminum alloys, other materials were recently tested for friction stir processing. Among them are copper and copper alloys which allow fabrication of ultrafine grains by friction stir processing.

Su et al. developed nanocrystalline structures in oxygen-free high thermal conductivity (OFHC) copper by using FSP with rapid cooling and a small diameter tool (Su et al., 2011). Crystallites having sizes of a few tens of nanometers were observed in the pin tool path, while the average grain size of newly formed nanolayer was 109 nm. In case of magnesium alloys, friction stir was not proved to be effective in producing nanograins as in case of other lightweight materials. During friction stir welding of magnesium alloys refined structures were obtained with improved characteristics and grain sizes between 10-16 μm (Yang et al., 2011).

B. FSP of Hard and High Melting Point Metals and Alloys

Recently, friction stir processing has progressed into high melting temperature and high strength materials. Low conductivity of these materials combined with high temperatures required to bring them in plastic domain, make heat to be partially conducted into the tool and partially reflected internally in the workpiece from the surface. In addition, high loads are applied to the friction stir tool producing accelerated tool wear or even tool breakage. Furthermore, tool life of friction stir tools is considerably reduced by overheating.

In case of steel FSP, for an actual temperature of approximately 900 °C in the workpiece material, tool tip temperature is expected to be higher than 1200 °C (Ohashi et al., 2013). To avoid rapid wear, generally for FSP of steel are used carbide and polycrystalline boron nitride tools (Choi et al., 2009, Hovanski et al., 2007). Their research demonstrated that for high carbon steels mechanical properties and the microstructure significantly depend on peak temperature and cooling rate. If friction stir temperature exceeds the A1 point (723 °C) and the cooling rate is too elevated, hard martensitic structure is obtained. A better refined microstructure, a two-phase structure of fine ferrite and globular cementite, was obtained for low cooling rate and peak temperature below A1 (Cui et al., 2007).

Very interesting results were obtained by Dehghani and Chabok (2010, 2011) who studied nanolayers obtained by friction stir processing on interstitial free steels. They established an optimal rotating speed for which very fine grains (50-120 nm) were produced and demonstrated that rotating speed above and below this rate increased the grain size.

Chen and Nakata (2009) applied friction stir processing on SKD61 tool steel and showed that low heat input is an effective method to produce a surface layer composed of relatively fine grain martensitic structure with good tensile properties and wear resistance. Other studies on tool steel AISI D2 (Sorensen, 2007), revealed that by FSP it is possible to refine microstructure to a grain size on the order of a micron or less, to enhance surface hardness, to increase wear resistance and to extend tool life. Several studies have been carried out for the benefit of gas and oil industry focusing on cavitation erosion of steel parts and revealed that resistance to cavitation erosion increases continuously with decreasing grain size (Kwok et al., 1998, Hattori and Ishikura, 2010, Pohl and Stella, 2002). In case of duplex stainless steel, (DSS) friction stir processing has proven to be a simple and economical way to produce surface layers with austenite nanograins (Escobar et al., 2013). Experimental results show a significant improvement of cavitation erosion resistance. The authors of this chapter studied friction stir processing of C48 carbon steel. Experimental studies were carried out in order to determine the influence of process conditions (rotational speed, translation speed and axial force) and to evaluate the performance in terms of size of newly formed nanocrystallites, depth change and hardness of transformed layer. It was demonstrated that in case of C48 the grain size was mostly influenced by the traverse speed (Langlade et al., 2012). Figure 5 shows two micrographs of C45 after FSP with the same rotational speed (2000rpm), same axial load (900N) and different traverse speeds (10mm/min and 70 mm/min). Smaller nanograins with average grain size of 0.8μm were produced with higher traverse speed, while decreasing traverse speed led to a thinner stirred layer with average grain size of 1.3μm.

Progress in tool materials recently enabled friction stir processing of superalloys and titanium alloys. Using polycrystalline cubic boron nitride (PCBN) tools (Yang et al., 2011) or boron nitride tools (Barabash et al., 2009), friction stir processing was applied to nickel based superalloys Inconel 600, respectively Inconel 738.

Although material properties were improved and microstructure was considerably refined down to an average of 5 μm, nanoscale grains could not be produced via friction stir.

Concerning titanium alloys, studies on friction stir welding (Farias A. et al., 2013, Mironov S. et al., 2008, Romulu M. et al., 2010, Su J. et al., 2013) or in friction stir process (Farnoush H. et al., 2012, Langlade C. et al., 2013, Li B. et al., 2013, Ma Z.Y., 2008, Pilchak L., Williams J.C., 2010, Zhou L. et al., 2010) confirmed that these alloys have a good potential to provide enhanced mechanical properties in comparison with other welding techniques or engineering surface treatments. Most of the published works are dedicated to friction stir of Ti-6Al-4V, the most currently used over one hundred known titanium alloys, and describe the microstructure and the mechanical behavior in terms of microhardness, tensile and fatigue. Ma (2008) in an interesting review dedicated to FS technology relates that concerning titanium alloys there is an increasing interest in the applications of near-net shape structural castings of titanium alloys. In this sense, via FSP the as-cast microstructure can be replaced by a fine superficial grained wrought microstructure. No information on stirred pure Ti exists in the consulted literature, however this material superficially hardened by such a treatment could be an alternative to titanium alloys which actually turn out to be harming to human body by spreading elements such as aluminium responsible for Alzheimer. The relationship between processing parameters, microstructure and mechanical properties is not fully established and understood (Su et al. 2013). The friction stir parameters concerning Ti-6l-4V are based on the friction welding know-how and are ranged in between 2.5 - 60 mm/min for the traverse speed and 500-1000 RPM rotational speed.

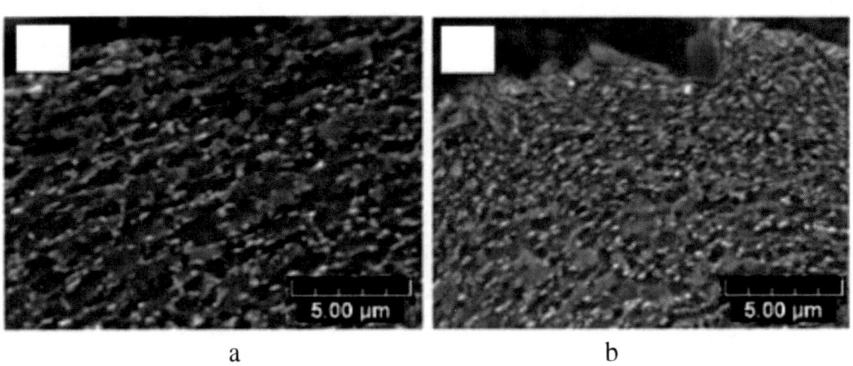

Figure 5. Nanograin layer obtained on C45 steel by friction stir processing at 2000rpm, 900 N, 10 mm/min (a) and 2000 rpm, 900 N,70mm/min (b) (Langlade et al., 2013).

The friction stir tools are the same as for friction stir welding and the materials for these tools are: W-Re alloys (Zhou L. et al., 2009), Mo-based alloys (Mironov et al., 2008), W -1% La_2O_3 (Su J. et al., 2013), WC-13%Co (Li B. et al., 2013) or alumina (Langlade C. et al., 2013). In the case of titanium alloys, a conical tool is needed because of the low thermal conductivity of titanium (Farias et al., 2013) and the cooling of the tool is recommended (Ma, 2008, Zhou L. et al., 2010).

Additionally, it has been found that during FSW or FSP on Ti-6Al-4V different microstructures can be created depending on the peak temperature in the stirred zones. The critical level of this temperature is the $\boldsymbol{\beta}$-transus temperature, approximately 880°C, relative to the allotropic transformation of titanium from alpha phase ($\boldsymbol{\alpha}$) to beta phase ($\boldsymbol{\beta}$). In the case of a subtransus processing, the stir zone microstructure consists of equiaxed $\boldsymbol{\alpha}$ grains, whereas for a supertransus processing, it consists of prior $\boldsymbol{\beta}$ grains containing $\boldsymbol{\alpha+\beta}$ colonies (Pilchak and Williams, 2010).

The microstructural refinement (Ma, 2008) of Ti-6Al-4V implies increased hardness, yield strength, crack initiation resistance and fatigue life. Recent studies conducted by Li et al. (2013) on FSP of Ti-6l-4V have related the possibility to obtain a nitriding superficial layer of stirred ultrafine grains under nitrogen atmosphere. The hardness of this layer lies around $1000HV_{0.2}$.

In order to synthetize data from recent work in the domain of friction stir, Table 2 presents results on most studied metals and alloys. To clearly demonstrate grain size reduction due to this process, histogram of Figure 6 presents data for aluminium and magnesium alloys, copper, titanium, stainless steels and two of most known superalloys.

Additionally, for better illustration grain refinement obtained by friction stir processing, the micrographs of Figure 7 exhibits different stirred structures. An interesting observation revealed (Figure 7c) is that stainless steel allowed the most significant gran size reduction (from tens of microns to a nanometer scale).

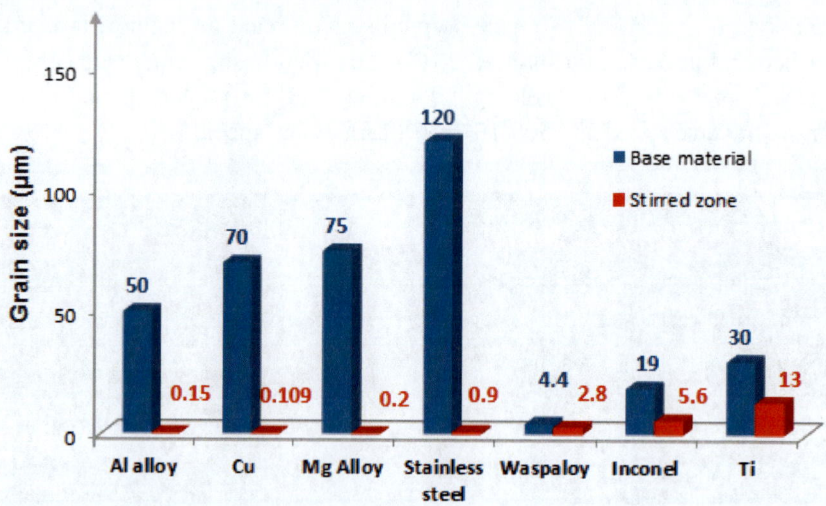

Figure 6. Grain size refinement via friction stir processing.

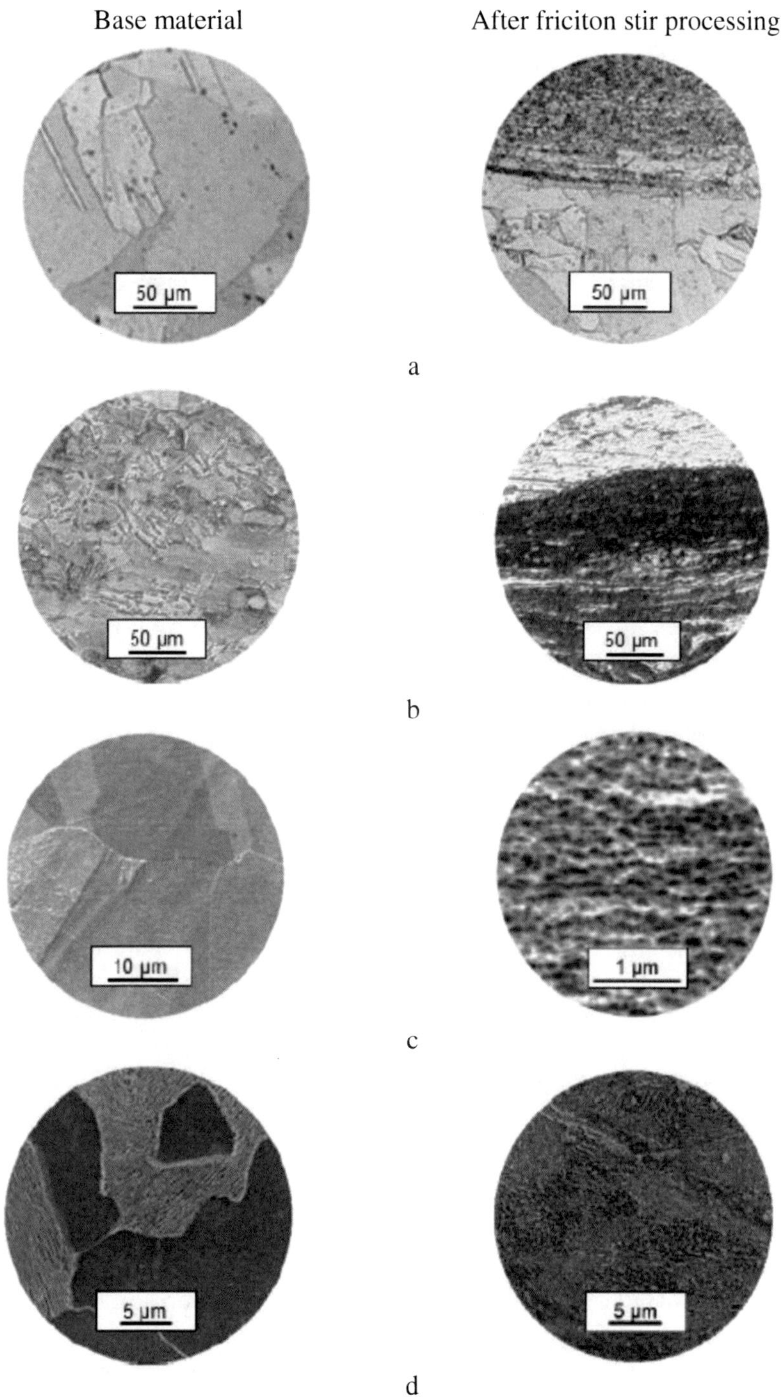

Langlade et al., 2013.

Figure 7. Microstructural modification induced by friction stir processing of Cu (a), Ti6AV (b), C316 stainless steel (c) and C45 carbon steel (d).

3. Processing Nanocomposite Layers by Friction Stir Processing

FSP for Fabrication of Metal Matrix Composites

Metal matrix composites are composite materials with generally two constituents, one being a metal and the other being reinforcement material. The latter is embedded into the metallic matrix to improve its properties.

Friction stir processing has been successfully used to obtain metal matrix composites (MMC) aiming to improve wear resistance and mechanical properties of surface layer. A thin composite layer is created on a metallic substrate by inserting added materials like graphite or ceramic materials (e.g., aluminum oxide, silicon carbide, cubic boron nitride, titanium carbide). The purpose of MMC is to improve surface quality, by increasing wear resistance or by enhancing lubrication properties, while preserving strength and toughness of the bulk material.

Friction stir has been proved to be an efficient technology when a composite surface was fabricated on 5083Al aluminum alloy plates by adding ceramic nanoparticles of silicon carbide SiC (Mishra et al., 2003). Ceramic powder of SiC was mixed with methanol and applied to workpiece surface before friction stir. Very good bonding was found between ceramic particles and aluminum alloy substrate. Remarkable results were obtained concerning hardness: at 27% volume of silicon carbide the microhardness doubled and a 44% increase of microhardness was observed at 13% volume of silicon carbide in the Al-SiC composite surface layer compared to the base material.

Shafiei-Zarghan et al. (2009) reinforced by multipass friction stir processing an aluminum alloy surface with ceramic nanosized particles of Al_2O_3. Experimental research showed that Al2O3 particles were not enough dispersed only after one pass, instead they formed clusters of 550 nm in average. Better results were obtained after three or four friction stir passes when cluster size was less than 150 nm. In addition, the grains of 6082 Al matrix, initially having 120 μm, were downsized to 890 nm and to 660nm after the third and fourth pass respectively.

A steel/titanium carbide (TiC) nano-composite surface layer with ultra-fine grains of less than 600 nm was successfully fabricated by friction stir processing (Ghasemi-Kahrizsangi and Kashani-Bozorg, 2012). High hardness, high strength, high rigidity, very good wear resistance, low friction coefficient, high melting point and high chemical stability make TiC a valuable material for coating steel parts.

In the field of composites, especially those obtained by powder metallurgy (PM), that suffer from microstructural problems like the presence of prior particle boundaries and porosity, friction stir can be used like a homogenization tool. The works of Ma, 2008 have shown the possibility of FSP to create homogeneous nanocomposites on a local scale from components obtained by PM (Anvari et al., 2013, Devaraju et al., 2013, Khayyamin et al., 2013, Khodabakhshi et al., 2013, Lee et al., 2013, Shafiei-Zarghan et al., 2009, You et al., 2013a). Table 3 provides a summary of recent works concerning different FSP conditions on various composites.

Table 2. A summary of friction stir processing conditions that led to nanograined surface layer on different metals and alloys

Material	Tool material and characteristics					Friction stir process parameters and conditions			Grain size (nm)	References
	Tool material	Tool shape	Shoulder diameter (mm)	Pin diameter (mm)	Pin length (mm)	Rotation speed (rpm)	Traverse speed (mm/min)	Special conditions		
Al-6061	Tool steel	Cylindrical pin	12.7 19.1	3.18	2.79	1000	60	FSP submerged under water	100-200	Hoffman and Vecchio, 2005
Al 5083	H13 steel	Cylindrical pin	15	4	1	1000 1300	100 150	Mixture of methanol and dry ice cooling	300-400	Yazdipour et al., 2009
Al 1050	SKD 61	Special design	7	3	3	560	155	-	500	Kwon et al., 2003
Al–4Mg–1Zr	*NA	Threaded pin	12	4	NA	600	25.4	-	700	Ma and Mishra, 2005
OFHC copper	H13 steel	Cylindrical pin	7.5	2.5	2	800	120	Dry ice cooling	50-300	Su et al., 2011
60/40 brass	NA	Cylindrical pin	12	4	2	1000-1500	500-2000	-	800-1500	Park et al., 2004
Mg–Al–Zn	NA	Cylindrical pin	10	3	3	1200	28-33	Liquid nitrogen cooling	100-300	Chang et al., 2007
interstitial free steel	WC	No pin	16	-	-	1600	31.5 63	Plunge depth: 0.5 mm water and ice cooling	50-120	Dehghani and Chabok, 2011
Super-austenitic steel	WC	No pin	16	-	-	2600	30	Workpiece preheated to about 200 °C Water and ice cooling	50-90	Mehranfar and Dehghani, 2011
Stainless steel	WC	Cylindrical pin	16	5	1.5	800	50	Plunge depth: 1.7 mm Argon shielded tool	900	Saeid et al., 2010
Low carbon steel	TiC-based cermet	No pin	14	-	-	400	50	Flowing water cooling	1000	Xue et al., 2013

*NA-not available.

Table 3. A summary of friction stir processing conditions that led to nanograined surface layer on different metals and alloys

Material	Tool material and characteristics					Friction stir process parameters and conditions			Nanocomposite grain size		References
	Tool material	Tool shape	Shoulder diameter (mm)	Pin diameter (mm)	Pin length (mm)	Rotation speed (rpm)	Traverse speed (mm/min)	Special conditions	Reinforcement particles size	Stirred zone grain size	
6082 Al/Al_2O_3	H13 tool steel	Specially designed	16	5	4	1000	135	groove filled with Al_2O_3 powder 4 overlapped passes	Al_2O_3: ~50 nm	~660 nm	Shafiei-Zarghan et al., 2009
6061 Al/Cr_2O_3	H13 tool steel	Conical threaded pin	18	6 to 4	4	630	100 mm/min	Cr_2O_3 powder applied on Al6061 plate by atmosphere plasma spray 6 overlapped passes	$Al_{13}Cr_2$: <100 nm $Al_{11}Cr_2$: <100 nm	~6 µm	Anvari et al., 2013
Al/Mg/CuO	NA	Standard tool with pin	18	6	1.2	500	15	Al/Mg/CuO powder metallurgy 4 overlapped passes	Al_2Cu: ~ 200nm Mg O < 1 µm	~600 nm	You et al., 2013a
5754 Al/Si_3N_4	NA	Tool with pin	NA	NA	NA	1400	100	groove filled with Si_3N_4 powder 2-4 overlapped passes	Si_3N_4: ~140 nm	~7µm	Moghaddas and Kashani-Bozorg, 2013
Al/Al-Mo	NA	Standard tool with pin	18	6	1.2	1400	30-45	Al/Mo powder metallurgy 4 passes	$Al_{12}Mo$: ~200 nm	~2 µm	Lee et al., 2013
Alumix 431D/SiC	PVD TiN/ZrN	Tapered threaded pin	12	5.1 base 4 end	NA	454	88	Al/ two types of SiC powder metallurgy	SiC: 610 nm SiC: 6.25 µm	NA	Izadi et al., 2013
Al/Al_2O_3	H13 tool steel	Tool with pin	12	3	2.8	800-1360	30-200	Al/Al_2O_3 powder metallurgy	Al_2O_3: ~92 nm	1.4-2 µm	Khodabakhshi et al., 2013
6061 Al /SiC/Gr	H13 tool steel	Tapered threaded pin	24	8	3.5	3000	40	groove filled with SiC/Gr powder Axial force 5KN	SiC: ~5 µm Gr: ~5 µm	NA	Devaraju et al., 2013
Al/SiO_2	NA	Tool with pin	16	6	1.2	500-2000	15-90	Al/SiO2 powder metallurgy	Al_2O_3: ~20 nm	~1.5µm	You et al., 2013b
5A06/ $Al_{84.2}Ni_{10}La_{2.1}$	NA	Tool with pin	16	4	NA	500-1000	40-150	groove filled with Al–Ni–La amorphous strip Plunge depth: 0.5 mm air cooling	$Al_{84.2}Ni_{10}La_{2.}$: 90–400 nm	NA	Liu et al., 2013
Al5083/ Gr/Al_2O_3	H13 tool steel	Threaded tool	18	6	5	1250	20	groove filled with Al_2O_3/Gr powder 3 passes	Al_2O_3: ~80 nm Gr: 10-15 µm	NA	Mostafapour Asl and Khandani, 2013

Material	Tool material and characteristics					Friction stir process parameters and conditions			Nanocomposite grain size		References
	Tool material	Tool shape	Shoulder diameter (mm)	Pin diameter (mm)	Pin length (mm)	Rotation speed (rpm)	Traverse speed (mm/min)	Special conditions	Reinforcement particles size	Stirred zone grain size	
Al/CuO	NA	Standard tool with pin	18	6	1.2	500	15	Al/CuO powder metalurgy 4 passes	Al_2Cu: <100 nm Al_2O_3: <1µm	~1µm	You et al., 2013c
Cu/Gr	H13 tool steel	cylindrical pin Threaded pin Square pin Triangular pin	20	6.5	3	1600 rpm	20	groove filled with Gr powder 4 passes	Gr: ~5 µm	< 44.2 µm < 43 µm < 27.3 µm < 21.9 µm	Sarmadi et al., 2013
Cu/SiC Cu/TiC, Cu/B_4C, Cu/WC Cu/Al_2O_3	H13 tool steel	cylindrical pin	20	5	2.7	800-1200	20-60	grooves filled with SiC, TiC, B_4C, WC and Al2O3powder	SiC ~ 8 µm TiC ~ 2 µm B_4C ~ 4µm WC~ 5 µm Al_2O_3~10 µm	NA	Sathis-kumar et al., 2014
Cu/Y_2O_3	Tool steel	Scrolled shoulder, threaded pin	20	M6	3	500	40	groove filled with Y_2O_3 powder covered with thin Cu foil 3 passes water spray cooling	Y_2O_3: 10 nm	3 µm	Avettand-Fènoël et al., 2014
Cu/TiC	NA	conical pin	16 mm	4 to3	3	1000	50	a net of fine holes filled with TiC	TiC:7-16 µm	NA	Sabbaghian et al., 2014
Cu/SiC	NA	conical pin	16 mm	4 to 3 mm	3	1000	50	a net of fine holes filled with TiC	SiC:25 µm	NA	Akramifard et al., 2014
$AZ91/SiO_2$	H13 tool steel	Squared pin	NA	4.5x4.5	4	1250	63	groove filled with SiO_2	SiO_2: 10 nm	8 µm	Khayyamin et al., 2013
$AZ31/Al_2O_3$/ CNTs	NA	Conical pin	20	5.5 to 3.5	5	1050	33.4	groove filled with n-Al_2O_3 and CNTs	Al_2O_3: ~ 50 nm CNTs (carbon nanotubes)	NA	Lu et al. (2013)
Mg/nHA	H13 tool steel	Conical pin	15	5 to 3	2.7	1200 rpm.	12 mm/min 5kN load	groove filled with nHA powder 5kN load	nano-hydroxy-apatite (nHA) particles	3.5 µm	Ratna Sunil et al. (2014)

4. NUMERICAL SIMULATION OF FRICTION STIR PROCESS

The modeling of the friction stir process is a big challenge and visualization and analysis of the material flow, temperature field, stresses, and strains generated during the friction stir process can be obtained using numerical simulation (Figure 8).

This process can be seen as a multiphysics problem (Figure 9) and many strong physical links appear. The FSP leads to several coupled and non-linear physical phenomena such as high plastic deformation, heat generation due to friction and plastic deformation, mechanical stirring of the material, material flow, tool-work piece surface friction definition and microstructural changes as for the FSW process.

Most models have been developed to simulate the FSW process (Chiumenti et al., 2013, Hamilton et al., 2010, Kim et al., 2010, Neto et al., 2013, Pan et al., 2013, Pashazadeh et al., 2013). Very few models only concern the FWP. However, the strategies and techniques used for modeling the FSW are similar for the FSP. Actually, the physical involved phenomena by both processes, FSP and FSW are similar.

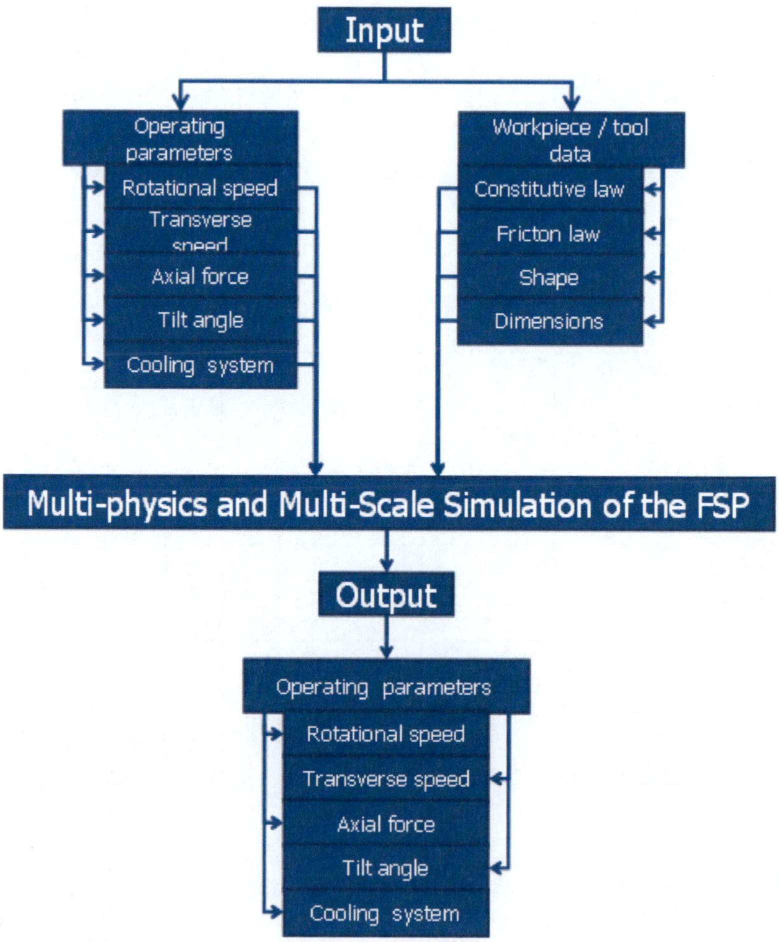

Figure 8. Numerical simulation of FSP.

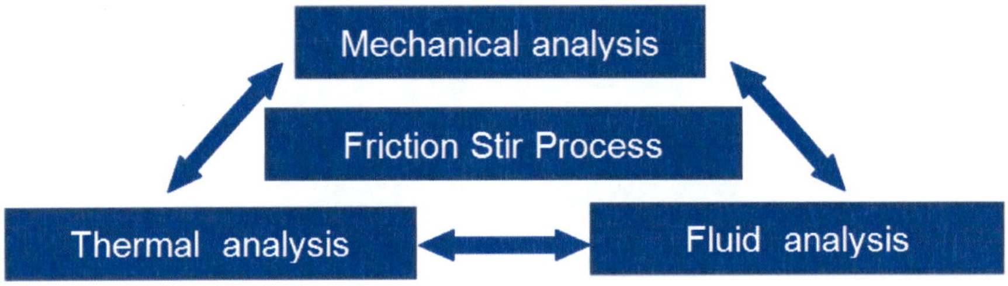

Figure 9. Multiphysics problem simulation.

Most often, two kinds of models are used: solid mechanics models and fluid dynamics models. Solid mechanics models allow the simulation of temperature distribution, stress, and strain. The tribological aspect of the process can also be taken into account.

Conventional formulations, such as Lagrangian and Arbitrary LagrangianEulerian (ALE) have been proposed in literature to study this type of problems. Lagrangian formulation is known for its capability to simulate material flow, including heat from friction but requires a complex remeshing. ALE formulation simulates flow and includes heat from friction and deformation, avoids degeneration of mesh. It allows for deformation of free surface. In (Tutunchilar et al., 2012), a fully coupled incremental Lagrangian thermo-mechanical simulation of FSP via employing commercial FEM software is presented. This study is able to predict the material flow pattern, temperature distribution.

More recently, discrete smoothed particle hydrodynamics have been used to model the FSW. The main advantage of this method over other classical numerical approaches is its ability to handle large deformations without failing due to excessive distortion, and be able to explicitly track the material history as in Lagrangian methods. Fluid dynamics models (CFD) allow the simulation of material flow around the tool and temperature distribution. Generally, the material flow is performed using an Eulerian formulation. The main drawbacks of CFD models are that residual stresses cannot be predicted and that free surface is neglected. Albakri et al. (2013) simulate FSP with in-process cooling through the backing plate using CDF analysis. In this paper, different aspects of the cooling process have been investigated to study their effects on the calculated temperature fields, flow stresses, grain size and hardness distributions. In the work of Yu et al. (2012), three-dimensional transient computational fluid dynamics (CFD) (fully coupled material flow and heat transfer simulation) models were developed to investigate the material flow and heat transfer during FSP. All these models can provide a better knowledge and understanding of the influence of the rotating speed and the normal load/penetration depth of the tool on the material flow, temperature and strains. From a practical point of view, Lagrangian, Eulerian, and ALE approaches have been used to numerically simulate the FSP, using software such as ABAQUS, Star-CCM+ (Albakri et al., 2013), Ansys/Fluent (Yu et al., 2012) and DEFORM-3D (Tutunchilar et al., 2012).

CONCLUSION

Friction stir processing is a novel technique with a multitude of benefits for surface engineering. Since friction stir was patented, a lot of industrial applications have been obtained due to numerous advantages.

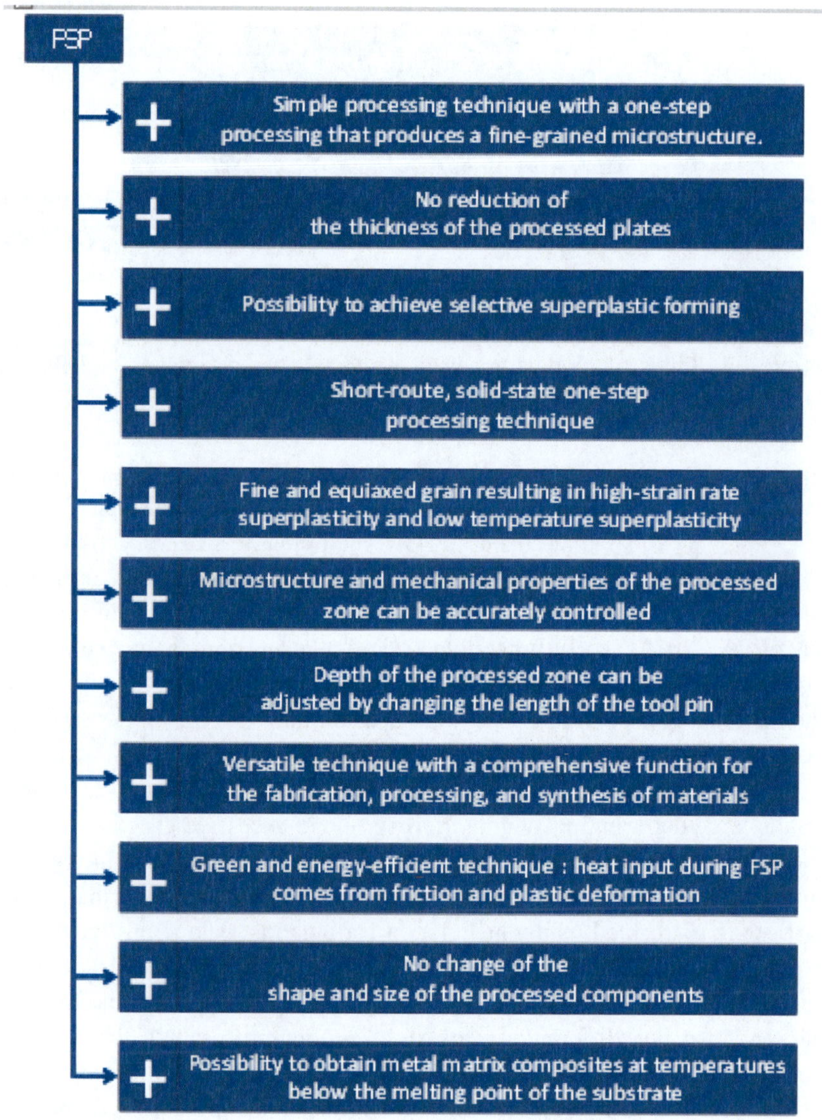

Figure 10. Advantages of the FS technique.

In this chapter, the major aspects and developments in friction stir processing are reviewed. It is shown that friction stir processing is an effective way to obtain surface nanolayers due to its numerous advantages summarized and illustrated in Figure 11 (Ma, 2008). Although friction stir processing is a promising technology, there are still important aspects that need further research. Fundamental research is required concerning the nature of material deformation and heat generation, the physics of FSP tool/processed material interface, the influence of process input parameters on the quality of processed surface etc. Particularly for high melting temperature or high strength alloy, reducing FSP tool wear is an important objective. Advancement in tool design and in tool material is expected in the near future. Regarding the machine-tools for FSP, developing machines with simultaneous dynamic control of several input parameters is a major challenge.

In spite of the challenges, studies clearly demonstrated the viability of friction stir processing. It is a step forward in surface engineering and many industries benefit and will continue to benefit from this innovative method.

REFERENCES

Akramifard, H. R., Shamanian, M., Sabbaghian, M., Esmailzadeh, M.: Microstructure and mechanical properties of Cu/SiC metal matrix composite fabricated via friction stir processing. *Materials and Design* 54, 838-844 (2014).

Albakri, A. N., Mansoor, B., Nassar, H., Khraisheh, M. K.: Thermo-mechanical and metallurgical aspects in friction stir processing of AZ31 Mg alloy - A numerical and experimental investigation. *Journal of Materials Processing Technology* 213(2), 279-290 (2013).

Anvari, S. R., Karimzadeh, F., Enayati, M. H.: A novel route for development of Al–Cr–O surface nano-composite by friction stir processing. *Journal of Alloys and Compounds* 562, 48-55 (2013).

Avettand-Fènoël, M.-N., Simar, A., Shabadi, R., Taillard, R., de Meester, B.: Characterization of oxide dispersion strengthened copper based materials developed by friction stir processing. *Materials and Design* 60, 343-357 (2014).

Babb, J. A., Steel, R., Packer, S. M., Williams, J.: Method for Using Modifiable Tool Control Parameters to Control the Temperature of the Tool During Friction Stir Welding. *US Patent* 7,909,231 B2, Mar. 22, 2011.

Barabash, O. M., Barabash, R. I., Ice, G. E., Feng, Z., Gandy, D.: X-ray microdiffraction and EBSD study of FSP induced structural/phase transitions in a Ni-based superalloy. *Materials Science and Engineering A* 524, 10-19 (2009).

Chabok, A., Dehghani, K.: Formation of nanograin in IF steels by friction stir processing. *Materials Science and Engineering A* 528, 309-313 (2010).

Chang, C. I., Du, X. H., Huang, J. C.: Achieving ultrafine grain size in Mg–Al–Zn alloy by friction stir processing. *Scripta Materialia* 57, 209-212 (2007).

Chen, Y. C., Nakata, K.: Evaluation of microstructure and mechanical properties in friction stir processed SKD61 tool steel. *Materials Characterization,* 60, 1471-1475 (2009).

Chiumenti, M., Cervera, M., de Saracibar, C., Dialami, N.: Numerical modeling of friction stir welding processes. *Computer Methods in Applied Mechanics and Engineering* 254 (0), 353-369 (2013).

Choi, D. H., Lee, C. Y., Ahn, B. W., Choi, J. H., Yeon, Y. M., Song, K., Park, H. S., Kim, Y. J., Yoo, C. D., Jung, S. B.: Frictional wear evaluation of WC-Co alloy tool in friction stir spot welding of low carbon steel plates. *International Journal of Refractory Metals and Hard Materials,* 27(6), 931-936, (2009).

Cruz, M. F., Freitas, G., Zanini, H., da Costa, J. A., de Oliveira Pereira, R. F., Pereira, E., Fernandez, F. F., Andena, M.: Friction Stir Welding Spindle Downforce and Other Control Techniques, Systems and Methods. *US Patent* 8,261,959 B2, Sep. 11, 2012.

Cui, L., Fujii, H., Tsuji, N., Nogi, K.: Friction stir welding of high carbon steel. *Scripta Materialia* 56(7), 637-640 (2007).

Darras, B. M., Khraisheh, M. K., Abu-Farha, F. K., Omar, M. A.: Friction stir processing of commercial AZ31 magnesium alloy. *Journal of Materials Processing Technology* 191, 77-81 (2007).

Dehghani, K., Chabok, A.: Dependence of Zener parameter on the nanograins formed during friction stir processing of interstitial free steels. *Materials Science and Engineering A* 528, 4325-4330 (2011).

Devaraju, A., Kumar, A., Kotiveerachari, B.: Influence of rotational speed and reinforcements on wear and mechanical properties of aluminum hybrid composites via friction stir processing. *Materials and Design* 45, 576-585 (2013).

Elangovan, K., Balasubramanian, V.: Influences of tool pin profile and tool shoulder diameter on the formation of friction stir processing zone in A6061 aluminium alloy. *Materials and Design* 29, 362-373 (2008).

Escobar, J. D., Velásquez, E., Santos, T. F. A., Ramirez, A. J., López, D.: Improvement of cavitation erosion resistance of a duplex stainless steel through friction stir processing (FSP). *Wear* 297(1-2), 998-1005 (2013).

Farias, A., Batalha, G. F., Prados, E. F., Magnabosco, R., Delijaicov, S.: Tool wear evaluations in friction stir processing of commercial titanium Ti–6Al-4V. *Wear* 302, 1327-1333 (2013).

Farnoush, H., Sadeghi, A., Bastami, A. A., Moztarzadeh, F., Mohandesi, J. A.: An innovative fabrication of nano-HA coatings on Ti-CaPnanocomposite layer using a combination of friction stir processing and electrophoretic deposition. *Ceramics International* 39, 1477-1483 (2013).

Ghasemi-Kahrizsangi, A., Kashani-Bozorg, S. F.: Microstructure and mechanical properties of steel/TiCnano-composite surface layer produced by friction stir processing. *Surface and Coatings Technology* 209, 15-22 (2012).

Hamilton, R., Mackenzie, D., Li, H.: Multi-physics simulation of friction stir welding process. *Engineering Computations* 27(8), 967-985 (2010).

Hattingh, D. G., Blignault, C., van Niekerk, T. I., James, M. N.: Characterization of the influences of FSW tool geometry on welding forces and weld tensile strength using an instrumented tool. *Journal of Materials Processing Technology* 203, 46-57 (2008).

Hattori, S., Ishikura, R.: Revision of cavitation erosion database and analysis of stainless steel data. *Wear* 268, 109-16 (2010).

Hofmann, D. C., Vecchio, K. S.: Submerged friction stir processing (SFSP): An improved method for creating ultra-fine-grained bulk materials, *Materials Science and Engineering A* 402, 234-241 (2005).

Hovanski, Y., Santella, M. L., Grant, G. J.: Friction stir spot welding of hot-stamped boron steel. *Scripta Materialia* 57(9), 873-876 (2007).

Izadi, H., Nolting, A., Munro, C., Bishop, D. P., Plucknett, K. P., Gerlich, A. P.: Friction stir processing of Al/SiC composites fabricated by powder metallurgy. *Journal of Materials Processing Technology* 213, 1900-1907 (2013).

Jana, S., Mishra, R. S., Baumann, J. A., Grant, G.: Effect of process parameters on abnormal grain growth during friction stir processing of a cast Al alloy, *Materials Science and Engineering A* 528, 189-199 (2010).

Jata, K. V., Subramanian, P. R., Reynolds, A. P., Trapp, T., Helder, E.: Friction Stir Welding of Titanium Alloys for Aerospace Applications: Microstructure and Mechanical

Behavior. *Proceeding of the Fourteenth International Offshore and Polar Engineering Conference*, Toulon, May 23-28, 2004.

Jayaraman, M., Sivasubramanian, R., Balasubramanian, V.: Effect of Process Parameters on Tensile Strength of Friction Stir Welded Cast LM6 Aluminium Alloy Joints, *Journal of Materials Science and Technology* 25 (5), 655-664 (2009).

Karthikeyan, L., Senthilkumar, V. S., Padmanabhan, K. A., On the role of process variables in the friction stir processing of cast aluminum A319 alloy. *Materials and Design* 31, 761-771 (2010).

Kennedy, D., Xue, Y., Mihaylova, M.: Current and Future Applications of Surface Engineering. *The Engineers Journal* (Technical) 59, 287-292 (2005).

Khayyamin, D., Mostafapour, A., Keshmiri, R.: The effect of process parameters on microstructural characteristics of AZ91/SiO2 composite fabricated byFSP. *Materials Science and Engineering A* 559, 217-221 (2013).

Khodabakhshi, F., GhasemiYazdabadi, H., Kokabi, A. H., Simchi, A.: Friction stir welding of a P/M Al–Al_2O_3 nanocomposite: Microstructure and mechanical properties. *Materials Science and Engineering A* 585, 222-232 (2013).

Kim, D., Badarinarayan, H., Kim, J., Kim, C., Okamoto, K., Wagoner, R. H., Chung, K.: Numerical simulation of friction stir butt welding process for AA5083-H18 sheets. *European Journal of Mechanics - A/Solids* 29(2), 204-215 (2010).

Kwok, C. T., Man, H. C., Cheng, F. T.: Cavitation erosion and damage mechanisms of alloys with duplex structures. *Materials Science and Engineering A* 242, 108-120 (1998).

Kwon, Y. J., Saito, N., Shigematsu, I.: Friction stir process as a new manufacturing technique of ultrafine-grained aluminum alloy, *Journal of Materials Science Letters* 21, 1473-1476 (2002).

Kwon, Y. J., Shigematsu, I., Saito, N.: Production of ultra-fine grained aluminum alloy using friction stir process. *Materials Transactions* 44(7), 1343-1350 (2003).

Langlade, C., Schlegel, D., Gete, E., Roman, A., Folea, M.: Formation of a TTS layer on Steel Samples by Friction Stir Processing. *Proceedings of the 26th International Conference on Surface Modification Technologies*, Ecully, France, June 20-22, 2012.

Langlade, C., Schlegel, D., Gete, E., Roman, A., Folea, M.: Friction stir tests on different materials. *Internal report of IRTES-LERMPS laboratory*, 2013.

Lauro, A.: Friction stir welding of titanium alloys, *Welding International* 26(1), 8-21 (2012).

Leal, R. M., Leitão, C. Loureiro, A. Rodrigues, D. M., Vilaça, P.: Material flow in heterogeneous friction stir welding of thin aluminium sheets: Effect of shoulder geometry. *Materials Science and Engineering A* 498, 384-391 (2008).

Lee, I. S., Kao, P. W., Chang, C. P., Ho, N. J.: Formation of Al-Mo intermetallic particle-strengthened aluminum alloys by friction stir processing. *Intermetallics* 35, 9-14 (2013).

Li, B., Shen, Y., Luo, Hu W.: Fabrication of TiCp/Ti–6Al–4V surface composite via friction stir processing (FSP): Process optimization, particle dispersion-refinement behavior and hardening mechanism. *Materials Science and Engineering A* 574, 75-85 (2013).

Li, B., Shen, Y., Hu, W.: Surface nitriding on Ti–6Al–4V alloy via friction stir processing method under nitrogen atmosphere. *Applied Surface Science* 274, 356-364 (2013).

Liu, P., Shi, Q., Zhang, Y.: Microstructural evaluation and corrosion properties of aluminium matrix surface composite adding Al-based amorphous fabricated by friction stir processing. *Composites: Part B* 52, 137-143 (2013).

Lu, D., Jiang, Y., Zhou, R.: Wear performance of nano-Al_2O_3 particles and CNTs reinforced magnesium matrix composites by friction stir processing. *Wear* 305, 286-290 (2013).

Ma, Z. Y., Mishra, R. S.: Development of ultrafine-grained microstructure and low temperature (0.48 Tm) superplasticity in friction stir processed Al–Mg–Zr. *Scripta Materialia* 53, 75-80, (2005).

Ma, Z. Y.: Friction Stir Processing Technology: A Review. *Metallurgical and Materials Transactions* 39 A, 642-658 (2008).

Mahmoud, T. S., Mohamed, S. S.: Improvement of microstructural, mechanical and tribological characteristics of Al alloys using friction stir processing. *Materials Science and Engineering A* 558, 502-509 (2012).

Mishra, R. S., Ma, Z. Y., Charit, I.: Friction stir processing: a novel technique for fabrication of surface composite. *Materials Science and Engineering A* 341, 307-310 (2003).

Mishra, R. S., Z. Y. Ma, Z. Y.: Friction stir welding and processing. *Materials Science and Engineering R* 50, 1-78 (2005).

Mehranfar, M., Dehghani, K.: Producing nanostructured super-austenitic steels by friction stir processing. *Materials Science and Engineering A* 528, 3404-3408 (2011).

Menon, S. K., Pierce, F. A., Rosemark, B. P., Oh-Ishi, K., Swaminathan, S., Mcnelley, T. R.: Strengthening Mechanisms in NiAl Bronze: Hot Deformation by Rolling and Friction-Stir Processing. *Metallurgical and Materials Transactions A* 43A, 3687-3702 (2012).

Mironov, S., Zhang, Y., Sato, Y. S., Kokawa, H.: Development of grain structure in b-phase field during friction stir welding of Ti–6Al–4V alloy. *Scripta Materialia* 59, 27-30 (2008).

Moghaddas, M. A., Kashani-Bozorg, S. F.: Effects of thermal conditions on microstructure in nanocomposite of Al/Si3N4 produced by friction stir processing, *Materials Science and Engineering A* 559, 187-193 (2013).

Mostafapour Asl, A., Khandani, S. T.: Role of hybridratioin microstructural, mechanical and sliding wear properties of the Al5083/$Graphite_p/Al_2O_{3p}$ a surface hybrid nanocomposite fabricated via friction stir processing method. *Materials Science and Engineering A* 559, 549-557 (2013).

Nakata, K.: Friction stir welding of copper and copper alloys. *Welding International* 19(12), 929-933 (2005).

Nakata, K., Kim, Y. G., H. Fujii, H. T., Tsumura, T., Komazaki, T.: Improvement of mechanical properties of aluminum die casting alloy by multi-pass friction stir processing, *Materials Science and Engineering A* 437, 274-280 (2006).

Neto, D., Neto, P.: Numerical modeling of friction stir welding process: a literature review. *The International Journal of Advanced Manufacturing Technology* 65(1-4), 115-126 (2013).

Ohashi, R., Fujimoto, M., *Friction Stir Welding Apparatus and Method: US Patent* 8,528,803 B2, Sep. 10, (2013).

Pan, W., Li, D., Tartakovsky, A., Ahzi, S., Khraisheh, M., Khaleel, M.: A new smoothed particle hydrodynamics Newtonian model for friction stir welding: Process modeling and simulation of microstructure evolution in a magnesium alloy, *International Journal of Plasticity, Article in Press*, (2013).

Park, H. S., Kimura, T., Murakamic, T. Naganod, Y., Nakata, K., Ushio, M.: Microstructures and mechanical properties of friction stir welds of 60% Cu–40% Zn copper alloy, *Materials Science and Engineering A* 371, 160-169, (2004).

Pashazadeh, H., Teimournezhad, J.: Numerical investigation on the mechanical, thermal, metallurgical and material flow characteristics in friction stir welding of copper sheets with experimental verification. *Materials and Design* (2013).

Pilchak, L., Williams, J. C.: Microstructure and Texture Evolution during Friction Stir Processing of Fully Lamellar Ti-6Al-4V. *Metallurgical and Materials Transactions* 42A, 773-794 (2011).

Pohl, M., Stella, J.: Quantitative CLSM roughness study on early cavitation-erosion damage. *Wear* 252(5-6), 501-511 (2002).

Ramulu, M., Edwards, P. D., Sanders, D. G., Reynolds, A. P., Trapp, T.: Tensile properties of friction stir welded and friction stir welded-superplastically formed Ti–6Al–4V butt joints. *Materials and Design* 31, 3056-3061 (2010).

Ratna Sunil, B., Sampath Kumar, T. S., Uday Chakkingal, V. Nandakumar, Mukesh Doble: Friction stir processing of magnesium–nanohydroxyapatite composites with controlled in vitro degradation behavior. *Materials Science and Engineering C* 39, 315-324 (2014).

Reshad Seighalani, K., Besharati Givi, M. K., Nasiri, A. M., Bahemmat, P.: Investigations on the Effects of the Tool Material, Geometry, and Tilt Angle on Friction Stir Welding of Pure Titanium. *Journal of Materials Engineering and Performance* 19(7), 955-962 (2010).

Sabbaghian, M., Shamanian, M., Akramifard, H. R., Esmailzadeh, M.: Effect of friction stir processing on the microstructure and mechanical properties of Cu–TiC composite. *Ceramics International, in press* (2014).

Saeid, T., Abdollah-zadeh, A., Shibayanagi, T., Ikeuchi, K., Assadi, H.: On the formation of grain structure during friction stir welding of duplex stainless steel. *Materials Science and Engineering A* 527, 6484-6488 (2010).

Sarmadi, H., Kokabi, A. H., Seyed Reihani, S. M.: Friction and wear performance of copper–graphite surface composites fabricated by friction stirp rocessing (FSP). *Wear* 304, 1-12 (2013).

Sathiskumar, R., Murugan, N., Dinaharan, I., Vijay, S. J.: Prediction of mechanical and wear properties of copper surface composites fabricated using friction stir processing. *Materials and Design* 55, 224-234 (2014).

Shafiei-Zarghan, A., Kashani-Bozorg, S. F., Zarei-Hanzaki, A.: Microstructures and mechanical properties of Al/Al2O3 surface nano-composite layer produced by friction stir processing. *Materials Science and Engineering A* 500, 84-91 (2009).

Sorensen, C. D., Nelson, T. W., Packer, S. M., Allen, C.: Friction Stir Processing of D2 Tool Steel for Enhanced Blade Performance. *Friction Stir Welding and Processing* 4, 409-418 (2007).

Su, J. Q., Nelson, T. W., McNelley, T. R., Mishra, R. S.: Development of nanocrystalline structure in Cu during friction stir processing (FSP). *Materials Science and Engineering A* 528, 5458-5464 (2011).

Su, J., Wang, J., Mishra, R. S., Xu, R., Baumann, J. A.: Microstructure and mechanical properties of a friction stir processed Ti–6Al–4V alloy. *Materials Science and Engineering A* 573, 67-74 (2013).

Tutunchilar, S., Haghpanahi, M., Givi, M. K., Asadi, P., Bahemmat, P.: Simulation of material flow in friction stir processing of a cast Al-Si alloy. *Materials and Design* 40(0), 415-426 (2012).

Wei, L., Nelson, T. W., Influence of heat input on post weld microstructure and mechanical properties of friction stir welded HSLA-65 steel, *Materials Science and Engineering A* 556(2012) 51-59.

Xue, P., Xiao, B. L., Wang, W. G., Zhang, Q., Wang, D., Wang, Q. Z., Ma, Z. Y.: Achieving ultrafine dual-phase structure with superior mechanical property in friction stir processed plain low carbon steel, *Materials Science and Engineering A* 575, 30-34 (2013).

Yang, Q., Li, X., Chen, K., Shi, Y. J.: Effect of tool geometry and process condition on static strength of a magnesium friction stir lap linear weld. *Materials Science and Engineering A* 528, 2463-2478 (2011).

Yazdipour, A., Shafiei, M. A., Dehghanib, K., Modeling the microstructural evolution and effect of cooling rate on the nanograins formed during the friction stir processing of Al5083. *Materials Science and Engineering A* 527, 192-197 (2009).

You, G. L., Ho, N. J., Kao, P. W.: Aluminum based in situ nanocomposite produced from Al–Mg–CuO powder mixture by using friction stir processing. *Materials Letters* 100, 219-222(2013a).

You, G. L., Ho, N. J., Kao, P. W.: In-situ formation of Al2O3 nanoparticles during friction stir processing of Al\SiO2 composite. *Material characterization* 80, 1-8 (2013b).

You, G. L., Ho, N. J., Kao, P. W.: The microstructure and mechanical properties of an Al–CuO in-situ composite produced using friction stir processing. *Materials Letters* 90, 26-29(2013c).

Yu, Z., Zhang, W., Choo, H., Feng, Z.: Transient Heat and Material Flow Modeling of Friction Stir Processing of Magnesium Alloy using Threaded Tool. *Metallurgical and Materials Transactions A* 43(2), 724-737 (2012).

Zhang, D., Suzuki, M., Murayama, K.: Microstructural evolution of a heat-resistant magnesium alloy due to friction stir welding. *Scripta Materialia* 52, 899-903 (2005).

Zhang, Y., Sato, Y. S., Kokawa, H., Park, S. H. C., S. Hirano, S.: Microstructural characteristics and mechanical properties of Ti–6Al–4V friction stir welds, *Materials Science and Engineering A* 485 (1-2), 448-455 (2008).

Zhang, Z., Zhang, H. W.: Numerical studies on controlling of process parameters in friction stir welding. *Journal of Materials Processing Technology* 209, 241-270 (2009).

Zhang, Y. N., Cao, X., Larose, S., Wanjara, P.: Review of tools for friction stir welding and Processing. *Canadian Metallurgical Quarterly* 51(3), 250-261 (2012).

Zhou, L., Liu, H. J., Liu, Q. W.: Effect of rotation speed on microstructure and mechanical properties of Ti–6Al–4V friction stir welded joints. *Materials and Design* 31, 2631-2636 (2010).

In: Comprehensive Guide for Nanocoatings Technology. Vol. 1 ISBN: 978-1-63482-447-7
Editor: Mahmood Aliofkhazraei

Chapter 15

SURFACE ENGINEERING

B. S. Saini and V. K. Gupta
Department of Mechanical Engineering,
Punjabi University, Patiala, Punjab, India

ABSTRACT

The surface acts as an interface between the bulk of a material and its environment. Therefore, a modification in the surface properties plays an important role in performance characteristics of a component. Surface engineering involves bringing about desirable changes in the properties at or just below the surface of a substrate. It is generally employed for improving performance in various applications, such as tribological, electrical, optical, electronic, chemical or magnetic. Through surface engineering, it becomes possible to select the bulk of a component to meet general requirements, such as strength, stiffness, formability, cost, etc. and thereafter modify the surface up to a certain depth, viz. through heat treatment, or by adding a thin layer of another material on the substrate surface through some suitable deposition technique in order to induce the desired functional properties, viz. hardness, corrosion resistance, electrical conductivity etc. With rapid advancement in vacuum technology and control systems over the past few decades, surface engineering processes involving deposition of thin films have gained widespread importance in various fields – especially metal cutting and semiconductor chip fabrication industries.

INTRODUCTION

Surface engineering techniques are quite native to nature and can be found at work almost everywhere in the world around us. They constitute the basic means of providing protection to the functional but fragile interiors of various organisms on micro as well as macro scale, viz. seeds, vegetables, fruits, roots – all have peels for providing protection against the harsh environment and pest attacks. Some plants have a complex surface geometry, forming a hydrophobic layer on their foliage, which comprises of nano-scale structures and aids in keeping the leaves clean and hence, improve their photosynthesis efficiency.

It seems very obvious that Darwinian evolution has led to proliferation of species, which have been exploiting surface engineering to enhance their "genetic fitness". A common example in this context is the presence of hard enamel coating on our teeth. Likewise, many insects and birds have thin transparent oily films on their plumage, which not only provide protection against environment, but also impart brilliant hues through interference of light – a characteristic useful for befooling predators as well as attracting mates. Chameleon and octopus are typical examples, where very advanced control over surface colour and texture constitutes an important survival mechanism.

Surface modification techniques have been finding widespread application in human civilization since ancient times – perhaps long before the term "Surface Engineering" was coined. Since times immemorial, humans have been applying various types of cosmetics on their skin, lining their living spaces with materials different from the ones used in construction, forming glazed layers on cookware, using polish for wooden artifacts, applying greasy substances on agricultural implements for prevention against rust etc.

In modern times, surface engineering has become all the more important as it lets an engineer make the most optimal selection of materials for bulk and surface in order to meet the price and performance requirements. For the past few decades, efforts have been on to develop techniques for achieving optimal properties at the surface and the core of materials, so as to meet the performance requirements in specialized engineering applications, ranging from bio-mechanics, civil structures, electronics, automobiles to space.

SURFACE ENGINEERING OF MATERIALS

The surface treatments can be broadly classified into the following categories:

Surface modification: The process involves modification of properties up to a certain depth of the material. Examples of this process include heat treatment (with or without the addition of chemicals) viz. carburizing, nitriding, carbonitriding, cyaniding, induction and flame hardening.

Overlay (coating deposition): It involves deposition of a thin layer on the substrate surface by processes such as cladding, electroplating, thermal spray coating and physical/chemical vapour deposition.

Combination of surface modification and overlay: Sometimes, a combination of two processes, as described above, is used for achieving the desired properties. As an example, the steel surface may be hardened by plasma nitriding prior to depositing hard coating through physical vapour deposition (PVD) process.

Modification of overlay deposit: It is yet another method that involves a surface modification process for altering properties of the deposit, viz. ion-bombardment of a sputter-deposited coating; or shot-peening of a sputter-deposit for introducing compressive residual stresses on the surface.

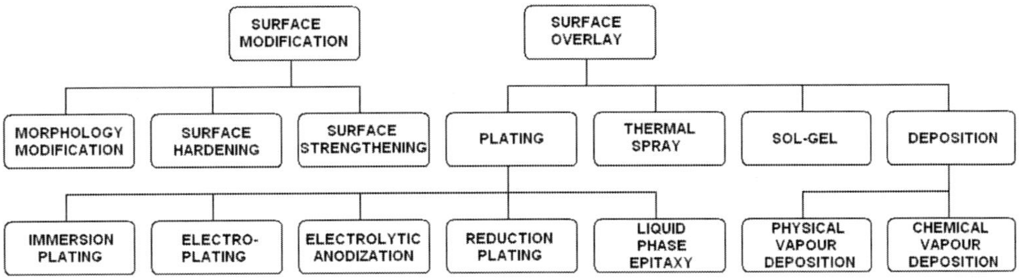

Figure 1.1. Surface engineering processes.

The commonly employed surface engineering techniques are depicted in Figure 1.1 and discussed in the following sections.

SURFACE MODIFICATION TECHNIQUES

Modification of Surface Morphology

Surface morphology refers to the degree of smoothness (or roughness) of the surface. Different degrees of smoothness could be desired for different applications.

Surface Smoothening

Smooth, polished surfaces, free from microscopic crests and valleys, lead to improvement in fatigue performance of machine elements. Such surfaces are also recommended for ensuring flawless deposition of coatings. Conventionally, surface smoothening is achieved through various types of dry or wet abrasive polishing methods. However, since abrasive polishing is always accompanied by plastic failure at the surface, it is often desirable to remove the resulting surface layer through chemical or electrochemical polishing, as carried out during sample preparation for estimation of residual stresses through X-ray diffraction technique. Combination of chemical and mechanical polishing is widely employed for planarizing surfaces in semiconductor device fabrication.

Surface Roughening

Conventional procedures of surface roughening are similar to the ones used for surface smoothening, viz. abrasive action and chemical etching – albeit with different process parameters than those employed for polishing. For example, the commonly used nital-etch of steel, if continued for prolonged duration of time, will lead to considerable roughening of the surface. Chemical etching works through preferential dissolution of different phases, grain boundaries, or different crystal facets.

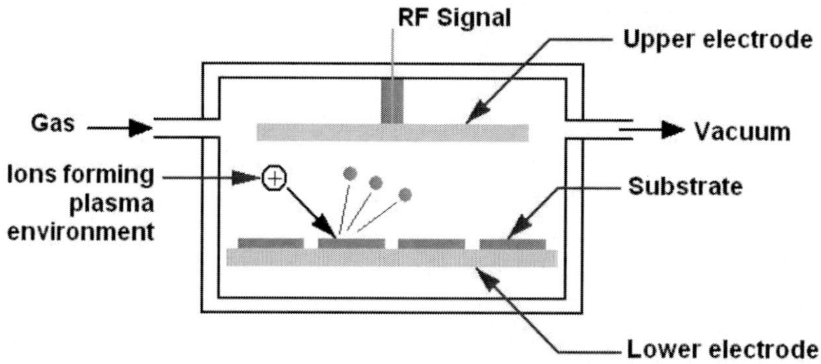

Figure 1.2. Schematic of sputter etching.

Sputter-etching [Refer Figure 1.2] is another technique for preferential etching of the surface on the basis of crystallographic orientation. In general, chemical sputter etching is largely isotropic in nature, while on the other hand, physical sputtering, viz. one employing Argon plasma, can be highly anisotropic for different crystallographic orientations. Sputter-texturing can be employed for developing micro or nano-sized topographic features with high aspect-ratios and very high surface areas – the characteristics, which can find novel applications, such as high optical absorption materials, or bio-medical implants for enhanced bone-growth and adhesion.

Surface Hardening

The most commonly used surface hardening technique with greatest engineering significance is heat treatment of metals and alloys – especially steels. A variety of heat treatment processes for different grades of steels have been used in the industry. For most of the medium and low carbon alloy steels that find widespread application in machine tool and automotive industries, the chemical diffusion heat treatment processes like nitriding and carburizing are widely employed for achieving the desired levels of hardness and strength. The process involves holding the components in a gaseous environment at a certain temperature for a specific period of time, followed by quenching in a suitable medium [Refer Figure 1.3]. Modern sealed quench furnaces offer higher productivity and better control over quality by avoiding contact with air.

Nitriding

Nitriding is possible on many grades of steels, but very high hardness is obtained in certain special grades of steels that contain nitride forming elements, viz. aluminium, chromium, molybdenum or vanadium, which help in maintaining a high concentration of nitrides near the surface and prevent their diffusion further deep into the core. Nitriding is generally performed with the help of ammonia, which provides nitrogen on steel surface through the following chemical reaction:

$$NH_3 \rightarrow N \downarrow + \tfrac{3}{2} H_2$$

Though nitriding of case achieves higher hardness in comparison to case carburization, the depth of nitrided case is shallow (25 – 750μm). Nitriding is also known to cause substantial improvement in fatigue performance [1-4]. However, owing to the associated shape distortions, the use of nitriding is generally restricted to tribological applications [5].

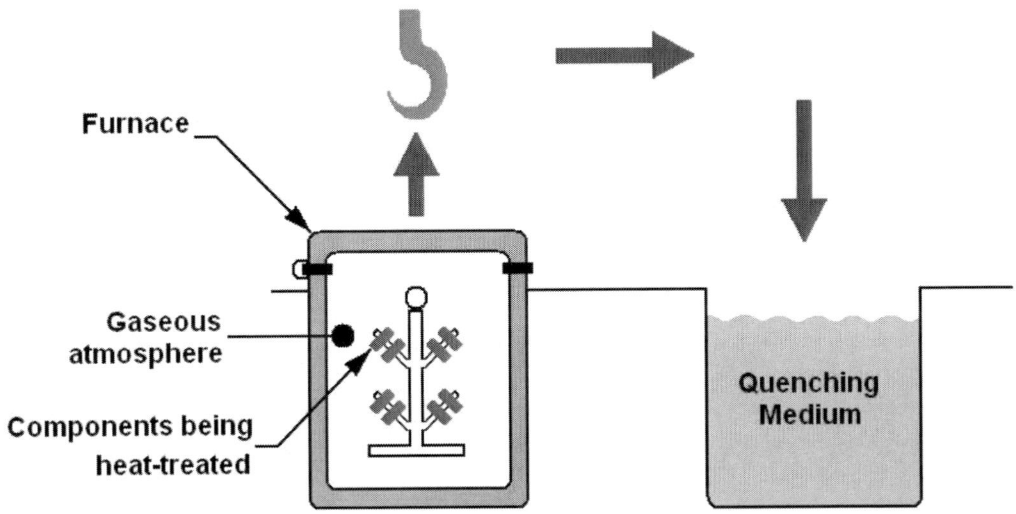

Figure 1.3. Schematic of chemical diffusion heat treatment process.

Case Carburization

Case carburization is the most common surface modification technique for steels [Refer Figure 1.4] and can provide an effective depth of case ranging from 50 to 3000 μm. The process finds widespread application as it leads to considerable improvement in mechanical properties of hardness, strength and endurance [6, 7]. The diffusion of carbon into the surface of steel component occurs by gas-metal reactions between various components of the gaseous environment and the solid solution austenite:

$$CO_2 + C \rightarrow 2CO$$
$$2CO \rightarrow C\downarrow + CO_2$$
$$CH_4 \rightarrow C\downarrow + 2H_2$$
$$CO + H_2 \rightarrow C\downarrow + H_2O$$

The carburizing potential of the gaseous atmosphere is a function of partial pressures of CO and CO_2. While high hardness of martensitic case is important from the standpoint of wear resistance, the enhancement in fatigue properties is attributed mainly to the presence of residual compressive stresses within the carburized case [8]. The compressive residual stresses arise as a result of interplay between varying carbon concentration and cooling rates encountered at different depths below the surface, which leads to a time lag in the occurrence of martensitic transformations and associated dimensional variations [9-11].

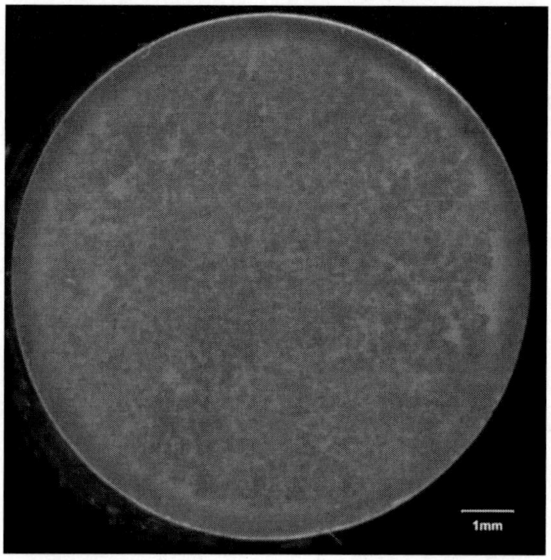

Figure 1.4. Metallograph showing cross-section of 6mm diameter SAE8620 steel specimen, case carburized to a depth of 0.7mm, etched in 2% nital for 8 seconds.

Surface Strengthening

Since most cracks generally initiate at the outermost surface, introduction of residual compressive stresses within the outermost surface layers (at the cost of tensile residual stresses in the core) helps improve fracture toughness and fatigue strength. Shot peening is a very popular process in this respect, which finds widespread application in industry. For ceramic materials, the similar effect is produced by *ion-peening* or *ion-implantation* while thermal stressing is employed for imparting residual stresses to glasses.

SURFACE DEPOSITION TECHNIQUES

The surface overlay techniques, as depicted in Figure 1.1, are broadly divided into four categories: plating, thermal spray processes, chemical vapour deposition (CVD) and physical vapour deposition (PVD). Among these, PVD and CVD constitute the basis for thin-film deposition technologies [Refer Table 1.1 for historical development of these techniques]. Both these processes are atomistic in nature – i.e., the overlay is deposited atom-by-atom. As a convention, overlay deposits having thicknesses ranging from a few atoms to several micrometers are termed as *thin films*, while those with higher thicknesses are called *coatings*. The actual thickness of coating is governed primarily by the type of application and its functional requirements, rather than the coating process followed, because the techniques for depositing coatings with thicknesses in the range of nanometers to micrometers are quite similar. Since there is no fundamental difference in the underlying processes, we shall be using these words without any discrimination throughout this chapter.

Table 1.1. Historical developments in surface engineering techniques

Year	Development
1852	The first sputtering experiments were reported by Grove [12].
1857	Faraday made early reports on evaporation [13].
1880s	First practical use of CVD was developed in the production of incandescent lamps to improve the strength of filaments by coating them with carbon or metal [14]
1892	First arc deposition was patented by Edison [15].
1893	First patent on industrial exploitation of CVD was granted to Lodyguine [16] for depositing W onto carbon lamp filaments through reduction of WCl_6 by H_2.
1911	Schoop [17] invented first thermal spray process and is now known as flame spraying.
1936	Magnetron sputtering was first proposed by Penning [18].
1938	Berghaus [19] was granted patent for the original Plasma assisted physical vapour deposition (PAPVD) system.
1953	Veszi [20] introduced the term reactive sputtering.
1955	Poorman et al. [21] invented major thermal spray processes like wire spraying and detonation gun deposition.
1958	Smith et al. [22] invented and patented high velocity oxyfuel (HVOF) process.
1960	Introduction of the terms *CVD* and *PVD* to distinguish "chemical vapor deposition" from "physical vapor deposition" [14].
1960	Introduction of CVD in semiconductor fabrication [14].
1960	CVD-TiC coating on cemented carbide tools was introduced and CVD-tungsten was developed [14].
1960s	Radio frequency (RF) was used to sputter material [19].
1962	George et al. [23] invented and patented plasma spray process.
1963	Mattox and McDonald [24] introduced simple ion plating method.
1963	Introduction of plasma CVD in electronics [25].
1967	Wasa [26] invented prototype of the planar magnetron.
1967	Pioneering work on the deposition of thin films by ion beam sputtering was reported by Chopra [27].
1968	Industrial use of CVD coated cemented carbides was reported [21].
1968	Morley [28] developed first ion plating system by using a hollow cathode discharge (HCD) gun and plasma activation.
Early 1970s	CVD was successfully used for manufacturing of electronic semiconductors and protective coatings for electronic circuits [16].
1970s	Ionized cluster beam deposition (ICBD), a modification of ion plating, was developed by Takagi [29]
1972	Plasma assisted physical deposition (PAPVD) of ceramic thin films was pioneered by Bunshah and Raghuram [30] and Raghuram and Bunshah [31].
1972	Bunshah proposed Activated reactive evaporation (ARE), commonly used for the deposition of metal oxides, carbides, and nitrides [32].
1974	Chapin [33] devised a practical planar magnetron system.
Mid-1970s	Development of reactively sputter-deposited hard coatings on tools [19].
1977	Balzers company [34, 35] introduced another version of the plasma-assisted ion plating process, which involves the use of an electron source in the form of a filament.
1979	Hot-Wire CVD (HWCVD) was first introduced and patented as thermal CVD [36].
Early 1980s	Evaporation systems came in to widespread use for the production of ceramic coatings, especially for cutting tools, most notably by the Balzers company [37].
1980s	Introduction of CVD diamond coatings [14].
1983	Thornton [38] introduced plasma-assisted or plasma enhanced deposition process.
1986	Windows and Savvides [39, 40] introduced "unbalanced magnetron sputtering".
1987-1988	Randhawa and Johnson [41,42] introduced cathodic arc plasma deposition (CAPD) process.
1990s	Rapid expansion of metallo-organic CVD (MOCVD) for ceramic and metal deposition [14].
1990s	Development of cluster tools combining CVD, PVD and other processing steps in a single tool for semiconductor fabrication [14].
1991	Brown et al. [43] were first to utilize dense *metal plasma* for immersion implantation.

Traditionally, PVD processes involve purely physical, such as evaporative methods, while CVD involves chemical reaction in gaseous, liquid, solid or plasma phases. A relatively new class of processes, based on glow discharges and reactive sputtering, involves a combination of both physical and chemical vapour deposition techniques.

With the exception of plating, each of the remaining deposition processes can be further classified on the basis of energy source used. Each process exhibits its own merits, demerits and associated range of applications. The resulting properties of the deposit are dependent upon the chemical and physical conditions prevalent during the deposition process. In other words, the properties required on the deposit will determine the coating process and parameters.

While planning deposition on a substrate, it should be borne in mind that the actual surface, which is going to participate in the coating process, generally differs from the bulk material due to contamination by adsorbed or chemically reacted material (viz. oxides) upto a certain depth. So, the surface may require cleaning or modification by mechanical, chemical, thermal or some other means, before any of these coating processes can be employed.

Plating

Plating involves deposition of inorganic coatings from liquid phase through some chemical process. Some commonly employed plating techniques are described below.

Immersion Plating

This process involves plating of a metallic substrate by immersion in electrolytic solution of the salt of a less electropositive metal through displacement reaction. While the metal from solution gets reduced to form film on the surface, some of the metal from the substrate will enter into the solution. As an example, if a zinc substrate is immersed in copper sulphate solution, the more electropositive zinc atoms will reduce ions of copper from the solution by the following redox reaction:

$$\text{Oxidation}: Zn \rightarrow Zn^{++} + 2e^{-}$$

$$\text{Reduction}: Cu^{++} + 2e^{-} \rightarrow Cu$$

As a result of this reaction, the surface of zinc substrate gets covered by a mosaic of oxidation (zinc) and reduction (copper) sites. As the plating progresses, the reduction sites eventually cover almost entire surface of the substrate and the reaction comes to a halt on its own. The process has limited application due to very small thickness of the deposits.

Electroplating

Electroplating is one of the oldest surface engineering techniques [44] that can produce deposits ranging from very thin films to thick coatings (electroforming). Norddeutsche Affinerie in Hamburg was the World's first electroplating plant that started its production in 1876.

In this process, the substrate to be plated is made as cathode in an electrolytic cell containing a suitable electrolyte of the specie that is to be plated. The deposition takes place as a result of reduction of ions to their elemental form. The process is commercially used for depositing Cr, Ni, Zn, Sn, In, Ag, Cd, Au, Pb and Rh; along with some alloy compositions such as Cu-Zn, Cu-Sn, Pb-Sn, Au-Co, Sn-Ni, Ni-Fe, Ni-P and Co-P. Owing to the useful physical and chemical properties, electroplating of chromium finds extensive application for steel components [45]. The coating provides good protection against corrosion, along with bright finish and high hardness.

Environmental concerns associated with the electroplating process have led to imposition of regulations [46], which have motivated a worldwide search for eco-friendly alternatives [47-48].

Electrolytic Anodization

In this process, the substrate gets oxidized through its reaction with negative ions from the electrolytic solution. This process is suitable for materials, whose oxides form continuous, non-porous, adherent and chemically inert coatings.

Since the formation of coating occurs as a result of chemical modification of surface layer, anodization can be termed as a surface modification process rather than surface overlay process, but is nevertheless placed under this section for the sake of logical continuity.

A commonplace application of this technique is anodization of aluminium cookware by the following reaction:

$$\text{Anodic reaction}: 2Al + 3H_2O \rightarrow Al_2O_3 + 6H^+ + 6e^-$$

$$\text{Cathodic reaction}: 6H^+ + 6e^- \rightarrow 3H_2$$

The process is also used for silicon, titanium, niobium and zirconium artifacts.

Chemical Reduction Plating

It is based on reduction of a metallic ion from the solution through addition of some suitable reducing agent (displacement reaction). Unlike electroplating, where deposition occurs only on cathode, the reaction in reduction plating is homogenous and leads to deposition everywhere in the solution. The prime merit of reduction plating is that unlike electroplating, this process can be used for plating metals onto non-conducting substrates, such as glass.

Historically, the method has been used for silvering of mirrors through reduction of silver from silver nitrate solution by addition of hydrazine or dextrose etc. The chemical reaction for reduction plating using dextrose as the reducing agent is as under:

$$C_6H_{12}O_6 + 2Ag(NH_3)_2^+ + 2OH^- \rightarrow C_6H_{12}O_7 + 2Ag + 4NH_3 + H_2O$$

Liquid Phase Epitaxy

Epitaxy refers to the growth of crystals on a crystalline substrate, so that the substrate and deposit have same structural orientation. If the deposited material is same as the substrate, the process is called *homoepitaxy*, while for dissimilar materials, it is termed as *heteroepitaxy*. Liquid phase epitaxy is often employed for deposition of epitaxial films of compound semiconductors onto a single crystal substrate [Refer Figure 1.5].

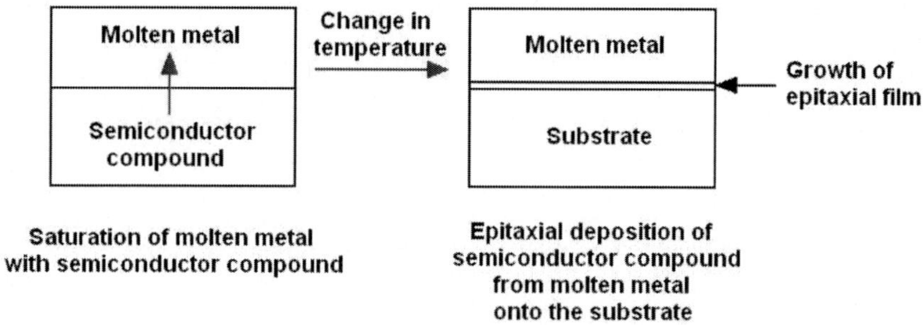

Figure 1.5. Schematic of liquid phase epitaxy.

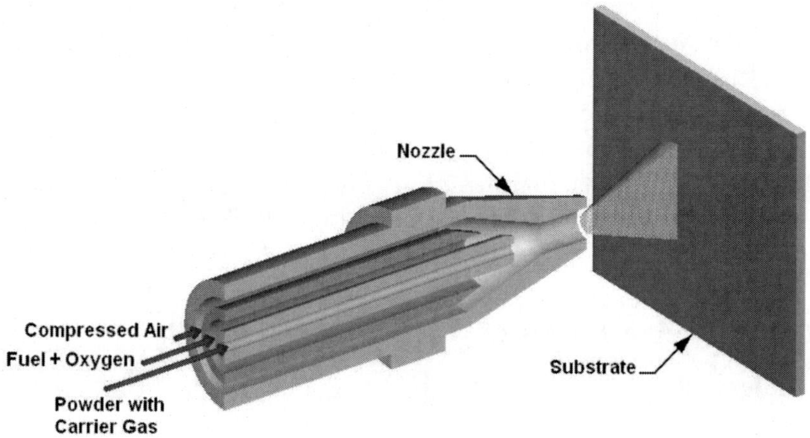

Figure 1.6. Schematic of thermal spray coating.

Thermal Spray Coating

In the thermal spray process, pre-alloyed powder of a given composition along with a carrier gas is injected through a high energy source [Refer Figure 1.6]. The intense heat rapidly melts the powder particles to form tiny droplets, which strike the substrate at high velocities and then rapidly solidify to form a mechanical bond with the surface being coated. The rate and quality of deposition depends upon the spray process, materials and processing parameters [49-52].

Major applications of thermal spray coatings include deposition of corrosion resistant coatings, hard-facing of surfaces and restoration of dimensions [53]. The process is fast and suitable for deposition of various metallic and oxide coatings; however, the coatings may lack quality and homogeneity.

Sol-Gel Deposition

Sol-gel deposition comprises of a class of processes involving compaction of molecules from some suitable solution to form a solid coating. Initially, the particles in the solution (*sol*) undergo partial bonding or compaction to form an integrated network (*gel*), which is subsequently dried and thermally treated to provide the required film. The process is suitable for depositing metal-oxide coatings on glass or metal substrates.

Chemical Vapour Deposition

The chemical vapour deposition (CVD) process involves passing a mixture of reactant gases in a high-temperature reactor to form a thin film deposit on the surface of heated substrate. Conventionally, the process was carried-out in *closed reactors*, in which the reaction chamber is closed after loading the reactants. However, the *open reactor design* [Refer Figure 1.7], in which chemicals are loaded and removed on a continuous basis, offers better performance and flexibility.

Table 1.2. CVD chemical reactions

Title	Generic Reaction	Examples
Thermal Decomposition Reactions	$MX(g)\rightarrow \underbrace{M(s)}_{Solid\,Coating}\downarrow + X(g)\uparrow$	$SiH_4 \rightarrow Si\downarrow + 2H_2\uparrow$ $B_2H_6 \rightarrow 2B\downarrow + 3H_2\uparrow$ $CH_4 \rightarrow C\downarrow + 2H_2\uparrow$
Reduction Reactions	$2MX(g)+H_2(g)\rightarrow \underbrace{2M(s)}_{Solid\,Coating}\downarrow + 2HX(g)\uparrow$	$2BCl_3 + 3H_2 \rightarrow 2B\downarrow + 6HCl\uparrow$ $SiCl_4 + 2H_2 \rightarrow Si\downarrow + 4HCl\uparrow$ $WF_6 + 3H_2 \rightarrow W\downarrow + 6HF\uparrow$
Displacement Reactions	$MX(g)+C(g)\rightarrow \underbrace{MC(s)}_{Solid\,Coating}\downarrow + X(g)\uparrow$ $MX(g)+CY(g)\rightarrow \underbrace{MC(s)}_{Solid\,Coating}\downarrow + XY(g)\uparrow$	$SnCl_4 + O_2 \rightarrow SnO_2\downarrow + 2Cl_2\uparrow$ $TiCl_4 + CH_4 \rightarrow TiC\downarrow + 4HCl\uparrow$
Redox Reactions	$2MX(g)\rightarrow \underbrace{M(s)}_{Solid\,Coating}\downarrow + MX_2(g)\uparrow$	$2TiCl_2 \rightarrow Ti\downarrow + TiCl_4\uparrow$ $2SiI_2 \rightarrow Si\downarrow + SiI_4\uparrow$

The chemical reactions taking place in the CVD process are usually endothermic and so, occur at the surface of the heated substrate. Depending upon their chemistry, the reaction products can either diffuse into or deposit on the surface of substrate. The overview of chemical reactions associated with CVD is provided in Table 1.2. A particular CVD process may involve one or more of such reactions in order to yield the desired coating.

On the basis of source of energy employed for initiating the CVD reaction, the vast conglomerate of CVD processes can be broadly classified into four major categories – thermal, photo, LASER and plasma. Each of these is discussed briefly in the following sub-sections.

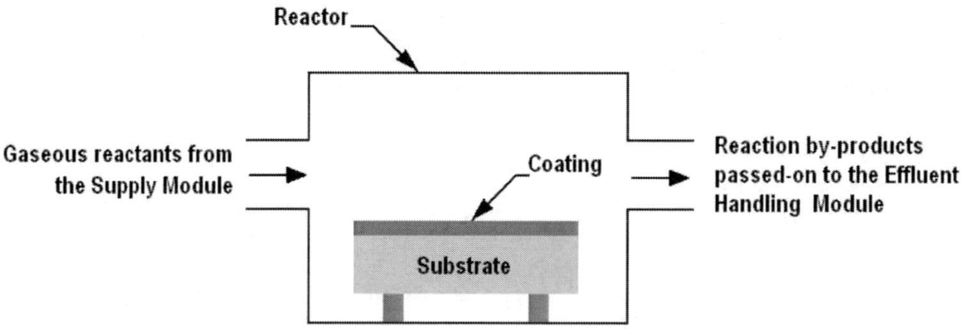

Figure 1.7. Schematic of open reactor Chemical Vapour Deposition setup.

Thermal CVD

Thermal CVD is a conventional CVD process and is generally carried out at temperatures in the range of 800 to 2000ºC. Process temperature and pressure, along with partial pressures of the reactant gases constitute the critical process parameters in any CVD process [54-56]. Although it is possible to achieve precise temperature control in furnace-type *hot wall* reactors, but the down side is that a major amount of deposition will take place on the reactor walls, which will not just call for frequent maintenance, but will also deteriorate deposition quality due to depletion of the reactive specie. *Cold wall* reactors are often employed in order to overcome these difficulties, wherein direct heating of the substrate takes place through conduction or induction. Optimal substrate temperature is important for ensuring that the reaction occurs on the substrate surface rather than within the gas phase itself. A variant of the process, in which CVD reaction is started and stopped through rapid heating or cooling of the substrate, is known as *rapid thermal* CVD (RTCVD).

Transportation of the reactant gases to the substrate surface is largely governed by partial pressures of the former, along with total pressure inside the reactor chamber. At higher pressures (viz. atmospheric pressures), the deposition process remains dependent on transport within the boundary layer. However, upon reducing the pressure, the rate of deposition gradually decouples from transportation hydrodynamics, thereby making it possible to exercise better control over film composition. On the basis of pressure within the reactor, the CVD processes are generally classified as *atmospheric pressure* CVD, *low pressure* CVD (1.33kPa to 0.01kPa) and *ultra high vacuum* CVD (< 0.1Pa). Lower pressures are of particular importance in cold wall reactors, as these reactors are susceptible to convection-related deposition problems encountered at higher pressures, which can have an adverse effect on uniformity of the deposit. Other major factors affecting transport are reactor geometry and substrate structure. The process can be employed for depositing high purity metallic and ceramic coatings, viz. oxides, carbides, nitrides etc.

Photo CVD

As the name suggests, in this process, the energy derived from photons (usually ultraviolet) is exploited for breaking chemical bonding within the reactant molecules, thereby initiating their reaction with the substrate material. Though the deposition rates are small, but the prime advantage of this process is that it can be carried out at room temperature.

LASER CVD

Unlike photo CVD, which does not require any heating of the substrate, LASER CVD involves local heating over a small portion of the substrate surface [Refer Figure 1.8]. The wavelength of the light is so chosen that it is not effectively absorbed by the gas molecules. By scanning the LASER beam over the surface at suitable speed, it is possible to form patterns of deposit on the substrate surface [57, 58].

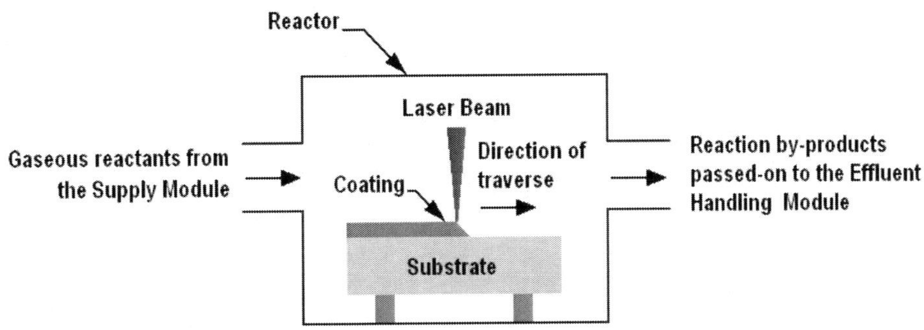

Figure 1.8. Schematic of LASER CVD.

Plasma CVD

The process of Plasma CVD had been developed in 1960s for deposition of silicon nitride for semiconductor application. The plasma serves to activate and partially decompose the precursor species in CVD reactors, thereby allowing deposition at a temperature lower than that required for thermal CVD [59, 60]. The lower processing temperatures render this process suitable for coating heat-sensitive materials, which would otherwise melt, degrade or undergo structural changes during thermal CVD. Also known by the names of plasma assisted CVD (PACVD), plasma enhanced CVD (PECVD) or glow-discharge CVD, the process has been successfully extended for producing various types of coatings, such as polymeric [61], diamond [62], TiN [53], amorphous carbon [63], cubic boron nitride [64] and silicon oxide [65].

The setup used in this process, as shown in Figure 1.9, is in principle similar to that employed for sputtering [Refer section on Sputter Deposition]. In this process, a microwave-frequency (2.45 GHz) or radio-frequency (13.56 MHz) AC voltage is applied to a gas held at low pressure (~200 Pa). The electric field ionizes the gas, leading to generation of ions and electrons. Due to their small mass, the electrons get accelerated to high energy levels, while the much heavier ions have less mobility under high frequency of applied electric field. As a result of this difference in kinetic energies, the glow-discharge plasma environment formed by the application of high frequency AC consists of two different temperatures in the same region, due to which it is also known as *non-isothermal* plasma. The high energy electrons perform various important tasks, such as ionizing the neutral gas atoms and dissociation of precursor gas molecules. During each half-cycle of the applied voltage, the ions and electrons alternately strike the target surface. While the ions serve to eject (sputter) particles from the target, the electrons serve to neutralize any charge buildup on the target – a particularly important step to prevent arcing in case of dielectric targets.

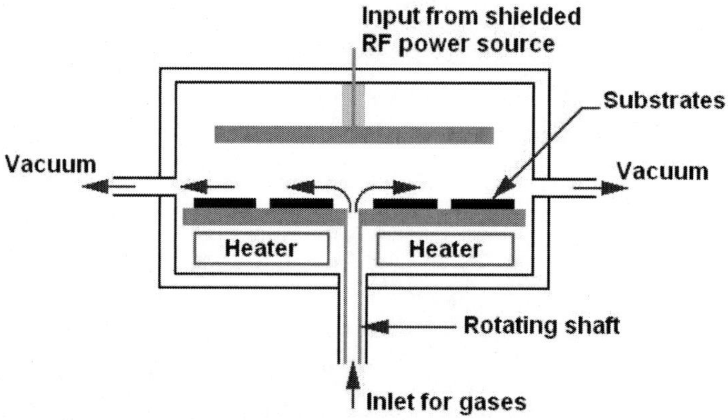

Figure 1.9. Schematic of plasma CVD.

For efficient sputtering (i.e., effective momentum transfer), the atomic mass of the gas in the process chamber should be similar to that of the target material (sputtering source), but the choice of gas is rather governed by the factors of cost and safety. As a result, argon is the most commonly used gas in PVD sputtering, while a combination of argon along with some other suitable gas (which has to supply the coating precursors) is generally employed in case of plasma CVD. The collisions between high energy electrons and gas molecules result in dissociation of the latter, which form the reactive chemical specie that bonds with the substrate surface to form coating.

The variation in applied voltage and other parameters, such as composition and pressure etc., affects the coating properties [66]. The three major types of PECVD reactor designs include parallel plate discharge, capacitive coupling tube and inductive coupling tube [67]. Another important variant of this process is metallo-organic CVD (MOCVD), which employs a metalorganic precursor in place of inorganic precursor. Details of other variants of CVD can be found in the review by Choy [16].

Most of the CVD processes are carried out in "open-loop" systems, where the reaction chemicals are injected into the reactor chamber from one side and removed from the other by effluent handling system. As the recovery of useful substance from among the effluents is generally subject to economic viability, the CVD process is usually accompanied by chemical waste. Thus, CVD is often accompanied by volatile by-products and unutilized precursor species. The prime demerits of the CVD process include high processing temperatures, coupled with chemical waste and low deposition rates.

Physical Vapour Deposition

Physical vapour deposition (PVD) is one of the oldest and widely used techniques in the laboratory as well as in industry for depositing thin films. Just like CVD, PVD is also a vapour deposition process, which can deposit atoms and/or molecules, but the difference is that in a purely PVD process, the deposition takes place through condensation - which is an exothermic process, while CVD takes place through an endothermic chemical reaction at the substrate surface. Modern coating equipment generally permits combination of PVD and CVD (viz. plasma CVD), so as to harness the merits of both the processes.

In order to ensure good adhesion of deposit, the substrates to be coated are first cleaned by mechanical, physical and chemical means, followed by degassing (heating in vacuum). Finally, the atomic contaminants on substrate surface can be removed by etching with Ar-ions [68]. The subsequent PVD process is generally carried out under high vacuum at substrate temperatures ranging from 150 to 500°C. The high-purity, solid coating material (viz. metals such as titanium, chromium, aluminium etc.) is first evaporated by application of energy. Depending on the material to be deposited, the energy can be provided to the target (source) through various techniques, viz. electrical resistance heating of filaments, conductive or radiative heating of crucibles, heating with electron beam, radio frequency inductive heating, arc heating etc. The evaporated material, which is in atomic or molecular form, gets transported to the substrate surface, where it condenses to form an adherent deposit.

The PVD process can also be used for depositing ceramic coatings by introducing a reactive gas (e.g., nitrogen or a gas containing carbon, such as methane) in the coating chamber. The reactive gas forms a compound with the metal vapour, which gets deposited on the substrate surface in the form of a thin coating.

In order to obtain a uniform coating thickness, the components being coated are rotated at uniform speed about various axes. A brief description of variants of PVD process is presented in the following sub-sections.

Vacuum Evaporation Deposition

As the name suggests, this deposition process is carried out in high vacuum (<0.01Pa). The coating material is evaporated by heating, viz. by passing current through the "boat" as shown in Figure 1.10. In the absence of scattering by any gaseous medium, the vapours of coating material essentially follow straight, "line of sight" paths to the substrate surface. High vacuum also ensures that the resulting deposit is free from any contamination. In order to avoid heating by radiative heat transfer, the substrate is usually held at some distance from the source. The transport process in vacuum evaporation is quite analogous to the transport of particles in solar wind, which traverse considerable distances through vacuum, before bombarding onto the bare surface of some object, such as moon or an artificial satellite.

The process is suitable for coating pure metals and is employed for metallization of compact disc glass masters with 400nm thick layer of Ni. Compounds, which can withstand high temperatures without getting decomposed, can also be deposited by this process, while the compounds having very high boiling/sublimation points at the depositing pressure (viz. certain oxides, carbides and nitrides), whose decomposition would occur before evaporation, can only be deposited through *reactive evaporation*, wherein the non-metallic component of the coating is introduced in gaseous phase, which reacts with the metal vapours to form a compound coating on the substrate.

The vacuum evaporation process yields high deposition rates, with the down side being that in the absence of scattering under high vacuum, the deposit will not reach inside cavities or obscured regions on the substrate. The process is generally used for making optical coatings and electrically conducting films.

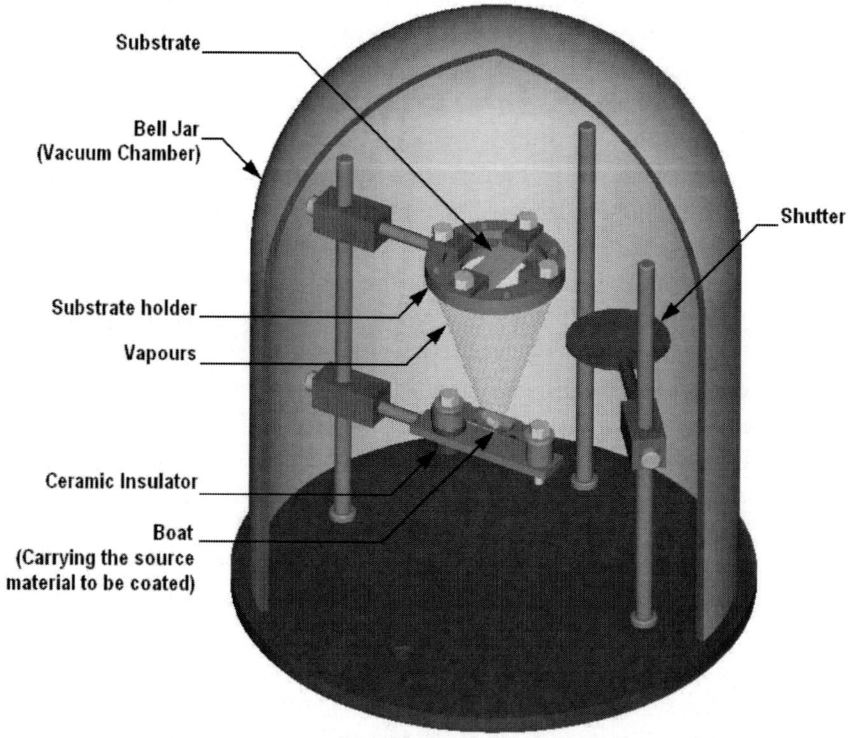

Figure 1.10. Schematic of Vacuum Evaporation Setup for laboratory use.

Electron Beam Physical Vapour Deposition (EB-PVD)

In this process, focused high-energy electron beams from electron guns are used for evaporating the coating material(s), as well as for preheating the substrate inside the vacuum chamber. The process offers possibilities for controlling variations in the structure and composition of condensed deposits, thereby facilitating the deposition of multilayer coatings as well as *functionally graded coatings* [69].

The preliminary steps in this process involve thorough cleaning of components (substrates) to be coated, by mechanical and chemical means, followed by their placement in the coating chamber and evacuation to 0.01Pa. The substrates are made anode and bombarded with electrons. This process removes surface contaminants (degassing) through heating. Subsequently, the substrates are made cathode, while the electrons are attracted towards an auxiliary anode placed inside the chamber. A noble gas (viz. argon) is introduced into the chamber, whose atoms get ionized by their collisions with the energetic electrons. These positively charged, heavy argon ions then strike the substrate surface (which is now made as anode) and clean it (ion-etching) to provide a contaminant-free surface for coating [70].

After completion of preparatory steps, the deposition of coating is done through thermal evaporation of the coating specie by striking electron beam onto the target (source material), as shown in Figure 1.11. The components are usually rotated about their own axes as well as the axis of chamber in order to obtain a uniform thickness of the coating.

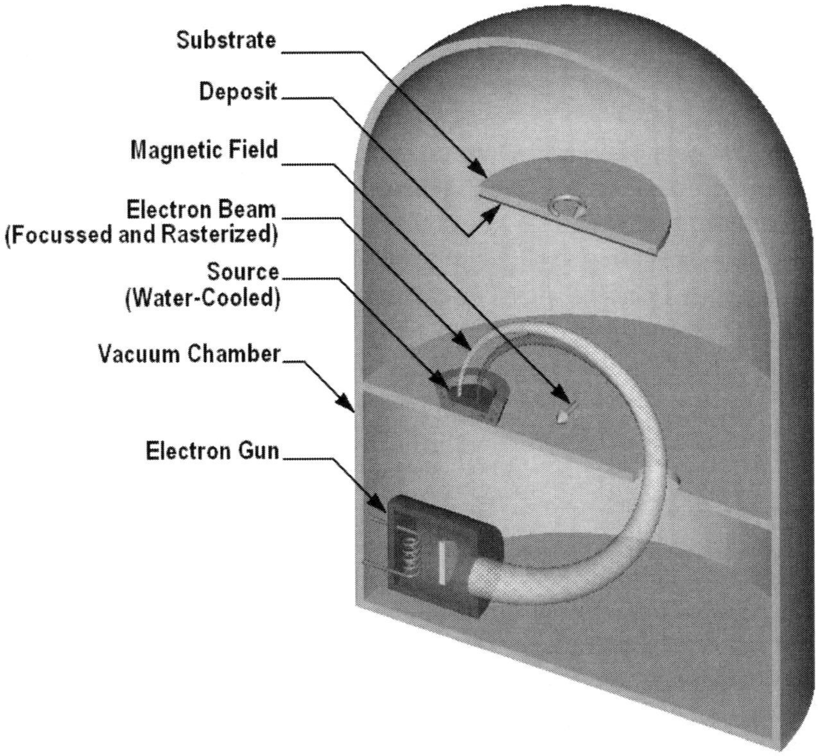

Figure 1.11. Schematic of EB-PVD Setup with focussed and rasterized beam.

Just like vacuum evaporation deposition, EB-PVD remains a line-of-sight process under low pressures and hence, the deposit will not reach inside cavities of the substrate. However, significant scattering may occur when the deposition is performed at pressures higher than 0.01Pa. EB-PVD is suitable for depositing pure metals (viz. Ti and Cr) as well as compounds, which can evaporate without undergoing decomposition in the vapour phase – viz. titanium carbide, titanium boride etc. Compounds, which are liable to undergo decomposition before their sublimation temperature, can be deposited either through *reactive evaporation* or *co-evaporation*. The former involves evaporation of pure metal (viz. tungsten) followed introduction of a reactive gas (viz. acetylene), which reacts with the metal vapours to provide non-metallic component (carbide) of the deposit (tungsten carbide). On the other hand, co-evaporation involves placing the coating source materials in separate ingots inside the chamber and heating them independently by means of electron beams of suitable energies. The energy of electron beam is thus used for controlling the rate of evaporation and in-turn, the composition of the deposit.

Sputter Deposition

Sputtering is one of the most versatile coating processes and is often employed as a combination of physical and chemical vapour deposition processes. Sputter deposition involves a non-thermal vaporization process, where atoms of the material to be deposited are physically ejected from solid surface of the target (source of coating material) by momentum transfer from atomic-sized energetic bombarding particles. The bombarding particles can be supplied by some ion-gun, but are usually in the form of gaseous ions (viz. Argon), accelerated from plasma. The plasma is typically formed around the target (i.e., source of deposition material) through application of high negative bias voltage to the target (cathode), while the chamber is maintained at low pressure [71] on the order of 1Pa. For efficient transfer of momentum, the bombarding particles should have similar atomic mass as the atoms of target material.

The most basic sputtering setup employs a DC *diode sputtering* configuration [Refer Figure 1.12 (a)] with electrically conducting hot or cold cathode. While the hot cathode exploits a thermo-electron emitting surface and can achieve very high plasma densities, the cold cathode configuration operates in *abnormal glow discharge* region. Positively charged ions from the plasma get accelerated towards the cathode, where they impinge with high energy and cause physical sputtering as well as emission of secondary electrons, which in-turn create further ions to sustain the plasma and so the process continues.

It is evident from the above description that sputter deposition follows a combination of ion-etching and vapour-deposition processes. The atomized particles (mostly neutral and a few in the form of ions) then get condensed on the substrate surface. The higher kinetic energy of sputtered particles results in better adhesion and higher density of deposit in comparison to that obtained in vacuum evaporation.

The DC diode sputtering process is not suitable for dielectric targets due to charge build-up problems. This limitation is often taken care of by using pulsed DC with various unipolar/bipolar waveforms.

A variant of this process is *triode sputtering*, wherein the plasma is established between a cathode (usually hot) and anode, while a third electrode is used to extract ions from the plasma using DC or RF potential [Refer Figure 1.12 (b)]. The non-uniform plasma density along longitudinal axis in this process is liable to cause non-uniform coating, which can be partially compensated by rotating the substrate.

Application of a magnetic field along the primary electrodes helps in arresting the electrons and hence, is useful for confining the plasma within a narrow zone close to the sputtering cathode [72-75]. Various types of axially symmetric *magnetron sputtering* configurations, viz. planar [Refer Figure 1.12 (c)], cylindrical, conical etc., can be designed to suit specific requirements. In addition to enhancing the deposition rate, the magnetic field serves to prevent thermal and radiation damage to the substrate surface from re-bombardment of secondary electrons, as they get trapped in spiral trajectories along magnetic field lines. This makes it possible to coat heat-sensitive materials, such as case-carburized steels, metal oxide semiconductors, or even plastics.

Films of metallic compounds can be deposited either through sputtering of compound targets or by means of *reactive sputtering*, wherein a reactive gas containing the non-metallic component of the coating is introduced; so that the atomized particles of metal react with the gas to form the compound to be deposited on the substrate surface. Reactive sputtering often requires the plasma to be extended away from the cathode surface, so as to interact and activate the reactive gas. This is usually achieved with the help of *unbalanced magnetron sputtering* wherein the magnetic field is so shaped that it allows some electrons to escape and form plasma away from the cathode. This region of plasma is useful for activating the reactive gas and providing ions for bombardment of the substrate. The process is commonly employed for deposition of insulators as well as diffusion barriers in semiconductor devices, metallization of semiconductor materials, glasses and compact discs and deposition of tribological coatings on cutting tools and machine elements [Refer Figure 1.12 (d)].

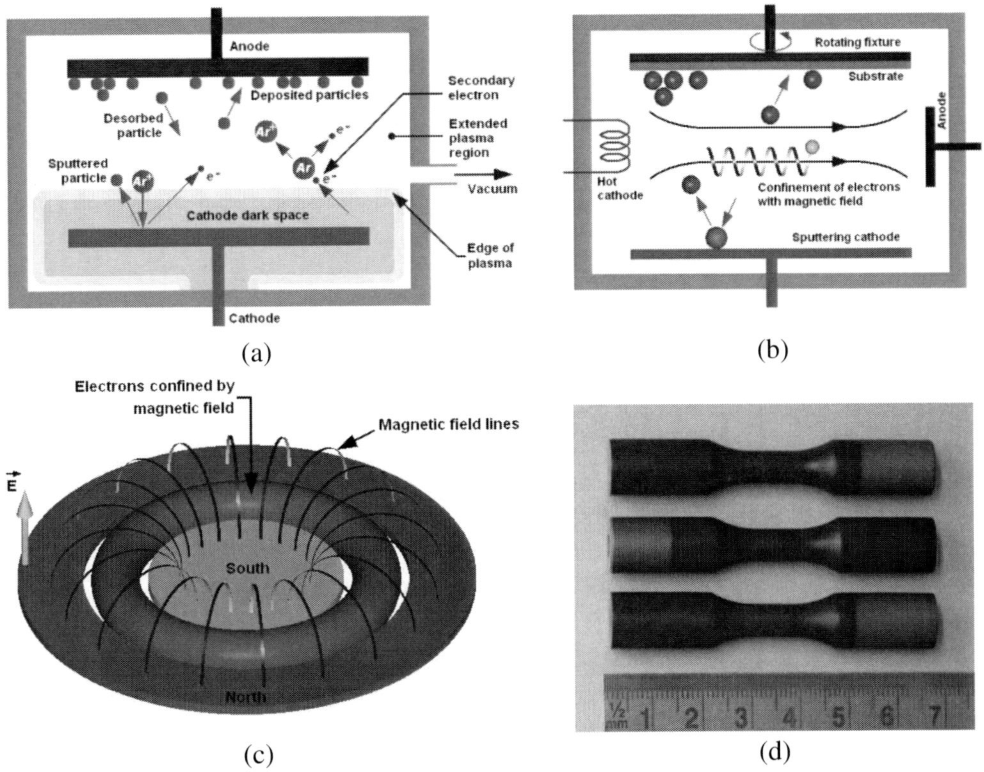

Figure 1.12: (a) Schematic of DC diode sputtering; (b) Schematic of DC triode sputtering; (c) Planar magnetron configuration; (d) Case carburized steel specimens coated with WC/C layer through PVD (Notice the lighter-coloured region on one end of each specimen, which remained uncoated due to masking by specimen holder).

Ion Plating and IBAD

Introduced by Mattox [76, 77] in 1960s, the ion-plating process was soon employed for protection of uranium reactor parts [78]. Subsequently, this method has often been used for coating of cutting tools, moulds, dies and other sliding elements. The two basic variants of this process are plasma based ion-plating and vacuum based ion-plating.

In ion-plating method, the substrate is made as cathode and is usually cleaned by means of ion bombardment prior to deposition. Subsequent to cleaning, the source material is evaporated before being partially ionized and accelerated to the substrate. Properties of the depositing film are controlled by means of concurrent or periodic bombardment by energetic particles derived either from an inert gas or a reactive gas. Ion bombardment during the course of deposition offers many benefits, such as resputtering of loosely-bonded film material, desorption of gases and conformal coverage of contoured surfaces. The process variables include energy, mass and flux of the bombarding species; besides ratio of impinging particles to depositing particles.

The vaporization of the depositing specie can be achieved by means of evaporation, sputtering, or arc erosion [79-82]. In case where plating is done in plasma environment, the energetic particles used for bombardment can originate from the plasma. The plasma is usually obtained by applying a high negative bias on the order of 5kV to the substrate (cathode) under low pressure. Application of reactive gas in plasma can be used for deposition of compounds of the coating specie. The carbides, nitrides, and oxides can be deposited by suitable combination of metal and reactive gas. Alternatively, if the plating is done in vacuum environment, then the ions can be produced in a separate ion-gun. In this case, the process is known as Ion Beam Assisted Deposition (IBAD).

If the process is carried out at relatively higher pressures, scattering by gas atoms can enhance surface coverage of the coating, thereby enabling deposition of hard coatings of metals as well as compounds over complex geometries. For further information, reader is referred to a detailed review by Mattox [83].

Arc Vapour Deposition

In this process, an electric arc of diameter on the order of a few micrometers is run over the solid, metallic coating material, causing it to evaporate. Because of the high currents and power densities used, the evaporated material is almost totally ionized and forms high-energy plasma. An electric bias is usually applied for accelerating the ions towards the substrate surface. The reader can refer different sources for further details of this deposition process [34, 84-86].

APPLICATIONS OF COATINGS

Among the various types of coatings developed and investigated, the simplest and the most widely used are the single layer deposits, while the more expensive, multilayer [87-91] and graded coatings [62] can be employed for tailoring certain properties to suit specific performance requirements.

Since a multitude of coatings of different materials can be deposited for a vast variety of applications; it is not possible to provide general guidelines regarding the most suitable deposition technique for a given application. Before selecting a combination of substrate, coating and processing parameters, it is important to investigate their impact on the desired properties of the component. It is likely that a considerable enhancement in one property could be accompanied by a detrimental effect on some other property of engineering importance. For example, an increase in hardness might be accompanied by loss of fracture toughness or fatigue resistance [92].

A brief outline of the broad application areas, which have been benefiting from advances in surface engineering, with particular reference to relatively new coating technologies (viz. PVD and CVD), is being presented in the following sub-sections. A summary of coating applications is provided in Table 1.3.

Manufacturing Industry

Rapid developments in new coating technologies have widely increased the use of coated components. The advances in surface engineering techniques have been of particular interest to the most demanding application areas, such as aerospace industry [93-97], followed by automobile and other manufacturing sectors. There are many commercially available coating techniques, like electro-chemical plating, chemical vapour deposition (CVD) and physical vapour deposition (PVD), which have gained widespread popularity in the manufacturing industry. The major application areas of surface coatings in manufacturing sector are generally based on their role in enhancing wear and corrosion resistance.

Wear Resistance Applications

Owing to their favourable tribological properties, such as wear-resistance, hot-hardness, oxidation resistance, residual compressive stresses, low coefficient of friction etc., the application of coatings on cutting tools, sliding bearings, seals and valves is now a well-established practice [98, 99]. Some of the well-established, standardized and commercially available coatings for such components are TiN, Al_2O_3, TiCN, TiAlN, AlCrN, Cr_7C_3, WC/C and diamond-like carbon (DLC). DLC films are of particular interest because of their unique blend of properties, viz. high hardness, high thermal conductivity, electrical insulation and resistance to heat and radiation.

In addition to the above-mentioned applications, there are many other machine elements, viz. automotive engine and driveline parts, which are subjected to vast variety of loads, contacts and environments during service. The performance as well as service life of these components remains dependent on the performance of lubricating oils. Since most of the high-performance lubricating oils are derived from petroleum products, their disposal poses a major environmental concern. There are many potential automotive components, which can benefit from thin coatings [100], thereby reducing dependence on lubricating oils, provided that the coating process is not detrimental to the favourable properties achieved during preceding heat treatment [101].

Corrosion Resistance Applications

Many types of coatings possess corrosion resistance against harsh chemicals even under elevated temperatures. In general, layers of corrosion resistant metals (metallization) or metal oxides (ceramics) are deposited on the substrates through evaporation or CVD respectively. Some common examples of corrosion resistant coatings for providing protection against atmospheric corrosion include aluminium metallization through evaporation and molybdenum silicide coating through CVD. The deposits of Cr_7C_3, Cr_3C_2 and Cr_2O_3 (chromia) are employed as intermediate layers in many corrosion resistance applications. The carbide coating of SiC is used for providing protection to marine components against sea water. Such coatings also offer good resistance against industrial waste, including sulfuric acid. Boron nitride coating, deposited through CVD process, provides protection to crucibles handling molten glasses or metals, while SiO_2 coating is employed on nuclear reactor components made of stainless steel for protection against oxidation.

Electronics Industry

1947 is generally perceived as the year marking birth of solid state electronics, when William Shockley, John Bardeen and Walter Brattain of Bell Telephone Laboratories discovered the transistor effect – a phenomenon, for which they were awarded Nobel Prize in

1956. They had used alloyed germanium for the purpose of making the first transistor; while silicon transistor was developed later on by Texas Instruments in 1954. The solid state diodes, transistors and other components gradually replaced the slow and power-hungry vacuum tubes. Following these early developments of discrete semiconductor components, Texas Instruments made a giant leap forward in 1959 by fabricating several components on a single chip; thereby paving the way for development of integrated circuits (ICs). Ever since, with the refinement in fabrication techniques, the number of components on a single chip has been gradually increasing – and today, we are in the era of Very Large Scale Integration (VLSI) or Ultra Large Scale Integration (ULSI) chip designs, wherein millions of components are packed in a single chip. The enhancement in design complexity, miniaturization, component reliability and switching speed would not have been accomplished without development of better understanding of the underlying physical and chemical deposition processes.

Electronic Components

The fabrication of discrete as well as integrated semiconductor electronic/optoelectronic components is greatly dependent on the technology behind formation of thin solid films of various semiconductor, dielectric and conducting materials from liquid, gas or vapour phases. The component fabrication process begins with the growth of epitaxial film of semiconductor material. For silicon epitaxy, this is usually done through hydrogen reduction of dichlorosilane vapours in a CVD reactor. Typical steps in the subsequent fabrication process may involve deposition of diffusion barriers (viz. TiN) in high aspect ratio (typically of 0.2 μm diameter and 2 μm deep) interconnection holes, followed by their filling with some suitable metal, such as tungsten or aluminium. The favourable characteristics of step-coverage and conformity exhibited by various CVD and PVD processes make them ideal choice for fabrication of semiconductor components [102]. CVD is also employed for deposition of dielectric films (viz. silicon dioxide, silicon nitride) in semiconductor component fabrication.

Further, as the fabrication of a typical semiconductor component can involve hundreds of processing steps, the preceding steps impose limitation on subsequent steps in term of temperature and holding time at that temperature – a restriction collectively known as *thermal budget* [103]. Precise control over processing parameters in PVD and CVD processes facilitates advance planning of thermal budget and so, helps meet the ever-increasing demands of circuit designers.

Data Storage Devices

Vacuum evaporation and sputter deposition processes for metallization of plastics and glasses are widely employed in fabrication of optical data storage media. PECVD coating of amorphous carbon finds applications in fabrication of sound heads and magnetic data storage devices.

Opto-electronic Components

Group II-VI compounds (chalcogenides) and group III-V compounds exhibit many interesting properties and their synthesis through CVD has led to a number of novel applications of these materials – especially in the area of opto-electronics. For example, GaAs and GaAlAs are used in making infrared light emitting diodes (LEDs), which find application in remote-control devices and covert-vision surveillance systems. CdS and CdSe are employed in fabrication of photoconductors and light detectors. CVD of ferroelectric materials, viz. AlN, $PbTiO_3$, ZnO, $Pb(Zr.Ti)O_3$, finds applications in various sensors and transducers, such as infrared sensing fire-alarms, electromechanical transducers etc.

Optical Applications

The optical applications of thin films are generally based on their properties of reflection, transmittance and interference. By exercising precise control on the composition and thickness of the deposited film, all these properties can be tailored according to requirement. While metallic films of aluminium, silver, gold or copper, generally deposited through evaporative metallization process, are essentially meant for obtaining high reflectivity; the interference effects in transparent coatings of other materials can be employed for selective transmission/reflection of a particular range of wavelengths, as creatively employed in "hot and cold mirrors" of film projectors.

Infrared Windows

Zinc selenide and zinc sulfide are transparent to infrared radiation. CVD grown ZnSe windows offer low absorption coefficient, which makes them suitable for thermal cameras. Owing to their capability to withstand thermal shocks, they also find application in high power CO_2 lasers. However, being a soft material, ZnSe is not suitable for harsh environments. The coating of ZnS, grown by CVD, exhibits high hardness and strength and is suitable for exterior infrared windows on aircrafts.

Antireflection Coatings

This class of coatings can comprise of a single transparent film having different refractive index than the substrate (viz. glass) or multiple films of high refractive index oxides (viz. TiO_2, ZrO_2, HfO_2, ThO_2), in combination with a low refractive index film (viz. SiO_2), deposited through evaporation or MOCVD. Antireflection coatings work on the principle of constructive or destructive interference of a selected wavelength of light (viz. visible or infrared) in transmitted and reflected beams respectively. These coatings are used on various components for enhancing transmissibility of light, viz. coatings on spectacles, camera lenses, photovoltaic panels, display devices and architectural glass.

Infrared reflective coatings are widely used on the inside of sodium vapour lamps (viz. indium tin oxide in SOX lamps) and automotive halogen lamps for retaining heat and in-turn enhancing the brightness of the lighting device. Reflective coatings on metallic surfaces find application in laser optics, where they have to withstand high beam intensities without degradation. Optical coatings with graded refractive indices find application in fabrication of filters for fibre optic communications.

Electrically Conducting Transparent Coatings

Transparent and electrically conducting oxides, such as indium tin oxide (ITO) offer a unique blend of properties. ITO is traditionally deposited through co-deposition of SnO_2 and In_2O_3 by sputtering. Such materials are used as antistatic coatings on television and computer monitors, as well as for providing protection to fragile electronic components against electrostatic build-up. They are also used for absorption or reflection of microwaves. Another important application of these transparent and electrically conducting coatings is for de-fogging and de-icing of automobile and aircraft windscreens. Due to poor scratch resistance of these sputter-deposited films, they are generally sandwiched beneath a second layer of glass.

Energy Sector

Due to ever-increasing demand for electricity, the energy sector has been growing steadily and adopting novel materials and techniques. While some of the new engineering materials help enhance efficiency of conventional, fossil-fuel based systems; others are aimed at developing renewable energy systems (viz. photovoltaics and wind turbines).

It is ironical that we have so far been able to tap just a tiny fraction of the huge potential of renewable energy resources (viz. wind and solar), while the supplies of many renewable energy materials are already beginning to become critical. For example, dysprosium, which is employed for enhancing life of permanent magnets in wind turbines, has become quite scarce; while the demand for silver, which is used for metallization of crystalline silicon photovoltaic cells, is soon going to surpass the total global production. So, any sustainable energy portfolio requires advancement of our engineering talent concerning renewable energy materials.

Power Plant Components

Vanes of gas and steam turbines operate under severe stress and harsh environment. The vane material deteriorates due to erosion, hot corrosion and sulfidication [104]. The application of thermal barrier coatings (TBCs) of ZrO_2 (zirconia), CoCrAlY and NiCrAlY through sputtering is known to significantly enhance the service life of turbines.

Photovoltaics

The first array of monocrystalline silicon was fabricated in 1958 for space application [14]. Ever since, the technology has proliferated considerably, but is yet to cover a lot more ground. A photovoltaic (PV) cell is essentially a semiconductor diode, in which one side of the *p-n* junction is made so thin that the light falling on it reaches the junction, where it is absorbed by electrons, following which, they cross the energy barrier.

Currently, the photovoltaic industry is dominated by crystalline silicon, whose wafers are generally sliced from crystalline silicon ingots, or sometimes grown by CVD. The subsequent steps in fabrication usually include diffusion of *n*-type doping, MOCVD of SiO_2 and deposition of aluminium through evaporation.

Amorphous silicon is the next candidate material for photovoltaics application and can be deposited as a large thin film through PACVD process. Its energy absorption characteristics are also superior to crystalline silicon; but currently, the process suffers from low yield, low efficiency and high cost.

The most efficient photovoltaic materials are the ones that respond to a wide range of wavelengths, viz. amorphous silicon, GaAs and other group III-V compounds. The CVD process has made it possible to fabricate photovoltaic cells of these materials. Further, multijunction photovoltaic cells, comprising of several thin layers of different optoelectronic materials, each tailored for absorbing a specific wavelength, are also being conceived at experimental level.

Table 1.3. Applications of coatings

Industry	Application Area	Coating Materials
Mechanical Manufacturing	Wear resistant coatings	TiN, Al_2O_3, TiCN, (TiAl)N, AlCrN, WC/C, DLC, W_3C+W2C+W, Cr_7C_3, SiC, Si_3N_4
	Corrosion resistant coatings	Cr, Ni, Zn, Sn, In, Ag, Cd, Au, Pb, Rh, CoCrAlY, NiCrAlY, Ni-B, Ni-P, Al_2O_3, TiB_2, SiC, Cr_2O_3, SiO_2
	Thermal barrier coatings for turbines	Al_2O_3, $Al_2O_3xTiO_2$, ZrO_2, Cr_2O_3
	Dry lubricating films	a-C, Ti(C,N), TiN
Semiconductor and other Opto-Electronic Applications	Semiconductors for microelectronics and energy conversion (Photovoltaics)	Si, Ge, Group III-V and II-VI compounds
	Interconnects and diffusion barrier coatings for semiconductors	TiN, In_2O_3, WSi_2, W, WN, Co, TiC
	Electrically conductive films	Indium Tin Oxide (ITO) Aluminium doped Zinc Oxide (AZO)
	Data storage	a-C layers for magnetic data storage; Al, Ni layers for optical data storage
	Films for selective transmission/reflection of radiation	TiO_2, ZrO_2, HfO_2, ThO_2, SiO_2
	Freestanding nano-structures	Ion-etched surfaces, viz. Cu

CONCLUSION

Surface engineering is widely employed in various fields, ranging from biomedical to aerospace. It is vital for industry, since it equips the design and manufacturing engineers with the capability of developing products, which can meet the ever-increasing expectations on performance and price fronts.

Surface coating technologies have come a long way from the classic, metal cladding, hot-dip or electroplating processes to the modern PVD and CVD techniques. The ability of these processes to precisely control the deposition parameters and employ *in-situ* chemical reactions of precursor species have led to the development of many advanced materials, whose properties are tailored to suit requirements of specific applications. Examples of well-established application areas of these coating techniques include deposition of protective coatings for steam/gas turbines and other components subjected to severe mechanical stress and harsh environments; tribological coatings for cutting tools; deposition of dielectric, semiconductor, metallic materials and diffusion barriers in semiconductor component fabrication; growth and surface deposition of optical components etc.

The replacement of existing surface overlay methods with more advanced deposition techniques has traditionally been governed by criticality of application (viz. solar photovoltaic cells for space probes), performance requirements and economic viability. The progress in coating technologies, along with mass-scale production of certain coated components, has led to a gradual lifting of economic viability barrier for many novel applications. Some candidate areas, which are likely to benefit from further scaling-up of coating technology in a foreseeable future, include replacement of lubricating oils in automotive engine and driveline components, fabrication of amorphous silicon and GaAs based photovoltaic systems and synthetic diamond heat-sinks for electronic components. With further enhancement in fundamental understanding of the process dynamics and improvements in process equipment, the coating processes are poised to open new vistas for providing solutions to the ever-deepening environment and energy crises.

REFERENCES

[1] Czerwiec, T., Michel, H. & Bergmann, E. (2000). Low-pressure, high-density plasma nitriding: mechanisms, technology and results. *Surface and Coatings Technology, 108-109*, 182-190.

[2] Czerwiec, T., Renevier, N. & Michel, H. (2000). Low-temperature plasma-assisted nitriding. *Surface and Coatings Technology, 131,* 267-277.

[3] Suh, C.M., Hwang, J.K. Son, K.S. & Jangb, H.K. (2005). Fatigue characteristics of nitrided SACM 645 according to the nitriding condition and notch. *Materials Science and Engineering A, 392*, 31 – 37.

[4] Ashrafizadeh, F. (2003). Influence of plasma and gas nitriding on fatigue resistance of plain carbon (Ck45) steel. *Surface and Coatings Technology, 173 –174*, 1196–1200.

[5] C. Godoy, C., Mancosu, R.D., Lima, M.M., Branda˜o, D., Housden, J. & Avelar-Batista, J.C. (2006). Influence of plasma nitriding and PAPVD $Cr_{1-x}N_x$ coating on the cavitation erosion resistance of an AISI 1045 steel. *Surface and Coatings Technology, 200*, 5370 - 5378.

[6] Park, Y.G., Wey, M.Y. & Hong, S.I. (2007). Enhanced wear and fatigue properties of Ti–6Al–4V alloy modified by plasma carburizing/CrN coating. *Journal of Materials Science: Materials in Medicine, 18*, 925–931.

[7] Hitoshi, I. (2005). Advanced acetylene vacuum carburizing. *IHI Engineering Review, 38*, 82-88.

[8] Preston, S. (1991). Bending fatigue strength of the carburizing steel SS 2506, *International Journal of Fatigue, 13*, 506.

[9] Ebert, L.J. (1978). The role of residual stresses in the mechanical performance of case carburized steels. *Metallurgical and Materials Transactions A, 9*, 1537 – 1551.

[10] Krauss, G. (1999). Martensite in steel: strength and structure. *Materials Science and Engineering A, 273–275*, 40–57.

[11] Yang, Q., Ren, X., Gao, Y., Li, Y., Zhao, Y. & Yao, M. (2005). Effect of carburization on residual stress field of 20CrMnTi specimen and its numerical simulation. *Materials Science and Engineering A, 392*, 240 – 247.

[12] Grove, W.R. (1852). On the electro-chemical polarity of gases. *Philosophical Transactions of the Royal Society of London, 142*, 87–101.

[13] Faraday, M. (1857). The bakerian lecture: experimental relations of gold (and other metals) to light. *Philosophical Transactions of the Royal Society of London, 147*, 145-181.

[14] Pierson, H.O. (1999). *Handbook of Chemical Vapor Deposition: Principles, Technology and Applications.* New York: Noyes Publications.

[15] Edison, T.A. (1892). Process of duplicating phonograms. *U.S. Patent No. 484582.*

[16] Choy, K.L. (2003). Chemical vapour deposition of coatings. *Progress in Materials Science*, *48*, 57–170.

[17] Tucker, Jr. R.C. (1994). Thermal Spray Coatings. In Cotell, C.M., Sprague, J.A. & Smidt, Jr. F.A. (Eds.). ASM Handbook Vol.5: Surface Engineering, (pp. 497-509). Ohio: ASM International.

[18] Penning, F.M. (1935). Coating by cathode disintegration. *U.S. Patent No. 2146025.*

[19] Berghaus, B. (1938). Improvements in and relating to the coating of articles by means of thermally vaporised material. *U.K. Patent No. 510993.*

[20] Jarratt, M. History of PVD coatings [online]. Nov. 7, 2010. Available from: http://www.pvd-coatings.co.uk/history-pvd-coatings/.

[21] Poorman, R.M., Sargent, H.B. & Headlee. L. (1955). Method and apparatus utilizing detonation waves for spraying and other purposes. *U.S. Patent No. 2714563 A.*

[22] Smith, G.H., Pelton, J.F. & Eschenbach, R.C. (1958). Jet plating of high melting point materials. *U.S. Patent No. 2861900.*

[23] Gage, R.M., Nestor, O.H. & Yenni, D.M. (1962). Collimated electric arc-powder deposition process. *U.S. Patent No. 3016447.*

[24] Mattox, D.M. & McDonald, J.E. (1963). *Interface formation during thin film* deposition. *Journal of Applied Physics, 34,* 2493-2494.

[25] Ruppert, W. (1960). Process for the production of titanium carbide coatings. *U.S. Patent No. 2962388.*

[26] Wasa, K. & Hayakawa, S. (1969). Low pressure sputtering system of the magnetron type. *Review of Scientific Instruments, 40*, 693-697.

[27] Chopra, K.L. & Randlett, M.R. (1967). Duoplasmatron ion beam source for vacuum sputtering of thin films. *Review of Scientific Instruments, 38*, 1147-1151.

[28] Morley, J.R. (1971). Vacuum vapor deposition utilizing low voltage electron beam. *U.S. Patent No. 3562141.*

[29] Takagi, T. (1988). *Ionized cluster beam deposition and epitaxy.* NJ: Noyes Publications.

[30] Bunshah, R.F. & Raghuram, R.C. (1972). Activated reactive evaporation process for high rate deposition of compounds. *Journal of Vacuum Science and Technology,* 9, 1385-1388.

[31] Raghuram, A.C. & Bunshah, R.F. (1972). The effect of substrate temperature on the structure of titanium carbide deposited by activated reactive evaporation. *Journal of Vacuum Science and Technology,* 9, 1389-1394.

[32] Bunshah, R.F. & Raghuram, A.C. (1972). Activated reactive evaporation process for high rate deposition of compounds. *Journal of Vacuum Science and Technology, 9,* 1385-1388.

[33] Wasa, K., Kitabatake, M., & Adachi, H. (2004). Thin film materials technology: sputtering of compound materials. New York: William Andrew Publishing.

[34] Moll, E. & Daxinger. H. (1980). Method and apparatus for evaporating materials in a vacuum coating plant. *U.S. Patent No. 4197175.*

[35] Pulker, H.K. & Daxinger. H. (1981). Method of producing gold-color coatings. *U.S. Patent No. 4254159.*

[36] Wiesmann, H., Ghosh, A.K., McMahon, T. & Strongin, M. (1979). a-Si:H produced by high temperature thermal decomposition of silane. *Journal of Applied Physics, 50*, 3752-3754.

[37] Schneider, J.M., Rohde, S., Sproul, W.D. & Matthews, A. (2000). Recent developments in plasma assisted physical vapour deposition. *Journal of Physics D: Applied Physics, 33,* R173–R186.

[38] Thornton, J.A. (1983). Plasma-assisted deposition processes: theory, mechanisms and applications. *Thin Solid Films, 107,* 3-19.

[39] Windows, B. & Savvides, N. (1986). Unbalanced dc magnetrons as sources of high ion fluxes. *Journal of Vacuum Science and Technology A, 4*, 453-456.

[40] Windows, B. & Savvides, N. (1986). Charged particle fluxes from planar magnetron sputtering sources. *Journal of Vacuum Science and Technology A, 4*, 196-202.

[41] Randhawa, H. & Johnson, P.C. (1987). Technical note: a review of cathodic arc plasma deposition processes and their applications. *Surface and Coatings Technology, 31,* 303-318.

[42] Randhawa, H. (1988). Cathodic arc plasma deposition technology. *Thin Solid Films, 167,* 175-185.

[43] Brown, I.G., Godechot, X. & Yu, K.M. (1991). Novel metal ion surface modification technique. *Applied Physics Letters, 58,* 1392-1394.

[44] Bancroft, W.D. (1905). The chemistry of electroplating. *The Journal of Physical Chemistry, 9*, 277–296.

[45] Barnhart, J. (1997). Occurrences, uses and properties of chromium. *Regulatory Toxicology and Pharmacology, 26*, 3–7.

[46] Baral, A. & Engelken, R.D. (2002). Chromium-based regulations and greening in metal finishing industries in the USA. *Environmental Science and Policy, 5*, 121–133.

[47] Smart, D., Such, T.E. & Wake, S.J. (1983). Novel trivalent electroplating bath. *Transactions of the Institute of Metal Finishing, 61*, 105 - 107.

[48] Brassard, J. (1988). Decorative chromium electroplating – a process in transition. *Metal Finishing, 86*, 105–108.

[49] Rabiei, A., Mumm, D.R., Hutchinson, J.W., Schweinfest, R., Rühle, M. & Evans, A.G. (1999). Microstructure, deformation and cracking characteristics of thermal spray ferrous coatings. *Materials Science and Engineering A, 269*, 152–165.

[50] Mostaghimi, J., Chandra, S., Ghafouri-Azar, R. & Dolatabadi, R. (2003). Modeling thermal spray coating processes: a powerful tool in design and optimization. *Surface and Coatings Technology, 163 –164*, 1–11.

[51] Paredes, R.S.C., Amico, S.C. & d'Oliveira, A.S.C.M. (2006). The effect of roughness and pre-heating of the substrate on the morphology of aluminium coatings deposited by thermal spraying. *Surface and Coatings Technology, 200*, 3049 – 3055.

[52] Kout, A. & Müller, H. (2009). Parameter optimization for spray coating. *Advances in Engineering Software, 40*, 1078 – 1086.

[53] Gargeda, T.W., Lanza, V. & Roebuck, A.H. (1986). Thermal spray coatings. *Journal of Protective Coatings and Linings, 3*, 44 – 48.

[54] Jensen, K.F. (1987). Micro-reaction engineering applications of reaction engineering to processing of electronic and photonic materials. *Chemical Engineering Science, 42*, 923–958.

[55] Gates, S.M. (1988). Adsorption kinetics of SiH_4, Si_2H_6 and Si_3H_8 on the Si(111)-(7x7) surface. *Surface Science, 195*, 307-329.

[56] Lee, P.W., Omstead, T.R., McKenna, D.R. & Jensen, K.F. (1987). In situ mass spectroscopy and thermogravimetric studies of GaAs MOCVD gas phase and surface reactions. *Journal of Crystal Growth*, *85*, 165 - 174.
[57] Van Hemert, R. L., Spendlove, L.E. & Sievers, R.E. (1965). Vapor Deposition of Metals by Hydrogen Reduction of Metal Chelates. *Journal of the Electrochemical Society*, *112* (11), 1123–1126 (1965).
[58] Larson, C. E., Baum, T. H., & Jackson, R. L. (1987). Chemical Vapor Deposition of Gold. *Journal of the Electrochemical Society*, *134* (1), 266.
[59] Musil, J. (1986). Microwave plasma: its characteristics and applications in thin film technology. *Vacuum*, *36*, 161 – 169.
[60] Rie, K.-T., Gebauer, A. & Woehle, J. (1993). Plasma assisted CVD for low temperature coatings to improve wear and corrosion resistance. *Surface and Coatings Technology*, *60*, 385 - 388.
[61] Memming, R., Tolle, H.J. & Wierenga, P.E. (1986). Properties of polymeric layers of hydrogenated amorphous carbon produced by a plasma-activated chemical vapour deposition process II: Tribological and mechanical properties. *Thin Solid Films*, *143*, 31 – 41.
[62] Bachmann, P.K., Leers, D. & Lydtin, H. (1991). Towards a general concept of diamond chemical vapour deposition. *Diamond and Related Materials*, *1*, 1 – 12.
[63] Meneve, J., Dekempeneer, E., Wegener, W. & Smeets, J. (1996). Low friction and wear resistant a-C:H/a-Si_{1-x},C_x:H multilayer coatings. *Surface and Coatings Technology*, *86-87*, 617-621.
[64] McKenzie, D.R., McFall, W.D., Reisch, S., James, B.W., Falconer, I.S., Boswell, R.W., Persing, H., Perry, A.J. & Durandet, A. (1996). Synthesis of cubic boron nitride thin films. *Surface and Coatings Technology*, *78*, 255—262.
[65] Zajickova, L., Janca, J. & Perina, V. (1999). Characterization of silicon oxide thin films deposited by plasma enhanced chemical vapour deposition from octamethylcyclotetrasiloxane/oxygen feeds. *Thin Solid Films*, *338*, 49-59.
[66] Köcker, G.M. & Habig, K. –H. (1996). Influence of different production parameters on the functional behaviour of tools and parts after coating. *Surface and Coatings Technology*, *82*, 294 - 304.
[67] Sherman, A. (1987). *Chemical vapor deposition for microelectronics*. Park Ridge (NJ): Noyes.
[68] Musket, R.G., McLean, W., Colmenares, C.A., Makowiecki, D.M. & Siekhaus, W.J. (1982). Preparation of atomically clean surfaces of selected elements: A review. *Applications of Surface Science*, *10*, 143—207.
[69] Movchan, B.A. (2002). Functionally graded EB-PVD coatings. *Surface and Coatings Technology*, *149*, 252–262.
[70] Bloyce, A. (2000). Coatings for high performance pumps and compressors. *World Pumps*, *(Jan)*, 43–45.
[71] Bogaerts, A., Neyts, E., Gijbels, R. & Mullen, J.V.D. (2002). Gas discharge plasmas and their applications. *Spectrochimica Acta Part B*, *57*, 609–658.
[72] Thornton, J.A. & Penfold, A.S. (1978). Cylindrical Magnetron Sputtering. In Vossen, J.L., & Kern, W., (Eds.). *Thin Film Processes* (pp. 76–113). New York: Academic Press.

[73] Fraser, D.B. (1978). The Sputter and S-Gun Magnetrons. In Vossen, J.L., & Kern, W., (Eds.). *Thin Film Processes* (pp. 115–129). New York: Academic Press.

[74] Waits, R.K. (1978). Planar Magnetron Sputtering. In Vossen, J.L., & Kern, W., (Eds.). *Thin Film Processes* (pp. 131–173). New York: Academic Press.

[75] Kelly, P.J. & Arnell, R.D. (2000). Magnetron sputtering: a review of recent developments and applications. *Vacuum*, *56*, 159-172.

[76] Mattox, D.M. (1964). Film Deposition Using Accelerated Ions. *Electrochemical Technology*, *2*, 295.

[77] Mattox, D.M. (1965). *Interface formation and the adhesion of deposited thin films*. (SC-R-65-852). Springfield, VA, USA: Sandia Corporation.

[78] Mattox, D.M. & Bland, R.D. (1967). Aluminum coating of uranium reactor parts for corrosion protection. *Journal of Nuclear Materials*, *21*, 349–352.

[79] Zega, B., Kornmann, M. & Amiguet, J. (1977). Hard decorative TiN coatings by ion plating. *Thin Solid Films*, *45*, 577-582.

[80] Weissmantel, C., Reisse, G., Erler, H.-J., Henny, F., Bewilogua, K., Ebersbach, U. & Schurer, C. (1979). Preparation of hard coatings by ion-beams methods. *Thin Solid Films*, *63*, 315-325.

[81] Fukutomi, M., Fujitsuka, M. & Okada, M. (1984). Comparison of the properties of ion-plated titanium carbide films prepared by different activation methods. *Thin Solid Films*, *120*, 283–292.

[82] Martin, P.J., Netterfield, R.P. & Kinder, T.J. (1990). Ion-beam deposited films produced by filtered arc evaporation. *Thin Solid Films*, *193-194*, 77–83.

[83] Mattox, D.M. (2000). Ion plating – past, present and future. *Surface and Coatings Technology*, *133-134*, 517-521.

[84] Lindfors, P.A. & Mularie, W.M. (1986). Cathodic arc deposition technology. *Surface and Coatings Technology*, *29*, 275 - 290.

[85] Sanders, D.M. & Anders, A. (2000). Review of cathodic arc deposition technology at the start of new millennium. *Surface and Coatings Technology*, *133-134*, 78 - 90.

[86] Martin, P.J. & Bendavid, A. (2001). Review of the filtered vacuum arc process and materials deposition. *Thin Solid Films*, *394*, 1-15.

[87] Rincon, C., Romero, J., Esteve, J., Martınez, E. & Lousa, A. (2003). Effects of carbon incorporation in tungsten carbide films deposited by R.F. magnetron sputtering: single layers and multilayers. *Surface and Coatings Technology*, *163-164*, 386-391.

[88] Wolfe, D.E., Singh, J. & Narasimhan, K. (2002). Synthesis of titanium carbide chromium carbide multilayers by the coevaporation of multiple ingots by electron beam physical vapor deposition. *Surface and Coatings Technology*, *160*, 206– 218.

[89] Knotek, O., Löffler, F. & Krämmer, G. (1993). Process and advantage of multicomponent and multilayer PVD coatings. *Surface and Coatings Technology*, *59*, 14– 20.

[90] Meletis, E.I., Erdemir, A. & Fenske, G.R. (1995). Tribological characteristics of DLC films and duplex plasma nitriding/DLC coating treatments. *Surface and Coatings Technology*, *73*, 39-45.

[91] Rie, K.-T. & Broszeit, E. (1995). Plasma diffusion treatment and duplex treatment – recent development and new applications. *Surface and Coatings Technology, 76-77*, 425-436.

[92] Wilson, A.D., Leyland, A. & Matthews, A. (1990). A comparative study of the influence of plasma treatments, PVD coatings and ion implantation on the tribological performance of Ti–6Al–4V. *Surface and Coatings Technology*, *114*, 70– 80.

[93] Schijve, J. (2009). Fatigue damage in aircraft structures, not wanted, but tolerated? *International Journal of Fatigue*, *31*, 998–1011.

[94] Zhang, X., Boscolo, M., Figueroa-Gordon, D., Allegri, D. & Irwing, P.E. (2009). Fail-safe design of integral metallic aircraft structures reinforced by bonded crack retarders. *Engineering Fracture Mechanics*, *76*, 114–133.

[95] Goldsmith, N.T., Clark, G. & Barter, S.A. (1996). A growth model for catastrophic cracking in an RAAF aircraft. *Engineering Failure Analysis*, *3*, 191–201.

[96] Maricocchi, A., Bartz, A. & Wortman, D. (1997). PVD TBC experience on GE aircraft engines. *Journal of Thermal Spray Technology*, *6*, 193–198.

[97] Miller, R.A. (1997). Thermal barrier coatings for aircraft engines: History and directions. *Journal of Thermal Spray Technology*, *6*, 35–42.

[98] Su, Y.L., Yao, S.H., Wei, C.S. & Wu, C.T., (1998). Evaluation on the tension and fatigue behavior of various PVD coated materials. *Thin Solid Films*, *322*, 218–224.

[99] Su, Y.L., Yao, S.H., Wei, C.S., Wu, C.T. & Kao, W.H. (1998). Evaluation on the wear, tension and fatigue behavior of various PVD coated materials. *Materials Letters*, *35*, 255–260.

[100] Vetter, J., Barbezat, G., Crummenauer, J. & Avissar, J. (2005). Surface treatment selections for automotive applications. *Surface and Coatings Technology*, *200*, 1962–1968.

[101] Saini, B.S. & Gupta, V.K. (2010). Effect of WC/C PVD coating on fatigue behaviour of case carburized SAE8620 steel. *Surface and Coatings Technology*, *205*, 511–518.

[102] Kern, W. & Schuegraf, K.K. (2002). Deposition Technologies and Applications: Introduction and Overview. In Seshan, K. (Ed.), *Handbook of Thin-Film Deposition Processes and Techniques: Principles, Methods, Equipment and Applications*, (pp. 11-36). New York: Noyes Publications.

[103] Jeng, J.C., Lee, B.C., Su, A.J. & Yu, C.C. (2010). Thermal Budget Control for Rapid Thermal Processing Systems with Spike-Shaped Temperature Profile. In Kothare, M., Tade, M., Wouwer, A.V. & Smets, I. (Eds.), *Proceedings of the 9th International Symposium on Dynamics and Control of Process Systems* (pp. 521-526), Leuven, Belgium.

[104] Kiameh, P. (2002). *Power Generation Handbook: Selection, Applications, Operation, and Maintenance*. (1). New York: McGraw-Hill Professional.

In: Comprehensive Guide for Nanocoatings Technology. Vol. 1 ISBN: 978-1-63482-447-7
Editor: Mahmood Aliofkhazraei

Chapter 16

THE SIGNIFICANCE OF ELECTROCHEMICAL TECHNIQUES IN NANO TECHNOLOGY

I. Gurrappa, I. V. S. Yashwanth* and I. Mounika
Defence Metallurgical Research Laboratory,
Kanchanbagh, Hyderabad, India

ABSTRACT

Electrochemistry plays a significant role in nanotechnology. The electrochemical techniques help not only in synthesizing a variety of nanomaterials and coatings starting from metals to tubes, rods or wires of different dimensions but also in characterizing them for various applications including preventing global warming, which is a major threat to the entire world. The chapter reviews the importance of electrochemical techniques in synthesizing nanostructured metals, alloys, coatings, composites, Titania nanotubes; rods, wires and carbon nanotubes that exhibit remarkably enhanced characteristics over other materials/coatings apart from their classification, synthesis, properties and applications. Subsequently, the advantages of electrochemically prepared nano metals, alloys, coatings, tubes, wires, rods and the necessity of their use in effectively preventing global warming and in modern industries that allow the systems to be used safely at varied environmental conditions, which not only would enhance the efficiency of industrial systems but also their life span and consequently the profitability, has been stressed.

INTRODUCTION

The Nature is a greatest gift by God to the entire living organisms. Among the living organisms, the human beings are mostly enjoying and utilizing it to a greater extent, thus causing air pollution, water pollution etc. It is the responsibility of scientists all over the world to focus on it and come out with appropriate technologies.

---

* Corresponding author: I. V. S. Yashwanth. Presently at Texas University in Elpaso, US. E-mail: igp1@rediffmail.com.

One of the biggest problems facing the world today is global warming. Many scientists believe that our production of carbon dioxide and other greenhouse gases are having a heating effect on the atmosphere, and this could be very dangerous for human life. Air pollution is one prime reason for this scourge.

Generally any substance that humans introduce into the atmosphere that has damaging effects on living organisms and the environment is considered air pollution. Smog hanging over cities is the most familiar and overt form of air pollution. There are different kinds of pollution—some visible, some invisible—that contribute to global warming. Carbon dioxide, a greenhouse gas, is the main pollutant that is warming the Earth. It is widely considered to be a pollutant when associated with transportation vehicles including trains, planes, power plants, and other human activities that involve the burning of fossil fuels such as gasoline and natural gas.

In the past 150 years, such activities have pumped enough carbon dioxide into the atmosphere to raise its levels higher than they have been for hundreds of thousands of years. Other greenhouse gases include methane—which comes from such sources as swamps and gas emitted by livestock—and chlorofluorocarbons (CFCs), which were used in refrigerants and aerosol propellants. Another pollutant associated with climate change is sulfur dioxide, a component of smog. Sulfur dioxide and closely related chemicals are known primarily as a cause of acid rain. Volcanic eruptions can spew massive amounts of sulfur dioxide into the atmosphere.

In fact, volcanoes used to be the main source of atmospheric sulfur dioxide but today human beings are responsible. In order to overcome these problems, the scientists across the globe have been trying to find the cordial solutions. One of the solutions is the development of Carbon and Titania nanotubes. In the resent exploration, it is shown that Titania nanotubes are very handy for environment purification and self-cleaning (Table I).

Significantly growing energy needs and increasing environmental concerns, alternatives to the use of non-renewable and polluting fossil fuels have to be investigated. One such alternative is solar energy. It is such an interminable energy that if we are able to covert and store it with advanced technologies then no one in the world will suffer with power dearth. Due to the nature of solar energy, two components are required to have a functional solar energy generator.

Table 1. The importance of nanomaterials

Applications	References
Global warming	[1]
Solarisation	[2-4]
Gas sensors	[5, 6]
Photocatalysis	[7, 8]
Photo electrochemical hydrogen generation	[9, 10]
Orthopedic implants	[11, 12]
Vaccum microelectronics	[13]
Energy storage	[14, 15]
Hydrogen storage	[16-19]
Templates	[20]

These two components are a collector and a storage unit. The collector simply collects the radiation that falls on it and converts a fraction of it to other forms of energy (either electricity and heat or heat alone).

The storage unit is essentially required because of the non-constant nature of solar energy -at certain times only a very small amount of radiation will be received. These components can be amalgamated by fabricating nanotubes and manufacture dye sensitized solar cells which have the required properties to produce superlative results.

According to the authors, nanostructured metals, alloys, composite coatings, Carbon and Titania nanotube fabrication by economically viable and environmentally friendly techniques are essentially required in preventing environmental pollution, solarisation of a variety of systems starting from household to space vehicles, sensors, hydrogen storage, biomedical applications etc. Though the efforts have been made for the past one decade, much research is highly needed in this direction.

NANOSTRUCTURED COATINGS

Nanostructured materials offer great potential for various applications due to their superior characteristics that are not typically found in conventional materials (Figure 1). Due to novel properties and varied potential applications, the nanomaterials with typical grain sizes < 100 nm, are attracting increasing attention from researches all over the world.

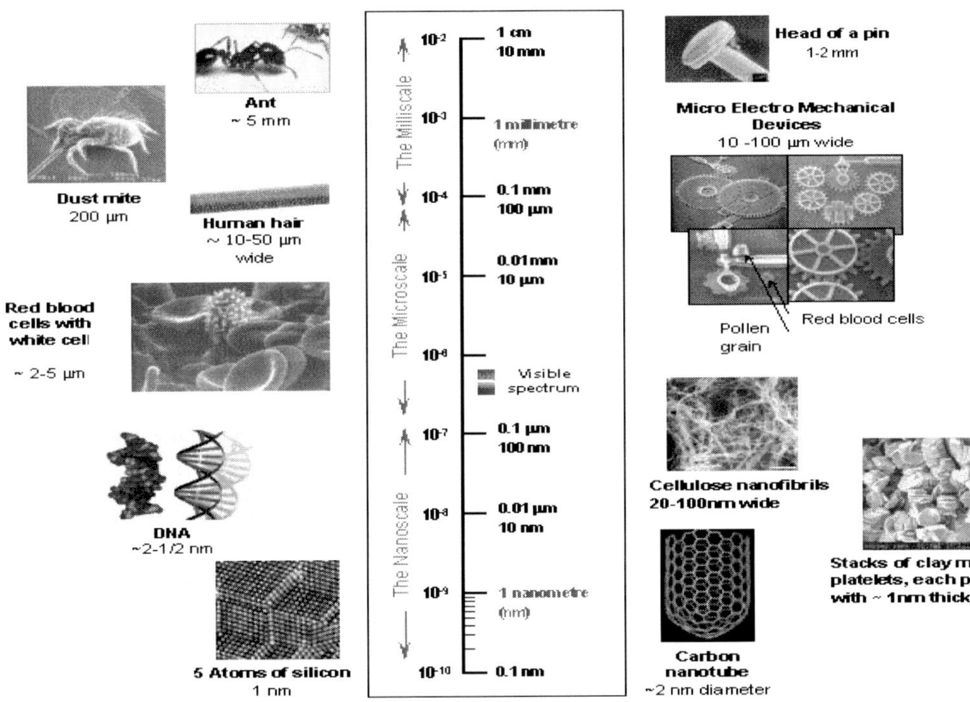

Figure 1. Comparison of nano with conventional materials.

Because of small grain size of these materials and consequently the large volume fraction of atoms in or near the grain boundaries, these materials exhibit properties that are often superior and sometimes completely new, in comparison to those of conventional coarse grained materials.

These include increased strength/hardness, enhanced diffusivity, improved ductility/toughness, reduced density, reduced elastic modulus, higher electrical resistivity, increased specific heat, higher coefficient of thermal expansion, lower thermal conductivity, increased corrosion and wear resistance and superior soft magnetic properties. However, it should be realized that a full understanding of the structures, thermal stability and properties of these novel materials and coatings is not yet available.

Classification

Nanocrystalline materials are single phase or multiphase polycrystals, the crystal size of which is of the order of a few (typically 1-100) nanometers in at least one dimension and they can be basically equiaxed in nature. They are classified as three-dimensional (3D), two-dimensional (2D), one-dimensional (ID) and zero-dimensional nanostructures (Table II).

The nanocrystalline materials may contain crystalline, quasicrystalline or amorphous phases and can be metals, ceramics or composites. Among the above, maximum attention has been paid to the synthesis, consolidation and characterization of the 3D-nanostructured crystallites followed by the 1D-layered nanostructures. While the formers are expected to find applications based on their high strength, improved formability, and a good combination of soft magnetic properties, the latter are intended for electronic and other applications.

Table 2. Classification of nanocrystalline materials

Dimensionality	Designation	Typical method(s) of synthesis
Zero dimensional	Clusters	Sol-gel method
One dimensional (1D)	Layered (lamellar)	Electrodeposition Vapour deposition
Two dimensional (2D)	Filamentary	Chemical vapour deposition
Three dimensional (3D)	Crystallites (equiaxed)	Gas condensation Mechanical alloying

Relatively few investigations have been carried out on the zero-dimensional clusters and 2D-filamentary nanostructures.

Preparation

Nanocrystalline materials can be synthesized by a number of techniques with the starting material existing either in the vapour, liquid, or solid state. Inert gas condensation, in which the evaporated material is quenched onto a cold substrate, was the first technique used to synthesize nanocrystalline metals and alloys. Subsequently, plasma processing and their physical and chemical vapour deposition techniques have also been used. Electrodeposition and rapid solidification processing use the liquid state as the starting material. Mechanical alloying, friction stir welding, severe plastic deformation, spark erosion, sliding wear and repeated cold rolling produce nanocrystalline materials in the solid state. While some of these techniques are commercially in use for large-scale production purposes (e.g., inert gas condensation, electrodeposition and mechanical alloying), many of other techniques are essentially used for laboratory investigations. Most of the early work (1975-90) dealt with materials produced by crystallization of amorphous precursor materials. The second group of materials that have been extensively studied since 1990 are nanostructures that are produced by electrodeposition technique. Table II provides the major findings of typical nanocrystaline nickel produced by electrodeposition. Nanostructured materials often exhibit outstanding properties from those of bulk materials as mentioned earlier. Therefore, these materials became popular worldwide to use in a variety of applications. Electrodeposition is a unique technique in which a variety of materials can be processed like metals, ceramics, polymers etc. Electrochemical strategies offer important advantages and unique possibilities in the development of nanomaterials and nanostructures. Recently, novel electrodeposition technologies have evolved into an important branch of nanotechnology. With this technique, it is possible to obtain porosity free nanocrystalline (NC) metal specimens in a single step process [28]. In order to have a better understanding of the nature of nanocrystalline state, it may be useful to establish similarities and differences in the microstructure of NC metals manufactured via different technological routes. It was reported that the microstructure of NC Ni and Cu obtained through electrodeposited (ED) and severe plastic deformation (SPD) techniques are different. For ED Ni, the grain structure is controlled by the parameters of the layer growth process and by the misfit of the crystal structure of deposit and substrate. On the other hand, the structure of SPD Ni is controlled by the movement of dislocations created by the applied deformation [29].

Table 2. Improved properties of nanocrystalline materials

Property	Observation	References
Hardness	5 times harder	[21, 22]
Friction	Reduction in coefficient of friction by 50%	[23]
Corrosion resistance	Improved significantly	[24]
Strength	3 to 10 times stronger	[24]
Wear Resistance	Increased by 170 times	[223]
Magnetic	Lower coercivity, increased resistivity by 3 times, Ms reduced by 5%	[25, 26]
Hydrogen Diffusion	Higher hydrogen diffusion	[27]
Electrocatalytic	Improved electrocatalytic activities for hydrogen evolution and hydrogen oxidation reactions	[27]

The co-deposition of ceramic materials with metals and polymers has created opportunities in the preparation of novel hybrid nanomaterials and nanostructures that cannot be obtained by other methods. Figure 2 presents general diagram for an electrodeposition process. For obtaining enhanced characteristics of any nanocoating, it is highly essential to optimize bath (electrolyte) composition as well as deposition parameters in addition to selection of appropriate anode and cathode materials with suitable finish. For example, Table

III provides the bath composition as well as deposition parameters for producing Ni-AlN nanocomposite coatings. Similarly, optimized parameters and bath composition are paramount in electrodeposition process of any new coating or material. Recent research has been focused on demonstrating the feasibility of co-deposition of various nanomaterials and the development of novel electrochemical strategies. The nanohybrides have combined and advantageous properties over that of individual materials and exhibit novel and advanced properties not achievable in single phase materials.

The choice of technique for the synthesis of nanocrystalline materials is determined by the ease of the process, its economic viability, its scalability and the purity of the end product desired. Most of the techniques referred to previously produce the nanocrystalline material in the powder form. The application of nanocrystalline materials requires that these powders are consolidated to full density with minimal no porosity and this is not a trivial matter. Full densification with complete inter-particle bonding requires exposure of the powder to high temperatures and pressures for extended periods of time, which is likely to coarsen the microstructural features. Retention of the material in the nanocrystalline state, however, warrants that the powder is not taken to high temperatures for longer periods of time; this may not result in complete densification.

Thus, successful consolidation to full density requires innovative methods of consolidation. However, since nanocrystalline materials have a large fraction of the atoms in the grain boundaries, the diffusivity of nanocrystalline materials is much higher than that of a coarse grained- material of same composition. This will facilitate achieving full consolidation of nanocrystalline materials at temperatures 300 °C to 400 °C lower than those required for the coarse grained materials. Successful consolidation of nanocrystalline powders has been achieved by electro-discharge compaction, plasma activated sintering, shock consolidation, hot isostatic pressing (HIP), ceracon processing, hydrostatic extrusion, strained powder rolling and sinter forging.

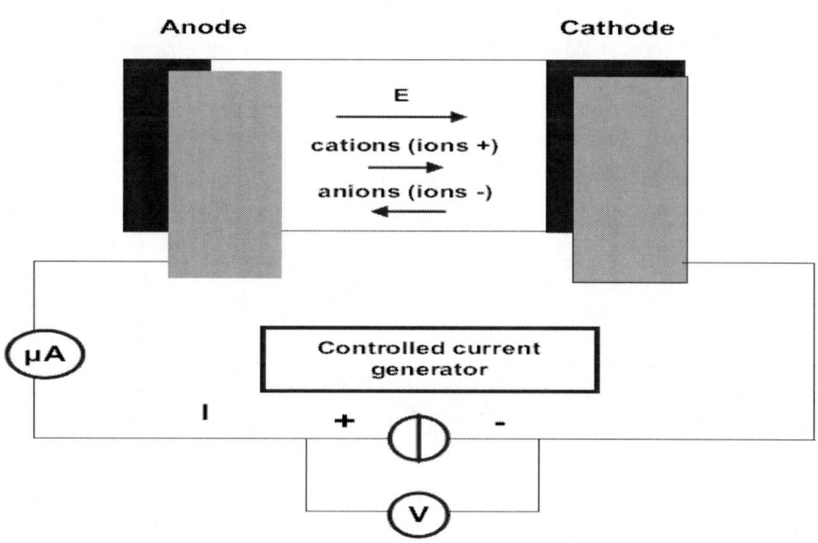

Figure 2. General diagram for electrodeposition process.

Table 3. Electrodeposition parameters for Ni-AlN nanocoating

Electrolyte and parameters	Composition and conditions
$NiSO_4.6H_2O$	300 g/l
$NiCl_2.6H_2O$	45 g/l
H_3BO_3	40 g/l
AlN powder	0-55 g/l
pH	2-6
Current density	0.5-5 A/dm^2
Temperature	55 °C

Table 4. Advantages of Electrochemical deposition technique

S.No	Advantage
1.	Rapid
2.	Low cost
3.	Free from porosity
4.	High purity
5.	Industrial applicability
6.	Potential to overcome shape limitations or allows the production of free standing parts of complex shape
7.	Offers higher deposition rates
8.	Produce coatings on widely differing substrates
9.	Ability to produce structural features with sizes ranging from nm to μm
10.	Easy to control alloy composition
11.	Ability to produce compositions unattainable by other techniques
12.	The possibility of formation of simple low cost multi layers in many different systems e.g., Cu/Ni, Ni/Ni-P etc.
13.	No post deposition treatments

Consolidation is not required if the powder could be used as it is (e.g., in slurries, as catalysts, or for coating purposes).

It is critical to synthesize nanosized powders in a clean environment and powders that are susceptible to contamination by oxygen and moisture must be protected during storage and transport. However, the nanostructured materials produced by electrodeposition and amorphous crystallization that lead directly to high-density bulk materials are not subject to these concerns.

Advantages of Electrochemical Deposition Technique

Synthesis of nanomaterials requires an atomistic deposition process and extreme control over deposition. Vapour deposition techniques have been used almost exclusively to produce these materials. The fact that electrochemical deposition being an atomic deposition process too, can be used to synthesize nanocomposites (Figures 3 and 4), has generated a great deal of interest in recent years.

The obvious advantages of this wonderful process of electrodeposition are provided in Table 4. The formation of nuclei on the electrode substrate was extensively studied on the basis of macroscopic thermodynamic considerations.

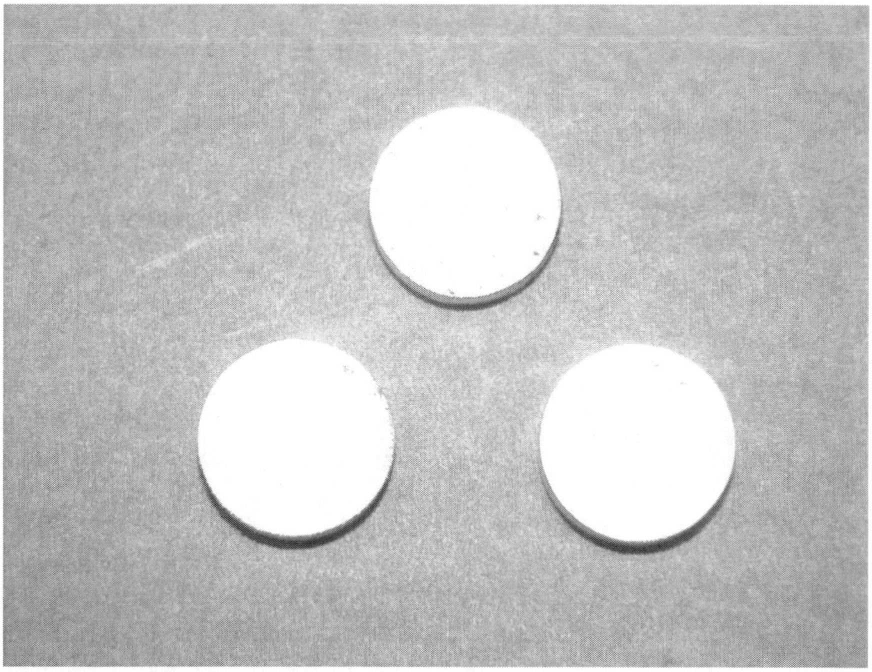

Figure.3. As coated different nanostructured coatings on DMR-1700 steel.

Figure 4. (Continued).

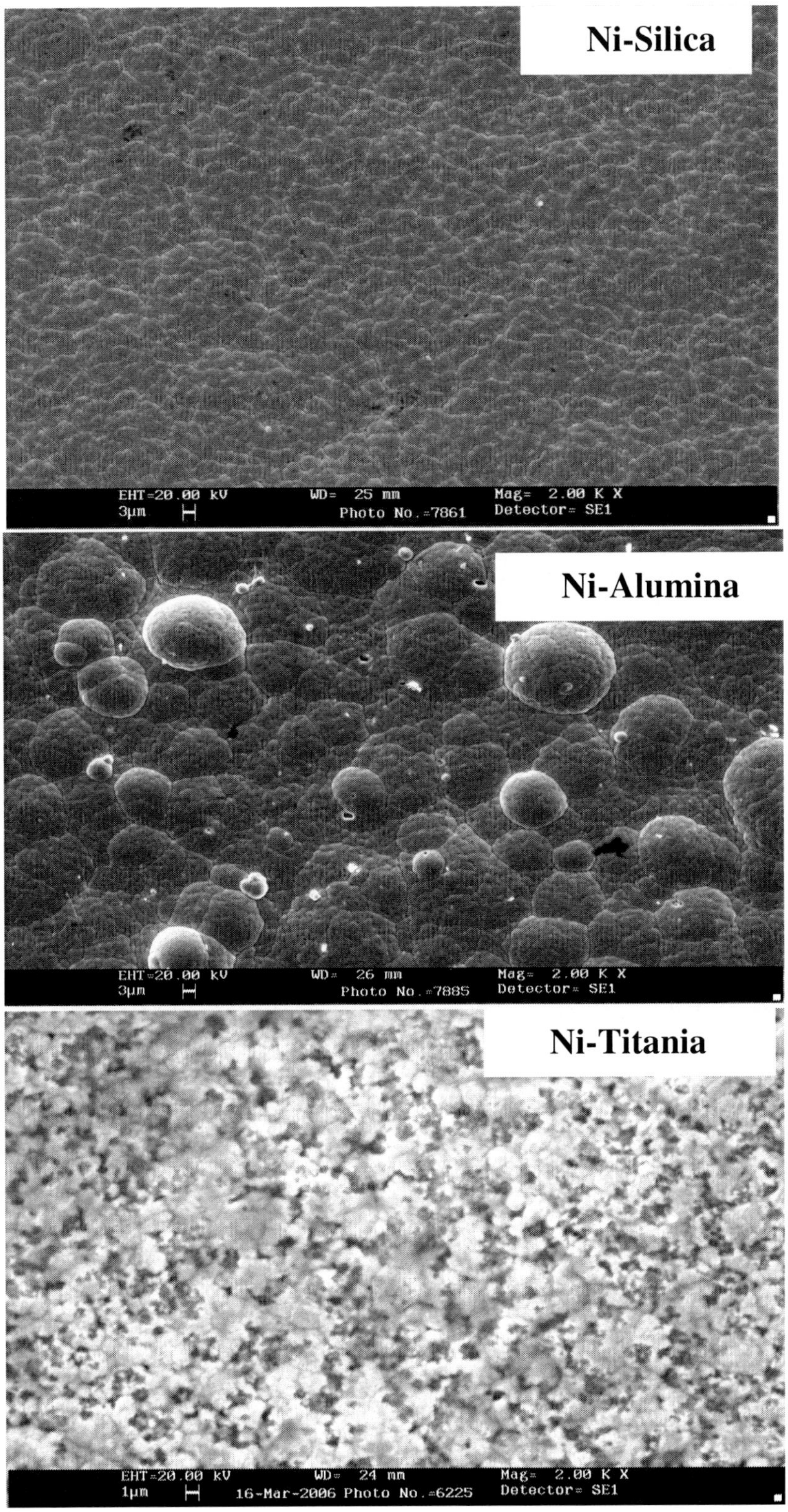

Figure 4. (Continued).

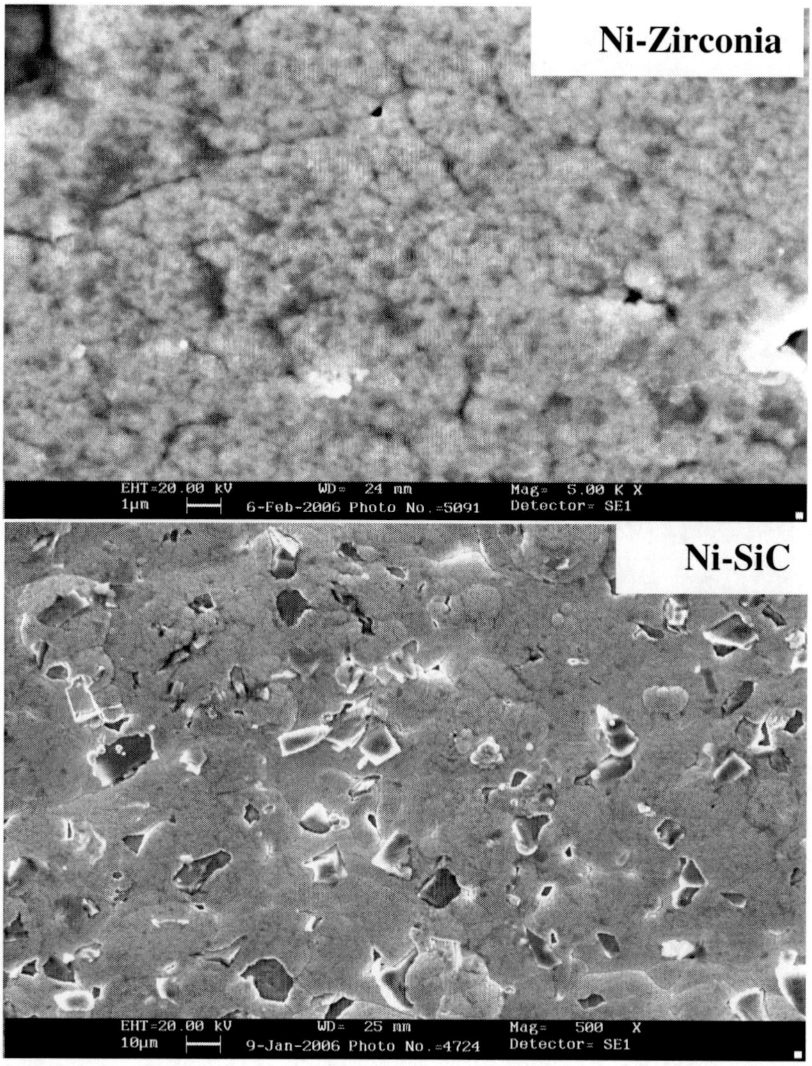

Figure 4. SEM of pure nickel,nanostructured Ni-silica, Ni-Alumina, Ni-Titania , Ni-Zirconia and Ni-SiC composite coatings.

The nucleation of nanostructures on the electrode substrate during electrodeposition is influenced by the crystal structure of the substrate, specific free surface energy, adhesion energy, lattice orientation of the electrode surface and crystallographic lattice mismatch at the nucleus-substrate interface boundary.

The final size distribution of the electrodeposits, however, strongly depends on the kinetics of the nucleation and growth. The electrodeposition process involves the formation of either an instantaneous or a progressive nucleation. In the case of instantaneous nucleation, all the nuclei from instantaneously on the electrode substrate, and subsequently grow with the time of electrodeposition. In contrast, the number of nuclei that formed is a function of time of electrodeposition in the progressive nucleation. These nuclei gradually grow and overlap; therefore, the progressive nucleation process exhibits zones of reduced nucleation rate around the growing stable nuclei.

The electrodeposition technique consists of an electrochemical cell and accessories for applying controlled current at a certain voltage. The cell usually contains a reference electrode, a specially designed cathode and an anode or counter electrode. The cathode substrate on which electrodeposition of the nanostructure takes place can be made of either nonmetallic or metallic material. Using the surface of the cathode as a template, various desired nanostructures or morphologies can be synthesized for specific applications. Such a template-assisted electrodeposition process can be broadly divided into two groups: active template-assisted and restrictive template-based electrodeposition.

The formation of nanostructures in active template –based synthesis results from growth of the nuclei that invariably nucleate the holes and defects of the electrode substrate. Subsequent growth of these nuclei at the template yields the desired surface morphology of the nanostructures, which can therefore be synthesized by choosing the appropriate surface of the electrode. Highly orientated pyrolitic graphite (HOPG), for example, is used extensively as an electrode substrate for the electrodeposition of silver, gold, molybdenum, palladium and platinum nanostructures. Electrodeposition initiates at the step edges, dislocations and defect sites of the electrode surface.

Another template-assisted synthesis of mostly metal nanowires involves the deposition of metal into the cylindrical pores or channels of an inert, non-conductive nanoporous electrode material.

Track-etch membranes, porous alumina and other nanoporous structures including conductive rubber polymers, metals, semiconductors, carbons and other solid materials were used as templates to prepare nanometersized particles, fibrils, rods and tubules.

Although a variety of templates and electrodeposition processes are reported in the literature, the universal application of such templates for depositing metallic nanostructures with controlled and well-defined morphologies is missing. The major problem with using electrodeposition to synthesize nanostructure is the preparation of proper templates.

Electrodeposition is highly influenced by the surface characteristics of the electrode substrate and shape and size of the deposit depend on the substrate. Therefore, further studies of electrode surfaces and subsequent surface engineering are essential to develop active templates for the successful synthesis of nanostructures using electrodeposition.

A number of advances have recently contributed to the use of electrodeposition for the fabrication of nanostructured ceramic materials and considerable attention has been given to the electrochemical deposition of ceramic coatings for biomedical applications. For such applications, thin films of hydroxyapatite (HA) were prepared using an electrodeposition technique (electrophoretic deposition (EPD) [30, 31] or electrolytic deposition (ELD) [32]. Hydroxyapatite is an established implants material because its chemical composition is similar to that of bone tissue. Electrodeposition technique was utilized for the preparation of coatings on Ti-6Al-4V surgical alloy substrates. In this process, the prepared coatings exhibited enhanced sinterability at relatively low temperatures. Nanostructured coatings of HA were also prepared by ED from aqueous solutions containing calcium and phosphate ions. The electrochemical method brings a novel approach to synthesis biomedical coatings and composites (Figures 2 and 3). Moreover, it provides an important contribution to the development of nanostructured compounds. Nanostructured ZnO obtained from electrodeposition has important applications in chemical sensors, semiconductor diodes, catalysis and solar cells [33-35]. Electrodeposited Titania is used in gas sensors, solar cells, biomedical implants, and electrochromic devices. Varying bath compositions and deposition parameters gives control over the morphology and porosity of the films. The nanostructured films have important advantages for applications in solar cells. The possibility of defect chemistry superlattices modulated on a nanometer scale is another exciting development. Fabrication of ceramic nanowires by using electrodeposition is another area of interest.

Cathodic electrodeposition enables the formation of ceramic particles in-situ in a polymer matrix, preventing particle agglomeration. An important development in this area is the electrodeposition of hybrid films based on Fe_2O_3 and poly-diallydimethyammonium chloride (PDDA). It was reported superparamagnetic properties of the hybrid films at room temperature. Hybrid polymer-ceramic films can be prepared on cathodic and anodic substrates by co-deposition of ceramic nanoparticles and polymers or by electrochemical intercalation of ceramic particles into the polymer matrix at the electrode surface. Various electrochemical techniques were developed for preparation of nanostructured hybrid metal/metal-oxide materials [36-39].

Ni-P deposits obtained from electrodeposition are important materials for the surface finishing of industrial materials owing to their electrocatalytic activity for hydrogen evolution, special paramagnetic properties, excellent microhardness and corrosion resistance [40, 41]. Electrodeposited amorphous Ni-P alloys posses a more regular amorphous structure and lower concentration of defects than melt spun Ni-P alloys [42].

It was reported that nanocomposite layers could be obtained by electrochemical deposition of silicon carbide nanoparticles (mean diameter 20 nm) in a nickel-plating bath. Benea et al. [43] showed that incorporation of 20 nm SiC particles into Ni deposition modifies the impedance and cathodic polarization diagrams in correlation with the modification of the surface morphology and crystallinity of nickel matrix. Multilayered structures of spinvalve type in the form of nanowires were deposited by Fedosyuk et al. [44] using pulsed electrolytic deposition from single electrolyte.

Among the diversified synthesis processes, the electrochemical method has been used for a wide range of nanostructured materials e.g., metal nanoparticles, nanowires, nonofilms, bullk nanocrystalline metals, laminated composites, multilayred coatings and nanoparticle-reinforced composite coatings. TiO_2 nanoparticle-reinforced Ni composite coatings obtained by electrodeposition exhibit unusual photoelectrochemical characteristics, which can be used to study electronic transitions in semiconductors [45]. The cycle performance of the nanosized Cu_6Sn_5 electrode (used for secondary lithium batteries) prepared by chemical reduction is significantly enhanced in comparison with that of the same material prepared by sintering or mechanical alloying [46]. Gao et al. [47] synthesized Cu nanowires, which are dense, continuous, with uniform diameter (60 nm) along the entire lengths of wires by potentiostatic electrochemical deposition of copper sulphate solution within the nanochannels of porous anodic alumina templates.

Silver nanorods were prepared by an electrochemical technique from an aqueous solution of $AgNO_3$ in the presence of polyethylene glycol (PEG) [48]. Electrochemically prepared Co nanofilms exhibit three to five times greater coercivity (H_c) than that of polycrystalline cobalt [49].

Fe nano wire was electrodeposited into the holes of porous anodic aluminium oxide (AAO). Magnetic measurements showed that its easy magnetization direction is perpendicular to the sample plane. This kind of nanowire array has potential applications in perpendicular magnetic recording [50]. Electrochemical reduction of metal oxides provides an appropriate technique for the production of nanocrystalline mesoporous metals at room temperature. It was demonstrated by producing Cu from polycrystalline sintered CuO and Cu_2O single crystals [51].

Some nanocrystalline alloys exhibit excellent resistance to corrosion and can be used as protective coatings [52]. The voltametric behaviour of gold nanowire arrays prepared by electrochemical method showed a significant enhancement of the ratio between the total electrolytically exposed surface area and the diffusionally accessible surface area in comparison with macroscopic flat gold electrodes [53].

Electrochemically deposited nanostructured platinum metals exhibit higher catalytic performances and stabilities for various catalytic reactions as well as having unique magnetic properties, compared with conventional metals [54]. The synthesis of Co and Fe nanowires and nanotubes electrochemically deposited through nanoporous membranes was reported [55]. The electrodeposition technique was reported to be an ideal technique for preparation of high performance tin-antimony nano-structured electrode alloys used in lithium-ion batteries [56, 57]. Nanocrystalline FeTi, $LaNi_5$, Mg_2Ni and other materials based on Mg were reported to have easy absorption of hydrogen at low temperatures with very good kinetics and are not sensitive when exposed to air [58]. Chen et al. reported that the nano Ni particles with Ce shell structure have higher hydrogen storage properties due to the synergistic effect of Ce and Ni in the shell structure [59]. Layered Cu-Ni nanocomposites prepared by electrodeposition showed enhanced hardness [60]. Nanocomposite Ni/Al_2O_3 coatings that produced by electrodeposition, exhibit considerable enhancements in their hardness in comparison to pure Ni (Figure 5).

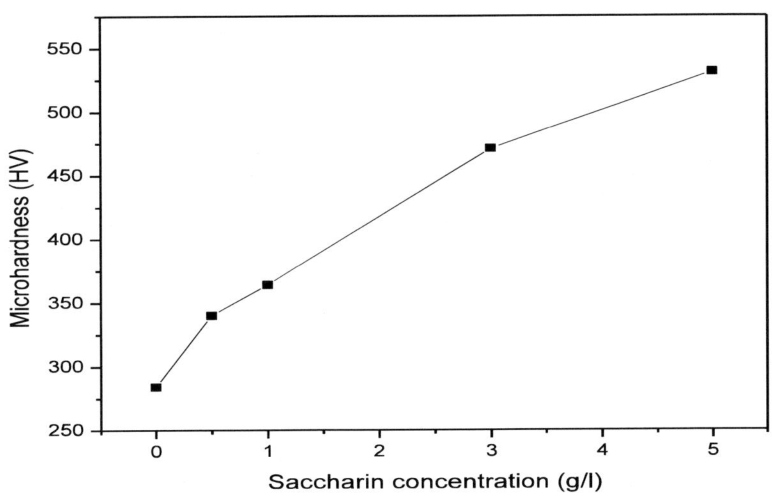

Figure 5. The microhardness of $Ni-Al_2O_3$ as a function of saccharin addition.

Alternative layers of two different materials with individual laminate thickness in the range of nanometers are possible with electrochemical deposition technique. This technique demonstrated the capability of relatively inexpensive process for the production of advanced materials with ultra fine microstructure [61]. Nanocrystalline nickel of about 17 nm size produced by electrodeposition significantly enhanced the electrocatalytic activity for the hydrogen evolution due to an increased density of active surface sites [62].

Electrodeposited Co-Ni-Cu/Cu superlattices showed a previously unreported transition from GMR (Giant Magnetoresistance) to AMR (Anisotropic Magnetoresistance) and back to GMR on annealing [63].

By controlling electrochemical deposition conditions adequately, a wide range of artificially structured coatings like epitaxial multilayered materials (Cu/Ni), amorphous material in combination with crystalline materials (NiPx/Sn) can be deposited [64].

Evans et al. and others [63, 65] opened the possibility of measuring the CPP-Mr as function of temperature on samples with resistance values reaching hundreds of ohms by making multilayered nanowires by electrodepostion. Multilayered Co/Cu wires produced by electrodeposition with layer thickness in the range of 3 to 100 nm exhibited GMR effect, which is similar to that of sputtered samples.

Electrochemistry is a key technology in nanoparticle science. For example electroplating offers novel routes to nanosized particles via arrested and templated electrodeposition. Electrochemistry is a suitable method for coupling particle activity to external circuitry in nanoparticle science in fields as optoelectronics and biological sensors. Electrochemistry is an enabling technology across the whole spectrum of nanoparticle research, from particle preparation to sensor applications and from band gap determination to solar cell operation [66]. Electrochemical methods for surface modification and materials processing play an important role in many modern technologies.

Electrochemical fabrication techniques are becoming increasingly attractive for the deposition of new materials and novel microstructures. Electrodeposition has advanced significantly over the past decade in syntheses of nanostructured materials to a practical industrial materials technology from a laboratory scale. The emerged materials for industrial applications are Ni, Pd, Cu, Ni-P, and Ni-Mo [67].

Structure

In order to understand the interrelationship between structure and properties, nanocrystalline materials need to be characterized on both atomic and nanometer scales. The microstructural features of importance include

1 Grain size, distribution and morphology
2 The nature and morphology of grain boundaries and interface phases
3 The perfection and nature of intragrain defects
4 Composition profiles across grains and interfaces and
5 Identification of residual trapped species from processing.

In the case of layered nanostructures, the features of importance are

a) The thickness and coherency of interfaces
b) Composition profiles across interfaces and
c) The nature of defects

There is gamut of characterization techniques that can yield structural information on nanocrystalline materials. These include direct microscopic techniques such as Transmission Electron Microscopy (TEM), Scanning Tunneling Microscopy (STM), Field Ion Microscopy (FIM) and the less direct electron, x-ray, and neutron diffraction techniques.

Indirect spectroscopic tools such as EXAFS, nuclear magnetic resonance, Raman and Mossbauer spectroscopies, and positron annihilation spectroscopy, have also been used. Other useful tools employed include differential scanning calorimetry (DSC), mass spectroscopy, x-ray fluorescence, atomic absorption spectroscopy, Auger electron spectroscopy and hydrogen absorption. Owing to the ultra fine scale of these materials, traditional characterization tools such as TEM and x-ray and neutron diffraction are both necessary and useful to understand the structure of nanocrystalline materials. However, for microchemical analysis on the requisite fine scale, further advances in the state of the art of instrumental capabilities will be necessary to obtain the desired lateral scale resolution. Only atom probe FIM or STM seem to offer the required lateral scale resolution for such chemical analysis at present.

The volume fraction of atoms residing in or near the grain boundaries increases as grain size decreases. Consequently, in materials with a small grain size of a few nanometers, the fraction of atoms in the grain boundaries is comparable to that in the grains. Therefore, nanocrystalline materials can be considered to consist of two structural components- the small crystallites with long-range order and different crystallographic orientations constituting the 'crystalline component', and a net work of intercrystalline regions (grain boundaries, triple junctions, etc.), the structure of which differs from region to region and is referred to as the interfacial component. The structure of the grain boundaries was investigated in great detail using high-resolution TEM techniques; the results are still inconclusive. There are two schools of thought regarding the grain boundary structures-one suggesting a gas–like disorder at the grain boundaries and the other suggesting that the structure of the grain boundaries is the same in both nanocrystalline and coarse-grained materials. The latter suggestion has gained widespread acceptance in recent years [68].

However, further research is needed to better understand the structure of nanomaterials produced by different techniques.

Properties

Because of very fine grain sizes, the nanocrystalline materials exhibit a variety of properties that are different and often considerably improved in comparison with those of conventional coarse-grained polycrystalline materials. These include morphological properties (Figure 4), increased strength/hardness (Figures 5-7) [21, 22, 69-72], enhanced diffusivity [24, 27], improved ductility/toughness [23], reduced density, reduced elastic modulus, increased specific heat, higher thermal expansion coefficient, lower thermal conductivity [25, 26], enhanced corrosion and wear resistance (Figures 8-9) [23, 69, 71] and superior reflectance (Figure 10) [73] and soft magnetic properties [25, 26] in comparison with conventional coarse grained materials. All these properties have to be investigated extensively for the new materials to be synthesized with a view to explore possible applications in different areas.

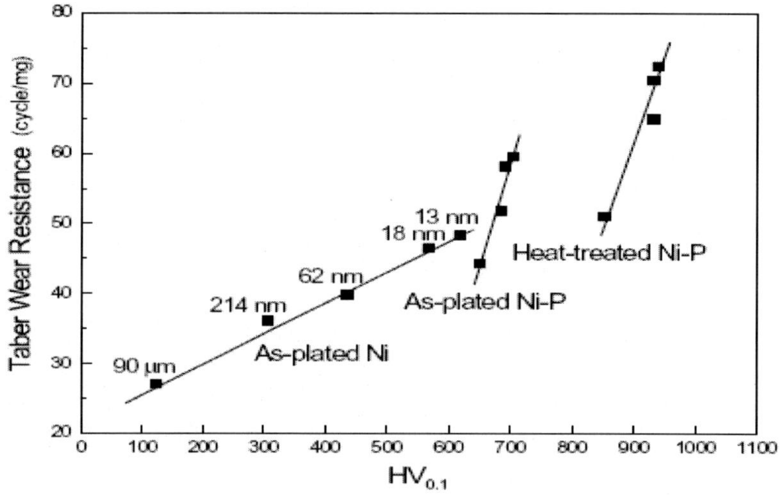

Figure 6. The relationship between hardness and Taber wear resistance of nanocrystialline Ni and Ni-P coatings [71].

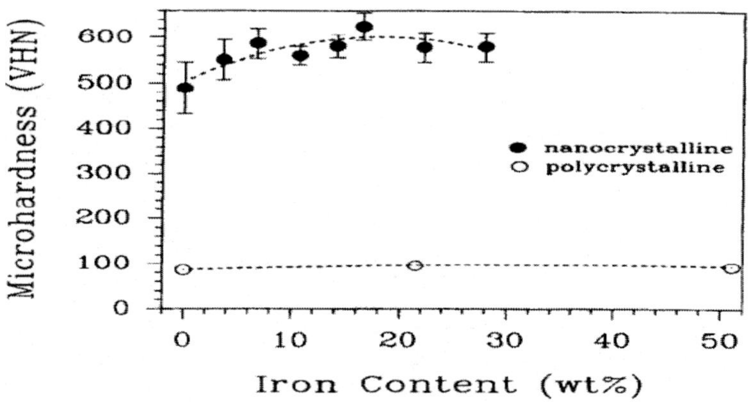

Figure 7. Microhardness of electrodeposited nanocrystalline and conventional polycrystalline Ni-Fe coatings as a function of iron content [72].

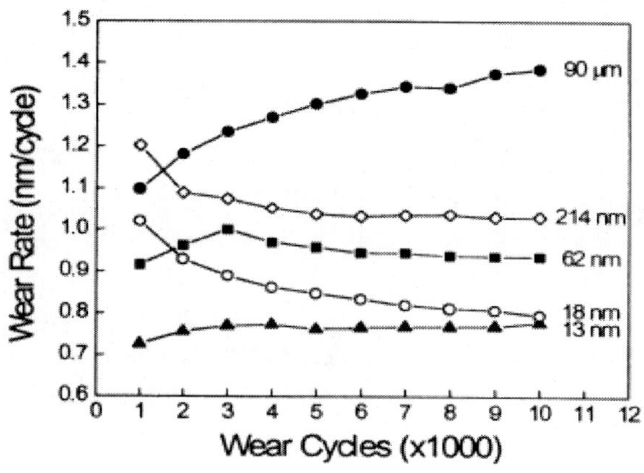

Figure 8. Average abrasive wear rate as a function of wear cycles [69].

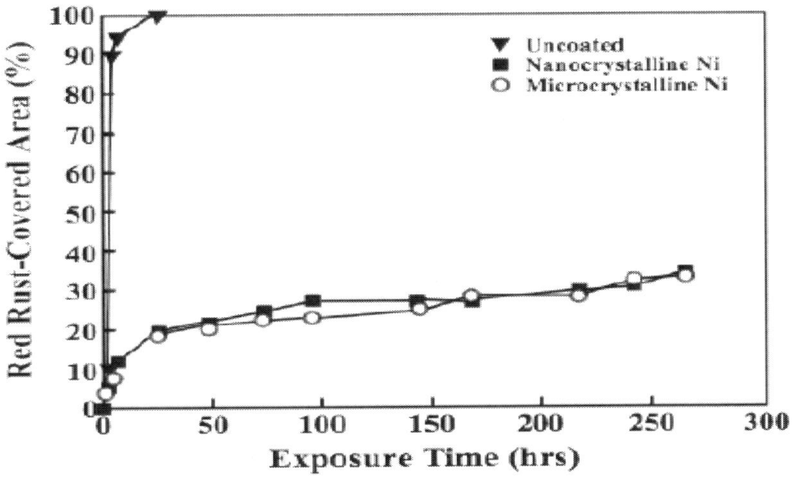

Figure 9. salt spray corrosion resistance for microcrystalline and nanocrystalline nickel on steel substrates [23].

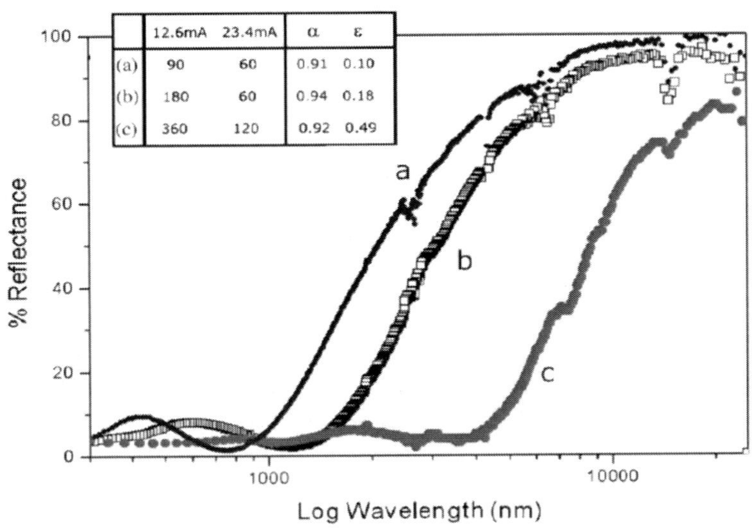

Figure 10. Percentage of total reflectance of nanostructured black nickel solar absorber coating under different conditions [73].

Industrial Applications

As a result of unique properties observed for electrodeposited nanomaterials, a number of industrial applications have emerged, some of which are presented in Table V.

For example, the combination of increased hardness/wear resistance and reduced localized corrosion results in improved protective coating performance.

The material could be used as hard facing on softer, less wear resistant substrates. For electroformed products such as printing mandrels and CD stampers, longer run life is expected. The magnetic and electrical properties make them attractive as soft magnets for high efficiency transformer, power supply and motor applications.

The future generation of recording head materials could also benefit from the excellent magnetic, electrical and wear properties.

The exceptional catalytic properties make them strong contenders for different applications like electrodes for hydrogen evolution reaction and fuel cells. Some nanocrystalline alloys exhibit very large solubility for hydrogen, making them suitable candidates for hydrogen storage materials.

As a result of enhanced properties, a wide range of materials can be synthesized and use them in a variety of applications for enhanced efficiency. In essence, effective commercialization of nanomaterials will enhance the production by several times by improving the efficiency of the systems significantly.

A growing number of companies are working to scale up their production capabilities and incorporate their nanostructured products into new and existing markets. In the United States alone, more than 70 companies are currently active at various levels in the development and production of nanostructured materials. More than a dozen of these businesses are involved in the manufacture of nanostructutred materials on an industrial scale. For example among other products, nanosized particles are incorporated into commercially available abrasive polishing slurries, fire retardant materials, magnetic fluids, magnetic recording tapes, sunscreens and transport wood stains. The bulk nanostructured material namely soft magnetic iron-based alloy is in industrial use among others. By the end of this century, nanostructured materials are expected to play an increasingly significant role in many major industries.

TITANIA NANOTUBES

Titania nanotubes are helpful not only in preventing environmental pollution but also highly efficient in solarisation of a variety of systems starting from household to space vehicles apart from other applications. Synthesis of titania nanostructures can be achieved by various routes including sol–gel methods, template-assisted methods, hydro/solvothermal approaches and by electrochemical techniques.

Table 5. Examples of electrodeposited nanomaterials and their potential applications

Materials	Uses
Pd, Ni-Pd	Hydrogen storage and purification, electrodes for hydrogen evolution reaction and fuel cells
Ni-Al_2O_3, Ni-SiC	Corrosion protection, wear resistance
Co, Co-W, Co-P	Hard coatings, magnetic materials
Ni, Ni-P, Ni-Mo, Ni-Zn, Ni-Fe, Ni-Fe-Cr	Corrosion protection, stress corrosion cracking resistance, soft magnets, catalysts

Self-organized TiO_2 nanotube layers can be formed by a simple electrochemical oxidation reaction of a metallic titanium substrate under a specific set of environmental conditions (Figure 11). This type of self-aligned oxide nanotube layers has attracted considerable interest over the past 10 years because of a unique combination of the highly functional features of TiO_2 with regular and controllable nanoscale geometry (length, tube diameter, and self-ordering can be adjusted over large length scales). The synthesis is carried out by an economically viable conventional electrochemical anodization technique. It is remarkable to mention that the self-ordering anodization approach is not limited to titanium or titanium alloys alone but can be applied to a large range of other transition metals or alloys to form highly aligned oxide nanotube or pore structures. The electrochemical anodization approach discussed herein is self-organizing: it leads to an array of oxide nanotubes aligned perpendicular to the substrate surface (Figure 11) and to a well-defined (and controllable) tube length. The tubes are attached to the metal surface and are electrically connected and thus easy to handle.

The use of an electrochemical anodization technique allows virtually any shape of titanium (and other metal) surfaces to be produced a defined nanotube layer and is thus an extremely versatile (easy to scale up) process.

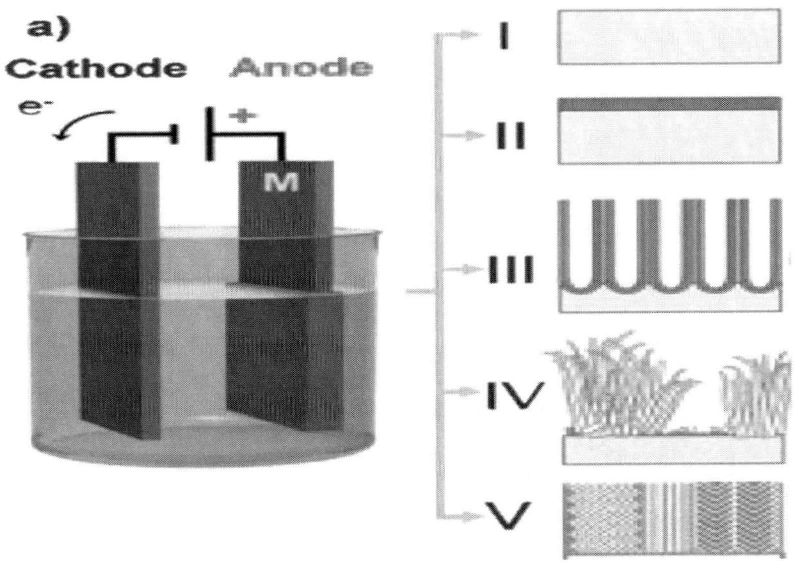

Figure 11. Schematic representation of the electrochemical anodization process and possible morphologies of titania nanotubes.

Rapid-breakdown anodization (RBA) technique is useful to produce comparably long nanotubes (several 100 mm) [within very short times some seconds to minutes. The major disadvantage is that the tubes are not well-defined regarding their length distribution, not well-organized over larger surface areas and they are hardly connected to the substrate. The anodic formation of several 10 mm thick ordered nanoporous TiO_2 layers with channels several nanometers to several tens of nanometers in diameter has been recently reported [74-77].

Self-organized oxide tube arrays can be obtained by an anodization process of a suitable material. When metals are exposed to a sufficiently anodic voltage in an electrochemical configuration (as in Figure 11), an oxidation reaction $M = M^{n+} + n\ e^-$ will be initiated. Depending mainly on the electrolyte and the particular anodization parameters, essentially three reactions takes place: i) The Mn^+ ions are solvatized in the electrolyte; that is, the metal is continuously dissolved (corrosion or electropolishing of the metal is observed) ii) the formed Mn^+ ions react with O^{2-} (provided by H_2O in the electrolyte) and form a compact oxide (MO) layer if MO is not soluble in the electrolyte; iii) under some electrochemical conditions, competition between solvatization and oxide formation is established (leading to porous MO). Under even more specific experimental conditions, a situation is established where self-organization during the growth of porous oxide takes place; furthermore, under some specific conditions, disorganized rapid growth of TiO_2 nanotube bundles or formation of thick self-organized meso-porous layers can be observed. With the modification and use of porous alumina either directly (such as in filters or for photonic crystals) or indirectly as a template for nanorod, nanowire, and nanotube synthesis by deposition of metal, semiconductor, or polymer materials, it is possible to synthesis different types of nano tubes, rods or wires. The first self-organized anodic oxides on titanium were reported for anodization in chromic acid electrolytes containing hydrofluoric acid by Zwilling and co-workers in 1999 [78, 79]. This work showed that organized nanotube layers of up to about 500 nm in thickness. However, the tube structure was not highly organized and the tubes showed considerable sidewall inhomogeneity.

It was recognized that small additions of fluoride ions to an acid electrolyte are the key to form these self-organized oxide structures [100-104].

Most crucial improvements to the geometry of the tubes were established by Macak et al. [85-87] firstly by altering pH and later non-aqueous electrolytes. Using organic electrolyte systems, such as ethylene glycol, almost ideal hexagonally arranged tube layers can be grown to a thickness of several hundreds of micrometers [88].

Furthermore, an optimized two-step anodization treatment, similar to the one reported for porous anodic alumina [89] can lead to virtually perfect hexagonally ordered arrays of TiO_2 nanotubes [90].

The nanotube layers, as they are grown in fluoride containing electrolytes, show a V-shaped sidewall thickness profile [91] (Figure 12) that is due to exposure of the formed tubes to the etching fluoride-containing electrolyte (permanently etching and thinning the tube walls). The general shape and composition distribution can therefore be summarized as in Figure 11. The outer part of the shell (OST) is typically dense very pure TiO_2, and the inner part (IST) of the tube is of a looser in nature and contains incorporated electrolyte components. Detailed SEM work showed that chemical dissolution of the inner TiO_2 layer is faster and thus contributes more to the fact that the walls become thinner towards the top of nanotubes. The tube layers overall have a considerable fluoride ion content but the fluoride (or carbon) content decays strongly upon annealing. The anodization time and etching rate define tube length. The diameter of the nanotubes is controlled linearly by the applied voltage (Figure 13) [92-94]. The fact that the voltage dependence has a different slope in non-aqueous electrolytes can be ascribed to a large extent to the low conductivity of organic electrolytes and resulting IR-drop effects.

As reaction products are formed with anodization time, the conductivity of the electrolyte changes [95] and thus variations in their diameter (larger diameters) for longer anodization times.

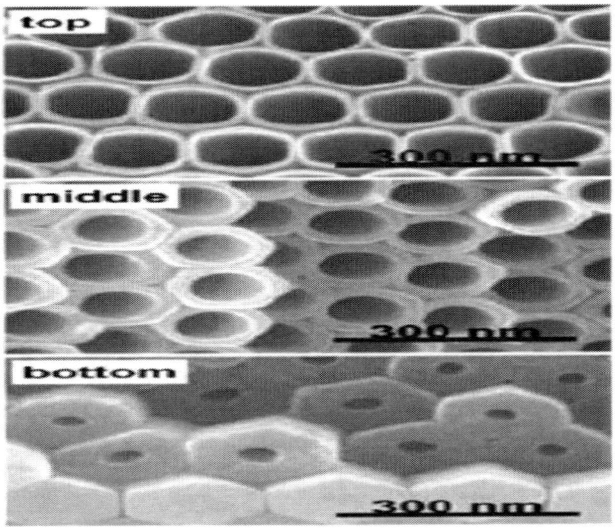

Figure 12. SEM of TiO_2 nanotubes indicating gradient in tube wall thickness [91].

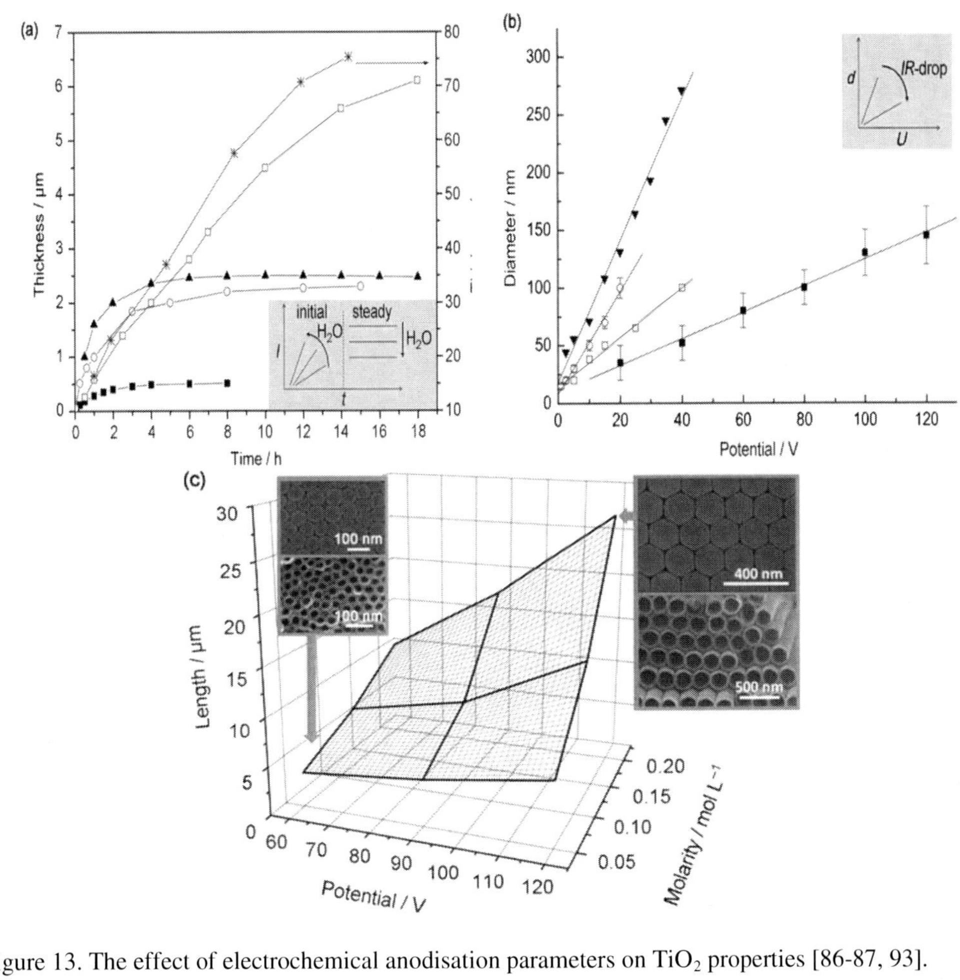

Figure 13. The effect of electrochemical anodisation parameters on TiO_2 properties [86-87, 93].

Investigations into optimized conditions for tube growth usually lead to results such as those shown in Figure 13c, where the tube length as a function of anodization voltage and fluoride concentration was investigated. It is remarkable that the fastest growth conditions typically also represent conditions for optimized ordering. The nanotubes grown from organic electrolytes, such as ethylene glycol, DMSO, glycerol, or ionic liquids, show some significant differences in terms of smooth wall tubes and in morphology and composition compared with nanotubes grown in aqueous electrolytes. Because of low water content, very long tubes and large diameters (up to 700 nm can be grown) [96].

Various modifications in the tube geometry can be achieved by changing the anodization voltage during the tube growth process (Figure 14). Applying voltage steps can be used to generate tube stacks [92, 97-99] or stratification layers (bamboo) [99, 100]. If the voltage is lowered during anodization, tube growth will be stopped or drastically slowed down, as the driving field across the bottom oxide is too low to maintain growth. Stratification layers can be generated when stepping first to a lower voltage while holding the lower voltage for a time that is sufficient to establish diffusion and field conditions corresponding to the lower applied voltage, and then stepping back to the original high voltage [99]. This process has of course to be carried out before tube bottoms are penetrated and a second tube layer grows. In this case, the result is the formation of a new initiation layer between the tubes. Regular tube structures with bamboo-like features can be fabricated by pulsing between two appropriate voltages [99, 100].

The distance between the bamboo rings can be altered by adjusting the holding times at different voltages. Bamboo-type nanotubes, with higher surface areas compared to normal TiO_2 nanotubes, are promising in the field of dye-sensitized solar cells [100].

The highest level of ordering in TiO_2 nanotube layers can be achieved in non-aqueous electrolytes, such as glycerol or ethylene glycol. Under optimized conditions (electrochemical and electrolyte) very closely packed tubes with a hexagonal symmetry can be obtained (Figure 14c). The key for a high degree of ordering is to grow under optimized electrochemical conditions, which are typically conditions under which maximum current is achieved but oxide dielectric breakdown (sparking) is avoided.

Even further improved ordering of the tubes can be achieved using pre-structured surfaces that provide geometrical guiding.

This was demonstrated by tube formation followed by a removal of the tube layer and a second anodization [90]. The organized dimpled titanium metal surface after removal of the first tube layer acts as initiation sites for tube growth in the second anodization step, which results in an optimized hexagonal arrangement. A very spectacular effect, two-size-scale self-organization was first observed when growing tubes on a complex biomedical titanium-based alloy [101, 102].

Properties

The main reason why TiO_2 nanotubes are currently attracting such a high interest is that some of the intrinsic properties of TiO_2 provide the basis for many outstanding functional features.

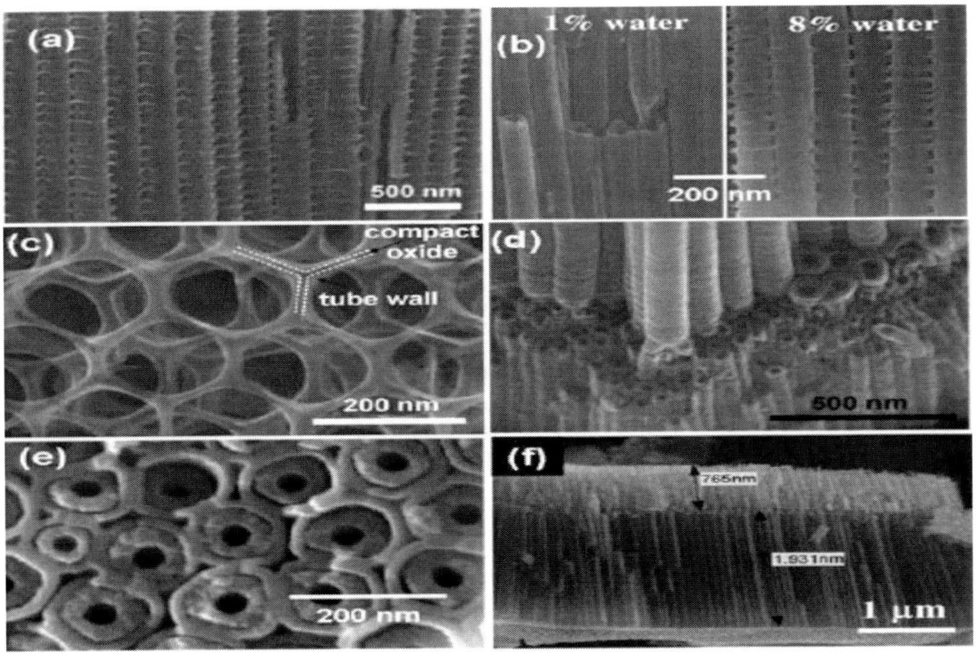

Figure 14. Morphologies of advanced titania nanotubes [99].

Crystal Structure

TiO_2 nanotubes, after electrochemical formation, are amorphous, and also the presence of nanocrystallites in the tube wall, particularly if anodization is carried out at higher voltages. For example, in acidic electrolyte (H_2SO_4), depending on the anodization conditions (sparking, potentistatic or galvanostatic), the as formed oxide film may consist of anatase, a mixture of anatase and rutile, or rutile crystallites [103-105]. In general, amorphous material (such as tubes and powders) can be converted into anatase (ca. 300–500 °C) or rutile (>550 °C) by a thermal treatment in air.

Optical and Electrical Properties

In TiO_2, the relevant energy levels that form the band edges and thus define the band gap are considered to be the Ti 3d states and O 2p levels. The lowest empty energy levels are Ti d_{xy} and thus they are representative of the conduction band (CB) edge, whereas full O 2p states (in particular, nonbonding pp states) define the valence band (VB).

Both anatase and rutile show this general distribution of states. TiO_2 has an indirect optical band gap for anatase of 3.2 eV, for rutile of 3.0 eV, and amorphous material is reported to have a mobility gap of about 3.2–3.5 eV [106, 107].

Reactivity

TiO_2 surfaces have been extensively studied regarding gas-phase adsorption and catalytic effects on various reactions such as CO oxidation [108], selective reduction of NO_x [109] and O_2 and water decomposition [110-112]. Rutile TiO_2 [90] in particular is the preferred substrate for UHV studies owing to the comparable ease with which these single crystal surfaces can be properly prepared. The most important reactive features of TiO_2 are: 1) its solubility in some complexing agents (HF, organic acids), 2) the possibility of modifying the surface with organic monolayers by surface hydroxyl group reactions 3) the feasibility of inducing apatite formation and its electrochemical properties. Electrochemically, anodically formed amorphous TiO_2 layers behave like a highly defective n-type semiconductor, and annealed to anatase or rutile like a highly doped n-type semiconductor [113, 114]. A very important electrochemical feature is the ability to reduce Ti^{4+} in the TiO_2 lattice to Ti^{3+} if a sufficiently negative potential is applied. The reduction reaction is accompanied by insertion of small cations into the TiO_2 lattice. There is a colour change associated with formation of the reduced Ti III material (usually dark blue to brown) that can be exploited for suitable electrochromic devices [115, 116]. For TiO_2 and other transition-metal oxides, the electrochemical reduction and re-oxidation are reversible with a switching time that is determined by the solid-state diffusion/migration of Y^+ into and out of the lattice. Therefore, nanotube surfaces with nanoscale wall thicknesses allow comparably fast switching in the millisecond to second range.

Modification of Tube Properties

It is possible to modify the nanotubes properties by using appropriate techniques. Modification of the TiO_2 nanotubes is mainly carried out by 1) heat-treatments 2) introducing other elements, or 3) by tube-wall decoration. The aim is to make the material suitable (or even more suitable) for various applications that rely on specific electrical, optical or chemical properties. In view of electronic properties, annealing to a crystalline structure mainly changes the conductivity and lifetime of charge carriers, while active doping or band gap engineering by introducing other elements targets decreasing the optical band gap, thus enabling a visible-light photo-response. Particle decoration is often used to increase surface catalytic effects and to create solid-state junctions.

Applications

Applications of TiO_2 nanostructures usually involve the exploitation of some unique feature of TiO_2 (electronic, ionic, or biocompatibility properties) and a significant enhancement of some reaction or transport rates that is obtained by using small scale dimensions (large surface area, short diffusion path, or size confinement effects). Using nanotubular assemblies also provides a preferred dimensionality to the system. Tubes grown on a metal substrate are vertically aligned to a back contact; that is, a direction for charge transfer towards the electrode is established. Nanotube layers provide well- defined top openings and are thus suitable for size-selective applications (filters) or for templating secondary material. The inside volume is well-defined and regular, thus applications such as nano test tubes are promising.

Prevention of Global Warming

As mentioned earlier, the Nature is a greatest gift by God to all the living organisms. Among them, the human beings are mostly enjoying and utilizing it to a greater extent. In the process, the human beings have started exploiting the natural resources to a maximum extent which in turn causing air pollution, water pollution etc. apart from their depletion over a period of time due to which global warming is taking place. It is the responsibility of scientists all over the world to focus on it and come out with appropriate technologies to protect the natural resources, prevent pollution and thus prevent global warming which in turn to pass it on to our future generations.

The titania nanotubes gravely help to prevent air pollution by converting CO_2 or CO present in the air to useful or harmless organic compounds (Figure 15). They are also quite useful as excellent filters during water purification process.

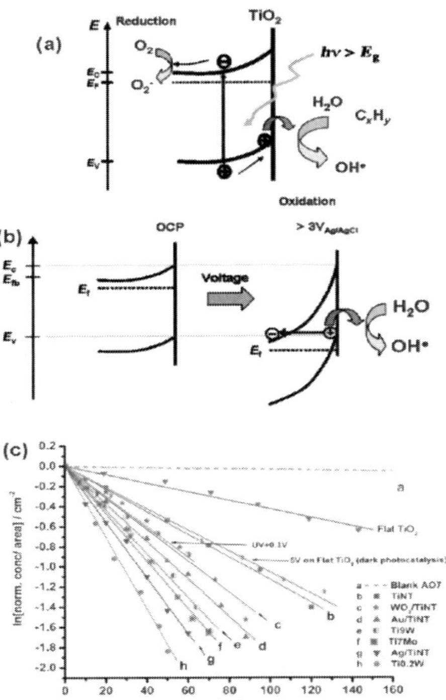

Figure 15. Photocatalytic effect of titania nanotubes [117].

Electrochemical Energy Devices (EEDs)

One of the most important applications of electrochemical energy devices (EEDs) is conversion and storage of energy. It remains a challenge for the EEDs in the 21st century to achieve high power density and long-term cycling life. In response to the needs of modern society and emerging ecological concerns, it is highly desirable to find new, low-cost and environmentally friendly electrochemical energy conversion and storage for powering an increasingly diverse range of applications, ranging from portable power for consumer electronics to potential transport applications. This allowed for rapid Research in electrochemical energy conversion and storage devices such as fuel cells, super-capacitors, and Li-ion batteries. As the performance of these EEDs depends intimately on the properties of their materials, nano tubes are the key materials.

Photocatalysis and Dark Photocatalysis

Titania nanotubes are the most photocatalytically active materials for decomposition of organic materials (degradation of organic pollutants) due to their band-edge positions relative to typical environments (such as water) (Figure 15) [117]. The basic principle involved in the photocatalytic mechanism is that UV light promotes electrons from the valence band to the conduction band; holes and electrons will be separated (under the field of the Schottky junction with the environment), reach the semiconductor–environment interface, and react with appropriate redox species in the environment. Several highly reactive species are generated from the surrounding water by charge exchange at the valence band (H_2O + h+ = OH^-) and at the conduction band (O_2 + e = O^{2-}). These radicals and peroxo ions are able to virtually oxidize all organic materials to CO_2 and H_2O. Furthermore, at the valence band, direct h+ transfer to adsorbed species to initiate decomposition also takes place. Nevertheless, all of the photocatalytic applications have in common the fact that a higher overall reaction rate is achieved using high-surface-area geometries. Thus, reactions are commonly performed using suspended nanoparticles or nanoparticulate electrodes. Ordered nanotube arrangements offer various advantages over nanoparticulate assemblies, as their defined geometry provides much defined retention times in nanoscopic photo-reactors. Moreover, the 1D geometry may allow a fast carrier transport and thus less unwanted recombination losses. A first investigation [118] has shown that TiO_2 nanotubes can indeed have higher photocatalytic reactivity than a comparable nanoparticulate layer.

Solarisation

Another very attractive application of TiO_2 is its use in solar cells. Dye sensitization of TiO_2 was extensively investigated.

The principle of DSSC involves a dye that absorbs light in the visible range and thereby excites electrons from the HOMO to the LUMO level, followed by a rapid injection of the excited electron into the conduction band of TiO_2 (Figure 16) [119].

The electrons travel through the TiO_2 layer to the back; meanwhile, the oxidized dye on the surface is regenerated by I^-/I_3^- electrolyte. Losses by recombination may happen principally in three ways: 1) After electron excitation from the HOMO to the LUMO level of the dye, de-excitation of the electron occurs in a radiative or non-radiative path 2) after injection of the electron in the conduction band of TiO_2, it may recombine with the oxidized dye and 3) the electron may recombine with the I_3^- ion in the electrolyte. A very interesting study has shown that a much higher electron diffusion length can be achieved in nanotubes than in nanoparticles [120].

In fact, from combined electron lifetime/diffusion measurements, it was deduced that electron diffusion lengths of 100 mm should be achievable. Therefore, it is possible to use these titania nanotubes for solarisation of a variety of systems starting from household to space vehicles.

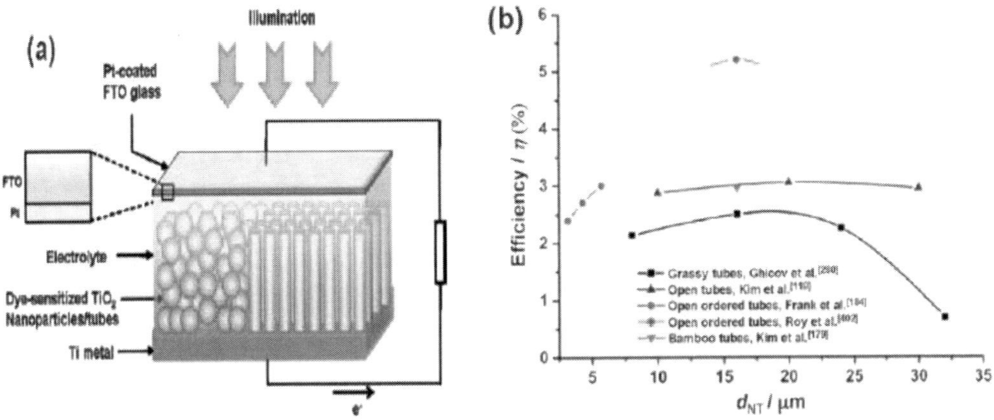

Figure 16. Titania naotubes based dye sensitized solar cell [120].

Biomedical

A very important application of titanium surface modification is in biomedical applications. About 40% of today's biomedical implant materials are based on titanium or titanium alloys. TiO_2 nanotubular surfaces are ideal for studying and applying size effects with living matter or biologically relevant species.

Not only can the diameter of the tube surfaces be accurately adjusted to virtually any value between 10-250 nm, but owing to the self-organizing nature, entire, even complex shaped surfaces (such as dental-implant screws or hip implants) can be coated easily with such nanotube layers.

The geometry of the TiO_2 nanotube arrays suggests that they can be used as a drug-delivery capsule if nanotube layers are separated (singled-out) and stabilized, or it can also be used as a drug-eluding coating on biomedical implant materials.

It was reported that TiO_2 nanotubes can be filled with magnetic Fe_3O_4 particles and thus be magnetically guided to desired locations. Such magnetic TiO_2 nanotubes can also be used directly for photocatalytic reactions with cells or tissue, for the site-selective killing of cancer cells.

Other Applications

TiO_2 layers have shown to have a high sensitivity to CO, H_2, and NO_x gases, and advantage of the nanotubular geometry, they can be used as nano test tubes. An excellent properties of their transparency for large range of light and self-cleaning, they can be used in static devices (MEMS). TiO_2 nanotubes decorated with co-adsorbed horseradish peroxidise (HRP) and thionine chloride helps them to use as biosensors, owing to their sensitivity to H_2O_2 concentration variations.

CARBON NANOTUBES

Presently, Carbon nanotubes have been increasingly used in commercial applications like composite materials, electrochemical devices, hydrogen storage, field emission devices, and nano scale electronic devices. These wide ranges of applications of carbon nanotubes are based on their unique physical and mechanical properties, which exhibit the high electrical and thermal conductivities and high mechanical strength. Carbon nanotubes are normally produced by either a carbon arc-discharge technique or a pyrolysis of hydrocarbon gases on particles of transition metals such as Fe, Ni, and Co in a chemical vapor deposition (CVD) reactor. In CVD technique, the produced carbon nanotubes diameters are controlled by the sizes of catalytic particles. The growth mechanism of the carbon nanotubes from a CVD process has been the subject of some controversy, particularly regarding the active catalytic state of the particles. Key steps include the absorption and decomposition of hydrocarbon molecules on the exposed metal surface to produce carbon species which dissolve into and diffuse through the underlying bulk, ultimately precipitating at the rear surface of the particle to form the carbon nanotubes.

At the same time, in recent years, more attention has been paid to electrochemical deposition technique for manufacturing thin films and devices due to its simplicity, its low capital equipment cost, and its ability to be scaled up for large production. Much work has been carried out over the world and came up with some techniques to produce carbon nanotubes. One of the profitable techniques includes producing carbon nanotubes from organic solvents at room temperature where a standard three-component electrochemical cell as the electrochemical reactor and a calomel electrode as reference electrode were used for the carbonaceous depositions. Two silicon wafers coated with equal ratios of Fe/Ni alloy nanoparticles were employed as counter electrode (anode) and working electrode (cathode), respectively.

A mixture of 40% methanol (CH_3OH) and 60% benzyl alcohol ($C_6H_5CH_2OH$) in volume was used as the electrolyte. The sizes of counter and working electrodes were about 25 mm - 20 mm. The distance between cathode and anode was kept at 5 mm, which was precisely controlled by a SiC spacer. The potential difference applied between the anode and cathode was kept nominally at 1000 V. The depositions were carried out at room temperature and magnetic stirring was employed to achieve a uniform distribution of carbonaceous deposit on the cathode. No carbonaceous deposits were found on the anode. After deposition, the samples were removed from the solution and allowed to dry in air. SEM, EDS and TEM analysis indicated that the complete surface is distributed with nanotubes (Figures 17-19) [121]. The carbonaceous deposits of the electrochemical reaction can be either in the form of films or nanotubes.

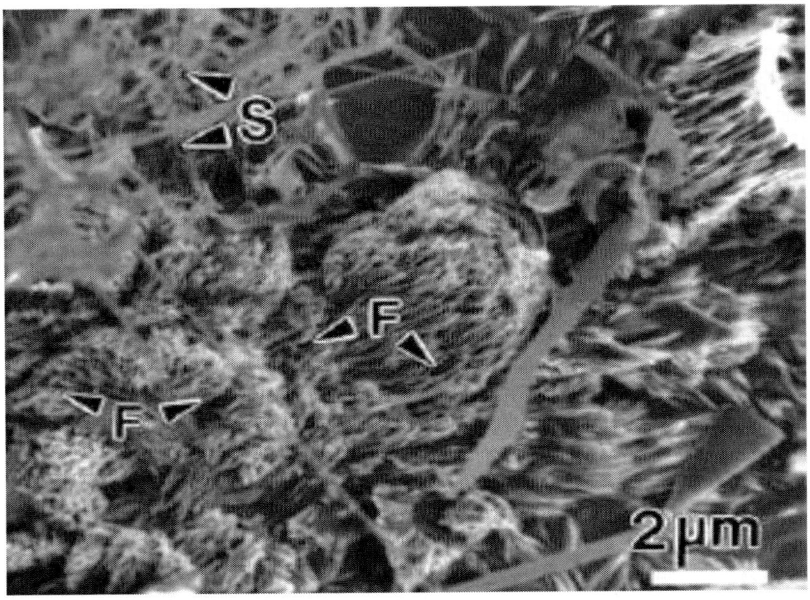

Figure 17. SEM of Carbon nanotubes prepared from Electrochemical process [121].

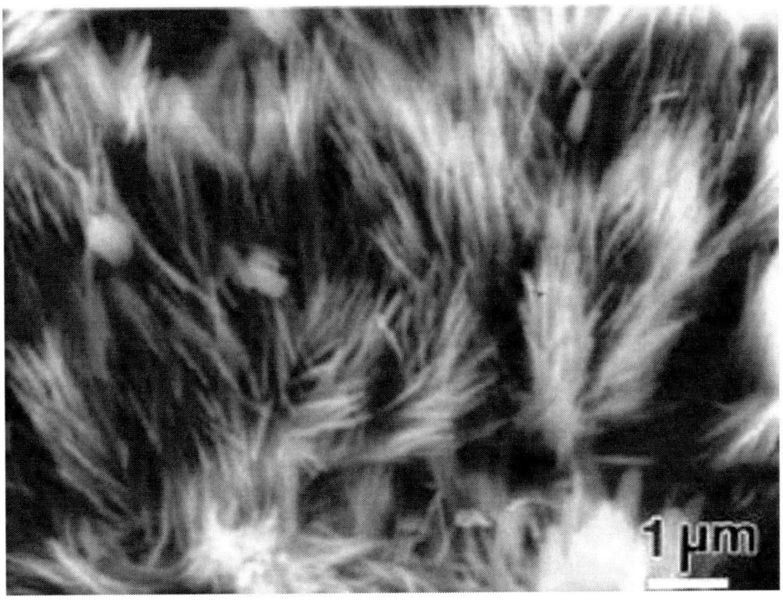

Figure 18. SEM of carbon nanotubes (bundles) prepared from Electrochemical process [121].

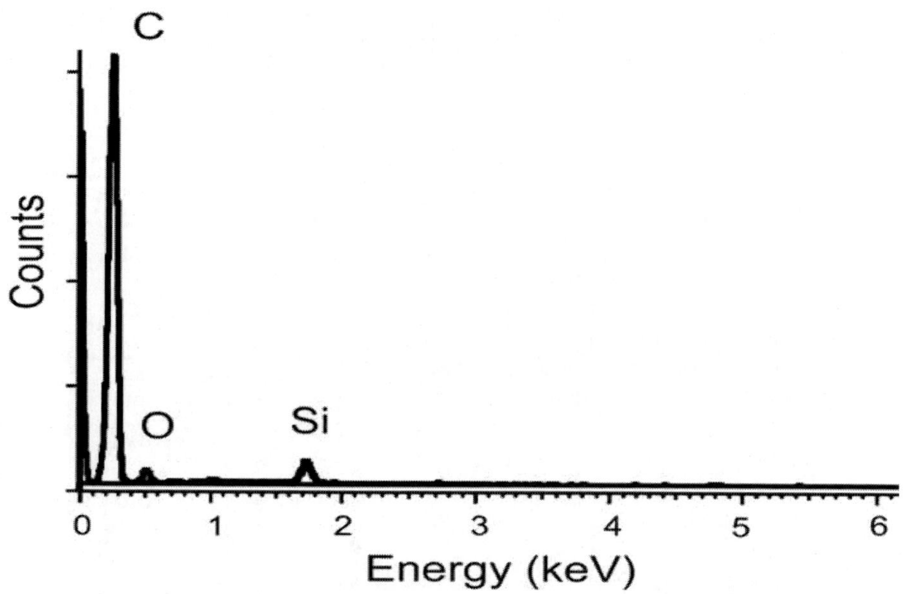

Figure 19. EDS spectrum of the carbon nanotube bundles. [121].

Properties and Applications

Vacuum Microelectronics

Carbon nanotubes have the apt combination of properties – nanometer-size diameter, structural integrity, high electrical conductivity and chemical stability – that make good electron emitters. Electron field emission from carbon nanotubes was first demonstrated in 1995 [13] and has since been studied intensively on various carbon nanotube materials. Compared to conventional emitters, carbon nanotubes exhibit a lower threshold electric field, as illustrated in Table VI.

Some of the electron emission devices in which carbon nanotubes are used to enhance the performance of the functionality are:

Cathode-Ray Lighting Elements

Cathode ray lighting elements with carbon nanotube materials as the field emitters have been fabricated by Ise Electronic Co. in Japan [123]. As illustrated in Figure 20, these nanotube-based lighting elements have a triode-type design.

Flat Panel Display

Prototype matrix-addressable diode flat panel displays have been fabricated using carbon nanotubes as the electron emission source [124].

One demonstration structure constructed at Northwestern University consists of nanotube-epoxy stripes on the cathode glass plate and phosphor-coated Indium-Tin-Oxide (ITO) stripes on the anode plate [124]. Pixels are formed at the intersection of cathode and anode stripes, as illustrated in Figure 21.

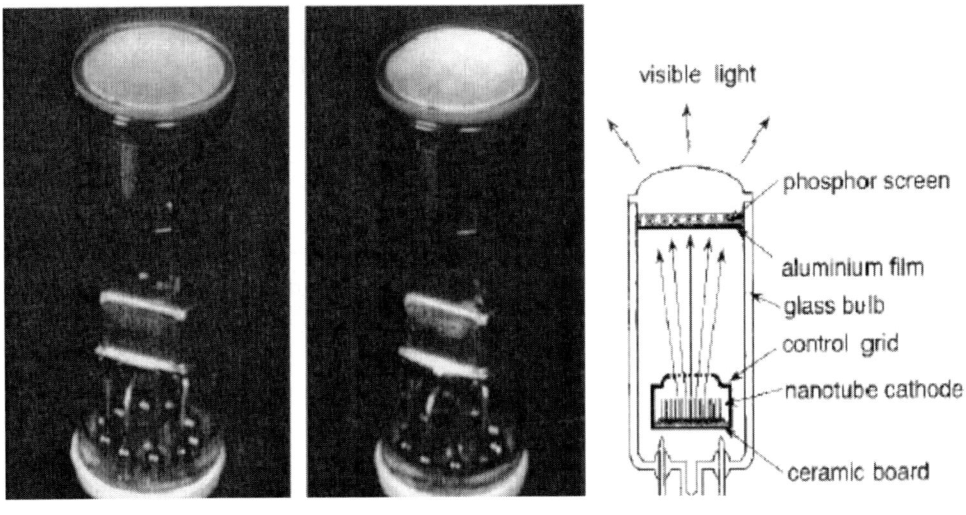

Figure 20. Schematic diagram of field emission light source using carbon nanotubes as cathodes [124].

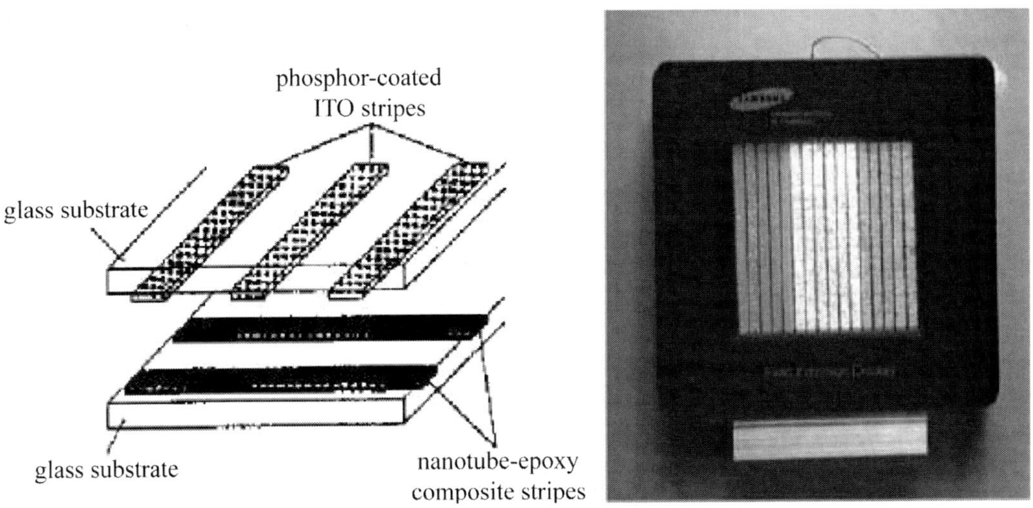

Figure 21. Schematic of a prototype field emission display using carbon nanotubes [14].

Energy Storage

Carbon nanotubes are being considered for energy production and storage. Graphite, carbonaceous materials and carbon fiber electrodes have been used for decades in fuel cells, batteries and several other electrochemical applications [14].

Several experiments have confirmed that compared to conventional carbon electrodes, the electron transfer kinetics take place fastest on nanotubes, following ideal Nernstian behavior [15].

Hydrogen Storage

The area of hydrogen storage in carbon nanotubes remains active and controversial. Extraordinarily high and reversible hydrogen adsorption in single walled nanotubes (SWNT) containing materials [16-19] and graphite nanofibers (GNFs) [125] has been reported and has attracted considerable interest in both academia and industry. Table VII summarizes the gravimetric hydrogen storage capacity reported by various groups.

Filled Composites

The mechanical behavior of carbon nanotubes is exciting since nanotubes are seen as the "ultimate" carbon fiber ever made.

Table 6. Threshold electrical field for different materials and carbon nanotubes (10 mA/cm^2 current density) [122]

Material	Threshold electrical field (V/μm)
Mo tips	50–100
Si tips	50–100
p-type semiconducting diamond	130
Undoped, defective CVD diamond	30–120
Amorphous diamond	20–40
Cs-coated diamond	20–30
Graphite powder(<1mm size)	17
Nanostructured diamond (Heat-treated in H plasma)	3–5 (unstable >30mA/cm^2)
Carbon nanotubes	1–3 (stable at 1 A/cm^2)

Table 7. Summary of reported gravimetric storage of H_2 in various carbon materials

Material	Max. wt% H_2	*T*(K)	*P* (MPa)
SWNTs(low purity)	5–10	133	0.040
SWNTs(high purity)	~4	300	0.040
GNFs(tubular)	11.26	298	11.35
GNFs(herringbone)	67.55	298	11.35
GNS(platelet)	53.68	298	11.35
Graphite	4.52	298	11.35
GNFs	0.4	298-773	0.101
Li-GNFs	20	473-673	0.101
Li-Graphites	14	473-674	0.101
K-GNFs	14	< 313	0.101
K-Graphite	5.0	< 313	0.101
SWNTs(high purity)	8.25	80	7.18
SWNTs(~50% pure)	4.2	300	10.1

The traditional carbon fibers have about *fifty times* the specific strength (strength/density) of steel and are excellent load-bearing reinforcements in composites. Therefore, Carbon Nanotubes are ideal candidates for structural applications.

Carbon fibers have been used as reinforcements in high strength, light weight, high performance composites; one can typically find these in a range of products ranging from expensive tennis rackets to spacecraft and aircraft body parts. NASA has recently invested large amounts of money in developing carbon nanotube-based composites for applications such as the futuristic Mars mission.

Nanoprobes and Sensors

The small and uniform dimensions of the nanotubes produce some interesting applications. With extremely small sizes, high conductivity, high mechanical strength and flexibility (ability to easily bend elastically), nanotubes may ultimately become indispensable in their use as nanoprobes. One could think of such probes as being used in a variety of applications, such as high resolution imaging, nano-lithography, nano-electrodes, drug delivery, sensors and field emitters.

Templates

Since nanotubes have relatively straight and narrow channels in their cores, it was speculated from the beginning that it might be possible to fill these cavities with foreign materials to fabricate one-dimensional nanowires. Early calculations suggested that strong capillary forces exist in nanotubes, strong enough to hold gases and fluids inside them [126]. The first experimental proof was demonstrated in 1993, by filling and solidification of molten lead inside the channels of MWNTs (Multi Walled Nanotubes) [20]. Wires as small as 1.2 nm in diameter were fabricated by this method. Thus, nanotubes can be been used as templates to create nanowires of various compositions and structures (Figure 22).

Figure 22. Schematic diagram showing the filling of empty one-dimensional hollow core of nanotubes with foreign substances [20].

Conclusion

The electrochemical techniques offer outstanding advantages over other techniques. It is expected that electrochemical techniques are expected to play an increasingly significant role in nanotechnology. Electrodeposition is an important technique for the fabrication of nanostructured materials, offering control over structure, composition and properties, and thus helps in preparation of novel materials with enhanced properties, which cannot be obtained with other techniques. Similarly, electrochemical anodization technique is a wonderful technique to produce carbon nanotubes, titania nanotubes, rods, wires with desired properties for different applications. Various electrochemical strategies were pioneered and are currently under investigation for fabrication of a variety of nanostructured materials.

Further developments of the electrochemical techniques are extremely essential not only for their preparation but also for characterization of a number of novel materials.

References

[1] I. V. S. Yashwanth and I. Gurrappa, *J. Nanomater. Molec. Nanotechnol.*, 3 (2014) 2 (http://dx.doi.org/10.4172/2324-8777.1000141).

[2] S. Lee, J. Park, D. H. Kim, W. M. Seong, and D. W. Kim, *Energy Environ. Sci.*, 5 (2012) 7989.

[3] J. Qiu, F. Zhuge, X. Li, X. Gao, and X. Gan, *J. Mater. Chem.*, 22 (2012) 3549.

[4] W. H. Ryu, D. H. Nam, Y. S. Ko, R. H. Kim, and H. S. Kwon, *Electrochim. Acta*, 61 (2012) 19.

[5] X. Zhang, J. Zhang, Y. Jia, P. Xiao, and J. Tang, *Sensors*, 12 (2012) 3302.

[6] O. K. Varghee, D. Gong, M. Paulose, K. G. Ong, and E. C. Dickey, *Adv. Mater.*, 15 (2003) 624.

[7] W. Y. Wang and B. R. Chen, *Intl. J. Photoenergy*, 168 (2013) 2013.

[8] S. Sreekantan, R. Hazan and Z. Lockman, *Solid State Sci. Technolo.*, 17 (2010) 132.

[9] K. S. Raja, V. K. Mahajan, M. Misra, *J. Power Sources*, 159 (2006) 1258.

[10] G. K. More, K. Shankar, M. Paulose, O. K. Verghese, and C. K. Grimes, *Nano Lett.*, 5 (2005) 191.

[11] K. S. Brammer, C. J. Frandsen and S. Jin, *Trends Biotechnol.*, 30 (2012) 315.

[12] E. Gongadze, D. Kabaso, S. Paur, J. Park, and P. Schmuki, *Mini Rev. Med. Chem.*, 13 (2013) 194.

[13] G. Rinzler, J. H. Hafner, P. Nikolaev, L. Lou, S. G. Kim, D. Tomanek, D. Colbert, R. E. Smalley, *Science* 269 (1995) 1550.

[14] R. L. McCreery, *Electroanal. Chem.*, 17, (ed. A. J. Bard) (Marcel Dekker, New York 1991).

[15] C. Dillon, K. M. Jones, T. A. Bekkedahl, C. H. Kiang, D. S. Bethune, M. J. Heben, *Nature* 386 (1997) 377.

[16] P. Chen, X. Wu, J. Lin, K. Tan, *Science* 285 (1999) 91.

[17] C. Liu, Y. Y. Fan, M. Liu, H. T. Cong, H. M. Cheng, M. S. Dresselhaus, *Science* 286 (1999) 1127.

[18] C. Nutenadel, A. Zuttel, D. Chartouni, L. Schlapbach, *Solid-State Lett.* 2 (1999) 30.

[19] A. Chambers, C. Park, R. T. K. Baker, N. M. Rodriguez, *J. Phys. Chem.* B 102 (1998) 4253.

[20] P. M. Ajayan, S. Iijima, *Nature* 361 (1993) 333.

[21] A. M. El-Sherik, U. Erb, G. Palumbo, and K. T. Aust, *Scripta Metall. et Mater.*, 27 (1992) 1185.

[22] C. Cheung, G. Palumbo and U. Erb, *Scripta Metall. et Mater.*, 32 (1994) 735.

[23] U. Erb, *Nanostr. Mater.*, 6 (1995) 533.

[24] N. Wang, Z. Wang, K. T. Aust, and U. Erb, *Acta Metall. et Mater.*, 43 (1995) 519.

[25] M. J. Aus. B. Szpunar, U. Erb, A. M. El-Sherik, G. Palumbo, and K. T. Aust, *J. Appl. Phys.*, 75 (1994) 3632.

[26] M. J. Aus, B. Szpunar, A. M. El-Sherik, U. Erb, G. Palumbo, and K. T. Aust, *Scripta Metall. et Mater.*, 27 1992) 1637.

[27] D. M. Doyle, G. Palumbo, K. T. Aust, A. M. El-Sherik, and U. Erb, *Acta Metall. et Mater.*, 43 (1995) 3027.

[28] I. Bakonyi, E. Toth-Kadar, T. Tarnoczi, L. K. Varga, A. Cziraki, I. Gerocs, and B. Fogarassy, *Nanostruct. Mater.*, 3 (1993) 155.

[29] A. Cziraki, I. Gerocs, E. Toth-Kada, and I. Bakonyi, *Nanostruct. Mater.*, 6 (1995) 547.

[30] I. Zhitomirsky and L. Gal-Or, *J. Mater. Sci. Mater. Med.*, 8 (1997) 213.

[31] I. Zhitomirsky, *Mater. Lett.*, 42 (2000) 262.

[32] M. Shirkhanzadeh, *J. Mater. Sci.*, 9 (1998) 67.

[33] M. Izaki and T. Omi, *J. Electrochem. Soc.*, 144 (1997) 1949.

[34] S. Peuloro and D. Lincot, *Adv. Mater.*, 8 (1996) 166.

[35] R. Liu, et al., *Chem. Mater.*, 13 (2001) 508.

[36] J. A. Switzer, et al., *J. Mater. Res.*, 13 (1998) 909.

[37] S. Kenane and L. Piraux, *J. Mater. Res.*, 17 (2002) 401.

[38] P. M. Vereecken, I. Shao and P. C. Searson, *J. Electrochem. Soc.*, 147 (2000) 2572.

[39] T. Yoshino, et al. *Jpn. J. Appl. Phys.*, 35 (1996) L1512.

[40] W. Sha and J. S. Pan, *J. Alloys and Comp.*, 182 (1992) L1-L3.

[41] B. P. Daly and F. J. Barry, *Intl. Mater. Rev.*, 48 (2003) 362.

[42] H. Habazaki, Y. P. Lu, K. Asami, and K. Hashimoto, *Corr. Sci.*, 32 (1991) 1227.

[43] L. Benea, P. L. Bonora, A. Borello, S. Martelli, F. Wenger, P. Ponthiaux, and J. Galland, *Solid State Ionics*, 151 (2002) 89.

[44] V. M. Fedosyuk, T. A. Tochitskij and S. A. Sharko, *Perspekt. Mater.*, 4 (2002) 74.

[45] J. Li, J. Jiang, H. He, and Y. Sun, *J. Mater. Sci. Lett.*, 21 (2002) 939.

[46] D. G. Kim, H. Kim, H. J. Sohn, and T. Kang, *J. Power Sources*, 104 (2002) 221.

[47] T. Gao, G. Meng, Y. Wang, S. Sun, and L. Zhang, *J. Phys. Condensed Mater.*, 14 (2002) 355.

[48] J.-J. Zhu, X.-H. Liao, X.-N. Zhao, and H.-Y. Chen, *Mater. Lett.*, 49 (2001) 91.

[49] P. N. Bartlett, P. N. Birkin, M. A. Ghanem, P. De Groot, and M. Sawicki, *J. Electrochem. Soc.*, 148 (2001) C119-C123.

[50] S. Yang, H. Zhu, D. Yu, Z. J. S. Tang, and Y. Du, *J. Magn. Magn. Mater.*, 222 (2000) 97.

[51] A. S. M. A. Haseeb, J. P. Celis and J. R. Roos, *Mater. Manuf. Processes*, 10 (1995) 707.

[52] W. Gunther, N. Pfander, G. Weinberg, R. Liedtke, J. Nissen, and R. Schlogl, *J. Mater. Sci.*, 35 (2000) 3963.

[53] C. T. Kunoshi, O. V. Correa, N. B. De Lima, and L. V. Ramanathan, *Ultrafine Grained Mater.*, 110 (2000) 99.
[54] P. Forrer, F. Schlottig, H. Siegenthaler, and M. Textor, *J. Appl. Electrochem.*, 30 (2000) 533.
[55] M. Ichikawa, *Platiunum Met. Rev.*, 44 (2000) 3.
[56] G. Tourillon, L. Pontonnier, J. P. Levy, and V. Langlais, *Electrochem. and Sold State Lett.,* 3 (2000) 20.
[57] E. Peled and A. Ulus, *Lithium Batteries*, (1998) 122.
[58] S. Villain, P. Knauth and G. Schwitzgebel, *J. Phys. Chem.*, B101 (1997) 7452.
[59] L. Zaluski, A. Zaluska, P. Tessier, J. O. Stromolsen, and R. Schulz, *Mater. Sci. Forum*, 225-227 (1996) 853.
[60] K. Z. Chen, Z. K. Zhang. Z. L. Cui, D. H. Zuo, and D. Z. Yang, *Nanostructured Mater.*, 8 (1997) 205.
[61] D. Simunovich, M. Schlesinger and D. D. Snyder, *J. Electrochem. Soc.*, 141 (1994) L10-L11.
[62] W. Schwarzacher, K. Attenborough, A. Michel, G. Nabiyouni, and J. P. *Meier, J. Magn. Magn. Mater.,* 165 (1997) 23.
[63] P. Evans, R. Hart, G. Nabiyouni, and W. Schwarzacher, *J. Magn. Magn. Mater.,* 148 (1995) 317.
[64] A. Blondel, B. Doudin and J.-Ph. Ansermet, *J. Magn. Magn. Mater.*, 165 (1997) 34.
[65] R. R. Oberle, M. R. Scanlon, R. C. Cammarata, and P. C. Searson, *Appl. Phys. Lett.,* 66 (1995) 19.
[66] D. J. Riley, *Current opinion in Colloid and Interface Science*, 7 (2002) 186.
[67] U. Erb, K. T. Aust and G. Palumbo, *Nanostructured Materials Processing, Properties and Technologies,* William Andrew publishing 2002, pp. 179-222.
[68] S. Ranganathan, R. Divakar and V. S. Raghunathan, *Scripta Mater.,* 22 (2001) 1169.
[69] D. H. Jeong, F. Gonzalez, G. Palumbo, K. T. Aust, and U. Erb, *Scripta Mater.,* 44 (2001) 493.
[70] A. F. Zimmerman, D. G. Clark, K. T. Aust, and U. Erb, *Mater. Lett.,* 52 (2002) 85.
[71] D. H. Jeong, U. Erb, K. T. Aust, and G. Palumbo, *Scripta Mater.,* 48 (2003) 1067.
[72] C. Cheung, F. Djuanda, U. Erb, and G. Palumbo, *Nanostruc. Mater.*, 5 (1995) 513.
[73] M. L. Cantu, A. M. Sabio, A. Brustenga, and P. G. Romero, *Solar Energy Mater. Solar Cells,* 87 (2005) 685.
[74] D. Kim, K. Lee, P. Roy, B. I. Birajdar, E. Spiecker, P. Schmuki, *Angew. Chem.* 121 (2009) 9490; *Angew. Chem. Int. Ed.* 48 (2009) 9326.
[75] K. Lee, D. Kim, P. Roy, I. Paramasivam, B. I. Birajdar, E. Spiecker, P. Schmuki, *J. Am. Chem. Soc.* 132 (2010) 1478.
[76] D. Kim, P. Roy, K. Lee, P. Schmuki, *Electrochem. Commun.*, 12 (2010) 574.
[77] P. Roy, T. Dey, K. Lee, D. Kim, B. Fabry, P. Schmuki, *J. Am. Chem. Soc.*, 132 (2010) 7893.
[78] V. Zwilling, E. Darque-Ceretti, A. Boutry-Forveille, D. David, M. Y. Perrin, M. Aucouturier, *Surf. Interface Anal.* 27 (1999) 629.
[79] V. Zwilling, M. Aucouturier, E. Darque-Ceretti, *Electrochim. Acta* 45 (1999) 921.
[80] D. Gong, C. A. Grimes, O. K. Varghese, Z. Chen, E. C. Dickey, *J. Mater. Res.* 16 (2001) 3331.
[81] R. Beranek, H. Hildebrand, P. Schmuki, *Electrochem. Solid-State Lett.* 6 (2003) B12.

[82] I. V. S.Yashwanth and I. Gurrappa, *J. Nat. Nanomater.* (submitted).
[83] I. V. S. Yashwanth and I. Gurrappa, *J. Nat. Nanotechnol.* (submitted).
[84] I. V. S. Yashwanth and I. Gurrappa, *J. Micro Bio. Nanotechnol.* (submitted).
[85] J. M. Macak, K. Sirotna, P. Schmuki, *Electrochim. Acta*, 50 (2005) 3679.
[86] J. M. Macak, H. Tsuchiya, P. Schmuki, *Angew. Chem.* 117 (2005) 2136; *Angew. Chem. Int. Ed.* 44 (2005) 2100.
[87] J. M. Macak, H. Tsuchiya, L. Taveira, S. Aldabergerova, P. Schmuki, *Angew. Chem.* 117 (2005) 7629; *Angew. Chem. Int. Ed.* 44 (2005) 7463.
[88] S. P. Albu, A. Ghicov, J. M. Macak, P. Schmuki, *Phys. Status Solidi RRL* 1 (2007) R65.
[89] H. Masuda, M. Satoh, *Jpn. J. Appl. Phys. Part* 1 35 (1996) L126.
[90] J. M. Macak, S. P. Albu, P. Schmuki, *Phys. Status Solidi, RRL* 1 (2007) 181.
[91] S. P. Albu, A. Ghicov, S. Aldabergenova, P. Drechsel, D. LeClere, G. E. Thompson, J. M. Macak, P. Schmuki, *Adv. Mater.*, 20 (2008) 4135.
[92] K. Yasuda, J. M. Macak, S. Berger, A. Ghicov, P. Schmuki, *J. Electrochem. Soc.*, 154 (2007) C472.
[93] J. M. Macak, H. Hildebrand, U. Marten-Jahns, P. Schmuki, *J. Electroanal. Chem.*, 621 (2008) 254.
[94] S. Bauer, S. Kleber, P. Schmuki, *Electrochem. Commun.*, 8 (2006) 1321.
[95] K. Lee, J. Lee, H. Kim, Y. Lee, D. Kim, P. Schmuki, Y. Tak, *J. Korean Phys. Soc.* 54 (2009) 1027.
[96] S. P. Albu, P. Schmuki, *Phys. Status Solidi, RRL* 4 (2010) 215.
[97] Y. Y. Song, P. Schmuki, *Electrochem. Commun.* 12 (2010) 579.
[98] Y. Y. Song, F. Schmidt-Stein, S. Bauer, P. Schmuki, *J. Am. Chem. Soc.* 131 (2009) 4230.
[99] S. P. Albu, D. Kim, P. Schmuki, *Angew. Chem.* 120 (2008) 1942; *Angew. Chem. Int. Ed.* 47 (2008) 1916.
[100] D. Kim, A. Ghicov, S. P. Albu, P. Schmuki, *J. Am. Chem. Soc.* 130 (2008) 16454.
[101] H. Tsuchiya, J. M. Macak, A. Ghlcov, P. Schmuki, *Small* 2 (2006) 888.
[102] H. Tsuchiya, T. Akaki, D. Terada, N. Tsuji, Y. Minamino, P. Schmuki, S. Fujimoto, *Electrochim. Acta* 54 (2009) 5155.
[103] J. C. Marchenoir, J. P. Loup, J. Masson, *Thin Solid Films*, 66 (1980) 357.
[104] J.-L. Delplancke, A. Garnier, Y. Massiani, R. Winand, *Electrochim. Acta*, 39 (1994) 1281.
[105] J. C. Marchenoir, J. Gautron, J. P. L. Metaux, *Corrosion-Industrie* (1977) 83.
[106] V. M. Naik, D. Haddad, R. Naik, J. Benci, G. W. Auner, *Mater. Res. Soc. Symp. Proc.* 775 (2002) 413.
[107] R. P. Lynch, A. Ghicov, P. Schmuki, *J. Electrochem. Soc.* 157 (2010) G76.
[108] S. Funk, U. Burghaus, *Catal. Lett.,* 118 (2007) 118.
[109] V. I. Prvulescu, P. Grange, B. Delmon, *Catal. Today* 46 (1998)233.
[110] S. Funk, B. Hokkanen, U. Burghaus, A. Ghicov, P. Schmuki, *Nano Lett.* 7 (2007) 1091.
[111] M. Kunat, U. Burghaus, C. Woll, *Phys. Chem. Chem. Phys.* 6 (2004) 4203.
[112] B. Hokkanen, S. Funk, U. Burghaus, A. Ghicov, P. Schmuki, *Surf. Sci.* 601 (2007) 4620.
[113] A. Ghicov, H. Tsuchiya, J. M. Macak, P. Schmuki, *Phys. Status Solidi A* 203 (2006) R28.
[114] J. W. Schultze, M. M. Lohrengel, *Electrochim. Acta*, 45 (2000) 2499.

[115] L. Kavan, M. Graetzel, S. E. Gilbert, C. Klemenz, H. J. Scheel, *J. Am. Chem. Soc.* 118 (1996) 6716.
[116] T. Oi, *Annu. Rev. Mater. Sci.* 16 (1986) 185.
[117] Y. Y. Song, P. Roy, I. Paramasivam, P. Schmuki, *Angew. Chem.* 122 (2010) 361; *Angew. Chem. Int. Ed.* 49 (2010) 351.
[118] J. M. Macak, M. Zlamal, J. Krysa, P. Schmuki, *Small* 3 (2007) 300.
[119] M. Gratzel, *J. Photochem. Photobiol.* C 4 (2003) 145.
[120] J. R. Jennings, A. Ghicov, L. M. Peter, P. Schmuki, A. B. Walker, *J. Am. Chem. Soc.* 130 (2008) 13364.
[121] Dan Zhou, Elvira V. Anoshkina, Lee Chow, Guangyu Chai, *Carbon* 44 (2006) 1013.
[122] W. Zhu, G. Kochanski, S. Jin, *Science* 282 (1998) 1471.
[123] Y. Saito, S. Uemura, K. Hamaguchi, *Jpn. J. Appl. Phys.* 37 (1998) L346.
[124] Q. H. Wang, A. A. Setlur, J. M. Lauerhaas, J. Y. Dai, E. W. Seelig, R. H. Chang, *Appl. Phys. Lett.* 72 (1998) 2912.
[125] M. S. Dresselhaus, K. A. Williams, P. C. Eklund, *MRS Bull.,* 24 (1999) 45.
[126] M. R. Pederson, J. Q. Broughton, *Phys. Rev. Lett.* 69 (1992) 2689.

EDITOR'S CONTACT INFORMATION

Mahmood Aliofkhazraei
Tarbiat Modares University,
Faculty of Engineering,
Department of Materials Engineering,
Tehran, Iran
E-mail: maliofkh@yahoo.com

INDEX

A

B

C

D

E

F

G

H

N

O

P

Q

R

S

T

U

V

W

X

Y

Z